AF818800

EXS 72

The Nervous Systems of Invertebrates: An Evolutionary and Comparative Approach

With a Coda written by T.H. Bullock

Edited by O. Breidbach
W. Kutsch

Birkhäuser Verlag
Basel · Boston · Berlin

Editors

PD Dr. Dr. O. Breidbach
Fakultät und Institut für Mathematik
Ruhr-Universität Bochum
Universitätsstr. 150, NA 5-26
D-44801 Bochum
Germany

Prof. Dr. W. Kutsch
Fakultät für Biologie
Universität Konstanz
Universitätsstr. 10
D-78464 Konstanz
Germany

Library of Congress Cataloging-in-Publication Data

The nervous systems of invertebrates : an evolutionary and comparative approach / with a coda written by T. H. Bullock ; edited by O. Breidbach, W. Kutsch.
p. cm. — (EXS ; 72)
Includes bibliographical references and index.

1. Invertebrates—Nervous system. I. Breidbach, Olaf. II. Kutsch, Wolfram, 1941– . III. Series.
QL364.N46 1995
596′.0188—dc20 94-35125
CIP

Deutsche Bibliothek Cataloging-in-Publication Data

The nervous systems of invertebrates : an evolutionary and comparative approach / with a coda written by T. H. Bullock. Ed. by O. Breidbach ; W. Kutsch. —Basel ; Boston ; Berlin : Birkhäuser, 1995.
(Experientia : Supplementum ; 72)

NE: Breidbach, Olaf [Hrsg.]

The publisher and editor can give no guarantee for the information on drug dosage and administration contained in this publication. The respective user must check its accuracy by consulting other sources of reference in each individual case.

The use of registered names, trademarks etc. in this publication, even if not identified as such, does not imply that they are exempt from the relevant protective laws and regulations or free for general use.

This work is subject to copyright. All rights are reserved, whether the whole or part of the material is concerned, specifically the rights of translation, reprinting, re-use of illustrations, recitation, broadcasting, reproduction on microfilms or in other ways, and storage in data banks. For any kind of use permission of the copyright owner must be obtained.

© 1995 Birkhäuser Verlag, PO Box 133, CH-4010 Basel, Switzerland

Softcover reprint of the hardcover 1st edition 1995

Printed on acid-free paper produced from chlorine-free pulp

ISBN-13: 978-3-0348-9949-9 e-ISBN-13: 978-3-0348-9219-3
DOI: 10.1007/978-3-0348-9219-3

9 8 7 6 5 4 3 2 1

Contents

Arthropoda

Chelicerata

Echinodermata

Methodological comment

Coda

The Nervous Systems of Invertebrates: An Evolutionary and Comparative Approach
ed. by O. Breidbach & W. Kutsch
© 1995 Birkhäuser Verlag Basel/Switzerland

Introductory remarks

Evolution is the unifying theme of biological thought.

Arbas et al. (1991)

Can it be doubted, from the struggle each individual has to obtain subsistance, that any minute variation in structure, habits, or instincts, adapting that individual better to the new conditions, would tell upon its vigour and health? Thus already in his first account on the evolution of species, communicated by C. Lyell to the Linnean Society, 1858, Charles Darwin pointed to the importance of a study of structure and behavior, for an understanding of the evolution of organisms.

More recent concepts, especially in sociobiology, strengthen this viewpoint that promotes analysis of animal behavior to understand the evolution of species. Since the nervous system controls behavior, for an approach to understand animal evolution it is essential to understand the evolution of the nervous system, in particular its most developed structure, the brain. Unfortunately, the vertebrate nervous system is composed of an enormous amount of highly complicated structures. Here, evolution of the nervous system can currently be described only at the level of centers and fiber tracts (Sarnat and Netsky, 1981), which themselves have developed already into very advanced structures. In invertebrates, however, we can describe several neuronal populations and even identify individual cells. Therefore, it might be of relevance to revert to the invertebrates in search of general rules controlling the evolution of nervous systems. Besides several practical reasons (see below), it is not our intention to artificially separate vertebrates and invertebrates. Modern developmental biology clearly demonstrates that essential features such as neuronal path finding or genetics of head formation are shared by both vertebrate and invertebrate species. For a further analysis of the wealth of information accumulated from vertebrates and concerning the evolution of certain brain areas (Sarnat and Netsky, 1981), it will be of significance to compare these data with results based on identified cells or cell groups collected from invertebrates.

Because of their, in general, less complicated organization of the nervous systems, invertebrates attracted such scientists from the last century ownards as Sigmund Freud, Hermann von Helmholtz, and Fridjof Nansen. In their description of especially the crustacean nervous systems these early neuroanatomists tried to detect general principles controlling the organization of all nervous systems (Florey and Breidbach, 1993). Until now, even though several nervous systems of invertebrates are of highly sophisticated structure (in this volume, for *Cephalopoda*, see Budelmann; arthropod brains, see Breidbach; Strausfeld et al.) the advantage of studying this large group is still attractive, even considering its apparent disparity.

1. In invertebrates the sheer smaller number and often apparent simplicity of the system enables to describe cell groups or even identify single neurons. Thus, an analysis of the invertebrate nervous system can be based on the very detailed description necessary when trying to adopt the morphological concept into modern neurobiological analysis. It allows to employ modern techniques of physiology, developmental biology and genetics on the level of the single modules a nervous system is formed of (Kutsch and Breidbach, 1994).
2. The invertebrates form a highly diversified group with a correspondingly diverse variation of their neural structures. Nevertheless, certain substructures, such as the visual neuropile or the neuroendocrine system, may display similarities. Such similarity might be explained either by a case of homology, depending on a common ancestor, or by a case of convergency, for which a common precursor is not existent. Cases of convergent evolution of the nervous systems suggest that certain computational tasks are performed best by a specific neural architecture (see Strausfeld et al., this volume). Therefore, the identification of such convergencies might give the opportunity to characterize common trends that are essential in the construction of patterns of the nervous systems even if they do not stem from a common ancestor. Nevertheless rules dominating the evolution of the nervous system should become more obvious when regarding possible homonomous or homologous structures (Boyan and Williams, this volume). Thereby, it might be possible to reconstruct the evolution of nervous systems utilizing the homology concept based on identified neurons (Kutsch and Breidbach, 1994; see also Kutsch and Heckmann, Bulloch and Ridgway, this volume).

In their outstanding work, Bullock and Horridge (1965) collected all the relevant information on the invertebrate nervous systems. This monumental study, concluded almost 30 years ago, was undertaken "to summarize the present state of knowledge and to bring together the scattered and growing literature" whereby – by refraining from additionally burdening this collection – "most phylogenetic comparisons are eschewed".

Supplementary knowledge accumulated during the following three decades and the introduction of new concepts and methods, such as provided by developmental biology and genetics, now enables us to tackle the phylogenetic implications. Neurophysiologists exploited the concept of the identified neuron introduced by Wiersma (1953) with the aim to understand nervous system morphology. Intensive anatomical studies supported by new staining methods such as single-cell dye injections or immunohistochemistry, or refinement of physiological methods such as single-cell recording or ablation in freely moving animals, and the introduction of optical recording techniques resulted in a considerable amount of new aspects. The accumulated data has enabled us, at least for a few "model" invertebrates, especially insects and gastropods, to outline some principal schemes of the functional organization of the nervous systems.

Furthermore, developmental biology and genetics have introduced new concepts into invertebrate neurobiology. Obviously, the developmental rules are coded genetically, but are expressed in cell morphologies (Lawrence, 1992). Accordingly, to understand the equivalent programs, one has to combine the genetic and the morphological perspectives. Since evolution of organisms relies on changes of such programs, to understand structural diversification in different species it is necessary to describe the processes of structural realizations of the equivalent species-specific genetic designs. One can follow the development of gene expression patterns and compare the developmental genetics of different species (Aisemberg et al., Meier and Reichert, Boyan and Williams, this volume). It has become possible to study the basic pattern of early embryogenesis with the goal to unveil common features throughout neurogenesis of different species (Whitington, Meier and Reichert, this volume). And one can study the genetics and expression of neurotransmitters or neuromodulators (Grimmelikhuijzen and Westfall, Agricola and Bräunig, this volume). A combination of all such data may result in some new vista of the invertebrate nervous systems and may give us some hints on possible evolutionary steps. Under these new perspectives the search for common principles controlling the organization of all nervous systems, such as envisaged previously by Freud, von Helmholtz, and Nansen, has become revitalized, supported by a strict phylogenetic methodology (Schmitt, this volume).

Therefore, in order to understand evolution, there must be a developmental morphological view which combines morphology, physiology and genetics (Breidbach, 1994). At the evolutionary level it has to be asked whether the diversifications in the nervous systems of certain invertebrate groups can be understood taking into consideration the variations in the appropriate developmental program. To analyze the basic mechanisms which control the fundamental lineage of developmental programs research has, inevitably, to focus on selected "model" animals. However, to understand the variation on which evolution bears, one has to consider the whole spectrum of diversity which is even found in the development of single species. The evaluation of the significance of a certain type of developmental program requires a comparative view. In this context, the phylogenetically based reconstruction of principle schemes of organization patterns, the "Bauplan", gives us the criteria to evaluate the evolutionary significance of a certain type of onto- genesis. Such a view even forms the basis to understand the functional morphology of the nervous systems. It has to be determined as to how far functional needs can alter the developmental programs (Walthall, Budelmann, this volume) as well as in which parts of the developmental programs they are conserved throughout different behavioral adaptations (Bulloch and Ridgway, Wegerhoff and Breidbach, Breidbach, Strausfeld et al., this volume).

If evolutionary implications are intended, then such an approach must be guided by phylogenetic methodology. Central for such an approach is a search for homologies that allows to identify common characters of a certain group (synapomorphy, *sensu* Hennig (1982)). They may enable to describe the ground plan, the principle structural and biochemical organization of the nervous system of a distinct invertebrate group (Reuter and Gustafsson, Sandeman and Scholtz, this volume). Following the reconstruction of such a general pattern for a certain group, deviation from this ground plan can be identified and the adaptational characteristics inducing or – at least – facilitating such an alteration can be disclosed. Hereby, it is to be determined how far functional needs have altered developmental programs, and which parts of the nervous system remained conservative in different organisms, even if different behavioral patterns have to be carried out (Strausfeld et al., Wegerhoff and Breidbach, Kutsch and Heckmann, Budelmann, this volume). The identification of convergence *versus* homology renders the tool for proceeding with such an analysis.

From the techniques available from neuroanatomy, neurophysiology, developmental genetics, and functional morphology, it might be possible to describe the evolution of the invertebrate nervous systems in such a perspective. The questions to answer in such an approach are:

1. Is it possible to describe a common "Bauplan" of the nervous system for any specific invertebrate group?
2. Are there any homologies among neighbour groups?
3. Is it possible to construct a basic nervous system developed by the ancestor of a group or groups?
4. Are there any rules dominating the change of the nervous system during evolution?
5. Are there conservative parts and faster evolving regions, such as the sensory neuropile?
6. Are there common trends of neural system changes correlated with behavioral pattern or invasion of new terrains?

7. Is it possible to correlate ontogenetic with phylogenetic alterations of specific neural populations?

The present volume tries to give some preliminary answers to some of the questions raised above. This is a first step towards an evolutionary neurobiology of invertebrates. The various contributions reflect the suitability of such an approach.

Since the publication of the fundamental treatise of Bullock and Horridge (1965) the knowledge on different views about the nervous systems of some invertebrate groups has profoundly enlarged. The present work tries to review certain aspects of such progressions. Therefore, it may serve as a frame of reference about recent attempts to understand the construction of different invertebrate groups. Clearly, the reader will notice a strong bias towards the arthropods, especially the insects. Only few studies are concerned with the neurobiology of Echinodermata (Cobb, this volume); important arthropod groups, like Arachnida (Wegerhoff and Breidbach, this volume) or Myriapoda (Kutsch and Heckmann, Whitington, this volume), are only briefly mentioned. Furthermore, supplementary studies concerning intermediate groups, such as Onychophora (Schürmann, this volume) are needed. This imbalance towards the other groups reflects the present attitude, to center research on specific animals, the "model" organisms. The evolutionary perspective presently outlined may demonstrate that such a restriction is inadequate for fully understanding the evolution of the nervous systems. One major outcome of evolution is species diversity!

Still, the descriptions collected in this book demonstrate the applicability of the comparative approach in invertebrate neurobiology. The authors not only describe the present state of the art for the various groups, but the contributions also demonstrate the potency of the different techniques available for evolutionary neurobiology: from identification of homeobox genes in different species (Aisemberg et al., this volume), parallel physiological and anatomical characterization of single, identified neurons (Boyan and Williams, this volume) to classical neuroanatomical studies (Sandeman and Scholtz, this volume). Thereby, the different chapters elucidate diversified facets of the comparative approach. They also emphasize that these different ventures to approach the invertebrate nervous systems form part of one basic attitude which, in its essential concepts and methodologies, is of morphological origin.

This compendium is based on contributions and discussions during an international symposium at the University of Constanze on "The Nervous Systems of Invertebrates: a Comparative Approach" (October 1993). We thank the *Freunde und Förderer der Universität*, the *Mainau-Stiftung* and the *Byk-Gulden* company, Constanze, for their generous financial support. We are also grateful to Dr. P. Gerlach and Ms. K. Metzger of the publisher

Birkhäuser Verlag for their efforts in compiling this volume. Especially, we thank Prof. Dr. T.H. Bullock for his continuous interest in our endeavors; it has been a great honor to receive his prolific and encouraging contribution.

Olaf Breidbach and Wolfram Kutsch

Altenberg, Germany, May 1994

References

Arbas, E.A., Meinertzhagen, I.A. and Shaw, S.R. (1991) Evolution in nervous systems. *Ann. Rev. Neurosci.* 14: 9–38.

Breidbach, O. (1994) Entwicklungsmorphologie - Ein neuer Ansatz zur Fundierung einer organismischen Biologie? *Jb. Geschichte u. Theorie der Biologie* 1: 21–43.

Bullock, T.H. and Horridge, G.A. (1965) *Structure and Function in the Nervous Systems of Invertebrates.* Two volumes. Freeman, San Francisco.

Darwin, C. and Wallace, A. (1858) On the tendency of species to form varieties and species by natural means of selection. *In:* P.H. Barrett (ed.): *Darwin, C. – The collected papers of Charles Darwin.* Univ. of Chicago Press, Chicago Vol. 3, p. 6. (1977).

Florey, E. and Breidbach, O. (1993) *Das Gehirn – Organ der Seele*? Akademie-Verlag, Berlin.

Henning, W. (1982) *Phylogenetische Systematik.* Parey, Berlin.

Kutsch, W. and Breidbach, O. (1994) Homologous structures in the nervous systems of Arthropoda. *Adv. Insect Physiol.* 24: 1–113.

Lawrence, D.H. (1992) *The Making of a Fly. The Genetics of Animal Design.* Blackwell, Oxford.

Sarnat, H.B. and Netsky, M.G. (1981) *Evolution of the Nervous System.* Oxford Univ. Press, New York, Oxford.

Wiersma, C.A.G. (1953) Neural transmission in invertebrates. *Physiol. Rev.* 33: 326–355.

The Nervous Systems of Invertebrates: An Evolutionary and Comparative Approach
ed. by O. Breidbach & W. Kutsch
© 1995 Birkhäuser Verlag Basel/Switzerland

The nervous systems of Cnidarians

C.J.P Grimmelikhuijzen[1,2] and J.A. Westfall[3]

[1]*Department of Cell Biology and Anatomy, University of Copenhagen, Universitetsparken 15, DK-2100 Copenhagen Ø, Denmark*
[2]*Centre for Molecular Neurobiology, University of Hamburg, Martinistraße 52, D-20246 Hamburg, Germany*
[3]*Department of Anatomy and Physiology, Kansas State University, Manhattan, Kansas 66506, USA*

Summary. Cnidarians have simple nervous systems and it was probably within this group or a closely-related ancestor that nervous systems first evolved. The basic plan of the cnidarian nervous system is that of a nerve net which, at some locations, has condensed to form nerve plexuses, or circular or longitudinal nerve tracts which may be syncytia. At the ultrastructural level, many cnidarian neurons have the combined characteristics of sensory, motor, inter- and neurosecretory neurons and thus appear to be multifunctional. We propose that these multifunctional neurons resemble the ancestors of the more specialized neurons that we find in higher animals today. The primitive nervous system of cnidarians is strongly peptidergic: from a single sea anemone species *Anthopleura elegantissima*, we have now isolated 16 different novel neuropeptides. These peptides are biologically active and cause inhibitions or contractions in muscle preparations or isolated muscle cells from sea anemones. The various peptides are located in at least six distinct sets of neurons showing that sea anemone neurons have already specialized with respect to their peptide content. Using immuno-electronmicroscopy, we have found that the peptides are located in neuronal dense-cored vesicles associated with both synaptic and non-synaptic release sites. All these data indicate that evolutionarily "old" nervous systems use peptides as transmitters. We have also investigated the biosynthesis of the cnidarian neuropeptides. These neuropeptides are made as large precursor proteins which contain multiple (up to 36) copies of immature neuropeptides. Thus, the biosynthesis of neuropeptides in cnidarians is very efficient and comparable to that of higher invertebrates, such as molluscs and insects, and vertebrates.

Introduction

Cnidarians are primitive animals, such as sea anemones, corals, and jellyfishes. During the 17th and 18th centuries many cnidarians were regarded as plants and later were given colorful names such as sea pansies, sea anemones, and sea roses. Even today, the beautiful "flora" of coral reefs is praised by traveling and diving guides, whereas cnidarians (especially corals) are meant. The phylum Cnidaria can be subdivided into four classes: the Hydrozoa (e.g., *Hydra*), Cubozoa (box jellyfishes), Scyphozoa (true jellyfishes), and Anthozoa (e.g., sea anemones and corals). The members of the first three classes frequently have a life cycle that includes polyp and medusa forms, whereas the Anthozoa have only polyp forms. Cnidarians can live as individuals or as a colony. The colony may consist of only polyp forms, or of mixed polyps and medusae. The colonies may be sessile or free-living, as is the case with siphonophores (e.g., "Portuguese man-of-war"), which are pelagic, swimming or floating, colonial hydrozoans consisting of a central stem or disk to which pneumatophores (float), swimming bells (medusae), and a variety of feeding, reproductive, defensive, and other modified polyps are attached.

Cnidarians are not only beautiful and impressive organisms, but they are also interesting for biologists as an experimental animal. This is for the following reasons. First, cnidarians have an anatomically simple nervous system (see below) and only a few, simple, behavioral patterns which, in the case of hydromedusae, include swimming, feeding, and a special type of defensive behavior ("crumpling"). This makes hydromedusae good model animals to study behavior (Anderson and Schwab, 1982; Spencer and Arkett, 1984; Anderson and Spencer, 1989). Second, *Hydra* and other hydrozoan polyps have an amazingly high regeneration capacity: from slices of *Hydra* or even dissociated and subsequently reaggregated cells, new animals develop within a few days (Gierer et al., 1972; Gierer, 1977). For this reason, *Hydra* and other hydropolyps are often used as a model organism by developmental biologists. Finally, nervous systems probably evolved in cnidarians or in a closely-related ancestor (Mackie, 1990). Therefore, cnidarians are interesting from an evolutionary point of view.

The anatomy of the cnidarian nervous system

The basic organization of the cnidarian nervous system is that of a nerve net. At some locations this nerve net has condensed to form longitudinal or circular nerve tracts. An example of a longitudinal nerve tract is the "giant axon" in the stem of physonectid siphonophores. This nerve tract consists of fused neurons and enables, by virtue of its enlarged diameter, fast signal transduction and, consequently, a fast escape reaction (Mackie, 1973, 1984; Grimmelikhuijzen et al., 1986). Examples of circular condensations are the inner and outer nerve rings at the bell margin of hydrozoan medusae. These rings, which consist of electrically coupled neurons, are capable of integrating a variety of sensory imputs and of transmitting these signals rapidly throughout the margin (Spencer and Arkett, 1984).

In addition to a nerve net, nerve rings, and nerve tracts, many hydrozoan, cubozoan, and scyphozoan medusae have developed light- and gravity-sensitive organs, which may have evolved in response to their free-living life style. The light-sensitive organs (ocelli) are found at the bases of the marginal tentacles and, in cubo- and scyphomedusae may be integrated into larger organs, the rhopalia. The ocelli consist of a cluster of light-sensitive neurons often associated with a cup of pigmented cells. In cubomedusae, the ocelli are sometimes equipped with lenses and light-adjustable irises and these organs can even be rotated towards the light source, thereby showing the properties of real eyes (G. Jarms, personal communication). The light-sensitive neurons in the ocelli of hydromedusae are connected by two or more nerve tracts to the outer nerve ring of the bell margin where the signals are integrated and transmitted to the motor neurons of the inner nerve ring (Spencer and Arkett, 1984; Grimmelikhuijzen and Spencer, 1984). Other sense organs integrated in the rhopalia are statocysts. These statocysts are pits or hollow structures filled with crystalline (calcareous) material surrounded by neurons that are connected to the nerve net of the bell. The statocyst neurons have a pacemaker function related to swimming behavior (Horridge, 1954, 1956). Specialized sense organs have not been found in polyps.

The organization of the cnidarian nervous system has been investigated by both light and electron microscopy. The visualization of neurons for light microscopy, however, has always been difficult, because, until recently, no good method has been available that stains cnidarian neurons well. The best stain used has been methylene blue, which was already applied about a century ago by Schneider (1890) and Hadži (1909). Using methylene blue staining, these investigators found that two types of neurons exist in *Hydra* and other cnidarians: (i) "sensory cells" which are long, slender neurons orientated perpendicular to the mesoglea (a layer of collagen located between the two cell layers: an outer ectoderm or epidermis and an inner endoderm or gastrodermis) and projecting to the surface of one of the cell layers. Near this surface the "sensory cell" forms a cilium which extends into the outer medium, or into the lumen of the gastric cavity; (ii) "ganglion cells" which are neurons with a more round perikaryon located in the basal parts of ecto- or endoderm. Westfall has reinvestigated the *Hydra* nervous system using electron-microscopy and found that there is no essential difference between "sensory" (Fig. 1a) and "ganglion cells" (Fig. 1b) in *Hydra*: both types of neurons contain a cilium (are "sensory"), store dense-cored vesicles at non-synaptic regions (are "neurosecretory"), and form chemical synapses with both epithelio-muscular cells and neighboring neurons (are both "motor-" and "interneurons") (Westfall, 1973a; Westfall and Kinnamon, 1978). Westfall proposes that these primitive, multifunctional nerve cells may resemble the ancestors of the more specialized neurons that we find in higher animals today. However, whether *all* cnidarian neurons have these multifunctional characteristics has yet to be established. It is possible that neurons exist which have three, two, or only one of the above-mentioned features available and, are, for example, only sensory-motor-, or interneurons. This is supported by recent scanning electron micrographs of dissociated neurons from the sphincter of the sea anemone *Calliactis parasitica*, showing that many ganglion cells lack the cilium observed in *Hydra* ganglion cells (Westfall, unpublished).

The existence of neurons in cnidarians having both sensory and motor functions suggests that these animals must have a reflex arc that is even simpler than the well-known monosynaptic reflex arc of mammals. Whereas in the mammalian nervous system at least two neurons are needed (one sensory and one motor neuron), one single sensory-motor-inter-neuron does the job in *Hydra* and probably many other cnidarians.

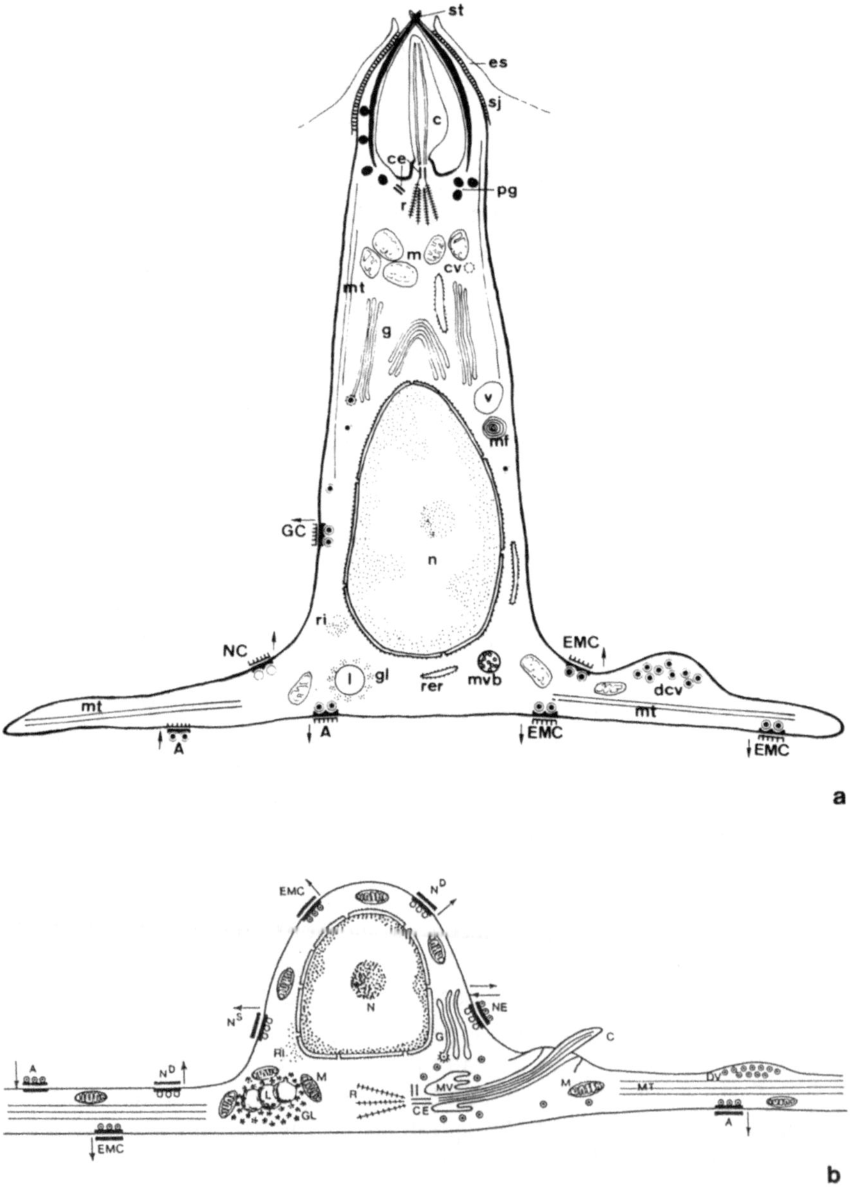

The function of this single, multifunctional neuron might be the control of local contractions that occur, e.g., during feeding (mouth opening, tentacle bending).

Neurotransmission

As mentioned earlier, dense-cored vesicles (90–150 mm) associated with synaptic and non-synaptic release sites have been found in all mature *Hydra* neurons (Westfall, 1973a; Westfall and Kinnamon, 1978, 1984). Similar observations have been made for other cnidarian species (Horridge and Mackay, 1962; Jha and Mackie, 1967; Westfall, 1973b; Quaglia and Grasso, 1986; see Westfall, 1987, for a review). The presence of chemical synapses and chemical (paracrine) neurotransmission in the cnidarian nervous system was confirmed by electrophysiological experiments. Using intracellular recordings at both pre- and postsynaptic neurons of hydro- and scyphomedusae, EPSPs were found with the expected, constant latency from the presynaptic spikes (Spencer, 1982; Spencer and Arkett, 1984; Anderson, 1985; Anderson and Spencer, 1989). Furthermore, neurotransmission was blocked after depletion of Ca^{2+} from the incubation medium, or after addition of excess of Mg^{2+}, indicating that classical, exocytotic release of neurotransmitter substances occurs (McFarlane, 1973; Satterlie, 1979; Spencer, 1982).

In addition to chemical synapses, electrical synapses have also been found in the cnidarian nervous system, but, so far, this has only been observed in members of the Hydrozoa. Electrical synapses have been demonstrated by electrical and dye coupling between neurons and by the presence of structures that were similar to gap junctions (Anderson and Mackie, 1977; Spencer, 1978, 1979; Spencer and Satterlie, 1980; Westfall et al., 1980; Spencer and Arkett, 1984). We have already

---

Figure 1. Schematic drawings of neurosecretory sensory-motor-interneurons of *Hydra*. **a:** Sensory cell from the tentacle showing neurosecretory dense-cored vesicles (dcv), a sensory cilium (c), and polarized synapses with epitheliomuscular cells (EMC), ganglion cells (GC), "en passant" axons (A), and nematocytes (NC). Other abbreviations are: paired centrioles (ce), coated vesicles (cv), epitheliomuscular cell sheath (es), Golgi complex (g), glycogen (gl), lipid droplets (l), mitochondria (m), myelin figure (mf), microtubules (mt), multivesicular body (mvb), nucleus (n), pigment granules (pg), rootlets (r), rough endoplasmic reticulum (rer), ribosomes (ri), septate junction (sj), stereocilia (st) and vacuoles (v). From Westfall and Kinnamon, 1978. **b:** Ganglion cell from the tentacle showing, again, neurosecretory dense-cored vesicles (DV), a sensory cilium (c) and polarized synapses with epitheliomuscular cells (EMC), axons (A), different types of nematocytes (N^S, N^P) and two-way somatic, interneuronal synapses (NE). Other abbreviations are: paired centrioles (CE), Golgi complex (G), glycogen particles (GL), lipid droplet (L), mitochondria (M), microtubules (MT), microvilli (MV), nucleus (N), rootlets (R) and ribosomes (RI). From Westfall, 1973a.

mentioned that in some cases, such as in the nerve net and "giant axon" of the stem of physonectid siphonophores, the neurons have apparently fused with each other to form true syncytia (Mackie, 1973; Grimmelikhuijzen et al., 1986). The presence of neuronal syncytia may be more common in the Cnidaria than has been realized so far and must be suspected wherever nerve tracts, nerve rings or "giant axons" are observed. The first step in the formation of a neuronal syncytium may be the formation of gap junctions, followed by fusion of the neuronal cell membranes (cf. Mackie et al., 1988; Mackie, 1989). Thus, the presence of electrical and dye coupling and gap junctions in, for example, the inner nerve ring of hydromedusae (Anderson and Mackie, 1977; Spencer, 1978, 1979; Spencer and Satterlie, 1980) does not exclude that this ring may be partially a syncytium.

Several chemical synapses in the cnidarian nervous system are bidirectional with synaptic vesicles found at both sides of the synapse (Horridge and MacKay, 1962; Westfall, 1973 a,b; Westfall and Kinnamon, 1984; Anderson and Grünert, 1988). Bidirectional or two-way synapses are especially frequent in the nerve nets of scyphomedusae where they transmit excitation in both directions: intracellular, transsynaptic recordings have shown that an action potential in one cell evokes an EPSP in the other cell with the same synaptic delay (1 ms), irrespective of which of the two cells is stimulated (Anderson, 1985; Anderson and Spencer, 1989). Bidirectionality could be responsible for the diffuse (nondirected) conduction that is often seen in cnidarian nerve nets, but bidirectional synapses are apparently not an absolute requirement and diffuse conduction can be explained by other means (Josephson et al., 1961; Anderson and Spencer, 1989).

Neuropeptides

It has generally been accepted that cnidarian neurons use neurotransmitters or locally acting (paracrine) hormones for signal transmission (see above). For a long time, however, the nature of these transmiter substances has remained unknown. We have been unable to demonstrate catecholamines, serotonin or acetylcholine in neurons of *Hydra* (Grimmelikhuijzen, 1986). These negative results have recently been confirmed by other researchers (Koizumi, personal communication). This means that the monoamines and acetylcholine are not generally occurring in the nervous systems of cnidarians and indicates that classical transmitters are not the evolutionarily "oldest" neurotransmitters.

Several years ago, we showed that antisera against the molluscan neuropeptide Phe-Met-Arg-Phe-NH_2 (FMRFamide; Price and Greenberg, 1977) and especially against its C-terminal fragment Arg-Phe-NH_2 (RFamide) stain neurons in all classes of Cnidaria (Grimmelikhuijzen et al., 1982, 1986, 1987, 1988a, 1989a,b, 1992a; Grimmelikhuijzen, 1983a, 1985; Grimmelikhuijzen and Spencer, 1984; Anderson et al., 1992). These results have been confirmed by other workers in the field (Mackie et al., 1985, 1988; Plickert, 1989; Weber, 1989; Koizumi et al., 1992; Kroiher and Plickert, 1992; Martin 1992). Staining with FMRFamide or RFamide antisera does not, of course, mean that FMRFamide is the transmitter substance in cnidarians; it only indicates that substances are present in cnidarian neurons that have structural similarities with the sequence FMRFamide or RFamide (see also below). Many cnidarians such as hydrozoan polyps and medusae are transparent and can be stained as whole mounts by the RFamide antisera (Grimmelikhuijzen and Spencer, 1984; Grimmelikhuijzen, 1985). This immunocytochemical staining method is superior to the methylene blue staining method of Schneider (1890) and Hadži (1909) and has given us a much clearer picture of the organization of the cnidarian nervous system than has previously been possible. Naturally, only a portion of all cnidarian neurons may be stained by the RFamide antisera (see also below), but this same drawback holds for any other staining method. After the classical work by Hadži (1909), it was generally believed that *Hydra* and other hydroid polyps have a diffuse, loosely interconnected nerve net with no forms of centralization. This misconception is still presented in most of the introductory textbooks of zoology (Figs 2a,b). In contrast to this picture, which held for over 70 years, staining with RFamide antisera shows that *Hydra vulgaris* (formerly called *Hydra attenuata*) has a strong agglomeration of sensory neurons and processes in the hypostome (around the mouth opening) and a densely packed collar of neurons in the peduncle (near the foot) (Figs 2c, 3a). In a related species, *Hydra oligactis*, there is an obvious nerve ring lying at the border of the hypostome and tentacles (Fig. 2d). These are clear examples of neuronal centralizations that have not been demonstrated to their full extent before by use of conventional histological techniques or electronmicroscopy (cf. Davis et al., 1968; Kinnamon and Westfall, 1981; Matsuno and Kageyama, 1984). Polyps of the marine hydroid *Hydractinia echinata* have a very dense neuronal plexus in the body column and a collection of sensory neurons around the mouth opening (Fig. 3b). Thus, the density and complexity of the hydroid nervous system has long been underestimated and this becomes even more true if one realizes that the RFamide-positive neurons are only a subpopulation of the total hydroid nervous system and that additional, peptidergic neurons exist (Grimmelikhuijzen, 1983b; Koizumi and Bode, 1991; Koizumi et al., 1992).

An account similar to that of the hydroid polyps can be given for other members of the Hydrozoa (Grimmelikhuijzen and Spencer, 1984; Grimmelikhuijzen et al., 1986), for Scyphozoa (Anderson et al., 1992), Cubozoa and Anthozoa (Grimmelikhuijzen, unpublished). The nervous

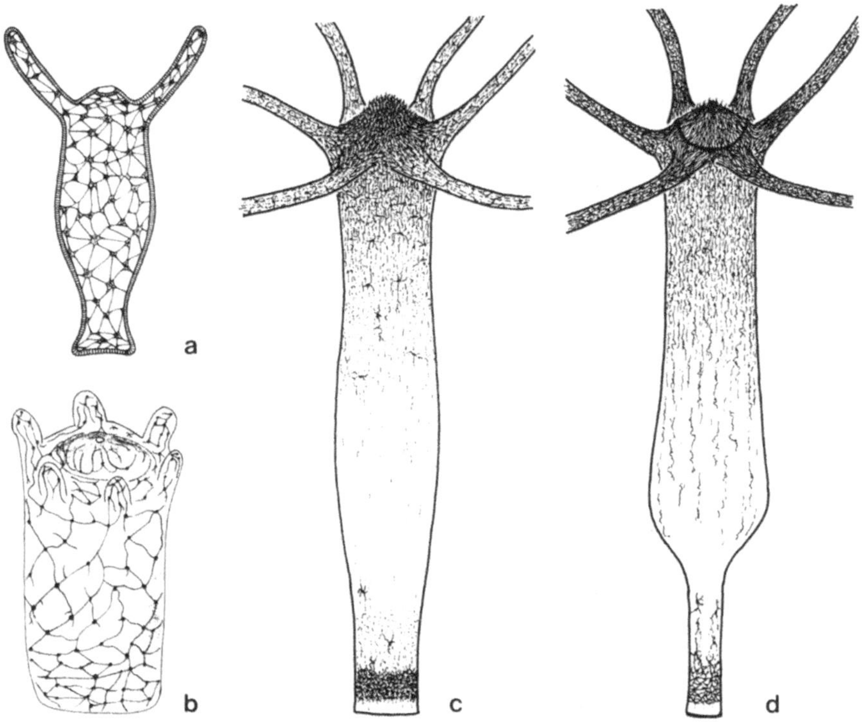

Figure 2. Drawings showing the nervous system of *Hydra*. (**a**) From a textbook of zoology by Marshall and Williams (1972); a similar picture is given in a textbook by Brusca and Brusca (1990). The nervous system is presented as a diffuse nerve net, with no regions of centralization. (**b**) From a zoology textbook by Barnes (1968). Again, only a diffuse, loosely interconnected nerve net is shown. (**c**) The nervous system of *Hydra vulgaris* stained with an antiserum against the sequence RFamide. Two centralizations occur: a strong agglomeration of sensory neurons and processes around the mouth opening and a densely packed collar of neurons in the peduncle. From Grimmelikhuijzen, 1985. (**d**) The nervous system of *Hydra oligactis* stained with an antiserum against RFamide. This species has a dense plexus of immunoreactive neurites in the hypostome, a cluster of sensory neurons around the mouth opening, and a nerve ring between hypostome and tentacle bases. The collar of neurons in the peduncle is less well developed. From Grimmelikhuijzen, 1985.

system of sea anemones, for example, has always been described as a diffuse, widely-meshed nerve net, even in the most respectable textbooks of zoology (Fig. 4a). The truth is, however, that such pictures are wrong, and that sea anemones have amazingly dense and highly organized nerve plexuses around the mouth and in many areas of their body column. An example of such a nerve plexus in the oral disk is given in Figure 4b.

Staining with RFamide antisera is not only an excellent technique to visualize a major portion of the cnidarian nervous system, but it also gives a clear indication of the neurotransmitter substances which may

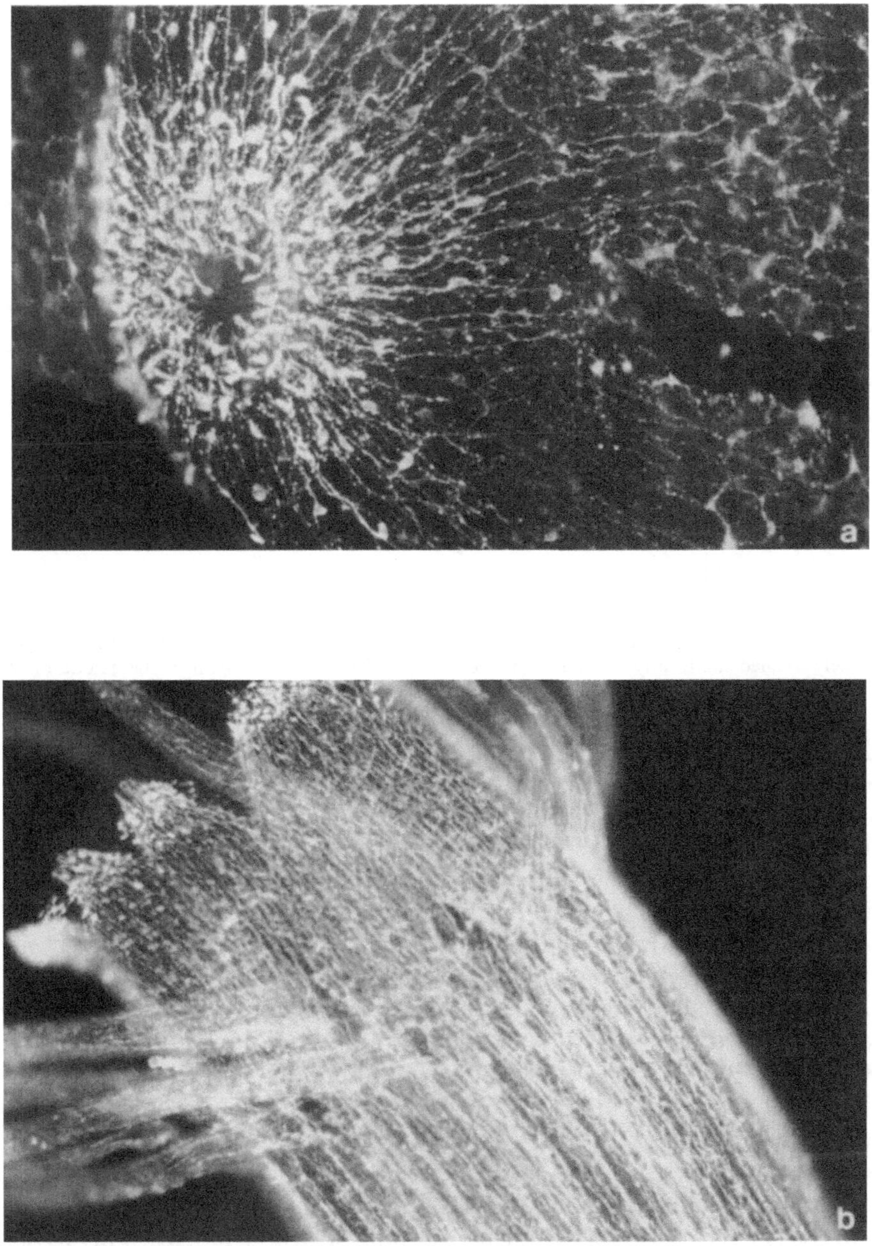

Figure 3. Whole-mount staining of hydroid polyps with an RFamide antiserum. (**a**) staining of the nervous system in the hypostome of *Hydra vulgaris*. Numerous sensory cells form a cluster around the mouth opening (× 180). (**b**) The nervous system in the head and body column of an adult gastrozooid (feeding polyp) of *Hydractinia echinata*. A very dense plexus of immunoreactive processes occurs throughout the polyp and numerous sensory neurons are present around the mouth opening (the mouth area has been broken by pressure of the cover slip) (× 120). Both figures are adapted from Grimmelikhuijzen, 1985.

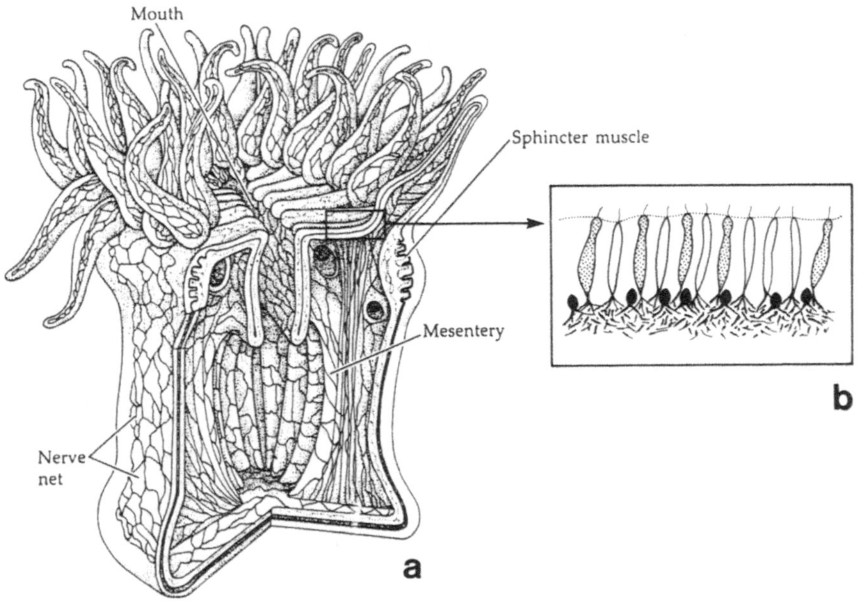

Figure 4. Drawings showing the nervous system of sea anemones. (**a**) From a textbook of invertebrate zoology by Brusca and Brusca (1990). The nervous system is presented as a widely-meshed, uniformly distributed nerve net. (**b**) The nervous system in the ectoderm of the oral disk as it appears after staining with three antisera: against Antho-RFamide (black ganglionic neurons), Antho-RWamides I and II (white sensory neurons), and the Antho-RPamides (spotted sensory neurons) (see also Tab. 1). The nervous system in the oral disk is a complex organization of different types of neurons forming a dense plexus of processes in the basal part of the ectoderm (Grimmelikhuijzen, unpublished).

be used. In order to isolate these substances, we have developed a radioimmunoassay against the sequence RFamide. Using this assay as a monitoring system, we have purified a novel neuropeptide, < Glu-Gly-Arg-Phe-NH_2 (Antho-RFamide), from extracts of the sea anemone *Anthopleura elegantissima* (Grimmelikhuijzen and Graff, 1986). Antho-RFamide has also been isolated from the sea pansy *Renilla köllikeri* and appears to occur generally in the Anthozoa (Grimmelikhuijzen and Groeger, 1987). Using our RFamide radioimmunoassay, we have also isolated four RFamide neuropeptides from *Hydra* (Moosler and Grimmelikhuijzen, unpublished), two RFamide peptides from the hydromedusa *Polyorchis penicillatus* (Grimmelikhuijzen et al., 1988b, 1992b), and three RFamide peptides from the scyphomedusa *Cyanea lamarckii* (Moosler and Grimmelikhuijzen, unpublished). The two *Polyorchis* peptides are given in Table 1. These and all other isolated cnidarian RFamide peptides are structurally related and have the C-terminal sequence Gly-Arg-Phe-NH_2 in common (Tab. 1). Thus, the Gly-Arg-Phe-NH_2 peptides appear to be ubiquitous in the Cnidaria.

Using Arg-Phe-NH_2, Arg-Ile-NH_2, Arg-Asn-NH_2, and Arg-Pro-NH_2 radioimmunoassays, we have isolated altogether 16 different neuropep-

Table 1. Neuropeptide families in cnidarians.

Species	Structure	Name
Anthopleura elegantissima	L-3-phenyllactyl-Phe-Lys-Ala-NH_2	Antho-KAamide
Anthopleura elegantissima	L-3-phenyllactyl-Tyr-Arg-Ile-NH_2	Antho-RIamide I
Anthopleura elegantissima	Tyr-Arg-Ile-NH_2	Antho-RIamide II
Anthopleura elegantissima	L-3-phenyllactyl-Leu-Arg-Asn-NH_2	Antho-RNamide I
Anthopleura elegantissima	Leu-Arg-Asn-NH_2	Antho-RNamide II
Anthopleura elegantissima	<Glu-Ser-Leu-Arg-Trp-NH_2	Antho-RWamide I
Anthopleura elegantissima	<Glu-Gly-Leu-Arg-Trp-NH_2	Antho-RWamide II
Anthopleura elegantissima	<Glu-Asn-Phe-His-Leu-Arg-Pro-NH_2	Antho-RPamide II
Anthopleura elegantissima	Leu-Pro-Pro-Gly-Pro-Leu-Pro-Arg-Pro-NH_2	Antho-RPamide I
Anthopleura elegantissima	Gly-Pro-Hyp-Ser-Leu-Phe-Arg-Pro-NH_2	Antho-RPamide IV
Anthopleura elegantissima	<Glu-Val-Lys-Leu-Tyr-Arg-Pro-NH_2	Antho-RPamide III
Anthopleura elegantissima	Tyr-Arg-Pro-NH_2	Antho-RPamide V
Anthopleura elegantissima	<Glu-Gly-Arg-Phe-NH_2	Antho-RFamide
Renilla köllikeri	<Glu-Gly-Arg-Phe-NH_2	Antho-RFamide
Polyorchis penicillatus	<Glu-Leu-Leu-Gly-Gly-Arg-Phe-NH_2	Pol-RFamide I
Polyorchis penicillatus	<Glu-Trp-Leu-Lys-Gly-Arg-Phe-NH_2	Pol-RFamide II

tides from the sea anemone *Anthopleura elegantissima* (Grimmelikhuijzen and Graff, 1986; Graff and Grimmelikhuijzen, 1988a,b; Grimmelikhuijzen et al., 1990; Nothacker et al., 1991a,b; Cartensen et al., 1992, 1993; see Grimmelikhuijzen et al., 1992a, for a review). Thirteen *Anthopleura* neuropeptides are given in Table 1. All peptides are structurally related and can be organized into six groups: Antho-RFamide, Antho-KAamide, the Antho-RIamides, -RNamides, -RWamides and -RPamides (Tab. 1). The peptides are biologically active and have characteristic inhibitory or excitatory actions on ecto- or endodermal, longitudinal or circular muscle preparations from sea anemones (McFarlane et al., 1987, 1991, 1992, 1993; McFarlane and Grimmelikhuijzen, 1991; Nothacker et al., 1991b; Carstensen et al., 1992, 1993). For some peptides, for example the Antho-RWamides, we have found that they act directly on isolated muscle cells, indicating that they are neuromuscular transmitters (McFarlane et al., 1991). The actions of the sea anemone peptides are summarized in Table 2.

We have raised antisera against the C and N termini of most of the peptides of Table 1. Staining with these antisera have shown that Antho-RFamide, Antho-KAamide, and the Antho-RIamides, -RNamides, -RWamides, and -RPamides are each produced by a different and characteristic group of neurons (Grimmelikhuijzen, unpublished). This means that there are at least six different sets of neurons in sea anemones. Thus, although many cnidarian neurons may be multifunctional, they are not identical and have already specialized with

Table 2. Excitatory (+) or inhibitory (−) actions of neuropeptides on different muscle groups of sea anemones.

Peptide	Tentacle	Body column	
	longitudinal	longitudinal	circular
Antho-RFamide	+	+	+
Antho-RWamide I, I	-	+	+
Antho-RNamide I	+	+	-
Antho-RIamide I	-	-	-
Antho-KAamide	-	-	-
Antho-RPamide I	+		+
Antho-RPamide II, III	-		+

respect to their peptide content. This, now, also implies a specialization in function (see Tab. 2).

We have started to investigate the ultrastructural localization of the cnidarian neuropeptides. Using RFamide antisera and gold-labeled secondary antibodies, we have found RFamide-like material in the granular cores of neuronal dense-cored vesicles of *Hydra* (Koizumi et al., 1989). These dense-cored vesicles were located in nerve processes paralleling the myonemes of epitheliomuscular cells and in nerve endings terminating on the myonemes, suggesting that the RFamide-like material is involved in neuromuscular transmission. Using antisera against the sea anemone neuropeptide Antho-RFamide, we found immunoreactive material in neuronal dense-cored vesicles that were associated with two-way neuro-neuronal synapses of sea anemones, and also with non-synaptic release sites (Westfall and Grimmelikhuijzen, 1993). Therefore, there are now morphological, physiological, and biochemical data indicating that primitive nervous systems use peptides as transmitters.

Biosynthesis of the cnidarian neuropeptides

We have also investigated the biosynthesis of the cnidarian neuropeptides that are listed in Table 1. However, as this topic is not fully covered by the scope of this volume, we prefer to give only a short overview. For more details, we would like to refer to a recent review (Grimmelikhuijzen et al., 1994) and to the original literature (Darmer et al., 1991; Schmutzler et al., 1992, 1994; Reinscheid and Grimmelikhuijzen, 1994).

In higher animals, neuropeptides are made as large precursor proteins (preprohormones) which consist of a hydrophobic, N-terminal signal sequence (for translocation across the endoplasmic reticulum membrane; the "pre"-part) and a prohormone part containing one or more copies of the immature neuropeptide. The immature neuropeptide sequences are coverted into mature neuropeptides by a series of processing steps (e.g., cleavage at basic amino acid residues) catalyzed by processing enzymes that are co-packaged with the neuropeptide precursor protein into the neurosecretory (dense-cored) vesicles. Maturation of the peptide precursor protein takes place during the transport of the neurosecretory vesicles from the Golgi system to their final release sites. From higher animals it is also known that the N-terminal <Glu group (see Tab. 1) orginates from an N-terminal Gln residue and that the C-terminal amide group (Tab. 1) is a conversion product from a C-terminal Gly residue. Thus, for the precursor of Antho-RFamide (<Glu-Gly-Arg-Phe-NH_2), for example, we expect a protein that should have an N-terminal signal sequence and one or more immature Antho-RFamide sequences (Gln-Gly-Arg-Phe-Gly) flanked by basic amino acid residues.

We have used two strategies to clone the Antho-RFamide precursor. First, we have raised antibodies against the internal, immature Antho-

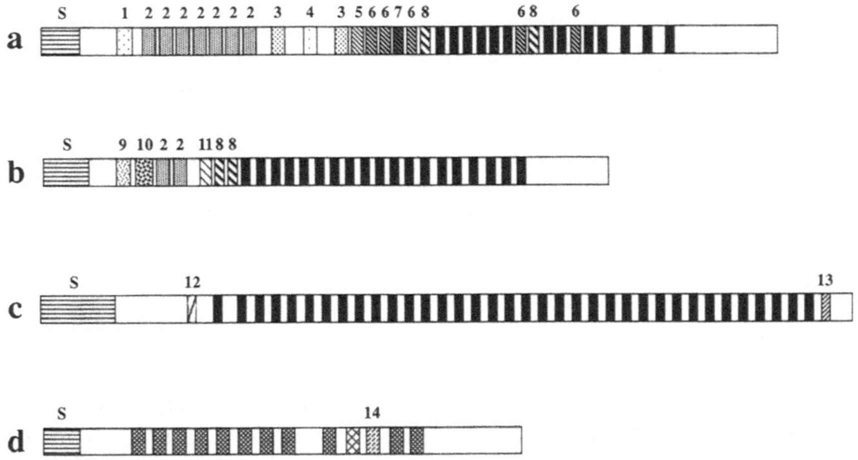

Figure 5. Schematic representation of the RFamide neuropeptide precursor proteins from two sea anemone species, a sea pansy and a hydromedusa. S, signal sequence. (**a**) The Antho-RFamide precursor from the sea anemone *Anthopleura elegantissima*. This precursor is 435 amino acid residues long and contains 13 copies of immature Antho-RFamide (black boxes) and 20 copies of other, putative neuropeptide sequences (numbered 1–8). Modified from Schmutzler et al., 1992. (**b**) The Antho-RFamide precursor from sea anemone *Calliactis parasitica*. This precursor is 334 amino acids long and contains 19 copies of immature Antho-RFamide (black boxes) and seven copies of other, putative neuropeptides (numbered 2, 8–11; the putative neuropeptides 2 and 8 are identical with the ones in the *Anthopleura* precursor). Modified from Darmer et al., 1991. (**c**) The Antho-RFamide precursor from the sea pansy *Renilla köllikeri*. This Antho-RFamide precursor is 487 amino acids long and contains 36 copies of immature Antho-RFamide and two putative neuropeptide sequences (numbered 12, 13). Modified from Reinscheid and Grimmelikhuijzen, 1994. (**d**) The Pol-RFamide precursor from the hydromedusa *Polyorchis penicillatus*. This precusor is 287 amino acid residues long and contains 11 copies of immature Pol-RFamide II (darkly hatched), one copy of immature Pol-RFamide I (lightly hatched), and one putative neuropeptide sequence (numbered 14). Modified from Schmutzler et al., 1994.

RFamide sequence that was coupled via two flanking Lys residues (Lys-Gln-Gly-Arg-Phe-Gly-Lys) to a carrier protein, thyroglobulin. The antibodies were subsequently affinity-purified and used to screen an expression cDNA (λgt11) library from sea anemones. Second, we have used a 384-fold degenerate pool of 15-mer oligonucleotide probes coding for the sequence Gln-Gly-Arg-Phe-Gly to screen the same λgt11 cDNA library. Both strategies were successful. Figure 5a shows a schematic representation of the Antho-RFamide precursor protein from *Anthopleura elegantissima*. This precursor contains 13 copies of immature Antho-RFamide and 20 other, putative neuropeptide sequences (Schmutzler et al., 1992). The Antho-RFamide precursor from the sea anemone *Calliactis parasitica* (Fig. 5b) contains 19 copies of immature Antho-RFamide and seven other, putative neuropeptides (Darmer et al., 1991). The Antho-RFamide precursor from the sea pansy *Renilla köllikeri* (Fig. 5c) is an extreme example of efficient neuropeptide production: it contains 36 copies of Antho-RFamide and two putative neuropeptide sequences (Reinscheid and Grimmelikhuijzen, 1994). This *Renilla* Antho-RFamide precursor protein is the most productive neuropeptide precursor protein known so far. The neuropeptides Pol-RFamide I and Pol-RFamide II from the hydromedusa *Polyorchis penicillatus* (Tab. 1) are produced by a common precursor protein (Fig. 5d) that contains one copy of Pol-RFamide I, 11 copies of Pol-RFamide II, and one putative neuropeptide sequence (Schmutzler et al., 1994). Thus, the biosynthesis of neuropeptides in cnidarians is very efficient and comparable to that of higher invertebrates, such as molluscs and insects, and vertebrates.

Conclusion

The nervous systems of cnidarians are "primitive", which, however, does not always mean "simple", as we can see, for example, from the high number of peptidergic systems present in a single species (Tab. 1), the complexity of neuropeptide precursors (Fig. 5), the complexity of single neurons (Fig. 1), or the existence of rather advanced specializations, such as the eyes in cubomedusae. The features that we find in the nervous systems of species of *all* cnidarian classes may be features that were present in the earliest cnidarians or in a closely related ancestor. To these features belong multifunctional neurons, chemical neurotransmission, and the use of peptides as neurotransmitters. Of all neuropeptides, the Gly-Arg-Phe-NH_2 peptides appear to be ubiquitous in the Cnidaria, so these types of peptides are good candidates for being the first transmitters.

Cnidarians are not on a side-line of neuronal evolution: action potentials, synaptic transmission, neurosecretion, neuropeptide biosynthesis, and many other neuronal mechanisms are as in higher animals.

Recently, Antho-RFamide and other Gly-Arg-Phe-NH_2 peptides have been isolated from neuronal tissue of crustaceans (Yasuda et al., 1993). This, again, suggests that cnidarians are near the main line of evolution and that many properties of the cnidarian nervous systems of today may reflect properties of the first nervous systems in evolution.

Acknowledgements
We thank Dr. G. Jarms (Zoological Institute, University of Hamburg) for access to unpublished material, Meike Bochow for typing the manuscript, and Elizabeth Beyerholm and Dörthe Claus for drawing Figures 4 and 5. This work was supported by the Danish Natural Science Research Council, Bundesministerium für Forschung und Technologie, Deutsche Forschungsgemeinschaft (Gr 762/10-1), NATO, Fonds der Chemischen Industrie (to C.J.P.G) and the NSF grant IBN-9120161 and NIH grant NS-10264 (to J.A.W.).

References

Anderson, P.A.V. and Mackie, G.O. (1977) Electrically coupled, photosensitive neurons control swimming in a jellyfish. *Science* 197: 186–188.

Anderson, P.A.V. and Schwab, W.E. (1982) Recent advances and model systems in coelenterate neurobiology. *Progr. Neurobiol.* 19: 213–236.

Anderson, P.A.V. (1985) Physiology of a bidirectional, excitatory, chemical synapse. *J. Physiol.* 53: 821–835.

Anderson, P.A.V. and Grünert, U. (1988) Three-dimensional structure of bidirectional, excitatory synapses in the jellyfish *Cyanea capillata. Synapse* 2: 606–613.

Anderson, P.A.V. and Spencer, A.N. (1989) The importance of cnidarian synapses for neurobiology. *J. Neurobiol.* 20: 435–457.

Anderson, P.A.V., Moosler, A. and Grimmelikhuijzen, C.J.P. (1992) The presence and distribution of Antho-RFamide-like material in scyphomedusae. *Cell Tissue Res.* 267: 67–74.

Barnes, R.D. (1968) *Invertebrate Zoology.* Second Edition, W.B. Saunders Company, Philadelphia (p. 83).

Brusca, R.C. and Brusca, G.J. (1990) *Invertebrates.* Sinauer Associates Inc., Sunderland (p. 245).

Carstensen, K., Rinehart, K.L., McFarlane, I.D., and Grimmelikhuijzen, C.J.P. (1992) Isolation of Leu-Pro-Pro-Gly-Pro-Leu-Pro-Arg-Pro-NH_2(Antho-RPamide), an N-terminally protected, biologically active neuropeptide from sea anemones. *Peptides* 13: 851–857.

Carstensen, K., McFarlane, I.D., Rinehart, K.L., Hudman, D., Sun, F. and Grimmelikhuijzen, C.J.P. (1993) Isolation of <Glu-Asn-Phe-His-Leu-Arg-Pro-NH_2 (Antho-RPamide II), a novel, biologically active neuropeptide from sea anemones. *Peptides* 14: 131–135.

Darmer, D., Schmutzler, C., Diekhoff, D. and Grimmelikhuijzen, C.J.P. (1991) Primary structure of the precursor for the sea anemone neuropeptide Antho-RFamide (<Glu-Gly-Arg-Phe-NH_2). *Proc. Natl. Acad. Sci. USA* 88: 2555–2559.

Davis, L.E., Burnett, A.L. and Haynes, J.F. (1968) Histological and ultrastructual study of the muscular and nervous system in *Hydra*. II. Nervous system. *J. Exp. Zool.* 167: 295–332.

Gierer, A., Berking, S., Bode, H., David, C.N., Flick, K., Hansmann, G., Schaller, H. and Trenckner, E. (1972) Regeneration of *Hydra* from reaggregated cells. *Nature New Biol.* 239: 98–101.

Gierer, A. (1977) Biological features and physical concepts of pattern formation exemplified by *Hydra. Curr. Top. Dev. Biol.* 11: 17–59.

Graff, D. and Grimmelikhuijzen, C.J.P. (1988a) Isolation of <Glu-Ser-Leu-Arg-Trp-NH_2, a novel neuropeptide from sea anemones. *Brain Res.* 442: 354–358.

Graff, D. and Grimmelikhuijzen, C.J.P. (1988b) Isolation of <Glu-Gly-Leu-Arg-Trp-NH_2 (Antho-RWamide II), a novel neuropeptide from sea anemones. *FEBS Lett.* 239: 137–140.

Grimmelikhuijzen C.J.P., Dockray, G.J. and Schot, L.P.C. (1982) FMRFamide-like immunoreactivity in the nervous system of *Hydra. Histochemistry* 73: 499–508.

Grimmelikhuijzen, C.J.P. (1983a) FMRFamide immunoreactivity is generally occurring in the nervous systems of coelenterates. *Histochemistry* 78: 361–381.

Grimmelikhuijzen, C.J.P. (1983b) Coexistence of neuropeptides in *Hydra*. *Neurosci.* 9: 837–845.
Grimmelikhuijzen C.J.P. and Spencer, A.N. (1984) FMRFamide immunoreactivity in the nervous system of the medusa *Polyorchis penicillatus*. *J. Comp. Neurol.* 230: 361–371.
Grimmelikhuijzen, C.J.P. (1985) Antisera to the sequence Arg-Phe-amide visualize neuronal centralization in hydroid polyps. *Cell Tissue Res.* 241: 171–182.
Grimmelikhuijzen, C.J.P. (1986) FMRFamide-like peptides in the primitive nervous systems of coelenterates and complex nervous systems of higher animals. *In:* G. Stephano (ed.): *Handbook of Comparative Opioid and Related Neuropeptide Mechanisms*. CRC Press, Boca Raton, pp. 103–115.
Grimmelikhuijzen C.J.P. and Graff, D. (1986) Isolation of <Glu-Gly-Arg-Phe-NH_2 (Antho-RFamide), a neuropeptide from sea anemones. *Proc. Natl. Acad. Sci. USA* 83: 9817–9821.
Grimmelikhuijzen C.J.P., Spencer, A.N. and Carré, D. (1986) Organization of the nervous system of physonectid siphonophores. *Cell Tissue Res.* 246: 463–479.
Grimmelikhuijzen C.J.P., Graff, D., Groeger, A. and McFarlane, I.D. (1987) Neuropeptides in invertebrates. *In:* M. A. Ali (ed.): *Nervous Systems in Invertebrates*. Plenum Press, New York, pp. 105–132.
Grimmelikhuijzen C.J.P. and Groeger, A. (1987) Isolation of the neuropeptide pGlu-Gly-Arg-Phe-amide from the pennatulid *Renilla köllikeri*. *FEBS Lett.* 211: 105–108.
Grimmelikhuijzen C.J.P., Graff, D. and Spencer, A.N. (1988a) Structure, location and possible actions of Arg-Phe-amide peptides in coelenterates. *In:* M.C. Thorndyke and G.J. Goldsworthy (eds): *Neurohormones in Invertebrates*. Cambridge University Press, Cambridge, pp. 199–217.
Grimmelikhuijzen, C.J.P., Hahn, M., Rinehart, K.L. and Spencer, A.N. (1988b) Isolation of <Glu-Leu-Leu-Gly-Gly-Arg-Phe-NH_2 (Pol-RFamide) a novel neuropeptide from hydromedusae. *Brain Res.* 475: 198–203.
Grimmelikhuijzen C.J.P., Graff, D. and McFarlane, I.D. (1989a) Neurones and neuropeptides in coelenterates. *Arch. Histol. Cytol.* 52 (Suppl.): 265–276.
Grimmelikhuijzen C.J.P., Graff, D., Koizumi, O., Westfall, J.A. and McFarlane, I.D. (1989b) Neurones and their peptide transmitters in coelenterates. *In:* P.A.V. Anderson (ed.): *Evolution of the First Nervous Systems*. Plenum Press, New York, pp. 95–109.
Grimmelikhuijzen, C.J.P., Rinehart, K.L., Jacob, E., Graff, D., Reinscheid, R.K., Nothacker, H.-P. and Staley, A.L. (1990) Isolation of L-3-phenyllactyl-Leu-Arg-Asn-NH_2(Antho-RNamide), a sea anemone neuropeptide containing an unusual amino-terminal blocking group. *Proc. Natl. Acad. Sci. USA* 87: 5410-5414.
Grimmelikhuijzen, C.J.P., Carstensen, K., Darmer, D., Moosler, A., Nothacker, H.-P., Reinscheid, R.K., Schmutzler, C., Vollert, H., McFarlane, I.D. and Rinehart, K.L. (1992a) Coelenterate neuropeptides: structure, action and biosynthesis. *Amer. Zool.* 32: 1–12.
Grimmelikhuijzen, C.J.P., Rinehart, K.L. and Spencer, A.M. (1992b) Isolation of the neuropeptide <Glu-Trp-Leu-Lys-Gly-Arg-Phe-NH_2(Pol-RFamide II) from the hydromedusa *Polyorchis penicillatus*. *Biochem. Biophys. Res. Commun.* 183: 375–382.
Grimmelikhuijzen, C.J.P., Darmer, D., Schmutzler, C., Reinscheid, R.K. and Carstensen, K. (1994). Biosynthesis of neuropeptides in the Cnidaria: new discoveries of old principles. *In:* K.G. Davey, R.E. Peter and S.S. Tobe (eds): *Perspectives in Comparative Endocrinology*. National Research Council Canada, Ottawa, pp. 97–108.
Hadži, J. (1909) Über das Nervensystem von *Hydra*. *Arb. Zool. Inst. Wien* 17: 225–268.
Horridge, G.A. (1954) Observations on the nerve fibres of *Aurelia aurita*. *Quart. J. Micr. Sci.* 95: 85–92.
Horridge, G.A. (1956) The nervous system of the ephyra larva of *Aurelia aurita*. *Quart. J. Micr. Sci.* 97: 59–74.
Horridge, G.A. and Mackay, B. (1962) Naked axons and symmetrical synapses in coelenterates. *Quart. J. Micr. Sci.* 103: 531–541.
Jha, R.K. and Mackie, G.O. (1967) The recognition, distribution and ultrastructure of hydrozoan nerve elements. *J. Morphol.* 123: 43–62.
Josephson, R.K., Reiss, R.F. and Worthy, R.M. (1961) A simulation study of a diffuse conducting system based on coelenterate nerve nets. *J. Theor. Biol.* 1: 460–487.
Kinamon, J.C. and Westfall, J.A. (1981) A three-dimensional serial reconstruction of neuronal distributions in the hypostome of a *Hydra*. *J. Morphol.* 168: 321–329.
Koizumi, O., Wilson, J.D., Grimmelikhuijzen, C.J.P. and Westfall, J.A. (1989) Ultrastructural localization of RFamide-like peptides in neuronal dense-cored vesicles in the peduncle of *Hydra*. *J. Exp. Zool.* 249: 17–22.

Koizumi, O. and Bode, H.R. (1991) Plasticity in the nervous system of adult *Hydra* III. Conversion of neurons to expression of a vasopressin-like immunoreactivity depends on axial location. *J. Neuroscience* 11: 2011–2020.

Koizumi, O., Itazawa, M., Mizumoto, H., Minobe, S., Javois, L.C., Grimmelikhuijzen, C.J.P. and Bode, H.R. (1992) Nerve ring of the hypostome in *Hydra*. I. Its structure, development, and maintenance. *J. Comp. Neurol.* 326: 7–21.

Kroiher, M. and Plickert, G. (1992) Analysis of pattern formation during embryonic development of *Hydractinia echinata*. *Roux's Arch. Dev. Biol.* 201: 95–104.

Mackie, G.O. (1973) Report on giant nerve fibres in *Nanomia*. *Publ. Seto Marine Lab.* 20: 745–756.

Mackie, G.O. (1984) Fast pathways and escape behavior in Cnidaria. *In:* R.C. Eaton (ed.): *Neural Mechanisms of startle behavior*, Plenum Press, New York, pp. 15–42.

Mackie, G.O., Singla, C.L. and Stell, W.K. (1985) Distribution of nerve elements showing FMRFamide-like immunoreactivity in hydromedusae. *Acta Zool.* 66: 199–210.

Mackie, G.O., Singla, C.L. and Arkett, S.A. (1988) On the nervous system of *Vellela* (Hydrozoa: Chondrophora). *J. Morphol.* 198: 15–23.

Mackie, G.O. (1989) Evolution of cnidarian giant axons. *In:* P.A.V. Anderson (ed.): *Evolution of the First Nervous Systems*. Plenum Press, New York, pp. 395–407.

Mackie, G.O. (1990) The elementary nervous system revisited. *Amer. Zool.* 30: 907–920.

Martin, V.J. (1992) Characterization of an RFamide-positive subset of ganglionic cells in the hydrozoan planular nerve net. *Cell Tissue Res.* 269: 431–438.

Marshall, A.J. and Williams, W.D. (1972) *Textbook of Zoology. Invertebrates*, Seventh Edition, The MacMillan Press, London (p. 12).

Matsuno, T. and Kageyama, T. (1984) The nervous system in the hypostome of *Pelmatohydra robusta*: the presence of a circumhypostomal nerve ring in the epidermis. *J. Morphol.* 182: 153–168.

McFarlane, I.D. (1973) Spontaneous contractions and nerve-net activity in the sea anemone *Calliactis parasitica*. *Mar. Behaviour Physiol.* 2: 97–113.

McFarlane, I.D., Graff, D. and Grimmelikhuijzen, C.J.P. (1987) Exitatory actions of Antho-RFamide, an anthozoan neuropeptide, on muscles and conducting systems in the sea anemone *Calliactis parasitica*. *J. Exp. Biol.* 133: 157–168.

McFarlane, I.D., Anderson, P.A.V. and Grimmelikhuijzen, C.J.P. (1991) Effects of three anthozoan neuropeptides, Antho-RWamide I, Antho-RWamide II and Antho-RFamide, on slow muscles from sea anemones. *J. Exp. Biol.* 156: 419–431.

McFarlane, I.D. and Grimmelikhuijzen, C.J.P. (1991) Three anthozoan neuropeptides, Antho-RFamide and Antho-RWamides I and II, modulate spontaneous tentacle contractions in sea anemones. *J. Exp. Biol.* 155: 669–673.

McFarlane, I.D., Reinscheid, R.K. and Grimmelikhuijzen, C.J.P. (1992) Opposite actions of the anthozoan neuropeptide Antho-RNamide on antagonistic muscle groups in sea anemones. *J. Exp. Biol.* 164: 295–299.

McFarlane, I.D., Hudman, D., Nothacker, H.-P. and Grimmelikhuijzen, C.J.P. (1993). The expansion behaviour of sea anemones may be coordinated by two inhibitory neuropeptides, Antho-KAamide and Antho-RIamide. *Proc. Roy. Soc. B.* (London) 253: 183–188.

Nothacker, H.-P., Rinehart, K.L. and Grimmelikhuijzen, C.J.P. (1991a) Isolation of L-3-phenyllactyl-Phe-Lys-Ala-NH_2(Antho-KAamide), a novel neuropeptide from sea anemones. *Biochem. Biophys. Res. Commun.* 179: 1205–1211.

Nothacker, H.-P., Rinehart, K.L., McFarlane, I.D. and Grimmelikhuijzen, C.J.P. (1991b) Isolation of two novel neuropeptides from sea anemones: the unusual, biologically active L-3-phenyllactyl-Tyr-Arg-Ile-NH_2 and its des-phenyllactyl fragment Tyr-Arg-Ile-NH_2. *Peptides* 12: 1165–1173.

Plickert, G. (1989) Proportion-altering factor (PAF) stimulates nerve cell formation in *Hydractinia echinata*. *Cell Diff. Develop.* 26: 19–28.

Price, D.A. and Greenberg, M. (1977) Structure of a molluscan neuropeptide. *Science* 197: 670–671.

Quaglia, A. and Grasso, M. (1986) Ultrastructural evidence for a peptidergic-like neurosecretory cell in a sea anemone. *Oebalia* 13: 147–156.

Reinscheid, R.K. and Grimmelikhuijzen, C.J.P. (1994) Primary structure of the precursor for the anthozoan neuropeptide Antho-RFamide from *Renilla köllikeri*: evidence for unusual processing enzymes. *J. Neurochem.* 62: 1214–1222.

Satterlie, R.A. (1979) Central control of swimming in the cubomedusan jellyfish *Charibdea rastonii*. *J. Comp. Physiol.* 133: 357–367.
Schmutzler, C., Darmer, D., Diekhoff, D. and Grimmelikhuijzen, C.J.P. (1992) Identification of a novel type of processing sites in the precursor for the sea anemone neuropeptide Antho-RFamide (<Glu-Gly-Arg-Phe-NH_2) from *Anthopleura elegantissima*. *J. Biol. Chem.* 267: 22534–22541.
Schmutzler, C., Diekhoff, D. and Grimmelikhuijzen, C.J.P. (1994) The primary structure of the Pol-RFamide neuropeptide precursor protein from the hydromedusa *Polyorchis penicillatus* indicates a novel processing proteinase activity. *Biochem. J.* 299: 431–436.
Schneider, K.C. (1890) Histologie von *Hydra fusca* mit besonderer Berücksichtigung des Nervensystems der Hydropolypen. *Arch. mikrosk. Anat.* 35: 321–379.
Spencer, A.N. (1978) Neurobiology of *Polyorchis*. I. Function of effector systems. *J. Neurobiol.* 9: 143–157.
Spencer, A.N. (1979) Neurobiology of *Polyorchis*. II. Structure of effector systems. *J. Neurobiol.* 10: 95–117.
Spencer, A.N. and Satterlie, R.A. (1980) Electrical and dye coupling in an identified group of neurons in a coelenterate. *J. Neurobiol.* 11: 13–19.
Spencer, A.N. (1982) The physiology of a coelenterate neuromuscular synapse. *J. Comp. Physiol.* 148: 353–363.
Spencer, A.N. and Arkett, S.A. (1984) Radial symmetry and the organization of central neurones in a hydrozoan jellyfish. *J. Exp. Biol.* 110: 69–90.
Weber, C. (1989) Smooth muscle fibers of *Podocoryne carnea* (Hydrozoa) demonstrated by a specific monoclonal antibody and their association with neurons showing FMRFamide-like immunoreactivity. *Cell Tissue Res.* 255: 275–282.
Westfall, J.A. (1973a) Ultrastructural evidence for a granule-containing sensory-motor interneuron in *Hydra littoralis*. *J. Ultrastruct. Res.* 42: 268–282.
Westfall, J.A. (1973b) Ultrastructual evidence for neuromuscular systems in coelenterates. *Amer. Zool.* 13: 237–246.
Westfall, J.A. and Kinnamon, J.C. (1978) A second sensory-motor-interneuron with neurosecretory granules in *Hydra*. *J. Neurocytol.* 7: 365–379.
Westfall, J.A. and Kinnamon, J.C. and Sims, D.E. (1980) Neuro-epitheliomuscular cell and neuro-neuronal gap junctions in *Hydra*. *J. Neurocytol.* 9: 725–732.
Westfall, J.A. and Kinnamon, J.C. (1984) Perioral synaptic connections and their possible role in the feeding behavior of *Hydra*. *Tiss. Cell* 16: 355–365.
Westfall, J.A. (1987) Ultrastructure of invertebrate synapses. *In:* M.A. Ali (ed.): *Nervous systems in invertebrates*. Plenum Press, New York, pp. 3–28.
Westfall, J.A. and Grimmelikhuijzen, C.J.P. (1993) Antho-RFamide immunoreactivity in neuronal synaptic and nonsynaptic vesicles of sea anemones. *Biol. Bull.* 185: 109–114.
Yasuda, A., Naya, Y. and Nakanishi, K. (1993) Isolation of Antho-RFamide related peptides from the eye stalks of blue crab. *Comp. Biochem. Physiol.* 104 B: 235–240.

The Nervous Systems of Invertebrates: An Evolutionary and Comparative Approach
ed. by O. Breidbach & W. Kutsch
© 1995 Birkhäuser Verlag Basel/Switzerland

The flatworm nervous system: Pattern and phylogeny

M. Reuter and M.K.S. Gustafsson

Department of Biology, Åbo Akademi University, Artillerigatan 6, FIN-20520 Åbo, Finland

Summary. The flatworms occupy a position at the base of the metazoan phylogenetic tree; they have a bilateral symmetric nervous system and an archaic brain. The following aspects, brought into focus by the use of new methods, will be dealt with in the present paper.

1. The high degree of diversity on all levels of the flatworm nervous system (NS).
2. The concept of *main nerve cords* is defined and the use of this concept in avoiding confusions in the terminology of nerve cords is stressed.
3. The archaic nature of the stomatogastric NS is reviewed.
4. The new data about neuronal celltypes implying advanced features at this low phylogenetic level.
5. The ultrastructural studies of neuronal cells indicating (A) that a common secretory cell type containing dense-core vesicles is archaic and a likely progenitor cell type for conventional neurons of advanced flatworms and (B) that an independent evolution of synaptic structures and glial cells has occurred inside the flatworm taxon.
6. The multitude of neuroactive substances demonstrated by light microscopic histofluorescence, immunocytochemistry, liquid chromatography, and HPLC. The cholinergic, aminergic, and peptidergic substances often occur in different neuronal compartments.

Introduction

The flatworms represent the most primitive Metazoan phylum with a true bilateral symmetry and a cellular mesoderm. They lack a circulatory system and their flattened body form is regarded as a consequence of their dependence of diffusion for the distribution of nutrients and gases. As no endocrine system occurs in flatworms all integration is taken care of by the nervous system (NS). The bilaterally symmetrical central nervous system (CNS) consists of an anterior concentration of nerve cells –"the brain"– and one or several pairs of longitudinal nerve cords. The cords are connected by commissures in a ladder-like configuration, "the orthogon" (Fig. 1). The peripheral nervous system consists of a meshwork of nerve fibres and bi- or multipolar nerve cells forming nerve plexuses interconnected with the CNS. The number and structure of plexuses vary, from infra- and/or subepithelial in lower flatworms to submuscular and specialized plexuses in higher flatworms.

The monophyletic or polyphyletic status of phylum Platyhelminthes is under discussion (Smith et al., 1986). However, according to the newest phyolgenetic system the stem form of Bilateria gave rise to both

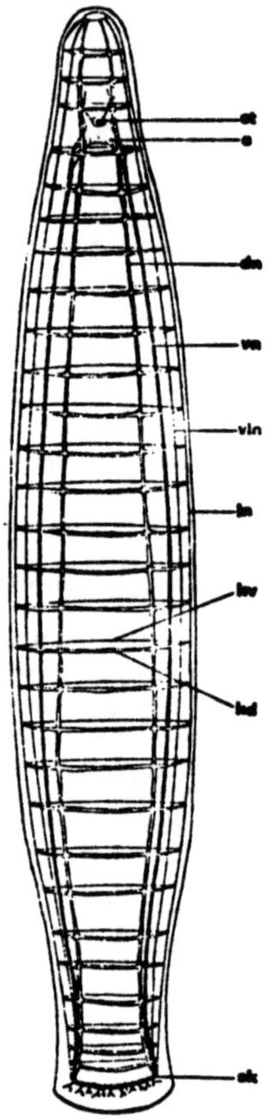

Figure 1. Orthogonal nervous system of *Minona trigonopora* (Proseriata) (Reisinger, 1972).

the Plathelminthes and the Eubilateria (Fig. 2) (Ax, 1984). According to Reisinger (1976) the *orthogonal pattern*, regarded as the basic plan for the protostome NS (Fig. 3), developed from the "Hautnervenplexus" and reflects a basic condition.

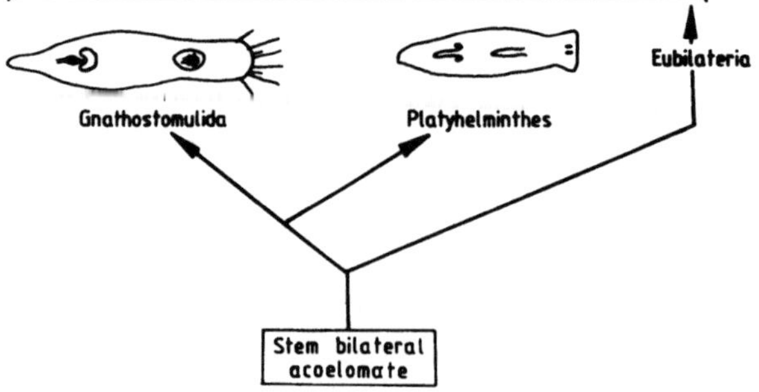

Figure 2. Phylogenetic relationship of the Bilateria (from Willmer, 1990 after Ax, 1985).

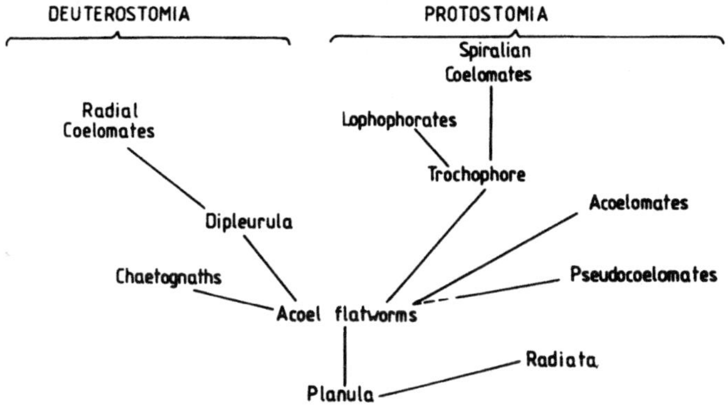

Figure 3. Traditional dichotomous phylogenetic tree of the invertebrates (Hyman, 1940).

In the present chapter, we will discuss the construction of the brain, the orthogon, the plexuses, the ultrastructure of the neuronal elements and the occurrence of neuroactive substances. The data of this paper are based on results obtained after the classic works of Hyman (1951) and Bullock and Horridge (1965). For recent reviews of the neuroanatomy of the turbellaria the reader is referred to Ehlers (1985) and Rieger et al. (1991), as for parasitic flatworms see Gustafsson (1988, 1992). The sensory structures of Plathelminthes have recently been discussed by Rohde (1990).

The central nervous system (CNS)

Brain

In the early bilateria an innovation took place that led to the development of a centralized NS with a cephalic ganglion to co-ordinate the input from the frontal sensory receptors and the activity of the two sides of the body. The cephalic ganglion is also the center for inhibiting and controlling reflexes in the peripheral nerve net (Koopowitz, 1970; Koopowitz et al., 1976). Flatworms are the most primitive animals in which we today can find this type of cephalic ganglion – *an archaic brain*. The structure of the brain varies. Lower flatworms have a few cells lateral to a fibrillar neuropile, while a distinct bilobed ganglion occurs in advanced flatworms. The lobes are connected by a commissure(s). Confocal microscopy (CSLM) shows that in *Microstomum lineare* (Macrostomida) two commissures occur, one thin dorsal and one thick ventral (Fig. 4). In the planarian brain several posteriorly located commissures occur. In *Diphyllobothrium dendriticum* (Cestoda) large neurons form the brain commissure (Fig. 5). In the majority of the

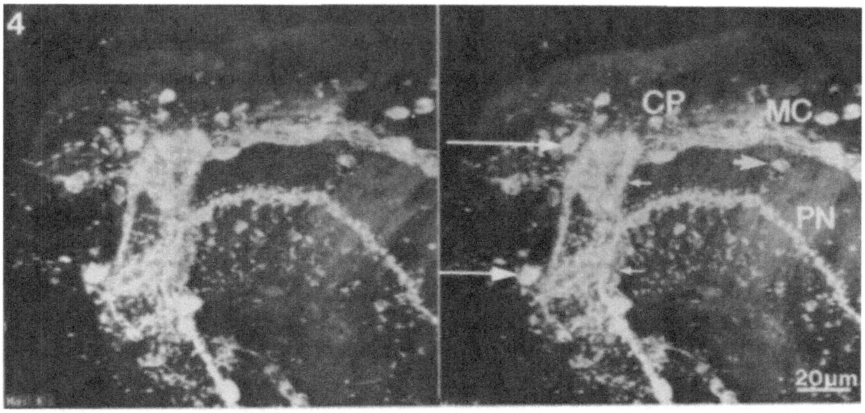

Figure 4. CSLM stereo pair showing IR to SP in brain of *Microstomum lineare*. Cells frontally to commissures (long arrows), neuropile (small arrows), pharyngeal nerve ring (PN), roots of main nerve cords (MC) (Reuter, 1994).

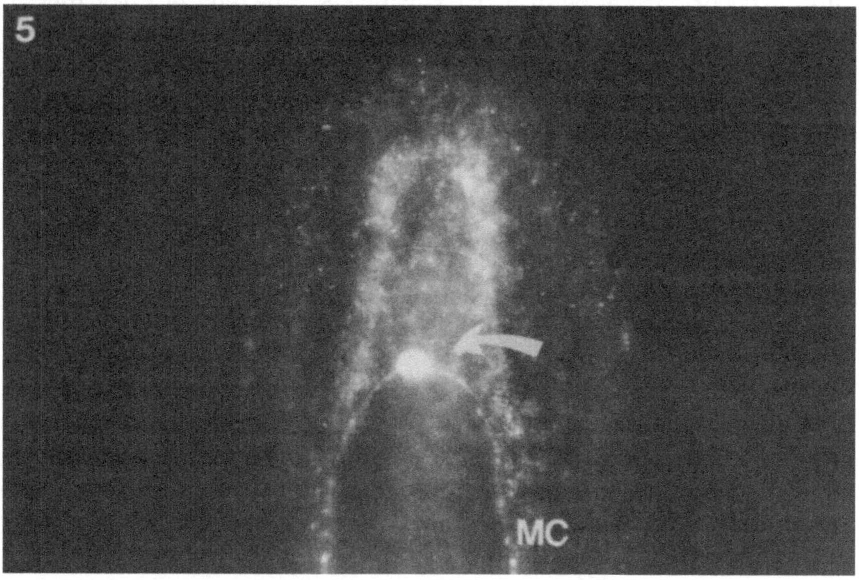

Figure 5. Whole mount showing head of *Diphyllobothrium dendriticum*. In the brain commissure, composed of large ganglion cells, a multipolar cell showing IR to RFamide occurs (bent arrow). Main nerve cord (MC); magnification ×250.

investigated flatworms the neurons are restricted to the brain periphery, while nerve fibers form a central neuropile. Muscles and necks of gland cells are observed crossing the neuropile, particularly in the stucturally spongy planarian brain (Baguñà and Ballester, 1978). An extracellular matrix layer forming a capsule around the brain occurs in species belonging to different flatworm taxa. In some worms, the neuropile, but

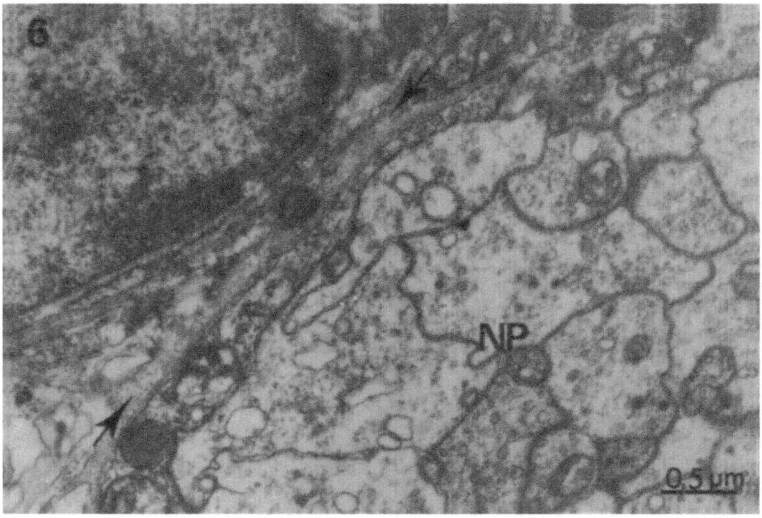

Figure 6. ECM (arrows) surrounding neuropile (NP), *Microstomum lineare*; magnification ×20,000.

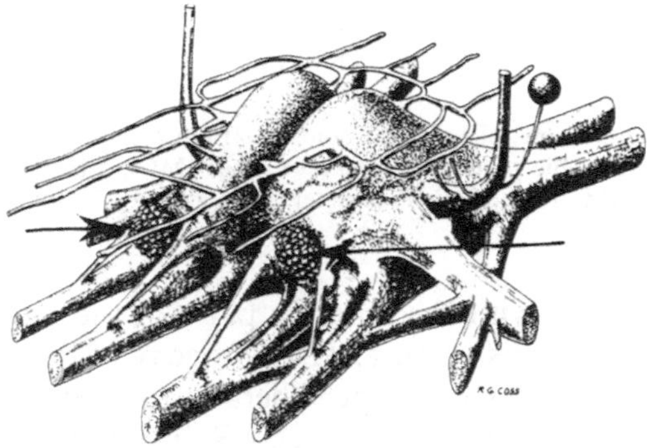

Figure 7. Serial reconstruction of the brain of *Notoplana acticola*. Clusters of globuli cells on the anterior side (arrows). Nerves I-VI. (Keenan and Koopowitz, 1981).

not the whole brain, is encircled by a sheath of extracellular matrix (ECM) continous with the matrix skeleton of the worm (Fig. 6). In polyclads specialized groups of cells, globuli cells, occur outside the capsule (Fig. 7).

Brain development

The formation of the flatworm brain from a nerve plexus and/or longitudinal cords will probably always remain an open question. It may have developed (1) from commissures in connection with the nerve

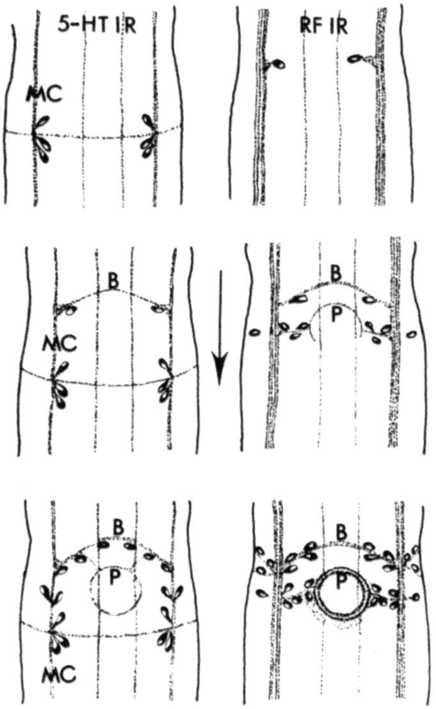

Figure 8. Development (arrow) of brain in zooids of *Microstomum lineare*. The formation of the new brain starts from the parental main nerve cords (MC), brain (B), pharyngeal nerve ring (P). Pattern of 5-HT-IR and RF-IR (Reuter and Palmberg, 1989).

net, (2) in connection with the anterior parts of the longitudinal cords or (3) as an independent structure. The ontogeny of the CNS is not well known. In *Bothrioplana semperi* peripherally located neuroblasts form brain primordia at a very early stage (Reinsinger, 1972). Later the connectives of the othogon are formed from other cells, but in contact with the primordia in an anterior-posterior direction. In the asexually reproducing and regenerating *M. lineare* the formation of the NS including the brain in new zooids is initiated by an accumulation of serotonin (5-HT) immunoreactive (IR) cells at the parental longitudinal cords (Fig. 8) (Reuter and Plamberg, 1989).

Orthogon

The basic orthogonal plan shows variation in the number of cords and in the total design. Bullock and Horridge (1965) describe nine different types of orthogons. They could neither be arranged in an evolutionary order nor could they be coupled to functional demands. Primitive organization plans occur side by side with derived plans in lower as well as in higher flatworm taxa. However, it is possible to discern evolutionary trends in the organization of the NS (Kotikova, 1986, 1991; Joffe 1990). The studies of Kotikova based on the acetylcholinesterase (AChE) and the glyoxylic-acid-induced fluorescence (GAIF) methods, show that all types of orthogons are basically homologous and derived from the "cord-type" observed in lower flatworms (Acoela and Catenulida) (Fig. 9). Some types develop directly from the "cord-type",

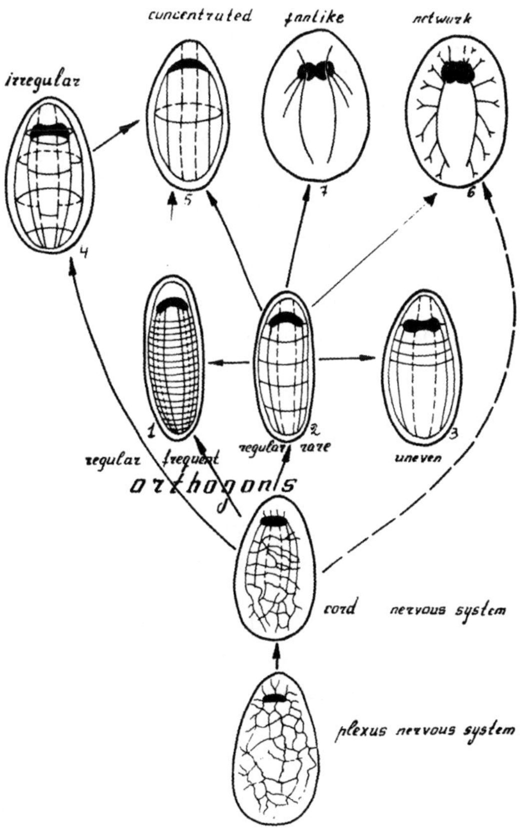

Figures 9–11. The orthogon

Figure 9. Orthogons in different groups of flatworms. Type 1: *Geocentrophora levanidorum*, *G. vagini* (Lecithoepitheliata), *Monocelis hopkinsi* (Proseriata), *Promesostoma paragracilis* (Typhloplanoidea), *Dendrocoelum lacteum* (Tricladida); 2. *Microstomum lineare* (Macrostomida) *Astratorynchuus robustus* (Typhloplanoidea) *Tetronchus monenteron* (Monogenea); 3. *Fellodistomum fellis*, *Spherostoma bramae* (Trematoda); 4. *Achrorynchides robustus* (Kalyptorynchia), *Coronpharynx pusilus* (Dalyellioida): *Friedmanniella bargusinica* (Prolecitophora); 5. *Macrostomum rostratum* (Macrostomida), *Gyratrix hermaphroditus*, *Castrada sp.* (Typhloplanoidea), *Grafilla graffi*, *Provortex karlingi Greysziroa expedita* (Dalyellioda); 6. *Acerotisa sp* (Polycladida) *Nitzschia sturionia* (Monogenea), *Facsiola hepatica* (Trematoda); 7. *Gnesioceros* (Polycladida) (Kotikova, 1986, 1991).

while others develop through transitional types. According to the author the main force determining the type of orthogon is the body shape, which in turn is connected to the life-style of the worm. Obviously, the same type of orthogon appears independently in distantly related orders (Kotikova, 1991).

Marker neurons

In connection with the study of the evolution of the orthogon within Plathelminthes neuroanatomical data have been little used (Reisinger, 1972; Sakharov, 1974; Kotikova, 1986; Joffe, 1990; Reuter, 1990; Joffe and Kotikova, 1991; Kotikova, 1991). Recently, attention has been paid to single neurons and groups of neurons, "marker neurons", identified by fluorescence methods (Hauser and Koopowitz, 1987; for review see Joffe and Kotikova, 1991; Kotikova, 1994). Morphologically similar neurons have been observed in the same position in closely related species (Fig. 10). In addition, homlogous neurons have been found in polyclads and rhabdocoels after Lucifer yellow dye injections (Keenan et al. 1981; Koopowitz, 1994). The use of marker neurons for nerve cords is discussed in a study of the proseriat *Bothriomolus balticus* (Joffe and Reuter, 1993).

Nerve cords

The nerve cords in flatworms have been named: dorsal, ventral, lateral, ventrolateral, and marginal. A confusion concerning the position of the main cords prevails. Main cords in lateral position characterizes Catenulida, Macrostomida, and the proseriat taxon Monocelididae, while a ventral position is typical for flatworms of higher taxa (Fig. 11) (Joffe and Reuter, 1993). Therefore, we want to introduce the term *main*

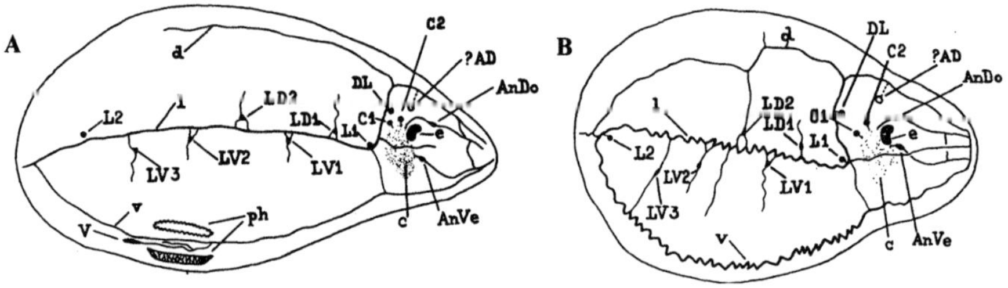

Figure 10. Pattern of CA positive neurons in two species, *Proxenetes flabellifer* (**A**) and *Trigonostomum breitfussi* (**B**). Individual marker neurons can be distinguished in the two worms. Brain (c) dorsal, ventral, marginal and lateral nerve cords (b, v, m, l), eye (e) pharynx (ph) (Joffe, 1991).

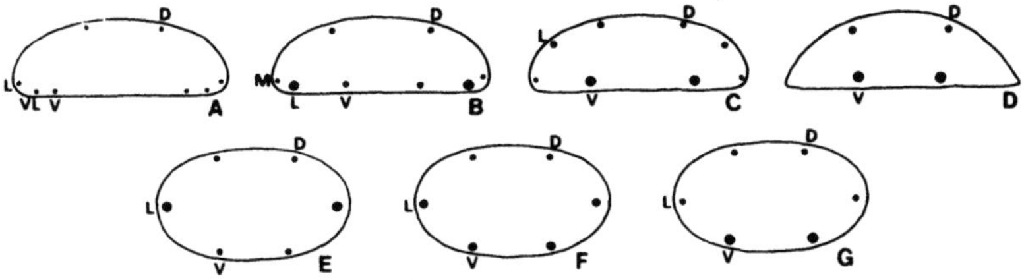

Figure 11. Number, position and relative width of longitudinal nerve cords in Seriata (A–D), Macrostomida (E), Rabdocoela (one variant) (F), and Neodermata (a typical pattern) (G). (A) *Minona trigonopora.* (B) *Monocelis* and *Archilopsis*, (C) *Bothriomolus balticus.* (D) *Bdellolura candida.* (E) *Microstomum lineare.* Cords: D, dorsal: L, lateral; M, marginal; V, ventral; VL, ventro-lateral (Joffe and Reuter, 1993).

cord as a common concept for the most prominent nerve cords in all flatworms. When distinguishing main cords form minor cords, the type of contact between them and the brain proved to be a useful characteristic (Kotikova and Timoschkin, 1987). Independent of lateral or ventral position, the main cords (1) start as multifiber outgrows (rootlets) from the brain ganglion, and (2) are distinguished by their size and 5-HT-IR and CA-positive marker neurons (Joffe and Reuter, 1993). Cords in other positions are thinner and have less pronounced contact with the brain.

The homology of main cords needs testing. The status of main cords as a stable character is, however, supported by the following observations. In *M. lineare* (Macrostomida) and *Promonotus schultzii* (Proseriata) the main cords (lateral in position) originate in the brain, while the ventral and dorsal cords have superficial contacts with the brain (see Fig. 8) (Reuter and Palmberg, 1989; Reuter, in preparation). In procercoids of *D. dendriticum* the NS consists of only two 5HT-IR main nerve cords connected by a commissure (Wikgren, 1986) (Fig. 12). In cercarians of the digenean *Schistosoma mansoni* and in adult cestodes a pair of central main cords project from the bilobed brain, while peripheral nerves occur associated to the peripheral musculature (Skuce et al. 1990; Gustafsson and Reuter, 1992).

The orthogonal pattern during the life-cycle

Parasitic flatworms go through more or less complicated life-cycles. The adaptations of NS to the changes in the external environment have been studied only in single species. In *D. dendriticum* the four 5-HT-IR ganglion cells in the brain commissure of the proceroid can be distinguished also at the plerocercoid and adult stage. From the plerocercoid stage onwards the aminergic and peptidergic NS develop strongly. The

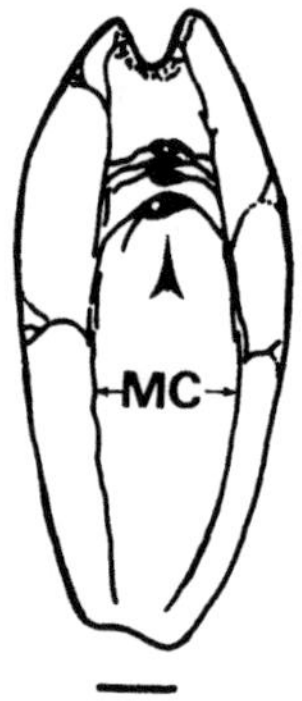

Figure 12. *Diphyllobothrium dendriticum* procercoid. Anti-serotonin positive cells and fibers revealed by whole mount ICC, brain commissure (arrowhead) and main nerve cords (MC). Bar = 20 μm (Wikgren, 1987).

main nerve cords enlarge and many minor nerve cords arise (Gustafsson et al., 1985; 1986). In connection with the final change of host, from cold fish to the warm seagull, the staining intensity of the NS increases and release take place from nerve terminals, indicating neuronal control over this important step in the worm's life (Gustafsson and Wikgren, 1981a,b,c; 1989). In the neck of the continously growing adult worm new neurons form from undifferentiated cells migrating into the main cords from the surrounding parenchyma at a rapid rate (Gustafsson, 1976). The development of the cholinergic NS during the life-cycle of the eye fluke *Displostomum pseudophataceum* has been followed. The transverse commissures increase in number during the change from the cercarial to metacercarial stage, but not when the worm reaches its final host; they just become dispersed (Fig. 13). The oral sucker of the adult worm becomes densely innervated, the hindbody develops and so does its NS (Niewiadomska and Moczon, 1982; 1984; 1987). The NS of miracidia and cercaria of *S. mansoni* has been studied by Skuce et al. (1990) with ICC. Neuropeptides of the pancreatic polypeptide family occur in the relatively simple NS of miracidia and in the more developed NS of cercaria. However, 5HT-IR was not observed in the larvae although it has been demonstrated in main and peripheral cords in the adult worm (Gustafsson, 1987).

The plexuses

In addition to nerve cords, nerve nets – plexuses – occur. Submuscular, subepidermal or subepithelial plexuses are observed in most flatworm taxa. Infraepithelial plexuses occur in the catenulid *S. leucops* (Reuter, unpublished), Acoelomorpha (Ehlers, 1985), polyclads (Koopowitz and

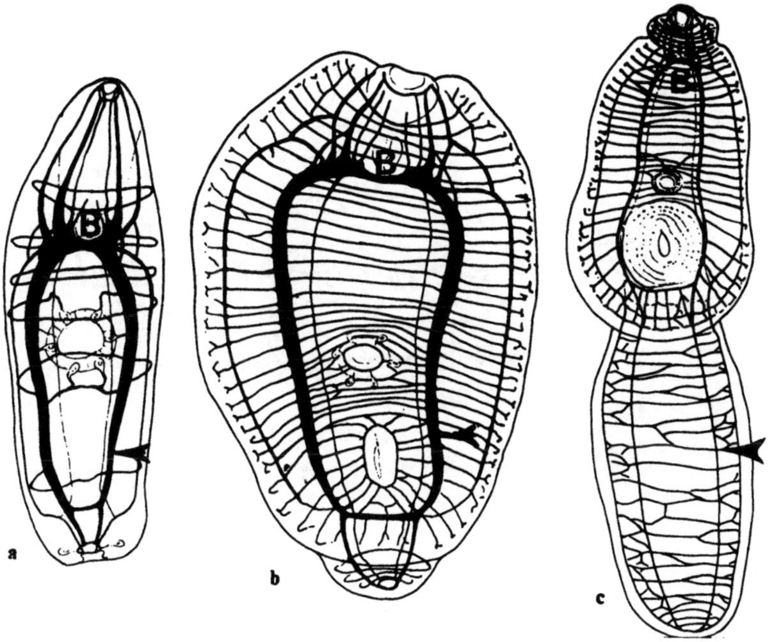

Figure 13. The orthogon during the life-cycle. NS of *Diplostomum pseudspathaceum*, main nerve cords (arrowheads): (a) cercaria (b) metacercaria (c) adult (Grabda-Kazubska et al., 1991). Brain (B), main nerve cord (arrow).

Chien, 1974, 1975) and cestodes (Gustafsson and Wikgren, 1981b,c; Gustafsson et al., 1985, 1986). Sensory nerves and terminals penetrating the epithelium are common to all flatworms. The nerve plexuses are in contact with both main and minor cords. In planarians the submuscular plexus beneath the muscular body wall forms a large meshed network usually close to or at the same level as the cords of the orthogon (Fig. 14) (Baguñà and Ballester, 1978). CSLM studies of lecithoepitheliats and proseriats have revealed that transverse orthogonal nerve fibers contact or constitute part of the submuscular plexus (Joffe and Reuter, 1993; Böckerman et al., in preparation). Thus the border between the submuscular plexus and minor orthogonal cords is often indistinct. The body musculature of tapeworms differs from that of other flatworms, by forming large longitudinal bundles. These muscles are surrounded by a nerve plexus on both sides. The cortical part of this plexus is in contact with the peripheral cords (Fig. 15). Specialized plexuses such as genital plexuses and plexuses for holdfast organs occur in advanced flatworms. The ventral nerve plate of landplanarians, lacking orthogonal longitudinal cords, holds an exceptional position (Reisinger, 1972, Rieger et al., 1991).

Cestodes differ from all other flatworms in having their nutrient-absorbing surface on the outside, in the tegument, which thus serves at least two purposes, protection and nutrients absorption. The innerva-

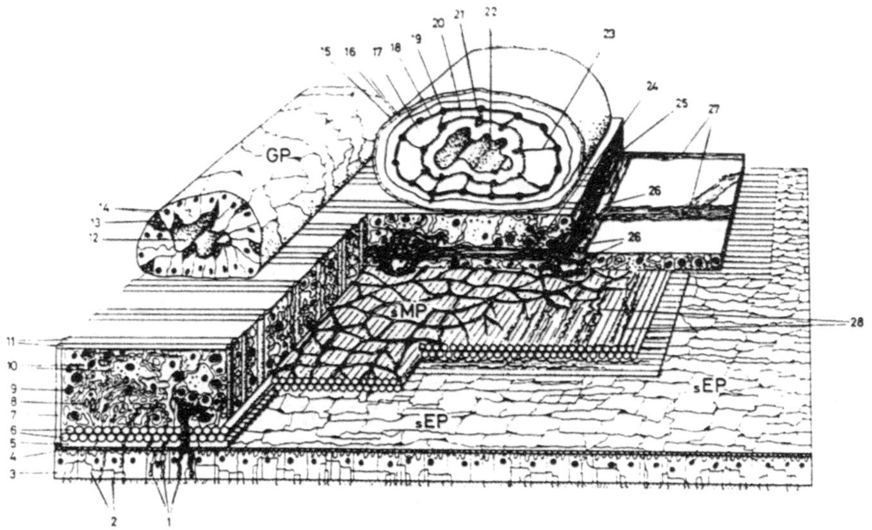

Figures 14–19. Nerve plexuses

Figure 14. Diagram showing the three nerve plexuses in *Dugesia lugubris*. Subepidermal nerve plexus (sEP) submuscular nerve plexus (sMP), gastrodermal plexus (GP) (for 1–28 see Baguñà and Ballester, 1978).

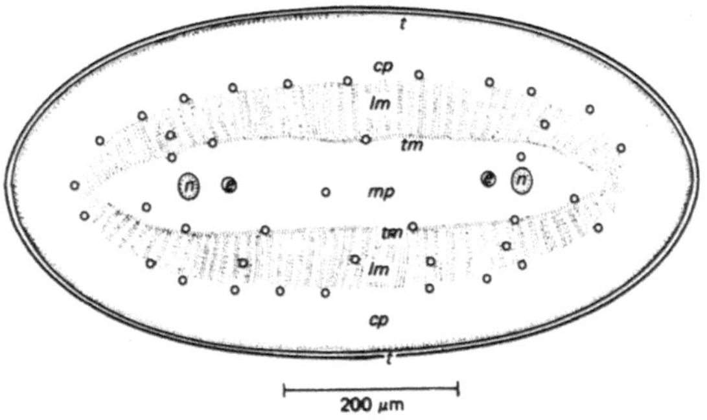

Figure 15. The circum-muscular pattern of nerve cells (o) in *Diphyllobothrium dendriticum*. Main nerve cords (n) tegument (t) cortical parenchyma (cp) longitudinal muscles (lm), transverse muscles (tm) medullary parenchyma (mp) (Gustafsson and Wikgren, 1981c).

tion of the subtegument is dense and varies along the body of the worm. Thus in *D. dendriticum* the most intimate contact between host and parasite is found along the inner border of the bothridia, which embrace the intestinal villi of the host (Fig. 16). The large number of especially peptidergic nerve terminals points to the importance of this surface. This becomes even more evident when the "inner" surface is compared with the "outer" surface of the scolex, where far fewer nerve terminals are found and with the subtegument of the strobila (Gustafsson, 1985, 1992).

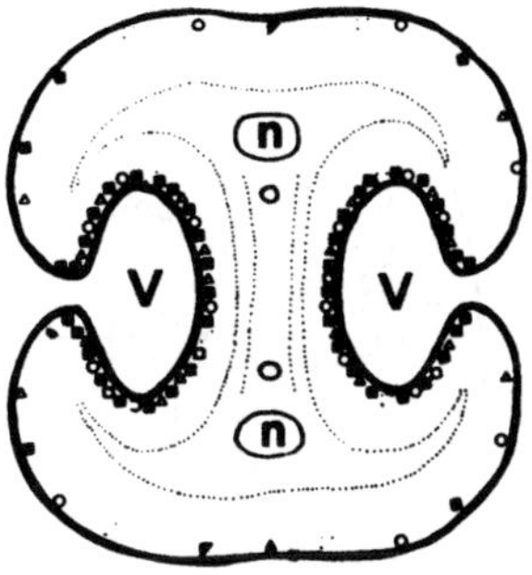

Figure 16. Innervation of nutrient absorbing inner bothridial surface of *Diphyllobothrium dendriticum*. Note abundance of nerve terminals IR to seven different neuronal mediators beneath basal lamina. V = place for host villus, n = main nerve cord (Gustafsson, 1992).

The pharyngeal and stomatogastric nervous system

Flatworms are the most primitive animals in which we can find an alimentary tract, with mouth, pharynx and intestine. An anus is lacking. The presence of an autonomous NS consisting of a pharyngeal and an intestinal part in most flatworms proves the relationship between Plathelminthes and Spiralia (Reisinger, 1976). The stomatogastric NS is of interest from an evolutionary point of view as it is regarded as a plesiomorphic character (Ehlers, 1985) and may show preserved archaic features. The pharyngeal NS of advanced flatworms consists of one or two pharyngeal nerve rings connected by a simple nerve net similar to the submuscular plexus. In addition, sensory nerves penetrating the pharyngeal epithelium occur (Reuter, 1975). The presence of two pharyngeal nerve rings connected to the main cords were recently shown by substance P antibodies already in the "primitive" catenulid *S. leucops* (Fig. 17) (Reuter, 1994).

The intermuscular intestinal nerve plexus is of an irregular and reticular structure without differentiated tracks (Reisinger, 1976; Baguñà and Ballester, 1978; Reuter and Palmberg, 1987; Reuter 1991). In the catenulid *M. lineare* nerve fibers and cells are observed among the intestinal cells, that is in an infraepithelial position (Fig. 18). ICC reveals that the contacts of the stomatogastric NS and the brain and/or the main cords is formed by 5-HT- and RF-IR fibres in different positions (Reuter and Palmberg, 1989). The absence of an intestinal NS in numerous typhloplanides lacking intestinal muscles and in all acoels has been interpreted as retrograde evolution (Reisinger, 1976).

The stomatogastric NS of the flukes is innervated by cholinergic, aminergic and peptidergic neurons (for refs see Gustafsson, 1992).

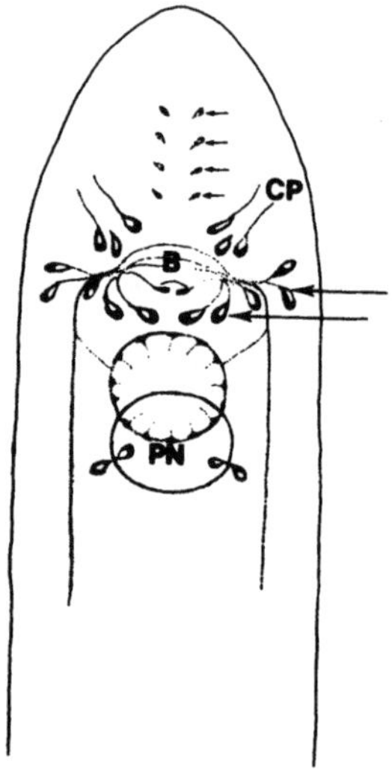

Figure 17. Pattern of SP-IR nerves in *Stenostomum leucops*. Note the two pharyngeal nerve rings (PN). Brain (B), ciliary pit (CP) (Reuter, 1994).

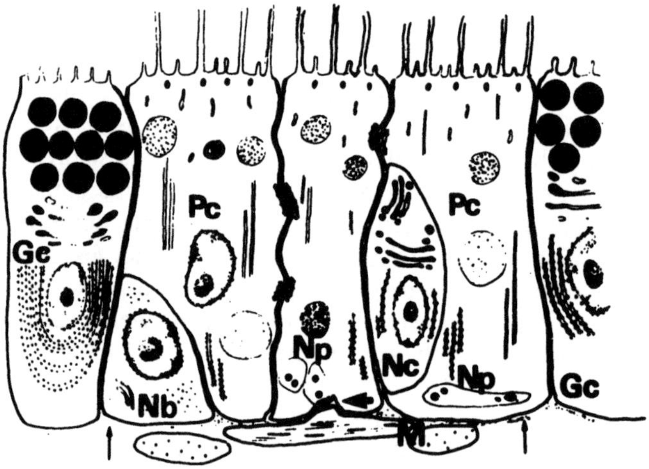

Figure 18. Neurons (Nc)and neuronal processes (Np) in gastroderm of *Microstomum lineare*. Granular club cell (Gc), phagocytic cell (Pc), muscle (M), synapse (thick arrow), basal lamina (thin arrow) (Reuter and Palmberg, 1987).

Genital plexus

An enormous production of eggs is characteristic of parasitic flatworms and represents an important target for chemotherapeutic attack. CSLM analyses of the NS of the reproductive organs of *Fasciola hepatica*, *S. mansoni*, *Haplometra cylindracea*, *Gorgoderina vitelliloba*, *Diclidophora*

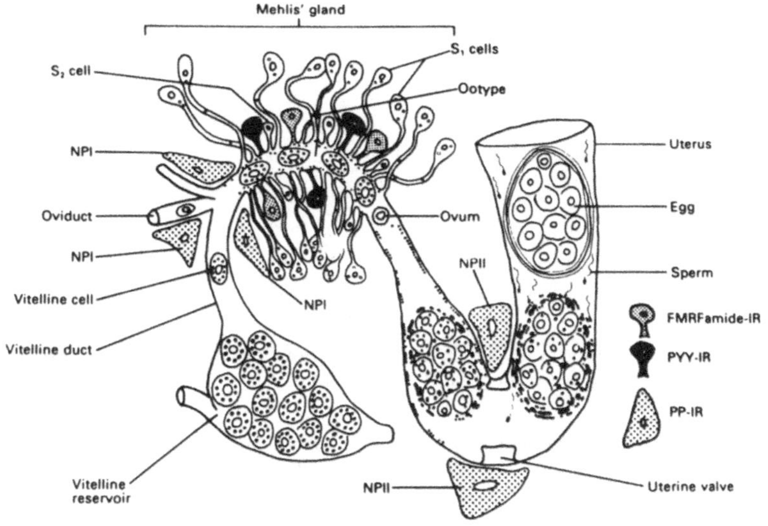

Figure 19. Distribution of peptidergic neurons in the ootype/Mehlis' gland complex of *Fasciola hepatica* (Magee et al., 1989).

merlangi and *D. dendriticum* have led to a new understanding of the innervation of these complex organs and new hope for the development of anthelminthics (Fig. 19). The complicated system of musculous ducts in the genital organs of parastic flatworms is innervated by aminergic and peptidergic neurons (for refs see Gustafsson, 1992). Contrary to the parasitic flatworms, the innervation of genital organs in free-living flatworms has gained little interest.

The neuronal cell types

In most invertebrates the majority of neurons has a unipolar organization. However, flatworms have a large variety of neuronal types such as uni-, bi-, and multipolar (see Rieger et al., 1991). Conventional microscopy was previously the only method for visualizing nerve cells in flatworms. Neither LM nor EM are entirely reliable in this respect. The introduction of fluorescence methods and CSLM was a great improvement. The most reliable results have been obtained by injecting dyes into individual cells (Keenan et al., 1981; Koopowitz, 1986; Koopowitz, 1994). This method has revealed corresponding morphologically distinct types in the rhabdocoel *Mesostoma Ehrenbergii* and the polycad *Notoplana actiola* (Koopowitz et al., 1994). Heteropolar bi- and multipolar, isopolar bi- and multipolar as well as unipolar cells have been found in these worms. In both worms similar bilaterally symmetrical dye-coupled

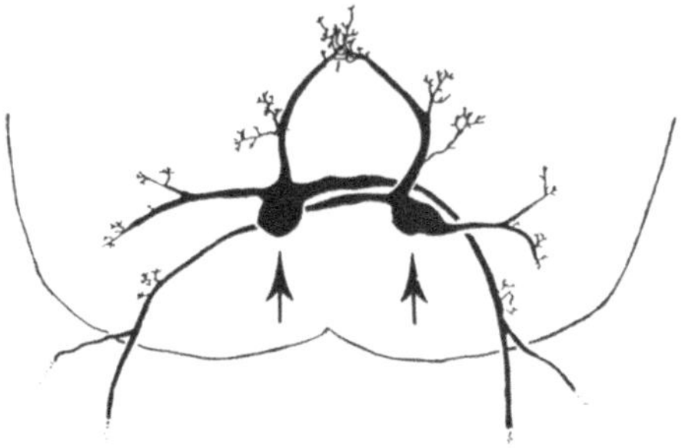

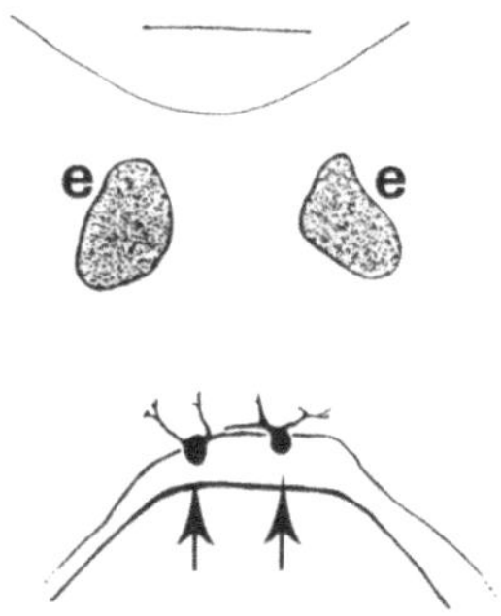

Figure 20. Dye-coupled multipolar heteropolar neurons (arrows) in the posterior part of the brain of *Notoplana acticola* (Polycladida) (Koopowitz, 1994). For magnification see bar in Figure 21.

Figure 21. Dye-coupled multipolar heteropolar neurons (arrows) in the posterior part of the brain of *Mesostoma Ehrenbergii* (Rhabdocoela), eyes (e). Bar = 100 µm (Koopowitz, 1994).

circuits are found in the posterior part of the brain (Figs 20, 21). The heteropolar multipolar nerve cell type is similar to nerve cells found in vertebrates. The stem bilaterian might have had a mixture of cell type similar to that in modern flatworms. Alternatively, the types might have evolved independently (Koopowitz, 1986).

The ultrastructure

Neurons

The most striking ultrastructural feature of the flatworm neurons is their high content of vesicles indicating a secretory nature. A single type of vesicle often dominates in the perikarya. More variation in vesicle types occurs in the fibers, where small clear vesicles are often seen alongside the dominating vesicle type. The vesicles fill the perinuclear cytoplasm and all the finer processes. They are formed in the active Golgi regions, found in the vicinity of the prominent nucleus. The nucleolus is usually distinct and free ribosomes occur abundantly. The amount of rough endoplasmic reticulum is usually moderate and found in the periphery. The large ganglion cells are characterized by their many outfoldings and cytoplasmic projections as well as their irregular nuclei (Lentz, 1976; Gustafsson unpubl.). The vesicle types have been used as markers for different cell types in the brain. The following types have been described:

1. Small clear vesicles (*sv*) (20–40 nm) of the "synaptic" type. *Sv* are regarded as cholinergic or as vesicles recapturing membranes or used for retrieval of Ca^{++}.

2. Dense-core vesicles (*dcv*) of varying types. Their core may be very dense, of medium density, amorphous with only a point-like center, or eccentric; *dcv* vary in size between 50–140 nm.
3. Large lucent vesicles or vacuoles (*llv*) (60–300 nm) are observed in sensory neurons, always together with large dense elementary granules.
4. Oval, flattened or drop-like vesicles (*ov*) with a content of varying density.
5. Large dense vesicles or elementary granules (*ldv*) (50-200 nm and larger).

Sv have often been regarded as cholinergic, *dvc* as aminergic, and *ldv* as peptidergic. However, the results from ICC on the EM level show that these broad categories no longer hold. Neuropeptides have been observed in *dcv* (Reuter et al., 1990) and several reports of coexistence of neuropeptides occur (Halton et al., 1992). A much broader variation in the content of the vesicles than expected seems to exist. Vesicle ultrastructure depends, in all probability, on developmental stage, processing of neuroactive substances and coexistence of neuroactive substances. A classification of neurons with respect to function and neuronal vesicle types is problematic or impossible on the basis of ultrastructure only. It is, however, rational and practical to group the neurons of the flatworms in main categories, all of which contain vesicles (Reuter and Gustafsson, 1989). Common to all investigated flatworms are : (1) the neurons containing *dcv* (Fig. 22 A). They are uni-, bi- or multipolar; (2) the neurons contaning *ldv* (Fig. 22 B). These are uni- or bipolar in microturbellarians, while the parasitic worms also have multipolar, heteropolar *ldv* containing neurons; (3) the sensory neurons which are usually bipolar, (Fig. 22 C); (4) in advanced large flatworms highly specialized ganglion cells containing solely small clear vesicles have been found (Morita and Best, 1966; Lentz, 1967).

The neurons of the catenulids *S. leucops*, *Catenula lemnae* and *C. leptocephala* and the macrostomid *M. lineare* seem to be structurally least differentiated. Their perikarya have a pronounced secretory character with the cytoplasm packed with abundant vesicles and conspicous Golgi complexes. They resemble the neuroblasts of higher animals (Moraczewski et al., 1977; Moraczewski, 1981). In more advanced flatworms a greater heterogeneity in population of vesicles is observed (Reuter, 1991). The cytoplasmic density and the design of the perikarya also differ. The observation of four neuronal cell types as well as "glial" cells in three different species of Acoela means that their NS can no longer be considered as primitive (Bendini and Lanfranchi, 1991) and contributes to the discussion of the phylogenetic position of the acoels (Ax, 1984; Ehlers, 1985).

The *dcv*-containing neuron dominates the ultrastructural picture of the flatworm NS. The core of the *dcv* varies in size and density. *Dcv*-containing neurons occur in the brain, the intestine, and as sensory

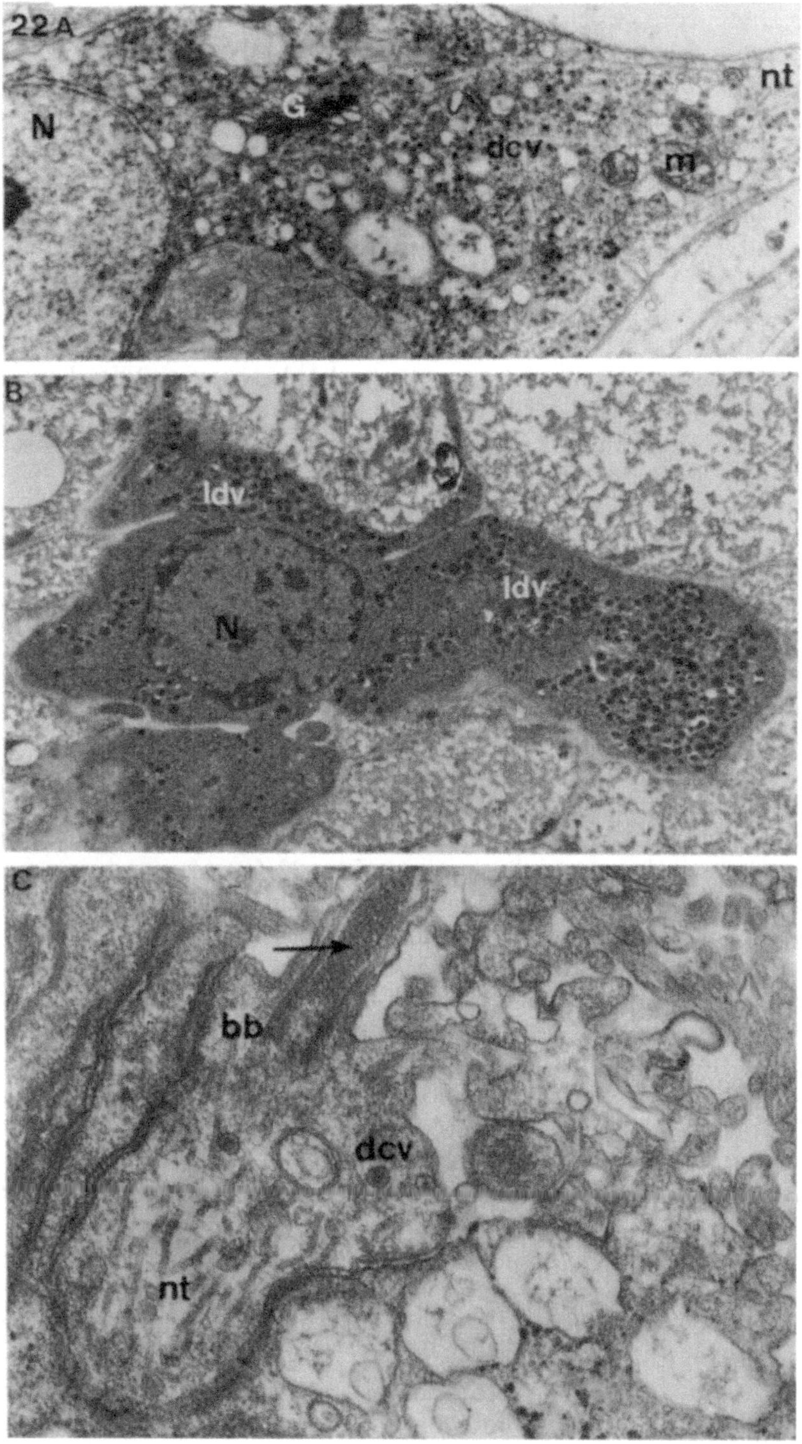
22A
N
G
dcv
m
nt
B
ldv
N
ldv
C
bb
dcv
nt

receptors. According to Reuter and Gustafsson (1989) the *dcv*-containing neuron may be considered an archaic cell type with synaptocrine and paracrine release – a neuroparaneuronal cell (Fujita et al., 1988). This cell type has been regarded as the progenitor of functionally different conventional neurons, in advanced flatworms and possibly in higher metazoans, too (Reuter and Gustafsson, 1989).

"Glial" cells

A second category of cells found in the NS of some flatworms are called neuroglia. This cell type lacks vesicles, has a thin cytoplasm with very few organelles and the main mass of cytoplasm concentrated in branches (Fig. 23) (Golubev, 1988). The cells are believed to perform supporting, isolating, and metabolic functions. Accessory and supporting cells lacking vesicles have been observed in planarians (Morita and Best, 1966; Palladini et al., 1983; Golubev, 1988; Trawicki et al., 1988), in Lecithoepitheliata (Böckerman et al., 1994a,b) and Acoela (Bedini and Lanfranchi, 1991). The precence of S-100-IR material, typical for vertebrate glia, has been demonstrated in the NS of *Crenobia alpina* (Tricladida) (Kerschbaum and Hermann, personal communication). Multilayered lamellae (Figs 24, 25) are found around neurons and/or nerve fibres of proseriats (Ferrero et al., 1985; Reuter, 1990), planarians, temnocephalids (Rhode and Webb, 1986; Trawicki et al., 1988), and in the polyclad *Notoplana acticola* (Koopowitz and Chien, 1974; Koopowitz, 1989). In the aspidogastrean *Multicotyle purvisi* nerve sheaths and deep invagiantions in the large axons occur (Rhode, 1970, 1971). In the lower flatworm taxa Catenulida and Macrostomida neither "glial" cells nor any form of wrapping can be seen. The observation of "glial" cells and wrappings of neurons and axons in scattered flatworm taxa indicates that glial cells in Plathelminthes may have originated independently from the glial cells in Eubilatheria.

ECM

Neurons in flatworms may also be separated by extracellular matrix (ECM) materials (for refs. see Lindroos and Reuter, 1991) of the same

---

Figure 22A–C. The secretory character of flatworm neurons, illustrated by examples. (**A**) Neuron from *Microstomum lineare* containing large amounts of dense core vesicles (dcv). Golgi (G), mitochondrion (m), neurotubules (nt), nucleus (N); magnification × 8000. (**B**) Neuron from *Diphyllobothrium dendriticum* containing large amounts of large dense vesicles (ldv). Nucleus (N); magnification × 5000. (**C**) Sensory neuron from *Microstomum lineare* containing dense core vesicles (dcv). Basal body (bb), neurotubules (nt), cilium (arrow); magnification × 40 000.

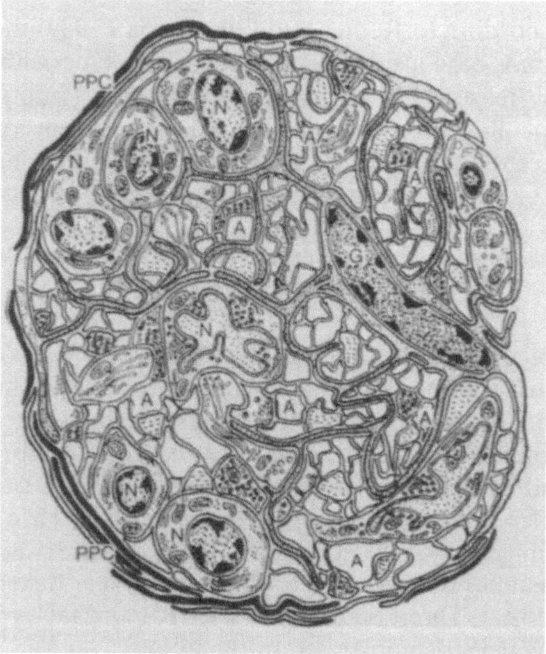

Figures 23–25. Neuroglia

Figure 23. Diagram of the fine structure of the ventral nerve cord in a planarian. Axons (A), glial cell (G), nerve cells (N), processes of parenchyma cells (PPC) (Gobulev, 1988).

nature as the ECM-forming brain capsules (see also section on *Brain*). In neurons of Lecithoepitheliats the ECM forms deep invaginations in in the perikaryon (Böckerman et al., 1994a,b). The morphology of the ECM in flatworms indicates a role as an internal medium for transport and exchange of substances (Lindroos and Reuter, 1991) (see also section on *Paracrine or nonsynaptic release*).

Synapses and non-synaptic release sites

A conspicous variation in the morphology of the release sites is observed (Figs 26, 27). It is difficult to draw a boundary between synaptocrine and paracrine release in flatworms (Reuter and Gustafsson, 1989).

Synaptocrine release

Several types of true chemical synapses have been observed in the worms. The chemical synapses have the morphological characteristics of

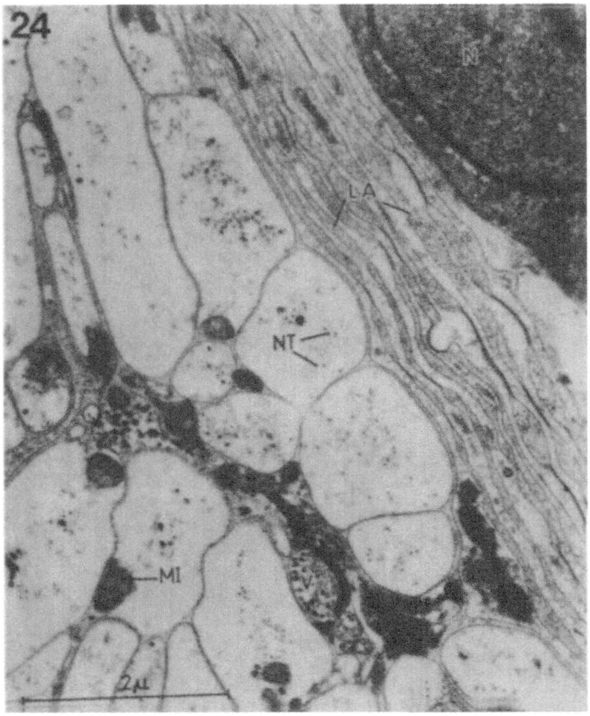

Figure 25. Section through nerve of *Multicotyle purvisi*. Note nerve sheat (LA). Mitochondrion (MI), neurotubule (NT) (Rhode, 1970).

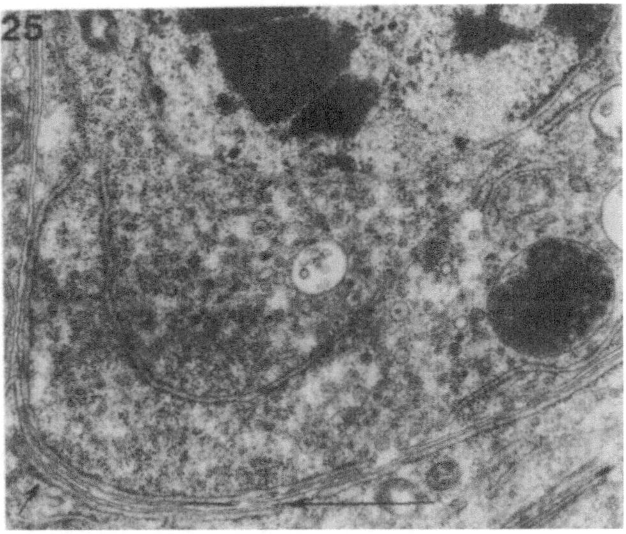

Figure 24. Wrapping (arrows) around nerve cell of *Archiloa unipunctata*; magnification × 30 000 (Reuter, 1990).

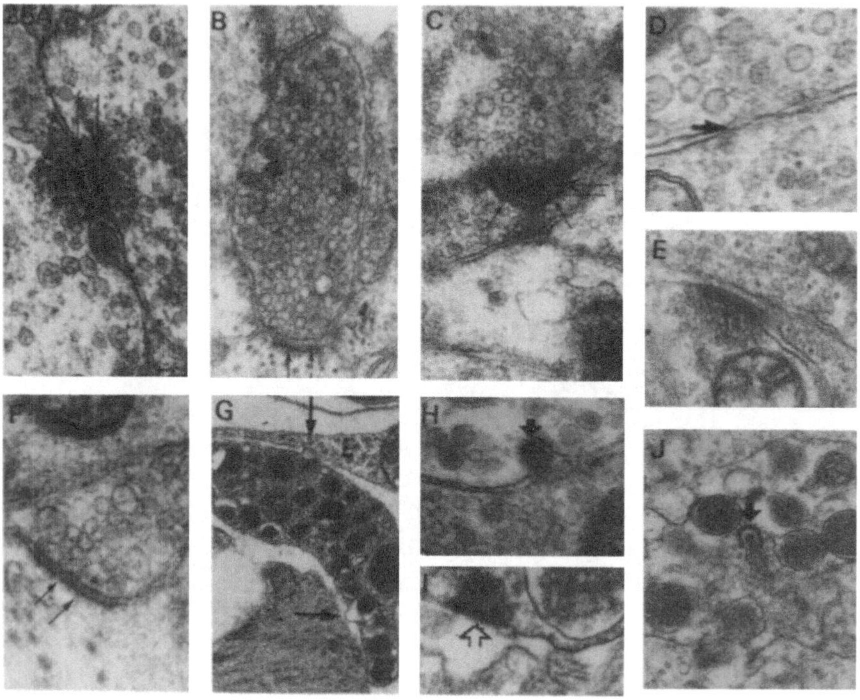

Figure 26. Types of synapses and release sites in flatworms. (A) Two-way synapse, *Stenostomum leucops* with paramembraneous densities (arrows); magnification ×50 000. (B) Single synapse, *Diphyllobothrium dendriticum* with postsynaptic density (arrows); magnification ×50 000. (C) Shared synapse, *Archiloa unipunctata* with paramembraneous density (large arrow), presynaptic density (thin arrow) and postsynaptic density (small arrows); magnification ×50 000. (D) Electrotonic junction, *Stenostomum leucops* (arrow); magnification ×50 000. (E) En passant synapse *Stenostomum leucops* with presynaptic density arrow); magnification ×40 000. (F) Sarcoplasmic junction, *Gyratrix hermaphroditus* with postsynaptic density (arrows); magnification ×40 000. (G) Protrusion towards muscle, *Diphyllobothrium dendriticum* (arrows), muscle (m); magnification ×15 000. (H) Omega profile towards nerve, *Microstomum lineare* (arrow); magnification ×60 000. (I) Non-synaptic release site, *Microstomum lineare* (empty arrow); magnification ×60 000. (J) Omega profile *Microstomum lineare* (arrow); magnification ×30 000.

conventional synapses. They are either unpolarized or polarized, i.e., the pre- and postsynaptic membranes are structurally similar or different. Unpolarized "two-way" synapses have been observed in microturbellarians and acoelans (Reuter and Palmberg, 1989; Bedini and Lanfranchi, 1991). The polarized synapses can be divided into the following categories: single, shared, en passant and multiple synapses, and neuromuscular junctions. The sarconeural junctions occur either at nonfilamentous areas (Reuter, 1990) or the cytoplasmic extensions of the muscle make contact with the axon (Chien and Koopowitz, 1972; Moraczewski, 1981). The types of presynaptic vesicles vary in size and density, but *sv* and *dcv* are the most common types. In the presynaptic

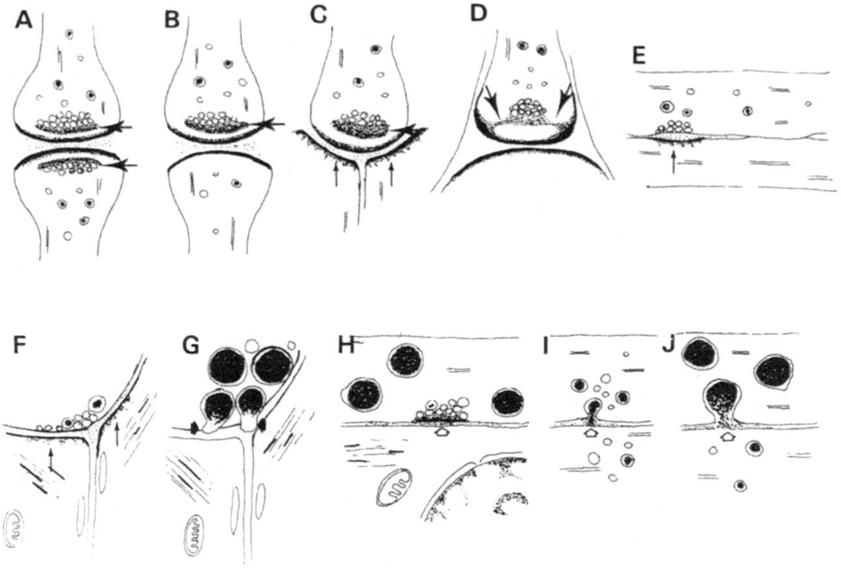

Figure 27. Diagram showing different types of synapses and release sites in flatworms. (A) Two-way synapse with paramembraneous densitities (arrows). (B) Single synapse with paramembraneous density (arrow). (C) Shared synapse with paramembraneous density (large arrow) and postsynaptic densities (small arrows). (D) Synapse with paramembraneous density connected by strings to presynaptic density. (E) En passant synapse and electrotonic junction. (F) Shared synapse-like structure towards muscle fibers, postsynaptic densities (arrows). (G) Release via protrusions (arrows) through axolemma towards muscles. (H) Synapse-like structure at cell body (empty arrow). (I) Release via omega figure of small dense vesicle (empty arrow). (J) Release via omega figure from large dense vesicle (empty arrows).

terminal a specialized structure, the "paramembraneous density", is often seen proximally from the presynaptic thickening of the membrane. In acoels a paramembraneous density resembling the presynaptic T-bar of arthropod synapses has been observed (Bedini and Lanfranchi, 1991). In geocentrophorans (Lecithoepitheliata) a paramembraneous density connected by strings of dense material to several thickenings of the presynaptic membrane has been observed (Böckerman et al., 1994a) (Fig. 27D). A greater electron density of the synaptic structures as well as serial or multiple chemical synapses occur more frequently in many advanced flatworms, indicating a greater specialization of their NS (Reuter and Lindroos, 1979; Gustafsson, 1984).

Paracrine or nonsynaptic release

(1) Synapse-like structures lacking the morphological characteristics of conventional synapses have been observed in lower turbellarians (Catenulida, Macrostomida) (for review, see Reuter, 1990). They are characterized by rows of vesicles close to the axolemma. The vesicles

vary in size and content from lucent vacuoles to *dcv*, and they are affilated with electron dense material. They occur at nerve fibers as well as close to muscles and extracellular stroma. Recently, the TARI method (Buma and Roubos, 1986) has been used to study the ultrastructure of the chemical release. The tannic acid used in this method intensifies the electron density of the interneuronal secretory substances, i.e., it freezes the cores. In different flatworms this method has revelaed that chemical messengers are extruded at non-specialized sites through the axolemma directly to neighboring nerve fibers (Golding, 1984; Gustafsson, 1990; Reuter and Palmberg, 1990; Reuter, 1991). (2) Omega profiles formed by *ldv* constitute another type of paracrine release of neuroactive substances from peptidergic neurosecretory processes observed between nerve processes and extracellular stroma, facing muscles and neoblasts (Reuter, 1981; Palmberg and Reuter, 1983; Reuter and Gustafsson, 1989; Gustafsson, 1990). (3) In addition, protrusions through the axolemma in neurites containing *ldv* occur in the cestode *D. dendriticum*. The release takes place towards the extracellular stroma beneath the basement membrane and towards muscle fibers in the parenchyma (Gustafsson and Wikgren, 1981c).

The types of release sites observed indicate that chemical signalling in flatworms is transmitted as neurotransmitters or neuromodulators, either directly to their targets at synaptic and nonsynaptic release sites or as local chemical mediators via intercellular matrix (ECM). Signalling by hormones in a classic sense has to be excluded in these worms without a bloodstream. The dominance of neuronal transmission in acoelomates is a feature that links them to the Spiralia, forming the prostostomian series (Fujita et al., 1988).

Neuroactive substances

The term neurosecretory cell has long been used in connection with the flatworm's NS. This term is misleading as the recipient of the released material is not a vascular system but the intercellular space. During the last decade the borderline between the nervous and the endocrine systems has become indistinct and the two parts are now dealt with as a continuum, "the neuroendocrine system". The mesenger substances involved are small molecules such as amino acids and acetylcholine, biogenic amines such as catecholamines, serotonin, histamine, and neuropeptides.

First demonstrated were cholinergic substances with the AChE method. Later, a broad spectrum of aminergic and peptidergic neuroactive substances was demonstrated in most flatworm taxa by LM flourescence, ICC and high-pressure liquid chromatographic (HPLC) methods.

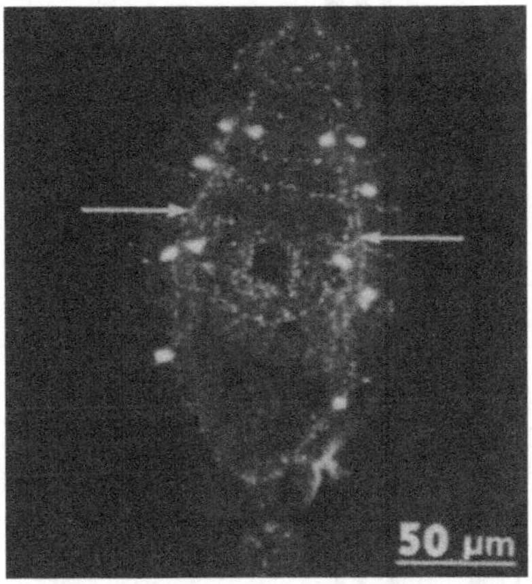

Figure 28. The aminergic nervous system as represented by 5-HT IR in the nervous system of *Cercaria emaculans* cercaria. Note the distinct orthogonal pattern and the paired arrangement of 5-HT-IR cell bodies associated to the main cords (arrows). Bar = 50 μm. (Pan et al., 1994).

Aminergic substances

The presence of IR to the indolamine 5-HT has been demonstrated in the NS of all flatworm taxa (Fig. 28) (Reuter et al., 1986; Reuter and Gustafsson, 1989; Gustafsson, 1982; Gustafsson and Reuter, 1992; Joffe and Reuter, 1993; Pan et al., 1994). Analyses with HPLC confirm the results (Eriksson et al., 1993a,b). Melatonin, a 5-HT derivative inhibiting fissioning of decapitated specimens, has been detected by HPLC and RIA in the head of *D. dorotocephala* (Morita and Best, 1984). In *D. japonica* it inhibits regeneration (Yoshizawa et al., 1991). The catecholamines dopamine (DA) and noradrenaline (NA) have been demonstrated by the Falck–Hillarp and GAIF methods (Gustafsson and Wikgren, 1981a; Buchman and Prento, 1989; Gustafsson and Eriksson, 1991; Reuter and Eriksson, 1991; Joffe and Reuter, 1993). The presence of DA, NA, and Dopa, the precursor of DA, in flatworms is also verified in different flatworms by HLPC (Fig. 29) (Welsh and King, 1970; Algeri et al., 1983; Gustafsson and Eriksson, 1991; Reuter and Eriksson, 1991; Eriksson et al., 1993a,b). Measurements of fluorescence spectra indicate substances of DA nature in the polyclad *Notoplana acticola* (Hauser and Koopowitz, 1987). Recently, IR to histamine has been detected both in lower and advanced flatworms (Wikgren et al., 1990; Panula et al., 1994).

Peptidergic substances

Cells staining like neurosecretory cells of higher metazoans have been identified by paraldehyde-fuchsin, paraldehyde-thionin, and chrome-

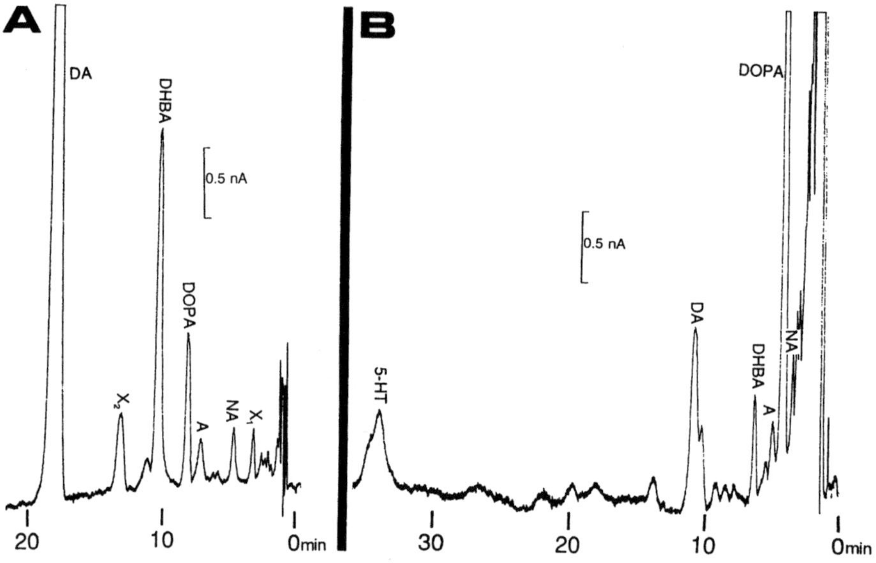

Figure 29. Chromatograms of monoamines in *Diphyllobothrium dendriticum*. (**A**) The catecholamine content in the scolex and neck region of adult worms after purification of alumina. DOPA, dopamine (DA), noradrenaline (NA), and adrenaline (A) are identified, and two unknown substances X_1 and X_2, can be seen. DA dominates this region. (**B**) The monoamine content in the proglottides of adult worms. Serotonin (5-HT), DOPA, DA, NA, and A are identified. DOPA is totally dominating this region (Eriksson et al., 1993).

haematoxylin phloxine methods (for review see Webb, 1988; Gustafsson and Wikgren, 1981a,b). The first authentic neuropeptide isolated from a flatworm (the ruminant tapeworm *Moniezia expansa*) is neuropeptide F (NPF) (Maule et al., 1991). NPF is a 39-amino acid peptide with a C-terminal phenylalaninamide. The primary structure of NPF from *M. expansa* and the terrestrial turbellarian *Artioposthia triangulata* has been determined (Maule et al., 1991; Curry et al. 1992) (Fig. 30). The flatworm NPF is regarded as the invertebrate equivalent or, indeed, the phylogenetic precursor of the NPY family peptides (Halton et al., 1992). The molecular structure of flatworm neuropeptides will certainly

▼▼▼ ▼ ● ▼▼▼▼▼

a) KVVHLRPRSSFSSEDEYQIYLRNVSKYIQLYGRPRFNH$_2$

b) PDKDFIVNPSDLVLDNKAALRDYLRQINEYFAIIGRPRFNH$_2$

Figure 30. Comparison of the complete primary structure of *Artioposthia triangulata* (a) and *Moniezia expansa* (b) neuropeptide F. The homologous residues (arrow) and the conservatively-substituted residue (circle) are indicated (Curry et al., 1992).

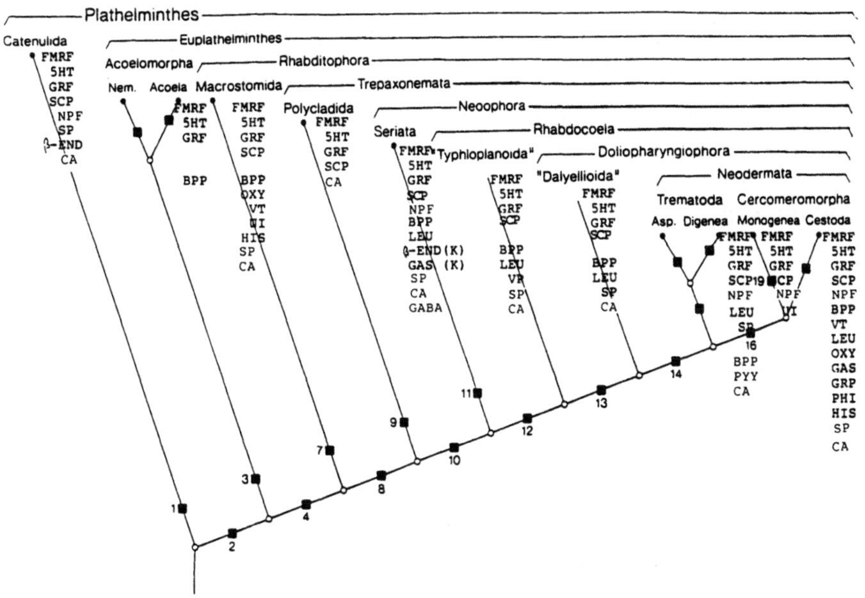

Figure 31. Neuropeptides and bioamines identified by ICC and histochemical methods introduced into the phylogenetic scheme of different taxa by Ehlers (1985). β-END β-endorphin; BPP bovine pancreatic polypeptide; CA catecholamines; FMRF FMRFamide; 5-HT serotonin; GABA γ-amino butyric acid; GAS gastrin; GRF growth hormone releasing factor; GRP gastrin releasing peptide; HIS histamine; LEU leu-enkephalin; NPF neuropeptide F; OXY oxytocin; PHI peptide histidine isoleucine; PYY peptide YY; SCP small cadiac peptide B; SP substance P; UI urotensin I; VP vasopressin; VT vasotocin.

be determined in the future. However, the small size of the worms makes it difficult to get sufficient quantities of material for the extraction. An alternative strategy is to employ DNA recombinant methods. To date, one such study has been documented by Duvaux-Miret et al. (1990). A β-endorphin-like peptide and a proopiomelanocortin-related gene were demonstrated in all stages of the life-cycle of *S. mansoni.* The peptidergic substances in flatworms are, with the exception of NPF, demonstrated only by ICC and mostly with antisera to vertebrate neuropeptides (Fig. 31) (Reuter et al., 1988; Halton et al., 1990; Gustafsson, 1992). Apart from NPF, few invertebrate neuropeptides have been identified in flatworms by means of ICC. Strong IR to the molluscan cardioactive peptide (FMRF-amide) and to the RF-amide from coelenterates, belonging to the same neuropeptide family, has been demonstrated in all flatworm taxa. Furthermore, IR to the molluscan small cardiac peptide (SCP_B) is demonstrated in *M. lineare* and *D. dendriticum* (Reuter and Palmberg, 1989; Gustafsson and Wikgren, 1989). SCP_B seems to colocalize with FMRF in brain neurons. The peptidergic substances demonstrated by antisera to the neuropeptides of vertebrates may be considered as substances like their counterparts in higher animals. IR to the vetebrate growth factors, basic fibroblast growth factor (bFGF), epidermal growth factor (EGF), and epidermal growth factor receptor (EGFr) have been demonstrated in both lower and higher flatworms (Gustafsson and Eriksson, 1992; Reuter and Kuusisto, 1992; Gustafsson et al., 1994). Western blot analysis shows that the molecular weight of bFGF in *D. dendriticum* is 47 kDa and in *D. tigrina* 48 kDa (Gustafsson and Eriksson, 1992; Eriksson, personal communications).

Compartments

Coexistence of neuroactive substances is rarely observed in flatworms (Gustafsson and Wikgren, 1989; Halton et al., 1991; Reuter and Palmberg, 1989) In *S. leucops* different neuroactive substances are confined to different neuronal compartments (Fig. 32) (Wikgren and Reuter, 1985). The same seems to be true, at least for aminergic and peptidergic substances, in the macrostomid species *M. lineare* (Reuter

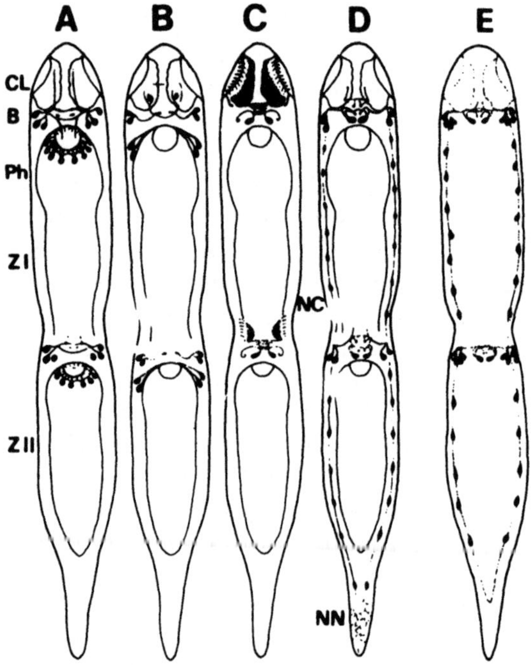

Figures 32–34. Compartments

Figure 32. IR pattern of five neuroactive substances in *Stenostomum leucops*. FMRFamide (**A**), β-endorphin (**B**), substance P (**C**), serotonin (**D**), catecholamines (**E**). Brain (B), cerebral lobes (CL), pharynx (Ph), main nerve cords (NC), nerve net (NN), first zooid (ZI), second zoid (ZII) (Wikgren and Reuter, 1985).

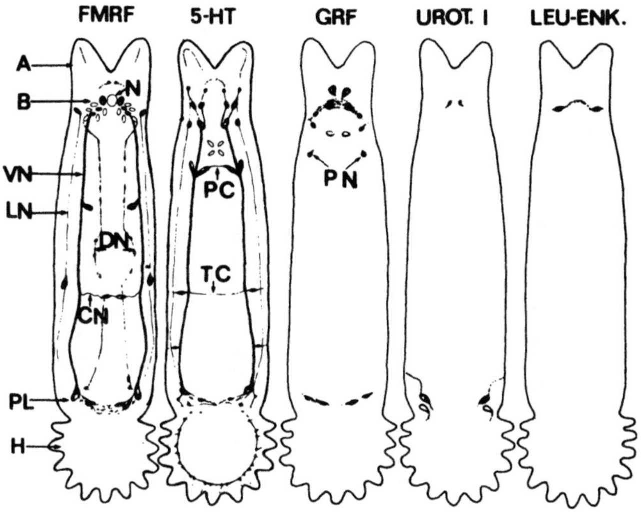

Figure 33. IR pattern of five neuroactive substances in *Gyrodactylus salaris*. Adhesive organ (A), brain (B), main nerve cord (VN), lateral nerve cord (LN), dorsal nerve cord (DN), nerve ring around mouth (N), connecting nerve (CN), postpharyngeal commissure (PC), transverse commissure (TC), latero-pharyngeal neurons (PN), haptor organ (H), nerve plexus (PL) (Reuter, 1987).

and Palmberg, 1989), and the trend is seen in more advanced flatworms, too (Figs 33, 34A–C) (Maule et al., 1990; Gustafsson, 1991; Gustafsson and Eriksson, 1992; Gustafsson and Reuter, 1992).

It is premature to draw general conclusions about the distribution of the substances. Some characteristic features of location are, however, apparent. Thus, the main cords are adjoined by 5-HT and CA-positive neurons (Fig. 34) (see also p. 33). Furthermore, if a positive staining result for a neuroactive substance is obtained, it will always also appear in the main cords. In the *stomatogastric* NS, neurons and fibers are intensely stained by NPF, FMRF/RF, and SP. In flatworms covered by cilia the *subepidermal* plexus shows intense 5-HT-IR. In the planarian *D. tigrina* antibodies to gamma-amino-butyric acid (GABA) stain the CNS, including the lateral branches of the nerve cords and the subepidermal plexus, while the transverse commissures and the submuscular plexus remain unstained (Eriksson and Panula, 1994). The distribution of the SP-IR nerves in association with *sensory structures* in microturbellarians, adult *S. mansoni*, and *D. dendriticum* makes this substance a likely candidate in sensory functions (Gustafsson, 1987; Gustafsson et al., 1992; Reuter, 1994). 5-HT and CA-positive reactions associated with *sensory structures* have been observed in lower as well as higher flatworms.

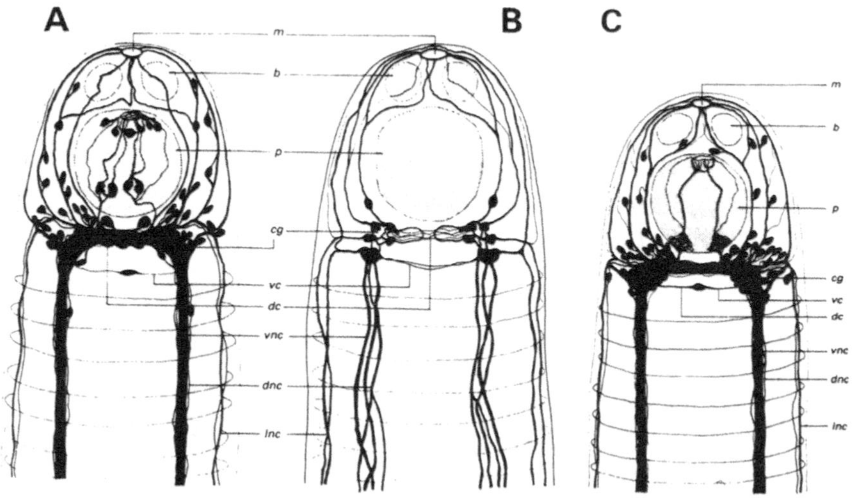

Figure 34. Diagram of cholinergic (**A**), serotoninergic (**B**) and peptidergic (**C**) nervous systems in forebody of *Diclidophora merlangi*. Mouth (m), buccal suckers (b), pharynx (p), cerebral ganglion (cg), ventral commissure (vc), dorsal commissure (dc), ventral nerve cord (vnc), dorsal nerve cord (dnc), lateral nerve cord (lnc) (Maule et al., 1990).

Conclusive remarks

The most characteristic feature of the flatworm nervous system is the high diversity on all levels. Multiplicity is found on the level of the orthogon, on the cellular level with uni-, bi-, and multipolar cell types, on the ultrastrucural level with different types of vesicles and release sites and, also regarding the neuronal substances. The high degree of morphological diversity reflects the versatility that shapes the basis for evolution and is typical for archaic groups of organisms (Mamkaev, 1986).

The studies of the organization of the NS shows A) that *main nerve cords* can be distinguished from minor cords by their multifibrilarity, their marker neurons positive for catecholamines and serotonin, and their contacts with the brain, B) that a stomatogastric infra- and/or subepithelial NS occurs in lower as well as higher flatworms and probably also in their stem form C) that regarding the evolution of the orthogon the following trends can be suggested: (1) a reduction in the number of nerve cords and a simultaneous strengthening of remaining cords (Reisinger, 1972), (2) a concentration of plexal fibers to longitudinal cords (Joffe and Reuter, 1993) and (3) an independent evolution of similar orthogonal types in distantly related groups (Kotikova, 1986, 1991).

The ultrastructural studies of neuronal cells indicate (A) that the secretory cell type containing *dcv* is archaic and a likely progenitor cell type for conventional neurons (Reuter and Gustafsson, 1989), (B) that an independent evolution of advanced neuronal cell types and "glial" cells has occurred inside the flatworm taxon, and (C) that an increase in density and a specialization of synaptic structures, multiple and serial synapses, as well as an increase in heterogenity of neuronal vesicle population have taken place.

The studies of neuroactive messengers show that, in addition to a multitude of peptidergic substances, aminergic and cholinergic substances have an important role already at this low phylogenetic level.

References

Algeri, S., Carolei, A., Ferretti, P., Gallone, C. Pallafini, G. and Venturini, G. (1983) Effects of dopaminergic agents on mono amine level and motor behaviour in planaria *Dugesia gonocephala*. *Comp. Biochem. Physiol.* 74C: 27–30.

Ax, P. (1984) *Das Phylogenetische System. Systematisierung der lebenden Natur aufgrund ihrer Phylogenese*. Gustav Fischer Verlag, Stuttgart, New York.

Baguñá, J. and Ballester, R. (1978) The nervous system in planarians: peripheral and gastrodermal plexuses, pharynx innervation, and the relationship between central nervous system and acoelomate organization. *J. Morphol.* 155: 237–252.

Bedini, C. and Lanfrachi, A. (1991) The central and peripheral nervous system of Acoela (Plathelminthes). An electron microscopical study. *Acta Zool.* 72: 101–106.

Buchman, K. and Prento, P. (1989) Cholinegic and aminergic elements in the nervous system of *Pseudodactylogyrus bini* (Monogenea). *Dis. aquat. Org.* 6: 89–92.

Bullock, T.H. and Horridge, G.A. (1965) *Structure and function in the nervous systems of invertebrates*. Vol. I., W.H. Freeman and Co., San Francisco.

Buma, P. and Roubos, E.W. (1986) Ultrastructural demonstration of nonsynaptic release sites in the central nervous system of the snail *Lymnaea stagnalis*, the insect *Periplaneta americana* and the rat. *Neurosci.* 17: 867–879.

Böckerman, I., Reuter, M. and Timoshkin, O. (1994a) Ultrastructural study of the central nervous system of endemic *Geocentrophora* (Prorynchidae, Plathelminthes). *Acta Zool.* 75: 47–55.

Böckerman, I., Raikova, O.I., Reuter, M. and Timoshkin, O. (1994b) Ultrastructure of the central nervous system of *Geocentophora baltica* (Lecithoepitheliata) and the surface sensillae in the *Geocentrophora* group. *Hydrobiologia*; *in press*.

Chien, P.K. and Koopowitz, H. (1972) The ultrastructure of neuromuscular system in *Notoplana acticola*, a free-living polyclad flatworm. *Z. Zellforsch.* 133: 277–288.

Curry, J., Shaw, C., Johnston C.F., Thim, L. and Buchanan, K.D. (1992) Neuropeptide F: Primąry structure from the turbellarian *Artioposthia triangulata*. *Comp. Biochem. Physiol.* 101C: 269–274.

Duvaux-Miret, O., Dissous, C., Gautron, J.P., Pattou, E., Kordon, C. and Capron, A. (1990) The helminth *Schistosoma mansoni* espresses a peptide similar to human β-endorphin and posesses proopiomelanocortin-related gene. *New Biologist* 2: 93–99.

Ehlers, U. (1985) *Das Phylogenetische System der Plathelminthes*. Gustav Fischer Verlag, Stuttgart, New York.

Eriksson, K.S. and Panula, P. (1994). Gamma-aminobutyric acid in the nervous system of a planarian. *J. Comp. Neurol.* 345: 528–536.

Eriksson, K., Gustafsson, M. and Åkerlind, G. (1993a) High-performance liquid chromatographic analysis on monomines in the cestode *Diphyllobothrium dendriticum*. *Parasitol. Res.* 79: 699–702.

Eriksson, K., Reuter, M. and Timoshkin, O. (1993b) Analysis of catecholamines in turbellarians of Lake Baikal. *Comp. Biochem. Physiol.* 105C: 509–511.

Ferrero, E.A., Lanfranchi, A. and Bedini, C. (1985) An ultrastructural account of otoplanid Turbellaria neuroanatomy. I. The cerebral ganglion and the peripheral nerve net. *Acta Zool.* 66: 63–74.

Fuijta, T., Kanno, T. and Kobayashi, S. (1988) *The Paraneuron.* Springer Verlag, Tokyo, Berlin, Heidelberg, New York, London, Paris.

Golding, D.W. and Bayraktaroglu, E. (1984) Exocytosis of secretory granules – a probable mechanism for the release of neuromodulators in invertebrate neuropiles. *Experientia* 40: 1277–1280.

Golubev, A.I. (1988) Glia and neuroglia relationships in the central nervous system of the Turbellaria (Electron microscopic data). *Fortschr. Zool.* 36: 31–37.

Grabda-Kazubska, B., Niewiadomska, K., Kanev, I. and Bayssade-dufour, C. (1991) Systéme nerveux des trématodes – Trematodes nervous system. *Ann Parasitol. Hum. Comp.* 66 (Suppl. I): 24–31.

Gustafsson, M.K.S. (1976) Observations on the histogenesis of nervous tissue in *Diphyllobothrium dendriticum* Nitzsch, (Cestoda, Pseudophyllidea). *Z. Parasitenkd.* 50: 313–321.

Gustafsson, M.K.S. (1984) Synapses in *Diphyllobothrium dendriticum* (Cestoda). An electron microscopic study. *Ann. Zool. Fennici* 21: 167-175.

Gustafsson, M.K.S. (1985) Cestode neurotransmitters. *Parasitology Today* 3: 72–75.

Gustafsson, M.K.S. (1987) Immunocytochemical demonstration of neuropeptides and serotonin in the nervous system of adult *Schistosoma mansoni. Parasitol. Res.* 74: 168–174.

Gustafsson, M.K.S. (1988) The helminth nervous system. *In*: H. Melhorn (ed.): *Parasitology in Focus. Facts and Trends*, Springer Verlag, Berlin, Heidelberg, New York, London, Paris, Tokyo, pp. 295–310.

Gustafsson, M.K.S. (1992) The neuroanatomy of parasitic flatworms. *Adv. Neuroimmunol.* 2: 267–286.

Gustafsson, M.K.S. and Eriksson, K. (1991) Localization and identification of catecholamines in the nervous system of *Diphyllobothrium dendriticum. Parasitol. Res.* 77: 498–502.

Gustafsson, M.K.S. and Erikson, K. (1992) Never ending growth and a growth factor. I. Immunocytochemical evidence for the presence of basic fibroblast growth factor in a tapeworm. *Growth Factors* 7: 327–334.

Gustafsson, M.K.S. and Reuter M. (1992) The map of neuronal signal substances in flatworms. *In*: R.N. Singh (ed.): *Nervous Systems. Principles of Design and Function.* Wiley Eastern Limited, New Delhi, Bangalore, Bombay, Calcutta, Guwahati, Hyderabad, Lucknow, Madras Pune, pp. 165–188.

Gustafsson, M.K.S. and Wikgren, M.C. (1981a) Peptidergic and aminergic neurons in adult *Diphyllobothrium dendriticum* Nitzsch, 1824 (Cestoda, Pseudophyllidea). *Z. Parasitenkd.* 64: 121–134.

Gustafsson, M.K.S. and Wikgren, M.C. (1981b) Activation of the peptidergic neurosecretory system in *Diphyllobothrium dendriticum* (Cestoda, Pseudophyllidea). *Parasitology* 83: 243–247.

Gustafsson, M.K.S. and Wikgren, M.C. (1981c) Release of neurosecretory material by protrusions of bounding membranes extending through the axolemma, in *Diphyllobothrium dendriticum* (Cestoda). *Cell Tissue Res.* 220: 473–479.

Gustafsson, M.K.S. and Wikgren, M.C. (1989) Development of immunoreactivity to the invertebrate neuropeptide small cardiac peptide B in the tapeworm *Diphyllobothrium dendriticum. Parasitol. Res.* 75: 396–400.

Gustafsson, M.K.S., Eriksson, K. and Hyden, A. (1994) Never ending growth and a growth factor. II. Immunocytochemical evidence for the presence of epidermal growth factor in a tapeworm. *Hydrobiologia*; *in press.*

Gustafsson, M.K.S., Lehtonen, M.A.I. and Sundler, F. (1986) Immunocytochemical evidence for the presence of "mammalian" neurohormonal peptides in neurones of the tapeworm *Diphyllobothium dendriticum. Cell Tissue Res.* 243: 41–49.

Gustafsson, M.K.S., Nässel, D. and Kuusisto, A. (1993) Immunocytochemical evidence for the presence of substance P-like peptide in *Diphyllobothrium dendriticum. Parasitology* 106;83–89.

Gustafsson, M.K.S., Wikgren, M.C., Karhi, T.J. and Schot, L.C.P. (1985) Immunocytochemical demonstration of neuropeptides and serotonin in the tapeworm *Diphyllobothrium dendriticum. Cell Tissue Res.* 240: 255–260.

Hanström, B. (1926) Über den feineren Bau des Nervensystems der Tricladen Turbellarien auf Grund von Untersuchungen an *Bdelloura candida. Acta Zool.* 7: 101–115.

Halton, D.W., Shaw, C., Maule, A.G. Johnston, C.F. and Fairweather, I. (1992) Peptidergic messengers: A new perspective of the nervous system of parastic platyhelminths. *J. Parasitol.* 78: 179–193.

Hauser, M. and Koopowitz, H. (1987) Age-dependent changes in fluorescent neurons in the brain von *Notoplana acticola*, a polyclad flatworm. *J. Exp. Zool.* 241: 217–225.

Hyman, L.H. (1951) *The Invertebrates. II Platyhelminthes and Rhynchocoela. The Acoelomate Bilateria.* McGraw-Hill, New York.

Joffe, B.I. (1990) Morphological regularities in the evolution of the nervous system in the Plathelminthes: anatomical variants of orthogon and their dependence upon the body form. (In Russian). *Proc. Zool. Inst. Leningrad* 221: 87–125.

Joffe, B.I. and Kotikova, E.A. (1991) Distribution of catecholamines in tubellarians (with discussion of neuronal homologies in the Platyhelminthes). *Stud. Neurosci.* 13: 77–113.

Joffe, B.I. and Reuter, M. (1993) The nervous system of *Bothriomolus balticus* (Proseriata) – a contribution to the knowledge of the orthogon in the Plathelminthes. *Zoomorphology* 113: 113–127.

Keenan, C.L., Coss, R. and Koopowitz, H. (1981) Cytoarchitecture of primitive brains: Golgi studies in flatworm. *J. Comp. Neurol.* 188: 647–679.

Koopowitz, H. (1986) On the evolution of central nervous systems: Implications from polyclad turbellarian neurobiology. *Hydrobiologia* 132: 79–87.

Koopowitz, H. (1989) Polyclad Neurobiology and the evolution of Central Nervous Systems. *In*: P.A.V. Anderson (ed.): *Evolution of the First Nervous Systems*. NATO ASI Series. Series A: Life Science 188: 315–328.

Koopowitz, H. (1970) Feeding behaviour and the role of the brain in the polyclad flatworm, *Planocera gilchrist. Anim. Behav.* 18: 31–35.

Koopowitz, H. and Chien, P.(1974) Ultrastructure of nerve plexus in flatworms. I. Peripheral organization. *Cell Tissue Res.* 155: 337–351.

Koopowitz, H. and Chien, P. (1975) Ultrastructure of nerve plexus in flatworms. II Sites of synaptic interactions. *Cell Tissue Res.* 157: 207–216.

Koopowitz, H., Elvin, M. and Bae, T. (1994) Comparison of the nervous system of the rhabdocoel, *Mesostoma Ehrenbergii* with that of the polyclad, *Notoplana acticola. Hydrobiologia*; *in press.*

Koopowitz, H., Silver, K. and Rose, G. (1976) Primitive nervous systems. Control and recovery of feeding behaviour in the polyclad *Notoplana actiola. Biol. Bull.* 150: 411–425.

Kotikova, E.A. (1986) Comparative characterization of the nervous system of the Turbellaria. *Hydrobiologia* 132: 89–92.

Kotikova, E.A. (1991) The orthogon of the plathelminthes and main trends of its evolution. (In Russian). *Proc. Zool. Inst. St. Petersburg* 241: 88–111.

Kotikova, E.A. (1994) Glyoxylic-acid-induced-fluorescence in the nervous system of *Gyratrix hermaphroditus* (Kalyptorynchia, Polycyctididae). *Hydrobiologia*; *in press.*

Kotikova, E.A. and Timoshkin, O.A. (1987) The structure of the nervous system of Lecithoepitheliata and Prolecithophora from the Baikal lake (Turbellaria). (In Russian). *Proc. Zool. Inst. Leningrad* 167: 97–110.

Lentz, T.L. (1976) Fine structure of nerve cells in a planarian. *J. Morph.* 12: 323–338.

Lindroos, P. and Reuter, M. (1991) Extracellular matrix in some microturbellarians. *Hydrobiologia* 227: 283–290.

Maule, A.G., Halton, DW., Johnston, C.F. Shaw, C. and Fairweather, I. (1990) The serotoninergic, cholinergic and peptidergic components of the nervous system in the monogenean parasite, *Diclidophora merlangi*: a cytochemical study. *Parasitology* 100: 255–273.

Maule, A.G., Shaw, C., Halton, D.W. Thim, L. Johnston, C.F., Fairweather, I. and Buchanan, K.D. (1991) Neuropeptide F: a novel parastic flatworm regulatory peptide from *Moniezia expansa* (Cestoda, Cyclophyllidea). *Parasitology* 102: 309–316.

Moraczewski, J. (1981) Fine structure of some Catenulida (Turbellaria, Archophora). *Zoologica Poloniae* 28: 67–415.

Moraczewski, J. Czubaj, A. and Bakowska, J. (1977) Organization and ultrastructure of the nervous system in Catenulida (Turbellaria). *Zoomorphology* 87: 87–95.

Morita, M. and Best, J.B. (1966) Electron microscopic studies of planaria. III. Some observations on the fine structure of planarian nervous tissue. *J. Exp. Zool.* 161: 391–413.

Morita, M. and Best, J. B. (1984) Effects of photoperiods and melatonin on planarian asexual reproduction. *J. Exp. Zool.* 231: 273–283.

Niewiadomska, K. and Moczon, T. (1982) The nervous system of *Diplostomum pseudospathaceum* Niewiadomska, (Digenea, Diplostomum). I. Nervous system and chaetotaxy in the cercaria. *Z. Parasitenkd.* 68: 296–304.

Niewiadomska, K. and Moczon, T. (1984) The nervous system of *Diplostomum pseudospathaceum* Niewiadomska, 1984 (Trematoda, Diplostomatidae). II. Structure and development of the nervous system in metacercaria. *Z. Parasitenkd.* 70: 537–548.

Niewiadomska, K. and Moczon, T. (1987) The nervous system of *Diplostomum pseudospathaceum* Niewiadomska 1984 (Trematoda, Diplostomatidae). III. Structure of the nervous system in the adult stage. *Parasitol. Res.* 73: 46–49.

Palladini, G., Margotta, V., Carolei, A., Chiarini, F., Del Piano, M., Lauro, G.M., Medolago-Albani, L. and Venturini, G. (1983) The cerebrum of *Dugesia gonocephala* s.l. Platyhelminthes, Turbellaria, Tricladida. Morphological and functional observations. *J. Hirnforsch.* 23: 165–172.

Palmberg, I. and Reuter, M. (1983) Asexual reproduction in *Microstomum lineare* (Turbellaria). I. An autoradiographic and ultrastructural study. *Int. J. Reprod.* 5: 197–206.

Pan, J.-Z., Halton, D.W., Shaw C., Maule A.G. and Johnston, C.F. (1994) Serotonin (5-HT)- and neuropeptide-immunoreactivities in the intramolluscan stages of three marine trematode parasites. *Parasitology Res.; in press.*

Panula, P., Eriksson, K., Gustafsson, M. and Reuter, M. (1994) An immunocytochemical method for histamine: application to the planarians. *Hydrobiologia; in press.*

Reisinger, E. (1972) Die Evolution des Orthogons der Spiralier und das Archicoelomatenproblem. *Z. Zool. Syst. Evolutionsforsch.* 13: 184–206.

Reisinger, E. (1976) Zur Evolution des stomatogastrischen Nervensystem bei den Plathelminthen. *Z. Zool. Syst. Evolutionsforsch.* 14: 241–253.

Reuter, M. (1975) Ultrastructure of the epithelium and the sensory receptors in the body wall, the proboscis and the pharynx of *Gyratrix hermaphroditus* (Turbellaria, Rhabdocoela). *Zool. Scripta* 4: 191–204.

Reuter, M. (1981) The nervous system of *Microstomum lineare* (Turbellaria, Macrostomida). II. The ultrastructure of synapses and neurosecretory release sites. *Cell Tissue Res.* 218: 375–387.

Reuter, M. (1987) Immunocytochemical demonstration of serotonin and Neuropeptides in the nervous system of *Gyrodactylus salaris* (Monogenea). *Acta Zool.* 68: 187–193.

Reuter, M. (1990) From innovation to integration. – Trends of the integrative systems in microturbellarians. *Acta Acad. Aboensis Ser. B.* 50(7): 121–136.

Reuter, M. (1991) Are there differences between proseriates and lower flatworms in ultrastructure of the nervous system? *Hydrobiologia* 227: 221–227.

Reuter, M. (1994) Substance P immunoreactivity in sensory structures and the central and pharyngeal nervous system of *Stenostomum leucops* (Catenulida) and *Microstomum lineare* (Macrostomida). *Cell Tissue Res.*; *in press.*

Reuter, M. and Eriksson, K. (1991) Catecholamines demonstrated by glyoxylic-acid-induced fluorescence and HPLC in some microturbellarians. *Hydrobiologia* 227: 209–219.

Reuter, M. and Gustafsson, M. (1989) "Neuroendocrine Cells" in flatworms – Progenitors to metaxoan neurons? *Arch. Histol. Cytol.* 52 Suppl. 253–263.

Reuter, M., Lehtonen, M. and Wikgren, M. (1988) Immunocytochemical evidence of neuroactive substances in flatworms of different taxa – a comparison. *Acta Zool.* 69: 29–37.

Reuter, M., Wikgren, M. and Lehtonen, M. (1986) Immunocytochemical demonstration of 5-Ht-like substances in whole mount of *Microstomum lineare* (Turbellaria). *Cell Tissue Res.* 246: 7–12,

Reuter, M., Gustafsson, M.K.S., Lang, J. and Grimmelikhuijzen, C.J.P. (1990) The release sites and targets of nerve cells immunoreactive to RFamide – an ultrastructural study of *Microstomum lineare* and *Diphyllobothrium dendriticum* (Plathelminthes). *Zoomorphology* 109: 303–308.

Reuter, M., and Kuusisto, A. (1992) Growth factors in asexually reproducing Catenulida and Macrostomida (Plathelminthes)? A confocal, immunocytochemical and experimental study. *Zoomorphology* 112: 155–166.

Reuter, M. and Lindroos, P. (1979) The ultrastructure of the nervous system of *Gyratrix hermaphroditus* (Turbellaria, Rhobdocoela). *Acta Zool.* 60: 153–163.

Reuter, M. and Palmberg, I. (1987) An ultrastructural and immunocytochemical study of gastrodermal cell types in *Microstomum lineare* (Turbellaria, Macrostomida). *Acta Zool.* 68: 153–163.

Reuter, M. and Palmberg, I. (1989) Development and differentiation of neuronal subsets in asexually reproducing *Microstomum lineare*. Immunocytochemistry of 5-Ht, RF-amide and SCP_B. *Histochemistry* 91: 123–131.

Reuter, M. and Palmberg, I. (1990) Synaptic and nonsynaptic release in *Stenostomum leucops*. A study of the nervous system and sensory receptors. *Acta Acad. Abonensis Ser. B.* 50(7): 121–136.

Rieger, R.M., Tyler, S., Smith III, J.P.S. and Rieger, G. (1991) Platyhelminthes: Turbellaria. *In*: F.W. Harrison and B.J. Bogitsh (eds): *Microscopic Anatomy of Invertebrates 3: Platyhelminthes and Nemertinea*. Wiley-Liss Inc., New York, pp. 7–140.

Rohde, K. (1970) Nerve sheath in *Multicotyle purvisi* Dawes. *Naturwissenschaften* 57: 502–503.

Rohde, K. (1971) Untersuchungen an *Multicotyle purvisi* Daves, 1941 (Trematoda: Aspidogastrea). III. Licht und elektronmikroskopischer Bau des Nervensystems. *Zool. Jahrb. Anat.* 88: 320–363.

Rohde, K. (1990) Phylogeny of Platyhelminthes, with special reference to parasitic groups. *Int. J. Parasitol.* 20: 979–1007

Rohde, K. and Webb, R. (1986) Ultrastructure of neuroglia in the peripheral nervous system of Temnocephala sp. (Turbellaria, Temnocephalida). *Zool. Anz.* 216: 53–57.

Sakharov, D.A. (1974) Genealogy of Neurons (in Russian). Nauka Moscow. Skuce, P.J. Johnston, C.F., Fairweather, I., Halton, D.W. and Shaw, C. (1990) A confocal scanning laser microscope study of the peptidergic and serotinergic components of the nervous system in larval *Schistosoma mansoni*. *Parasitology* 101: 227–234.

Smith, J.P.S., Tyler, S. and Rieger, R. (1986) Is the Turbellaria polyphyletic? *Hydrobiologia* 132: 13–21.

Trawicki, W., Czubaj, A. and Moraczewski, J. (1988) The brain ultrastructure of *dendrocoelum lacteum* (O.F. Muller). *Fortschr. Zool.* 36: 195–200.

Van Houten, J. and Preston, P.R. (1987) Chemoreception in single-celled organisms. *In*: T.E. Finger (ed.): *Neurobiology of Taste and Smell*. John Wiley & Sons, New York, pp. 11–33.

Webb, R.A. (1988) Endocrinology of Acoelomates. *In*: H. Laufer and R.G.H. Downer (eds): *Endocrinology of Selected Invertebrate Types*. Alan R. Liss, New York, pp. 31–62.

Welsh, J.H. and King, E.C. (1970) Catecholamines in planarians. *Comp. Biochem. Physiol.* 36: 683–688.

Venturini, G., Carolei, A., Palladini, G., Margotta, V. and Lauro, M.G. (1983) Radioimmunological and immunocytochemical demonstration of met-enkephalin in planaria. *Comp. Biochem. Physiol.* 74C: 23–25.

Wikgren, M. (1987) *Neuroendocrine systems in flatworms*. Thesis. Åbo Akad. Univ.

Wikgren, M.C. and Reuter, M. (1985) Neuropeptides in an microturbellarian – wholemount immunocytochemistry. *Peptides* 6 (Suppl. 3): 471–475.

Wikgren, M., Reuter, M., Gustafsson, M. and Lindroos, P. (1990) Immunocyto-chemical localization of histmaine in flatworms. *Cell Tissue Res.* 260: 479–484.

Willmer, P. (1990) *Invertebrate relationships. Patterns in Animal Evolution*. Cambridge Univ. Press, Cambridge.

Yoshizawa, Y., Wakabayashi, K. and Shinozawa, T. (1991) Inhibition of planarian regeneration by melatonin. *Hydrobiologia* 227: 31–40.

The Nervous Systems of Invertebrates: An Evolutionary and Comparative Approach
ed. by O. Breidbach & W. Kutsch
© 1995 Birkhäuser Verlag Basel/Switzerland

Repeating patterns of motoneurons in nematodes: The origin of segmentation?

W.W. Walthall

Department of Biology, Georgia State University, Atlanta, Georgia, USA

Summary. Evolutionarily diverse groups of animals share numerous similarities as individual neurons are assembled into functional neural circuits. One example is the hierarchical sequence of events that individual nerve cells follow during morphological development. In the initial step a presumptive neuron is generated and positioned appropriately. Second, the undifferentiated cell elaborates a growth cone capable of interacting with extrinsic cues and leading the presumptive axonal process as it is guided into areas where potential synaptic targets reside. Finally, the differentiating nerve cell selects among appropriate and inappropriate target cells as it completes the process of selective synaptogenesis. The extracellular matrix molecule laminin provides a second example, this time at the molecular level. Biochemical and genetic studies have shown that this molecule directs process guidance of neurons in vertebrates, annelids, and nematodes. In both examples an interest in neural development has provided a window through which evolutionarily processes have been revealed. The free-living soil nematode *Caenorhabditis elegans* possesses several features that collectively place it in a rather unique position among metazoans and has allowed genetic and cellular studies to be integrated at the level of identified neurons and neural circuits. This review will focus on developmental studies of *C. elegans* locomotory neural circuits. General issues that will be addressed are the similarities and differences among different taxa regarding: the relationship between cell lineage and cell fate determination in generating reiterative neural patterns; pioneer cells and the molecular basis for process guidance and finally genetic and epigenetic events involved in sculpting highly specific synaptic patterns.

Introduction

The large number of similarities in metazoan nervous systems did not arise by chance; either they are vestiges of a common evolutionary origin, or they represent convergence upon a parsimonious solution to a common problem. Protostomes and deuterostomes represent the two most successful evolutionary lines. In spite of the fact that the two groups diverged several hundred million years ago, shared similarities suggest that the origin of critical molecules and mechanisms of pattern formation predate this separation. But what is it that we hope to accomplish by unlocking these relationships? The revelation of a common past allows one to marvel at the conservative nature of nature and adds arrows to a well-stocked quiver supporting neoDarwinian theories of evolution. More important is the potential to gain insight into the orchestration of more complex organizational relationships, such as the multifaceted problem associated with the assembly of neurons into functional neural circuits mentioned in the abstract. A critical question, which remains opaque, is how a finite number of genes directs the seemingly limitless number of decisions necessary to assemble a nervous system? Nematodes represent an ancient phylum with a number of striking idiosyncracies. However, the phylum diverged before the split between deuterostomes and protostomes and is therefore strategically positioned to allow comparative studies searching for unifying organizational principles among metazoans. In this review, I will emphasize unique aspects of these animals that presumably hold keys to conserved evolutionary mechanisms.

The nematode *Caenorhabditis elegans* is a self-fertilizing hermaphrodite with a reproductive cycle of 3.5 days. It was these properties, in part, that led Sydney Brenner to choose this animal as a model system for studying the genetic basis of development and behavior. Together these properties have nurtured the establishment of a sophisticated genetic system for generating and characterizing mutants (Brenner, 1974). Other important features of *C. elegans* include the small size, the limited number of somatic cells, and the transparent nature of the animal. Collectively, these properties have contributed to the accomplishment of two major goals initially set by Sydney Brenner. *First*, the charting of the complete somatic cell lineage from the fertilized egg to the adult (Sulston and Horvitz, 1977; Sulston et al., 1983), and *second* the reconstruction of the synaptic pattern for each of the neurons that comprise the animal's nervous system (White et al., 1986). A recent review has examined the genetic and cellular aspects of behavior (Bargmann, 1993).

The revelation of the invariant lineage has made it possible to assign a fate to each somatic cell in the animal. This fate map allows one to know either the progeny that a particular cell will produce or the terminally differentiated cell type that a cell will assume. The fertilized egg, P_0, divides to form a daughter, P_1, and a founder cell, AB. After six rounds of division six founder cells are present. These founder cells or their progeny continue to divide for approximately 500 min, at which time all but four of the embryonic complement of cells will have been generated. The animal hatches at 800 min when grown at 20°C. The AB cell, generated at the first division, produces primarily ectodermal and neural cells. Nematodes are pseudocoelomates, their body cavity is derived from the blastocoel; whereas the body cavity of eucoelomates is a byproduct of the invagination process associated with gastrulation. The spatial relationship between the muscles, the locomotory neurons and the pseudocoelomic cavity is illustrated in Figure 1. With the exception of the neurons associated with the pharynx, the nervous system is excluded from the coelomic cavity by a basal lamina produced from the basal face of the hypodermal cells (Hedgecock et al., 1990).

The establishment of the connectivity pattern for each of the 302 neurons in the adult hermaphrodite has provided a wealth of informa-

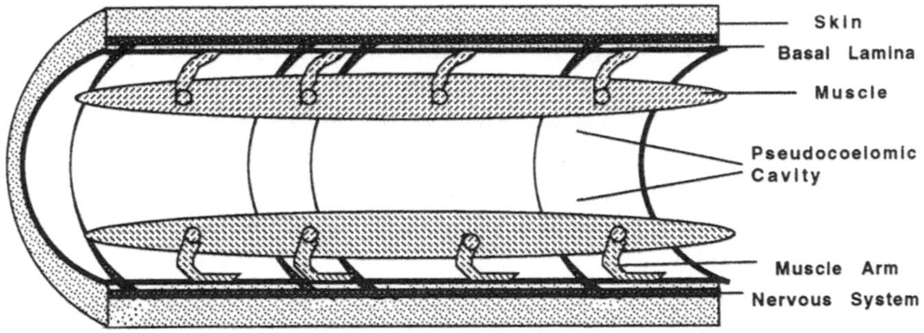

Figure 1. The physical relationship between the ectodermal and mesodermal tissue type in pseudocoelomates. The illustration is a cut-away from the left lateral perspective. Skin cells (light stipple) and locomotory neurons (dark stipple) are separated from the coelomic cavity by the basement membrane (thick black lines) the longitudinally running body muscles (diagonal lines) lie within the coelomic cavity. Only two of the four muscle groups are illustrated. Five of the seven classes of locomotory neurons have processess in both the dorsal and ventral nerve cords, which are connected via a circumferentially running commissure (four are shown). Muscle arms extend from the body of each muscle cell to the basement membrane adjacent to the nearest nerve cord. Neuromuscular junctions are composed of *en passant* presynaptic elements from the mns and the postsynaptic elements from the muscle arms.

tion regarding potential circuits for mediating many of the animals stereotyped behaviors (White et al., 1986). Pulsed dye lasers compatible with microscopes have made it possible to kill selected cells in the living, unstained embryo or animal. In many instances this has provided important confirmation regarding the role of neurons within a circuit. However, the connectivity patterns based solely on reconstructions cannot provide functional information regarding whether postsynaptic potentials are excitatory or inhibitory, or the role of modulatory mechanisms in setting the gain on particular sets of synapses. Unfortunately, the small size of *C. elegans* (1.5 mm in adult hermaphrodites) that was so important for reconstructing the entire nervous system, has been a major impediment in the electrophysiological characterization of the nervous system. Fortunately, comparative studies between neurons of the distantly related parasitic nematodes *Ascaris suum* and *C. elegans* reveal a surprising amount of structural similarity especially in the locomotory neurons (Walrond et al., 1985, Walrond and Stretton, 1985). The large size of *A. suum* has facilitated the electrophysiological analysis of homologous motoneurons and provided complementary data regarding neural function.

Locomotion in *C. elegans*: A model for cellular and genetic studies

C. elegans moves forward and backward in a sinuous pattern that is the product of synchronous waves of muscle contraction that propagate along the length of the animal. There are four groups of longitudinally running body muscles, one pair located to the right and left of the dorsal hypodermal ridge, and a second pair located to the right and left of the ventral hypodermal ridge; three of the groups contain 24 mononucleate muscle cells and the fourth contains 23 cells (Sulston et al., 1983). The dorsal pair of muscles acts in unison as do the ventral pair, however, the two pairs act antagonistically to one another. Typically, the animal lies on its side and body movement is restricted to the dorsal/ventral plane.

Each dorsal and ventral muscle sends axon-like processes, the muscle arms, to the centrally located dorsal and ventral nerve cords, respectively. The apposition of the muscle arms to the lateral sides of the two nerve cords means that it is unnecessary for the motoneurons to form complex branches to contact the muscle cells. All of the locomotory neurons possess unbranced cylindrical processes that run longitudinally in the cords and form *en passant* neuromuscular junctions with the muscle arms as illustrated in Figure 1.

There are a total of 69 locomotory neurons innervating body muscle, and they comprise seven different classes (White et al., 1986). One representative neuron illustrating the generalized anatomy and synaptic pattern of each class is shown in Figure 2. The pre- and postsynaptic processes of these motoneurons reside exclusively in the ventral nerve cord or in both the ventral and dorsal nerve cords. The cell bodies of the locomotory neurons are located ventrally in the retrovesicular ganglion, the ventral nerve cord and the preanal ganglion (Fig. 3). In addition to initiating forward or backward movement, the neural networks also insure that contractile waves in the dorsal and ventral muscle maintain the proper phase relationship as they propagate along the animal's body. Five of the classes possess dorsal and ventral processes, which are connected by single circumferential commissures. Of the five

---

Figure 2. The structures of representative neurons from each of the seven classes of locomotory neurons in the body of *C. elegans* as viewed from the left side. Embryonic (A and B) and postembryonic (C and D) neural circuits. The upper and lower stippled areas represent dorsal and ventral nerve cords, respectively. Inputs on the postsynaptic branches are indicated by triangles entering the branch. Outputs from the presynaptic branches are indicated by triangles exiting the branch and protruding into the coelomic cavity where the muscles would be located. The AVA and AVB interneuronal processes each represent groups of interneurons that drive backward and forward movement, respectively. Hollow shapes depict synapses or interneuronal processes that drive forward locomotion and black triangles and shapes drive backward locomotion. The hollow tubes between the interneuron AVB and the two classes of B mns represent gap junctions. (B) The stage specific respecification of the synaptic pattern of the DD mns is illustrated. The left panel illustrates the pattern in newly hatched animals and the right panel illustrates the DD mn pattern in the adult. (C) The structures of the two postembryonic mns with branches in both the dorsal and ventral nerve cords. (D) The structures of the two postembryonic classes of excitatory mns that are restricted to the ventral cord. (A-D) Diagrams are derivative of reconstructions from White et al., 1986; however, the synaptic patterns, although representative, are not intended to show the complete connectivity patterns for these mns.

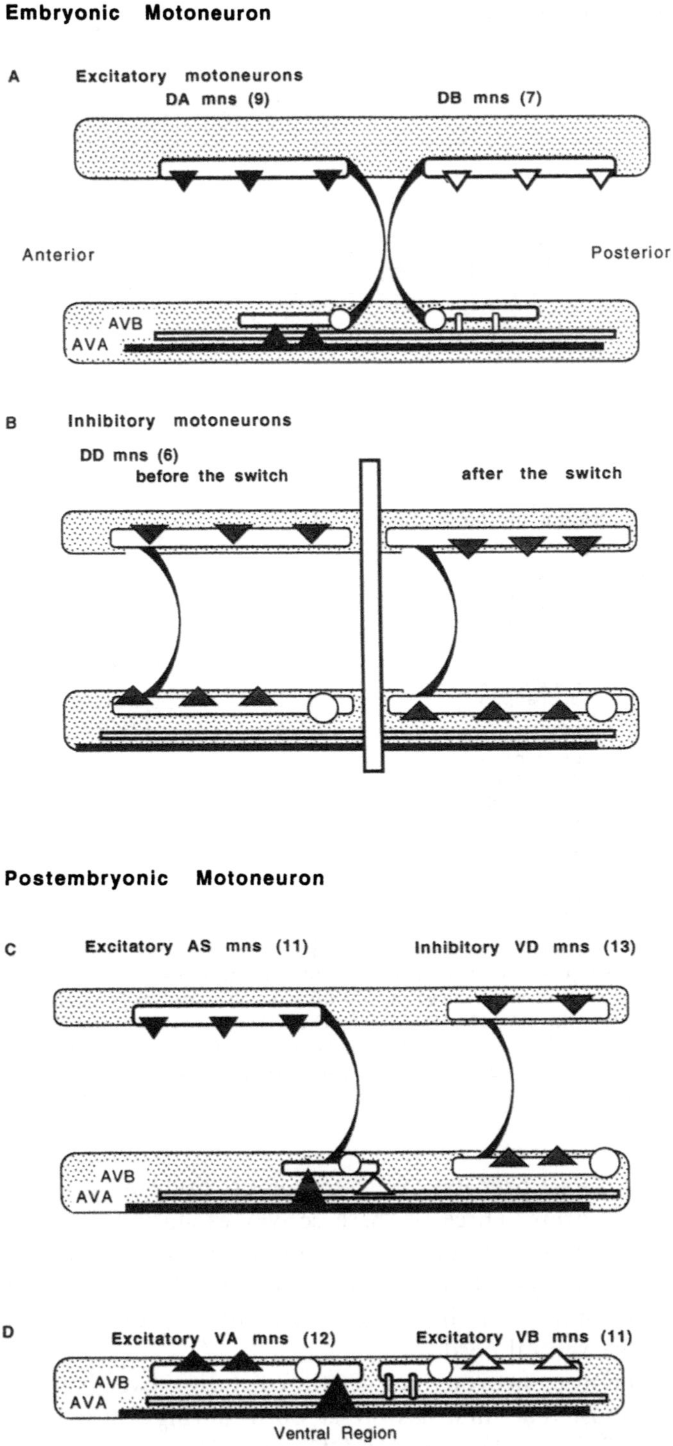
Embryonic Motoneuron
A
Excitatory motoneurons
DA mns (9)
DB mns (7)
Anterior
Posterior
AVB
AVA
B
Inhibitory motoneurons
DD mns (6)
before the switch
after the switch
Postembryonic Motoneuron
C
Excitatory AS mns (11)
Inhibitory VD mns (13)
AVB
AVA
D
Excitatory VA mns (12)
Excitatory VB mns (11)
AVB
AVA
Ventral Region

Newly Hatched L1--Locomotory Circuits--

A

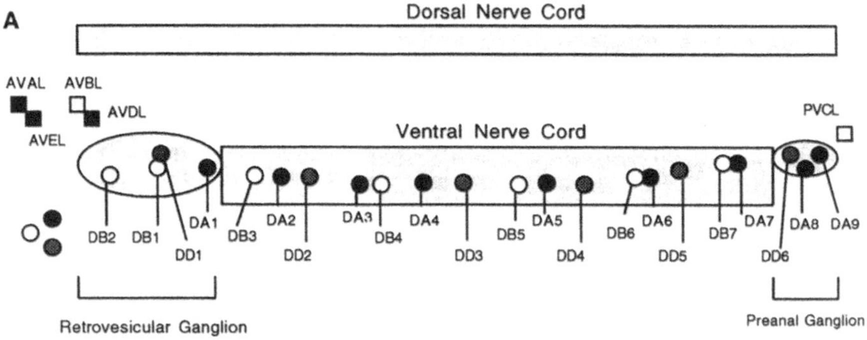

B

Lineal Origins of the Embryonic Motoneurons--

DD 1	AB	plppappap	DB1/3	AB	plpaaaapp
DD 3		plppapppa	DB 5		plpapappp
DD 5		plppapppp	DB 6		plppaappp
DD 2		prppappap	DB 2		arappappa
DD 4		prppapppa			
DD 6		prppapppp	DB 3/1		prpaaaapp
			DB 4		prpappapp
DA 1	AB	prppapaap	DB 7		prppaappp
DA 3		prppapapa			
DA 5		prppapapp			
DA 7		prpppaaap			
DA 8		prpapappp			
DA 2		plppapapa			
DA 4		plppapapp			
DA 6		plpppaaap			
DA 9		plpppaaaa			

Figure 3. The positions (A) and lineal origins (B) of the embryonic motoneuron cell bodies. Blackened shapes represent excitatory neurons involved in backward movement, clear shapes represent excitatory neurons involved in forward movement and cross-hatched shapes represent inhibitory motorneurons (Durbin, 1987). The lineages are taken directly from Sulston et al., 1983. AB represents one of the six founder cells. Most of the subsequent cell divisions occur along the anterior posterior axis and are denoted by an "a" or "p". Occassionally, a cell divides perpendicular to the longitudinal axis and then they are denoted as right (r) or left (l).

classes with dorsal processes four are almost exclusively presynaptic to dorsal muscle arms. Three of these convey excitation to dorsal muscle from interneurons whose processes run in the ventral nerve cord and the fourth class conveys inhibition from the ventral cord. The fifth has a postsynaptic branch in the dorsal cord and a presynaptic inhibitory branch in the ventral cord.

The locomotory neurons are generated during two distinct developmental periods. The first group is born during embryogenesis and consists of 9 DA, 7 DB, and 6 DD mns. With these 22 mns newly hatched animals are capable of moving forward and backward. The designations were originally based upon their distinctive morphologies and synaptic patterns (White et al., 1976). It was determined later using a laser microbeam and killing selected combinations of neurons that each of these classes makes specific contributions to locomotory behavior (Chalfie et al., 1985). It now appears that the DA and DD mns are necessary for backward movement and the DB mns for forward movement (Walthall et al., in press).

The connectivity pattern between interneurons and locomotory neurons suggested a subset of interneurons as likely candidates for contributing to locomotion (Chalfie et al., 1985). Ablation of selected interneuron pairs or their precursors and the observation of resulting locomotory defects has allowed their contributions to be inferred. The absence of the AVAR/L and AVDR/L resulted in animals unable to move backward; whereas PVCR/L and/or AVBR/L appear to be necessary for forward movement. The experiments have for the most part been performed and the behavior analyzed in L1 animals. However the interneuronal input onto the VA and AS mns is very similar and they are thought to be analogs to the DA cells and the interneuronal input onto the VB mns is identical to that of the DB mns, and the VB mns are thought to be ventral analogs (White et al., 1986).

Twenty-two locomotory neurons form a functional locomotory network at hatching. It is interesting to note that the two classes of excitatory mns innervate dorsal body muscle, exclusively; whereas the inhibitory DD mns innervate ventral body muscle, exclusively (Figs 2A and B). Inspite of the absence of excitatory input contractile waves propagate in both directions through the ventral muscle.

The second wave of locomotory neurons is born during the latter half of the first larval stage (L1) and probably becomes functional following the molt to the L2 stage (Figs 2C and D). These consist of the VA, VB, AS, and VD mns. The VA, VB, and AS mns are excitatory, however, of the three only the AS mns possess branches in both the dorsal and ventral cord. The presynaptic branch of each VA mn projects anteriorly from the cell body, whereas the shorter postsynaptic branch projects posteriorly. Conversely, the presynaptic branch of each VB mn projects posteriorly from the cell body and the postsynaptic branch projects anteriorly. The 13 VD mns are structurally very similar to the DD mns, in fact they assume the same synaptic pattern as was initially described for the DD mns (Fig. 2C). While the postembryonic mns are completing their morphological differentiation the DD mns undergo a complete respecification of their synaptic patterns (Fig. 2B; White et al., 1978). When this respecification is completed the DD mns receive ventral input from the postembryonic excitatory mns VA and VB and make inhibitory neuromuscular junctions with dorsal muscle.

The lineal origin of the locomotory neurons

When John Sulston and his colleagues analyzed the lineal relationships among the cells in *C. elegans*, numerous examples were revealed in which cells produced by mirror symmetrical lineal patterns assumed the same fates. ABplp and ABprp represent the earliest example and together they generate 21 of the 22 embryonic locomotory neurons. The spatial and lineal relationships among these neurons are illustrated in Figure 3B. Closer examination of the lineages that produce the three classes of embryonic locomotory neurons revealed that the DD mns are clonally related to one another. The six DD mns along with two interneurons. RIGL and R arise from mirror-symmetric lineages. The three DD cells derived from ABpl are 1, 3, and 5; the three from ABpr are 2, 4, 6. This is a rare occurrence among the lineages of the nervous system. Five of the nine DA mns share the same grandmothers as the DD mns; these are DA mns 1, 3 and 5 from the precursor on the right and DAs 2 and 4 from the left. The remaining four DA mns are closely related, but a number of different neuronal cell types are also generated. The lineal origins of the DB mns are diverse with no obvious structural relationships emerging among the cells generated from these lineages.

Although the embryonic mns assume fixed longitudinal positions, not unlike that observed in annelids and insects, the underlying lineal relationships show marked differences. Longitudinal organization in annelids is established early and systematically as longitudinal bands of cells are initiated by individual founder cells representing different germ bands. Neighbouring cells then establish skin, nerve and muscle for a particular segment (Shankland, 1991). In insect embryos CNS neuroblasts are systematically recruited from a population of ventral ectodermal cells, all of which have the same developmental potential (Doe and Goodman, 1985). Once designated, these cells undergo segmentally determined lineal patterns to produce the segmented ventral nerve cord. The underlying similarity between annelids and insects is the iterated pattern in which each segmental repeat is generated. A systematic pattern for generating embryonic mns is not apparent in nematodes. Instead, it appears that autonomous lineal mechanisms are primarily responsible for determining the fates for these cells (Hedgecock, 1990).

Conversely, the generation and positioning of the postembryonic locomotory mns, bears a greater resemblance to the types of mechanisms observed in leeches and insects. Twelve bilaterally symmetric blast cells located sublaterally on either side of the newly hatched animal migrate into the ventral nerve cord (Fig. 4A) where each follows the same stereotyped pattern of cell divisions (Fig. 4B) to produce the complement of postembryonic mns (Sulston and Horvitz, 1977). Although the pattern of cell divisions is stereotyped, the fates of the five daughter cells vary according to their position along the longitudinal

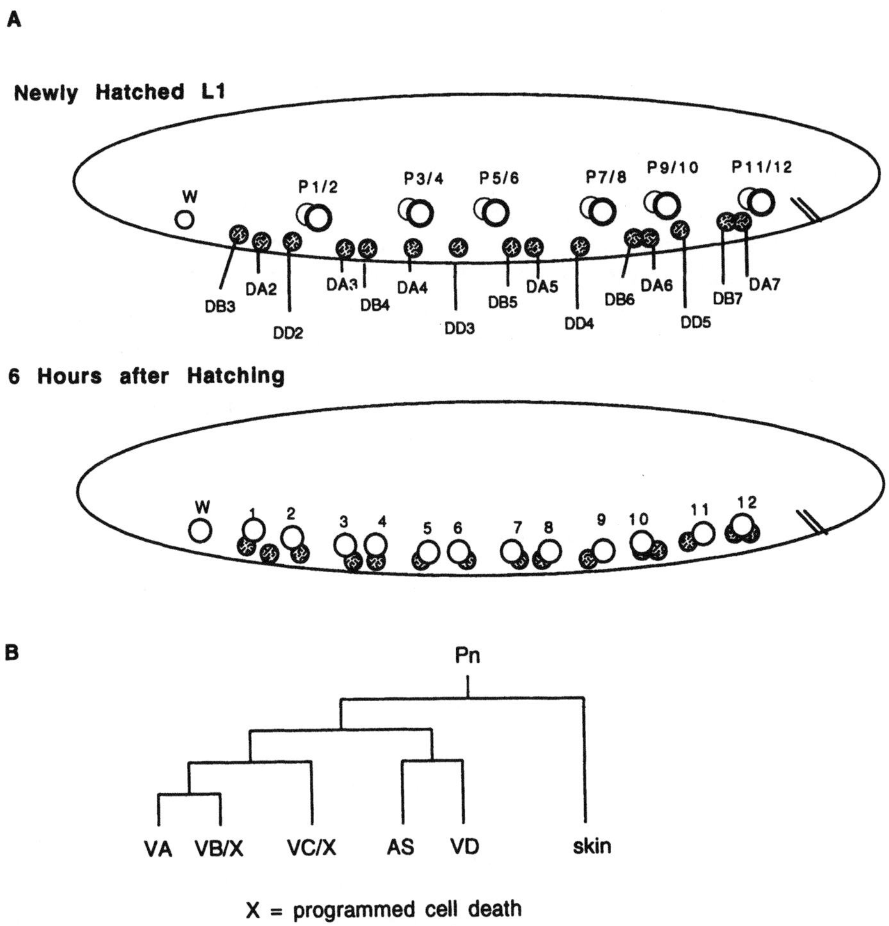

Figure 4. (A) Positions of the postembryonic blast cell bodies before and after their migration and entry into the ventral nerve core. (B) The pattern of cell division that produces the postembryonic motorneurons.

axis. Despite not showing overt segmental organization, a feature common in annelids, arthropods and vertebrates, a number of internal structures exhibits regional specification along the anterior-posterior axis. The five classes of postembryonic locomotory neurons, in fact, form periodic repeats and like their analogs from other phyla, these neurons are lineally more closely related to their neighbours than similar class members in different segmental repeats.

Genetic and molecular analyses have focused on mutations that disrupt pattern formation in *C. elegans* and have identified a cluster of genes which contain homeoboxes and which participate in the establishment of pattern along the animal's longitudinal axis (Wang et al., 1993). These investigations have revealed that the genomic organization of the homeotic complex is related to that first described in *Drosophila*, and then mammals. The similarities extend to establishing that the regional expression of these genes corresponds to their spatial location along the chromosome. Taking advantage of the precise fate maps for *C. elegans* cells, it has been possible to demonstrate that these genes create regionally distinctive patterns among the locomotory neurons (Salser et al., 1993). As was mentioned earlier, there are 11 VB mns. The posterior blast cells P11 and P12 undergo the stereotyped pattern of cell divisions, but the daughters (P11/12aaap) that would be expected to differentiate into the VB mns undergo programmed cell death instead. The homeotic gene that is expressed in this particular region is *mab-5* (*m*ale *ab*normal). Experiments using *mab-5* mutants have shown that in the absence of this gene's product the P11 and P12 derivatives normally programmed to die become VB mns instead.

Establishing the dorsal and ventral nerve cords (process guidance in *C. elegans*)

The guidance of neuronal processes has been an intensive area of study for the last decade and a conserved group of molecules, (members of the immunoglobulin supergene family), i.e., extracellular matrix molecules such as laminin have been characterized independently in at least three different phyla. In *C. elegans* several recent studies have investigated the role of three genes involved in circumferential migrations. Mutations in these genes were originally isolated as uncoordinated (*unc*) by Sidney Brenner. Later, a number of investigators devised staining techniques that could be used effectively in the light microscope and demonstrated that a subset of these mutants is involved in the guidance of both migrating cells and processes. Mutations have identified three genes that affect longitudinal process guidance and fasciculation (McIntire, 1992). Hedgecock et al. (1985) studied a second group composed of three genes involved in circumferential guidance. The *unc-6* gene product is required for migrations in both the dorsal and ventral halves of the animal, whereas the genes *unc-5* and *unc-40*, are primarily involved in dorsal and ventral migrations, respectively. Recent molecular studies have characterized two of these genes (Leung-Hagesteijn et al., 1992; Ishii et al., 1992). The product of the gene *unc-6* is a laminin-related protein involved in establishing a substrate suitable for guiding motile cells. The product of the gene *unc-5* is a transmembrane protein in the immunoglobulin supergene family. It is expressed in a number of different cell types that share one feature, the cell or an associated process migrates circumferentially in the dorsal half of the animal. Included are all of the mns with processes in both dorsal and ventral nerve cords (Leung-Hagesteijn et al., 1992). It has recently been shown by Culotti and his group that ectopic expression of this gene in a late-developing sensory neuron causes the neuronal process to completely redirect its direction of outgrowth (Hamelin et al., 1993). In mutants that disrupt neuronal process guidance similar disruptions are seen for both embryonic and postembryonic mns. This suggests that either the same guidance cues are present during the development of both classes of D mns, or alternatively, the postembryonic mns fasciculate onto the embryonic mns and use them as scaffolds. If so, guidance of the postembryonic mns might be independent of the circumferential cues just described, but might be indirectly affected due to the misguidance of the embryonic neurons. The processes of the DD and VD mns are tightly fasciculated in both nerve cords. Ablation of the two grandmothers in the embryo completely removed the six DD mns. Observation of the anatomy of the VD mns that had grown in the absence of the DD mns and the locomotory behavior in the ablated adults showed that the structural development of the VD mns was independent of the DD mns and suggests that the same guidance system is used by both embryonic and postembryonic mns (Walthall et al., in press).

Synaptic specificity and locomotory neurons

The sinuous movement of *C. elegans* requires precise regulation balancing excitation and inhibition in both the dorsal and ventral body muscles. A number of mutants has been isolated that exhibit asymmetric locomotory patterns. The cellular and molecular analyses of these genes have revealed that they often affect the synaptic specificity of a subset of motoneurons.

unc-4

The pair of interneurons AVBR and L normally form gap junctions exclusively with seven DB and 11 VB motoneurons and they along with a second pair of interneurons PVCR and L provide excitation for forward locomotion (Fig. 2A). Animals with a mutation in the *unc-4* gene exhibit an asymmetric locomotory pattern when attempting to move backward. The dorsal contractile waves are stronger than the ventral ones and the animal forms an "omega" figure with the dorsal aspect of the animal oriented centrally. Reconstructions of a serially sectioned *unc-4* mutant revealed an alteration in the pattern of synapses on the postembryonic VA mns so that their input had become identical to that of their sister cells, the VB mns (White et al., 1992). The replacement of synapses from the AVA and AVD interneurons with gap junctions from the AVB interneurons produced changes that impact on circuits for both forward and backward movement. The loss of synapses between the AVA, AVD and AVE interneurons and the VA mns removed the only source of excitation for the ventral mns during backward movement and appears to be responsible for the distinctive aspects of the *unc-4* mutant phenotype. The novel gap junctions from the AVB interneurons cause additional excitation of ventral muscle during forward movement but the resulting impairment is relatively mild.

The locomotory pattern of *unc-4* mutants suggested and the reconstructions confirmed that the synapses of VA mns 1, 11 and 12 are not altered in the *unc-4* mutants. It was noted by White and colleagues that of the 12 VA mns, 9 have VB mns as siblings. The three that do not are VA mns 1, 11, and 12, i.e., the subset of VA mns that lack a VB mn as a sibling. Although the synaptic pattern of the VA mns was altered to become identical to that of the VB mns, the VA mns retain their distinguishing anatomical features.

The *unc-4* gene has been molecularly cloned. Analysis of its sequence revealed a homeodomain, which suggests that the gene is involved in regulating the expression of target genes (Miller et al., 1992). The

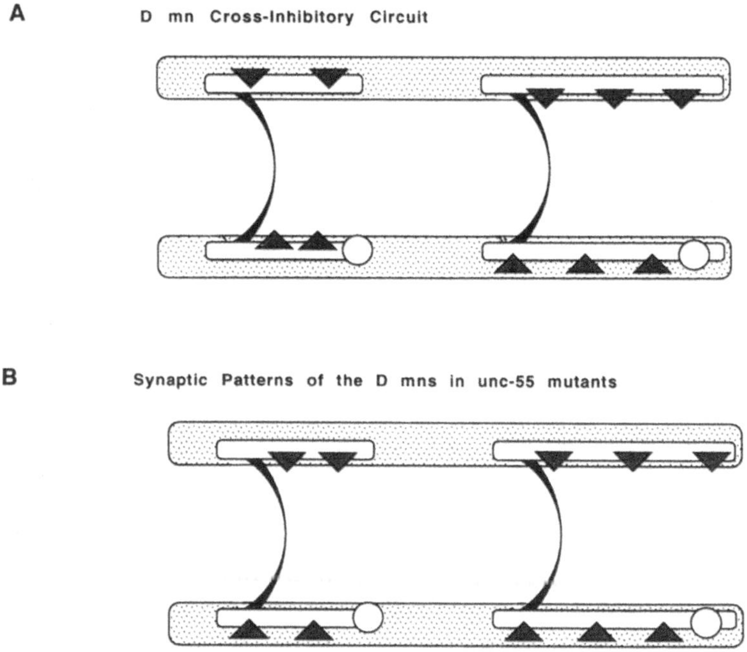

Figure 5. Homeotic transformation of the synaptic patterns of the VD mns in *unc-55* mutants. (A) The adult wild-type DD and VD mn patterns of synapses. Together the two classes of mns comprise a cross-inhibitory network that insures that simultaneous contraction of dorsal and ventral muscle at the same transverse level does not occur. (B) This schematic illustrates the altered connectivity of the VD mns in the *unc-55* mutant. Instead of a cross-inhibitory circuit both DD and VD mns conduct information that originates ventrally and results in the inhibition of dorsal muscle.

UNC-4 protein is expressed in the VA mns and raises the intriguing possibility that *UNC-4* may regulate downstream genes directly involved in choosing appropriate synaptic partners. In addition to the VA mns the gene product is also expressed in the DA mns but the synaptic pattern of these mns is not altered.

unc-55

Mutations in a second gene, *unc-55*, have been identified that transform the pattern of synapses of the VD mns so that they become identical to that of the DD mns (Fig. 5; John White and Leon Nawrocki, of the MRC, Cambridge U.K.; Walthall and Plunkett, personal communication). The VD and DD mns possess a number of structural and biochemical similarities, but differ from one another in their lineal origin and their pattern of synapses (White et al., 1986; Walthall, 1990). These similarities and differences raise a number of questions regarding their genetic specification. Mutations in several different genes cause similar defects in the DD and VD mns and result in a symmetrical locomotory defect in which both dorsal and ventral muscles contract when the animal initiates backward movement. The same locomotory pattern is observed if the majority of the DD and VD mns is ablated (McIntire et al., 1993). These mutations have identified genes involved in two features critical for the establishment of functional DD and VD mns: morphological development and intercellular communication. These similarities are apparently due to the activation of similar genetic programs. In *unc-55* mutants the sole functional distinction between the two cell types is removed as the VD mns adopt the synaptic pattern of the DD mns.

Conclusion

The question posed in the title as to whether the repeating pattern of mns in *C. elegans* represents the origin of segmentation is impossible to answer conclusively. However, examination of the lineal and pattern forming mechanisms involved in the generation of the embryonic and the postembryonic mns reveal interesting differences between the two groups. Although the positions of the embryonic mns are invariant there is not a true repeating pattern. In addition, there is no systematic pattern relating the lineages that generate the three mn classes. Conversely, the postembryonic mns do create a repeating pattern. More importantly, they are generated by a group of blast cells that undergoes an iterative pattern of cell divisions to produce the four classes of postembryonic mns. The generation of the postembryonic mns is reminiscent of the generation of segmental repeats in animals exhibiting more overt segmentation.

Although the lineal mechanisms used to generate the embryonic and postembryonic mn classes are different, mutants have identified a number of critical genes that relate the DAs to the VAs and possibly the ASs, the DBs to the VBs and the DDs to the VDs. As more evidence is gained indicating that identical genetic programs are used repeatedly to generate neurons with similar fates, then the critical question becomes, how are the differences generated? Single genes, such as *unc-4* and *unc-55*, exist whose products are necessary to produce different synaptic patterns among subsets of related neurons. In addition, a large body of evidence has characterized the critical contribution of epigenetic events, such as neural activity or hormonal changes, in producing distinguishing features among neurons.

Living organsims exhibit an immense amount of complexity at the molecular, cellular and behavioral level. A plan for efficiently generating this complexity is proposed in which the same genetic program is used in different places and at different times to generate neurons that have the potential to realize different fates. Determining how genetic and epigenetic information is then combined to provide unique instructions for the differentiation of individual nerve cells will have universal implications.

References

Brenner, S. (1974) The genetics of *Caenorhabditis elegans*. *Genetics* 77: 71–94.

Bargmann, C.I. (1993) Genetic and cellular analysis of behavior in *C. elegans*. *Annu. Rev. Neurosci.* 16: 47–71.

Chalfie, M., Sulston, J.E. White, J.G. Southgate, E., Thompson, J.N. and Brenner, S. (1985) The neural circuit for touch sensitivity in *Caenorhabditis elegans*. *J. Neurosci.* 5: 956–964.

Doe, C.Q. and Goodman, C.S. (1985) Early events in insect neurogenesis. II. The role of cell interactions and cell lineage in the determination of neuronal precursor cells. *Dev. Biol.* 111: 206–219.

Durbin, R.M. (1987) *Studies in the development and organization of the nervous system of Caenorhabditis elegans*. Ph.D. thesis, King's College, Cambridge, England.

Hamelin, M., Zhou, Y., Su, M-W., Scott, I.M. and Culotti, J.G. (1993) Expression of the *UNC-5* guidance receptor in the touch neurons of *C. elegans* steers their axons dorsally. *Nature* 364: 327–330.

Hedgecock, E.M., Culotti, J.G., Thomson, J.N. and Perkins, L. (1985) Axonal guidance mutants of *Caenorhabditis elegans* identified by filling sensory neurons with fluorescein dyes. *Dev. Biol.* 111: 158–170.

Hedgecock, E.M., Culotti, J.G. and Hall, D.H. (1990) The *unc-5*, *unc-6*, and *unc-40* genes guide circumferential migrations of pioneer axons and mesodermal cells on the epidermis in *C. elegans*. *Neuron* 4: 61–85.

Ishii, N., Wadsworth, W.G. Stern, B.D., Culotti, J.G. and Hedgecock, E.M. (1992) *UNC-6*, a laminin-related protein, guides cell and pioneer axon migrations in *C. elegans*. *Neuron* 9: 873–881.

Leung-Hagesteijn, C., Spence, A.M., Stern, B.D., Zhou, Y., Su, M.W., Hedgecock, E.M. and Culotti, J.G. (1992) *UNC-5*, a transmembrane protein with immunoglobulin and throm-

bospondin type 1 domains, guides cell and pioneer axon migrations in *C. elegans*. *Cell* 71: 289–299.
McIntire, S.L., Garriga, G. and Horvitz, R. (1992) Genes necessary for directed axonal elongation or fasciculation in *C. elegans*. *Neuron* 8: 307–322.
McIntire, S.L. Jorgensen, E. and Horvitz, R. (1993) The GABAergic nervous system of *Caenorhabditis elegans*. *Nature* 364: 337–341.
Miller, D.M., Shen, M., Shamu, M., Burglin, T., Ruvkun, G., Dubios, M. Ghee, M. and Wilson, L. (1992) *C. elegans unc-4* gene encodes a homeodomain protein that determines the pattern of synaptic input to specific motor neurons. *Nature* 355: 841–845.
Salser, S. Loer, C. and Kenyon, C. (1993) Multiple Hom-C interactions specify cell fates in the nematode central nervous system. *Genes and Development* 7: 1714–1724.
Shankland, M. (1991) Leech segmentation: Cell lineage and the formation of complex body patterns. *Dev. Biol.* 144: 221–231.
Sulston, J.E., Schierenberg, E., White, J.G. and Thomson, J.N. (1983) The embryonic cell lineage of the nematode *Caenorhabditis elegans*. *Dev. Biol.* 100: 64–119.
Sulston, J. and Horvitz, H.R. (1977) Postembryonic cell lineages of the nematode *Caenorhabditis elegans*. *Dev. Biol.* 56: 110–156.
Walthall, W.W. (1990) Metamorphic-like changes in the nervous system of the nematode *Caenorhabditis elegans*. *J. Neurobiol.* 21: 1085–1091.
Walrond, J.P., Kass, I., Stretton, A.O.W. and Donmoyer, J.E. (1985) Identification of excitatory and inhibitory motoneurons in the nematode *Ascaris* by electrophysiological techniques. *J. Neurosci.* 5: 1–8.
Walrond, J.P. and Stretton, A.O.W. (1985) Reciprocal inhibition in the motor nervous system of the nematode *Ascaris:* Direct control of ventral inhibitory motoneurons by dorsal excitatory motoneurons. *J. Neurosci.* 5: 9–15.
Wang, B.M., Muller-Immergluck, M., Austin, J., Rovinson, A., Chisolm, A. and Kenyon, C. (1993) A homeotic gene cluster patterns the anteroposterior body axis of *C. elegans*. *Cell* 74: 29–42.
White, J.G., Albertson, D.G. and Anness, M.A.R. (1978) Connectivity changes in a class of motoneurone during the development of a nematode. *Nature* 271: 764–766.
White, J.G., Southgate, E., Thomson, J.N. and Brenner, S. (1986) The structure of the nervous system of the nematode *Caenorhabditis elegans*. *Philos. Trans. R. Soc. Lond. Biol. Sci.* 314: 1–340.
White, J.G., Southgate, E. and Thomson, J.N. (1992) Mutations in the *Caenorhabditis elegans unc-4* gene alter the synaptic input to ventral cord motor neurons. *Nature* 355: 838–841.

The Nervous Systems of Invertebrates: An Evolutionary and Comparative Approach
ed. by O. Breidbach & W. Kutsch
© 1995 Birkhäuser Verlag Basel/Switzerland

Genesis of segmental identity in the leech nervous system

G.O. Aisemberg, V.Y. Wong and E.R. Macagno

Department of Biological Sciences, Columbia University, New York, NY 10027, USA

Summary. The *Antennapedia*-class homeobox genes are likely to play a role in the specification of neuronal identities in invertebrates. The leech *Hirudo medicinalis*, a species well-suited for the analysis of these genes at the level of identified neurons, contains homologs of many *Antennapedia*-class and related genes. The expression pattern in the central nervous system of four leech homebox genes was examined in detail. *Lox1* is expressed during early gangliogenesis in one pair of transient neurons present in every segment and, at later stages, in 15–20 pairs of neurons per segment. *Lox2* is expressed in 25–30 pairs of neurons repeated in the posterior two-thirds of the midbody. *Lox4* is present in 20–30 pairs of iterated neurons in the posterior half of the midbody, and in smaller subset of them in more anterior ganglia. *Lox6* is expressed in 15–20 pairs of neurons of the third subesophageal neuromere and in fewer cells of more posterior ganglia. The subsets of neurons that express these homeobox genes are different but overlapping. Combinations of *Lox* genes could in theory generate enough variability to specify all central neurons in a leech ganglion.

Introduction

The central nervous systems (CNS) of arthropods and annelids share a basic morphological plan, including an anterior dorsal brain or supraesophageal ganglion, a ventral nerve cord that consists of a string of segmental ganglia linked by longitudinal connective nerves, and nerve roots that exit form the lateral aspects of the ganglia and subdivide to form numerous minor branches in the periphery. In some groups, fusion of several ganglia takes place during development, and the separate neuromeres are in evidence only at early stages. This common basic plan, nonetheless, suggests that many developmental factors that determine this primary design, as well as many functional properties that are derived from it, are homologous among these groups. Within this context of similarities, a crucial question is how each individual neuron in an organism acquires its unique properties. The answer to this question will likely help us understand not only its development, but also how the nervous system works. An interesting aspect of this question in metameric animals like arthropods and annelids is how segmentally homologous central neurons from different ganglia of the ventral nerve cord obtain singular identities.

Studies of *Antennapedia-(Antp-)* class homebox genes in *Drosophila melanogaster* and in other systems suggest that these transcription factors, conserved in all metazoans examined thus far, play a role in defining the identity of specific neurons. In early embryogenesis, they seem to specify positional information along the antero-posterior axis (McGinnis and Krumlauf, 1992). Later in development, they are expressed in specific cells of certain regions of the fly nervous system and they are likely to specify the identities of these neurons (see, for example, Doe et al., 1988a). We are examining the expression of *Antp*-class genes in the leech *Hirudo medicinalis*, a well-characterized model organism for neurobiological research that allows us to examine the pattern of expression and function of these genes at the level of single, identified neurons.

The metameric pattern of the leech embryo develops from five pairs of teloblasts (Weisblat, 1981; Stent et al., 1982) that divide asymmetrically to produce a chain of cells called primary blast cells. These chains of cells become parallel and join laterally to form a germinal plate that generates the 32 embryonic neuromeres of the ventral nerve cord as well as all other segmental structures. The supraesophageal ganglion, however, originates from the micromere cap and is considered to be, as a consequence, of non-metameric origin. Although the neuromeres are very similar to one another, they show segment-specific differences (Levine and Macagno, 1990). For example, the first four and the last seven neuromeres fuse to form specialized structures of the leech central nervous system, namely the subesophageal ganglion and the tail ganglion. In addition, many segmental ganglia differ in the presence or fate of certain iterated neurons. An extreme case of segmental differentiation is the generation of approximately 300 extra neurons in the two segments that contain the genitalia (all other midbody ganglia contain about 400 neurons in the adult leech; see Macagno, 1980). During gangliogenesis, about 20% more neurons are born than survive in each neuromere (Stewart et al., 1986).

In *Hirudo medicinalis*, we can recognize neurons that express each of the *Antennapedia*-class genes and, in principle, identify traits that may be under their control. A first stage in such studies includes the isolation of *Antennapedia*-class leech homeobox (*Lox*) genes and the analysis of their patterns of expression, followed by the identification of neurons that express individual members of this group of genes. The second stage would require altering the normal expression of each of these genes, in order to gain insight into their function. In *Droposphila* and a few other insects, this can be done genetically, and in mice by targeted disruption in embryonic stem cells. In other arthropods and annelids, however, only indirect means, such as microinjection of mRNAs, antisense nucleic acids (RNAs or oligonulceotides) or purified antibodies, are presently available.

CNS Expression of Lox genes

The expression patterns of four *Lox* genes were examined in the embryonic CNS of the leech using *in situ* hybridization to probes specific for the different genes and, in one case, staining with antibodies raised against the *Lox1* protein expressed in *E. coli*. The results indicate that these genes define specific, overlapping subpopulations of central neurons within defined regions of the CNS.

Both transient and stable neurons express Lox1 in most ganglia

Lox1 is most homologous to *Sex combs reduced* (*Scr*) and *Antp of Drosophila*, although it also shares some sequence features with the *Antp*-class homeobox gene *fushi tarazu* (*ftz*) (Aisemberg and Macagno, 1994). It is expressed in 15–20 bilateral pairs of central neurons that are present in most segments. Although some of these cells are found in every segment, the full complement of *Lox1*-positive neurons can be detected only in the two posterior-most neuromeres of the subesophageal ganglion and in all midbody ganglia. These cells arise in approximately the same positions from ganglion to ganglion and are thought to be an iterated set of segmentally homologous neurons (Fig. 1).

Lox1 is expressed in a long but spatially restricted region of the nerve cord, like mammalian Hox genes and fly segment identity genes. However, *Lox1* covers a larger fraction of the total length of the CNS than any of these fly genes, thus approximating the CNS pattern of segmentation genes, which are turned on in every neuromere (see, for example, Doe et al., 1988b). On the basis of sequence homology and pattern of expression, *Lox1* may be the cognate of *Drosophila Scr*, mouse *Hoxc-6* (Kongsuwan et al., 1988) and human *HOXC-6* (Simeone et al., 1988). However, there are significant differences between *Scr* and *Lox1*: although expression of Scr has been detected in posterior neruomeres, high levels are present only in two anterior segments (Riley et al., 1987), while *Lox1* is expressed at a constant level in many segments. These differences notwithstanding, both genes are similar in having an anterior boundary in the head. An alternative and quite interesting possibility is that *Lox1* is a gene ancestral to fly segmentation genes like *ftz* and *even-skipped* (*eve*), considering that the neurogenetic function is likely to have preceded during evolution the involvement of these genes in segmentation (Patel et al., 1992).

With the help of the marker monoclonal antibody Lazl-1 (Stewart et al., 1987), we were able to identify two pairs of neurons that express both *Lox1* and Lazl-1 antigen. The first neurons to express *Lox1* during embryonic development are the Bipolar cells, a Lazl-1-positive pair of

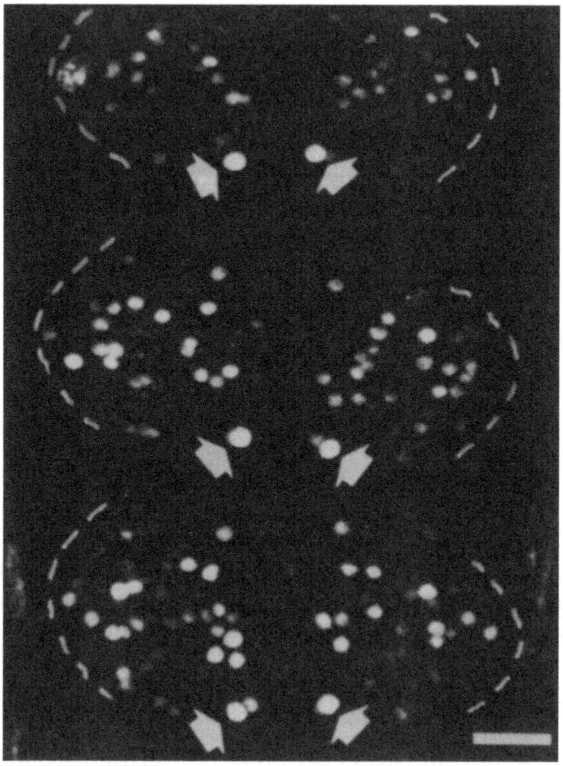

Figure 1. *Lox1* expression in the CNS at embryonic day 9. In this and the following figures, dorsal views of dissected leech embroys are shown. The embryos were stained with affinity purified anti-*Lox1* antibody. Serial optical sections of midbody ganglia 19–20 were taken at stage increments of 3 μm with a confocal microscope and superimposed. Twenty to 30 *Lox1*-positive neurons can be counted (Aisemberg and Macagno, 1994). The Bipolar cells are indicated with arrows and the outline of the ganglia with dashed lines. Anterior is to the top and the scale bar is 25 μm.

dorsal, postero-medial neurons present in all segments. The Bipolar cells can be detected with either antibody at very early stages of CNS development, when the ganglionic primordia start to form (at 6 days of embryonic development—E6—in *Hirudo medicinalis* developing at 23°C) and before most neurons differentiate. Each one of these spindle-shaped cells has two ipsilateral processes that run longitudinally for 3–5 anterior and 2–3 posterior segments, into the developing connective nerves, forming two continuous, parallel tracks that extend for the entire length of the embryo (Stewart et al., 1987). When the process of gangliogensesis ends (starting at E9), the Bipolar cells begin to degenerate (they loose their projections and their somata fragment) and quickly die. An individual Bipolar cell lives for about 3 days. These cells are particularly fascinating because they appear to play a role in marking pathways for the establishment of the longitudinal connective nerves. If this is the case, experimental alteration of *Lox1* expression in these cells is expected to produce major changes in the normal pattern of the leech CNS.

The L1 neurons, another pair of *Lox1*-positive bilateral neurons, can also be identified with the Lazl-1 antibody. They project ipsilateral primary axons to the posterior segemental roots, and secondary branches that, like the processes of the Bipolar cells, enter the longitudinal connective nerves without crossing the midline (Stewart, 1985). The L1 neurons, like all *Lox1*-positive neurons examined thus far (other than the Bipolar cells), are first detected at the end of gangliogenesis and are maintained throughout embryongenesis. The L1 neurons could not be detected with either the Lazl-1 or the anti-*Lox1* antibodies in the two anterior-most neuromeres of the subesophageal ganglion, where it is not known whether they are missing or they just fail to express these antigens. In addition, the Bipolar cells express very low levels of *Lox1* and the Lazl-1 antigen in these same two anterior segments, indicating that *Lox1* could control the expression of the Lazl-1 protein in these neurons. The existence of other neurons that contain only one of these two proteins suggests that, if there is any regulatory connection between *Lox1* and the Lazl-1 antigen, *Lox1* is neither necessary nor sufficient in all cells for the expression of the Lazl-1 protein. In *Drosophila*, homeobox genes that are expressed in different but overlapping subsets of cells show this kind of complex behavior. For example, *ftz* seems to activate *even-skipped* (*eve*) in the RP2 neurons but not in other neurons, and *eve* function is in turn required for normal axonal morphology in the RP2 and aCC neurons, but not in the pCC neurons (Doe et al., 1988a,b).

Neurons in medial and posterior ganglia express Lox2

Lox2, a homolog of the fly gene *Ultrabithorax* (*Ubx*), is expressed in the posterior two-thirds of the central nervous system, in an iterated set of neurons that includes some identified motor neurons (Wysocka-Diller et al., 1989; Nardelli-Haefliger and Shankland, 1992). Its expression is first detected during gangliogenesis (at E7) and it is still found at late stages of embryonic development. *Lox2*-positive neurons are present in the sixth midbody ganglion and in all ganglia posterior to it. Most of these ganglia contain 25–30 pairs of *Lox2*-positive neurons. However, *Lox2*, like *Lox1*, is expressed in gradually fewer neurons in the two anterior segments of its expression domain (Fig. 2). *Lox2* is detected in a cluster of only three or four bilateral pairs of neurons in the sixth midbody ganglion. In the seventh midbody ganglion, the number of *Lox2*-positive neurons is intermediate (15–20 pairs). The expression pattern of *Lox2* is consistent with its identification as a *Ubx* homolog because this

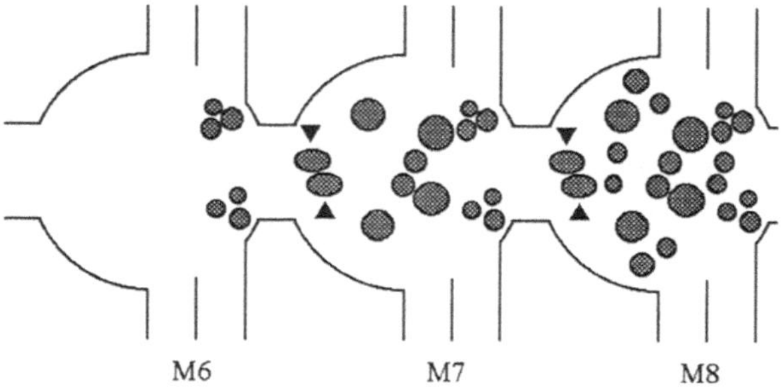

Figure 2. Anterior boundary of *Lox2* expression. Diagrammatic representation of the ventral aspect of midbody ganglia six (M6) through eight (M8). *Lox2* expression increases from anterior to posterior in this region (Wysocka-Diller et al., 1989). Arrowheads: RPE-like neurons. In this and the following figures, anterior is to the left.

gene is also expressed in approximately the posterior two-thirds of the *Drosophila* CNS, although the highest levels are found in a single embryonic parasegment (Akam and Martínez-Arias, 1985).

On the basis of their distinctive location, two pairs of neurons that express *Lox2* can be tentatively identified. These are the CV and the RPE-like neurons. These neurons should be easy to identify and study when normal *Lox2* expression is altered. The CV motor neurons, located antero-ventrally and close to the ganglionic midline, excite circular muscles. The RPE-like neurons, located at the antero-ventral margin of all midbody ganglia, are believed to be the segmental homologs of the RPE (rostral penile evertor) motor neurons of the sixth midbody ganglion (the true RPE neurons do not express *Lox2*). The RPE-like neurons do not innervate the male organ and have a divergent aborization pattern. Other neurons that can be identified by position and large size, such as the Bipolar cells, the Retzius cells, the AP neurons or the P, T and N mechanosensory neurons, do not seem to express *Lox2*.

The RPE neurons stop expressing Lox4 when they differentiate

Lox4, a probable homolog of the fly gene *abdominal*-A (abdA), has a main domain of expression posterior to the one of *Lox2*, but is also expressed in a few anterior central neurons (V.Y. Wong, G.O. Aisemberg and E.R. Macagno, unpublished results). Likewise, abdA is expressed in a domain that is posterior to the one of *Ubx* in the CNS of *Drosophila* (Karch et al., 1990). The main domain of *Lox4* expression starts in the 13th midbody ganglion and extends up to the last (21st) midbody ganglion, where about 20–30 pairs of neurons per ganglion are *Lox4*-positive. However, a smaller set of neurons also expresses *Lox4* in more anterior ganglia. In the third midbody ganglion only one

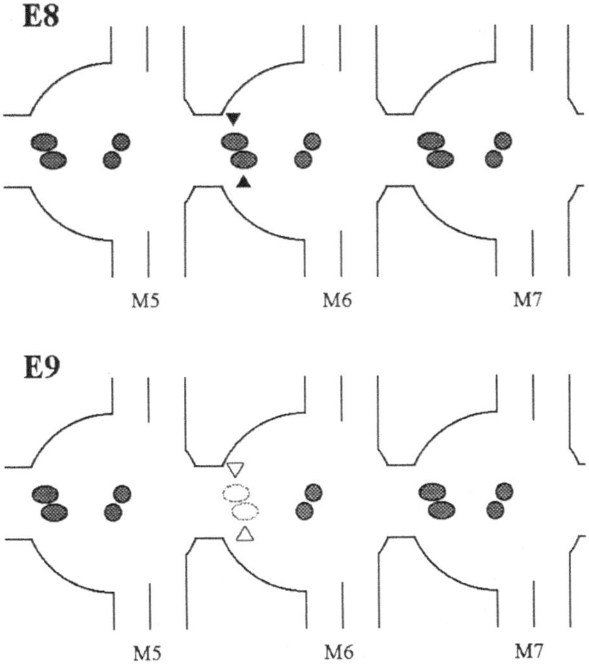

Figure 3. Changing *Lox4* expression in the anterior ventral nerve cord. Diagrammatic representation of midbody ganglia five (M5) through seven (M7), at 8 days (E8) and 9 days (E9) of embryonic development. *Lox4* is turned off in the RPE neurons of M6 between E8 and E9. Arrowheads: RPE motor neurons.

medial pair of unidentified neurons expresses *Lox4* and, in all midbody ganglia posterior to the third one, *Lox4* is also expressed in the RPE-like neurons.

Surprisingly, the expression of *Lox4* is very finely tuned in these cells: during gangliogenesis (at E8), *Lox4* is in both the RPE-like and the true RPE neurons, but 1 day later, when the neurons start to differentiate, *Lox4* is specifically turned off in the RPE neurons of ganglion 6, but not in their segmental homologs (Fig. 3). We interpret these results as meaning that *Lox4* is involved in the control of a developmental switch that changes the fate of these neurons, perhaps specifying the phenotype of the RPE-like neurons. If this is true, the fate of the RPE neurons should be changed by intracellular injections of *Lox4* mRNA or protein.

The identification of *Ubx* and *abdA* homologs in leeches contradicts a model for the evolution of *Antp*-class homeobox genes. Quite logically, it was postulated that posterior genes like *Ubx* and *abdA* were absent in a putative, myriapod-like ancestor of arthropods with very similar trunk segments. According to this model, these genes arose later, by duplication of an *Antp*-like gene, to suppress the action of anterior genes within the presumptive abdomen (Lewis, 1978; Akam, 1989). The finding of *Lox2* and *Lox4* in leeches indicates that homologs of *Ubx* and *abdA* were present much earlier, at least before the split of arthropods and annelids.

Neurons of the subesophageal ganglion and eye photoreceptor cells express Lox6

Lox6, a homolog of the fly gene *Deformed* (*Dfd*), was detected during gangliogenesis in 15–20 pairs of neurons of the third neuromere of the subesophageal ganglion and in 2–3 pairs of cells of more posterior neuromeres (V.Y. Wong, G.O. Aisemberg and E.R. Macagno, unpublished results). A main site of expression in a single segment of the leech subesophageal ganglion (Fig. 4) is consistent with the high sequence homology to *Drosophila Dfd* found in the *Lox6* gene (*Dfd* is expressed in a single head parasegment of the head; Martínez-Arias et al., 1987). We find significant that this homolog of *Dfd*, one of the homeobox genes best conserved throughout the animal kingdom, is more restricted in its CNS pattern of expression (and, therefore, more similar to its fly cognate) than the other *Lox* genes. It is possible that this conservation is actually due to true homology of the region affected by this gene between arthropods and annelids.

Double-label experiments with marker antibodies will be required to identify neurons in the fused neuromeres of the subesophageal ganglion where most *Lox6*-positive central neurons lie. In addition to its CNS expression, the *Lox6* mRNA is also found in photoreceptor cells of the five pairs of eye primordia in the *Hirudo* embryo, where it could be involved in specifying these structures of the nervous system.

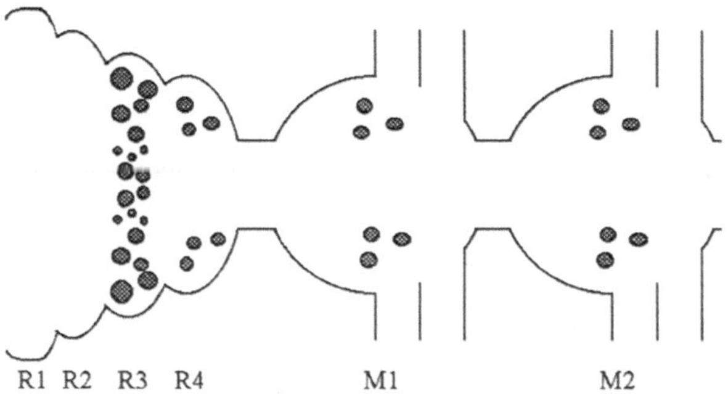

Figure 4. *Lox6* expression at the anterior end of the embryonic CNS. Diagrammatic representation of the ventral aspect of the subesophageal ganglion (segments R1–R4) and the first two midbody ganglia (M1 and M2). The main site of expression of *Lox6* is in R3.

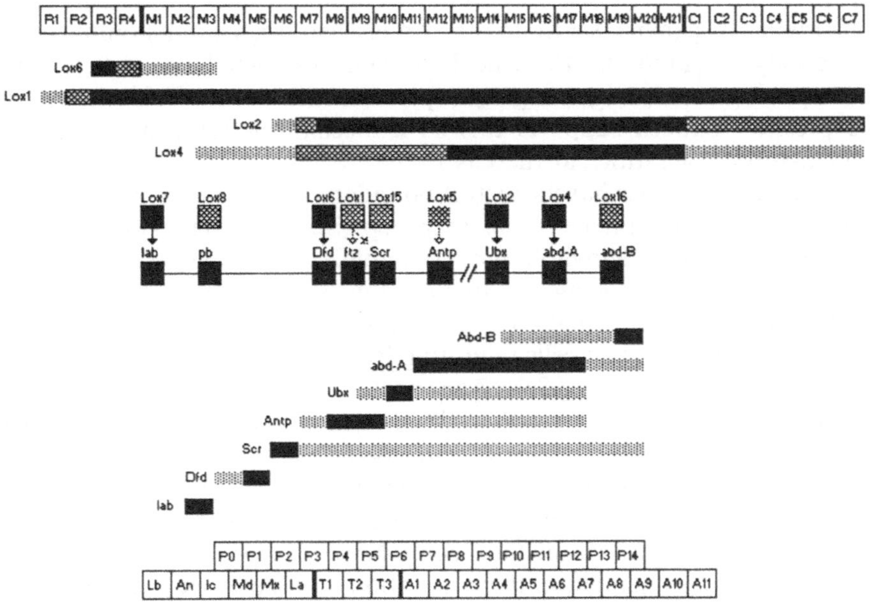

Figure 5. Comparison of leech *Lox* and fly HOM complex genes. Center: *Lox* genes that have specific homologs in *Drosophila* are represented by black squares. In cases where there is no certainty about gene homology, the genes are represented by shaded squares. The arrows are meant to indicate only sequence homology and not evolutionary relationship. The *Drosophila* genes *zen1*, *z2*, and *bicoid* (*bcd*) are generally thought not to have homologs in the homeobox complexes of the other species and are represented here by white squares. The actual order of *Lox* genes in the leech chromosome is presently unknown. *Lox5*, although expected to exist also in *Hirudo*, has been found only in *Helobdella*. *Lox2*, *Lox*6 and *Lox7* have been found in both *Helobdella* and *Hirudo*. All the other *Lox* genes shown here are from *Hirudo*. The CNS expression domains of some of these genes are represented with horizontal bars (darkness represents level of expression). Top: segmental organization of the leech CNS. R1–R4: rostral (or head) neuromeres, M1–M21: midbody ganglia, C1–C7: caudal (or tail) neuromeres. Bottom: segmental organization of the fly CNS. P0–P14: parasegments, Lb: labral, An: antennal, Ic: intercalary, Md: mandibular, Mx: maxillary, La: labial, T1–T3: thoracic segments, A1–A11: abdominal segments.

Conclusions

A comparison of the *Lox* genes of leeches to the *Drosophila* HOM complex is shown in Figure 5. The patterns of expression of the *Lox* genes examined thus far are different from one another at two levels: (1) the expression domains of individual *Lox* genes along the antero-posterior axis are not the same, although they may overlap, and (2) when two or more *Lox* genes are found within the same ganglion, they are specifically detected in different but overlapping subsets of neurons. Segmentally homologous neurons express the same *Lox* gene or genes in adjacent ganglia within these antero-posterior domains, but different *Lox* genes outside those domains. For example, the RPE-like neurons express only *Lox4* in midbody ganglia four and five, but *Lox4* and *Lox6* in midbody ganglia seven to 21, while their segmental homologs of midbody ganglion six, the true RPE neurons, that have a distinctive morphology and target organ, express neither gene. This variety in expression pattern supports this as a mechanism for generating segment-specific identities in the CNS.

Several *Lox* genes have a rostral boundary of expression at which the number of expressing cells increases stepwise in the posterior direction from ganglion to ganglion. For example, *Lox2* is expressed in 6–8 neurons in the sixth ganglion, in 30–40 in the seventh ganglion, and in 50–60 in the eighth ganglion of the midbody. In all leech ganglia some neurons originate from cells migrating from an adjacent segment (Weisblat and Shankland, 1985). The postero-ventral clusters of *Lox2*-expressing neurons of the sixth midbody ganglion derive from a primary blast cell that also produces neurons in segment seven (Nardelli-Haefliger and Shankland, 1992). The mixing of cells at the boundary of the expression domains of *Lox* genes could be an additional source of diversity among ganglia in the leech CNS.

In a leech hemiganglion, there are about 200 neurons. If each one of them expresses between one and four *Lox* genes, and assuming that there are ten *Lox* genes expressed in this way in the leech CNS, there will be about 400 different combinations of *Lox* genes in a ganglion, making it possible to have a unique "*Lox blend*" in every neuron of a given hemiganglion. It is tempting to propose that the *Lox* genes are controlling neuronal identities in this way. We recognize that this hypothesis must be oversimplified, but it lets us test possible molecular origins of neuronal diversity in the leech at the level of single, identified neurons, as a complement to genetic and embryological studies from other species.

Acknowledgments
Original research presented in this review was partially supported by NFS grant DCB-89-17096 and NIH grant HD-20954. We thank Nicholas Necles for photographic work.

References

Aisemberg, G.O. and Macagno, E.R. (1994) *Lox1*, an *Anntennapedia*-class homeobox gene, is expressed during leech gangliogenesis in both transient and stable central neurons. *Dev. Biol.* 161: 455–465.

Akam, M. (1989) Hox and HOM: Homologous clusters in insects and vertebrates. *Cell* 57: 347–349.

Akam, M.E. and Martínez-Arias, A. (1985) The distribution of *Ultrabithorax* transcripts in *Drosophila* embryos. *EMBO J.* 4: 1689–1700.

Doe, C.Q., Smouse, D. and Goodman, C.S. (1988a) Control of neuronal fate by the *Drosophila* segmentation gene *even-skipped*. *Nature* 333: 376–378.

Doe, C.Q., Hiromi, Y., Gehring, W.J. and Goodman, C.S. (1988b) Expression and function of the segmentation gene *fushi tarazu* during *Drosophila* neurogenesis. *Science* 239: 170–175.

Karch, F., Bender, W. and Weiffenbach, B. (1990) *abdA* expression in *Drosophila* embryos. *Genes Dev.* 4: 1573–1587.

Konguswan, K., Webb, E., Housiaux, P. and Adams, J.M. (1988) Expression of multiple homeobox genes within diverse mammalian haemopoietic lineages. *EMBO J.* 7: 2131–2138.

Levine, M. and Macagno, E.R. (1990) Segmentation and segmental differentiation in the development of the central nervous systems of leeches and flies. *Annu. Rev. Neurosci.* 13: 195–225.

Lewis, E.B. (1978) A gene complex controlling segmentation in *Drosophila. Nature* 276: 565-570.

Macagno, E.R. (1980) Number and distribution of neurons in leech segmental ganglia. *J. Comp. Neurol.* 190: 283–302.

Martínez-Arias, A., Ingham, P.W., Scott, M.P. and Akam, M.E. (1987) The spatial and temporal deployment of *Dfd* and *Scr* transcripts throughout development of *Drosophilia. Development* 100: 673–683.

McGinnis, W. and Krumlauf, R. (1992) Homeoboxes and axial patterning. *Cell* 68: 283–302.

Nardelli-Haefliger, D. and Shankland, M. (1992) *Lox2,* a putative leech segment identity gene, is expressed in the same segmental domain in different stem cell lineages. *Development* 116: 697–710.

Patel, N.H., Ball, E.E. and Goodman, C.S. (1992) Changing role of *even-skipped* during the evolution of insect pattern formation. *Nature* 357: 339–342.

Riley, P.D., Carroll, S.B. and Scott M.P. (1987) The expression and regulation of *Sex combs reduced* protein in *Drosophila* embryos. *Genes Dev.* 1: 716–730.

Shankland, M., Martindale, M.Q., Nardelli-Haefliger, D., Baxter, E. and Price, D.J. (1991) Origin of segmental identity in the development of the leech nervous system. *Development* Suppl. 2: 29–38.

Simeone, A., Pannese, M., Acampora, D., D'Esposito, M. and Boncinelli, E. (1988) At least three human homeoboxes on chromosome 12 belong to the same transcription unit. *Nucleic Acids Res.* 16: 5379–5390.

Stent, G.S., Weisblat, D.A., Blair, S.S. and Zackson, S.L. (1982) Cell lineage in the development of the leech nervous system. *In:* N.C. Spritzer (ed.): *Neuronal Development*, Plenum Press, New York, pp. 1–44.

Stewart, R.R. (1985) *The genesis of segmental differences in the nervous system of two species of leeches.* Ph. D. Thesis, Columbia University.

Stewart, R.R., Spergel, D. and Macagno, E.R. (1986) Segmental differentation in the leech nervous system: The genesis of cell number in the segmental ganglia of *Haemopis marmorata. J. Comp. Neurol.* 253: 253–259.

Stewart, R.R., Gao, W.-Q., Peinado, A., Zipser, B. and Macagno, E.R. (1987) Cell death during gangliogenesis in the leech: Bipolar cells appear and then degenerate in all ganglia. *J. Neurosci.* 7: 1919–1927.

Weisblat, D.A. (1981) Development of the nervous system. *In:* K.G. Muller, J.G. Nicholls and G.S. Stent (eds): *Neurobiology of the leech*, Cold Spring Harbour Laboratory, pp. 173–195.

Weisblat, D.A., and Shankland, M. (1985) Cell lineage and segmentation in the leech. *Phil. Trans. R. Soc. Lond. B* 312: 39–56.

Wysocka-Diller, J.W., Aisemberg, G.O., Baumgarten, M., Levine, M. and Macagno, E.R. (1989) Characterization of a homologue of bithorax-complex genes in the leech *Hirudo medicinalis. Nature* 341: 760–763.

The Nervous Systems of Invertebrates: An Evolutionary and Comparative Approach
ed. by O. Breidbach & W. Kutsch
© 1995 Birkhäuser Verlag Basel/Switzerland

Comparative aspects of gastropod neurobiology

A.G.M. Bulloch[1] and R.L. Ridgway[2]

[1]*Neuroscience Research Group, University of Calgary, Faculty of Medicine, 3330 Hospital Drive N.W., Calgary, Alberta, T2N 4N1, Canada*
[2]*Department of Biology, Seattle Pacific University, 3469-3rd Avenue W., Seattle, WA 98119, USA*

Summary. In this chapter, we compare the development, morphology, and behavioral function of identified neurons across a number of gastropod species. We first examine the origins of the earliest identifiable neurons and their corresponding behavioral roles. Several adult behaviors are then considered. We compare the transmitter phenotype and neuroanatomy of identified neurons known to be involved in three rhythmic behaviors: feeding, locomotion and respiration. Horizontal themes that are re-examined throughout this discussion are (1) the requirements for the neuroanatomy of a gastropod ancestor, (2) methods to identify neuronal homologues, and (3) the utility of identified gastropod neurons for studies at the morphological, physiological, biochemical, and molecular levels.

Introduction

Molluscs constitute a large, diverse phylum that includes about 100 000 living species distributed among seven classes: Monoplacophora, Aplacophora, Polyplacophora, Scaphopoda, Bivalvia, Gastropoda, and Cephalopoda. The latter three are the major classes, with representatives of the cephalopods being the most highly evolved. The earliest molluscs are thought to have arisen in the shallows of pre-Cambrian oceans from nonsegmented flatworm-like ancestors (see Runnegar and Pojeta, 1974), although signs of segmentation present in living Monoplacophora and Polyplacophora suggest the possibility of annelid-like ancestors (Lemche, 1958). Comparative anatomical and developmental studies of living species have been combined with information from the fossil record to reconstruct the body plan of a hypothetical ancestral mollusc which may also serve as a morphotype for modern molluscs. The basic features of this body plan are: (1) a *head* including eyes, tentacles, and mouth; (2) a *foot*, which may be continuous with the head; (3) a *mantle* or pallium, which secretes the protective shell and provides an external cavity for respiratory, excretory, and reproductive functions; and (4) a *visceral mass*, containing the heart, digestive gland, kidneys, and reproductive organs (Kandel, 1979; Barnes, 1980). This relatively simple yet adaptable body plan has been a key factor in the evolutionary success of molluscs and, more specifically, of the Gastropoda, which has the largest number and greatest diversity of molluscan species. The fact that gastropods have colonized every major habitat except the air is evidence of this diversity (Willows, 1973; Purchon, 1968).

Any review concerning the comparative aspects of gastropod neurobiology must acknowledge the contribution made by Bullock and Horridge (1965) in their treatise "*Structure and Function in the Nervous Systems of Invertebrates*". Their landmark chapter on the Gastropoda reviewed most of the early anatomical and physiological studies in the field and set the stage for subsequent research. A more recent comparative review by Dorsett (1986) has also been invaluable in this context. In the present chapter, we have, by necessity, limited our scope to a discussion of neural correlates underlying three rhythmic behaviors exhibited by most gastropods: feeding, locomotion, and respiration. If we assume common ancestry (see Fig. 1), we can hypothesize that at least some basic neural elements of these behaviors are shared among the various extant gastropod groups, even though significant diversification has occurred over time. In seeking evidence to support this hypothesis, we have examined these behaviors to the level of identified neurons. What constitutes homology at the cellular level is problematical in itself, for it is impossible to *directly* show common ancestry between two cells in different species (see Croll, 1987). However, among the criteria we (and others) have used to infer such homologies are: soma position and size, axon trajectory, neuritic branching patterns, synaptic connectivity, electrophysiological parameters, biochemical parameters, and developmental lineage.

Overview of nervous system structure and development

Structure

The diversity of the Gastropoda is reflected in the structure of the nervous systems within the three subclasses (see Fig. 1): Prosobranchia, Opisthobranchia, and Pulmonata. The more primitive prosobranchs, such as the abalone *Haliotis*, have two long parallel nerve cords running anteroposteriorly and crosslinked periodically by commissures (Fig. 2). Nerve cell bodies, however, are primarily restricted to paired ganglia of the head region. The typical gastropod nervous system includes paired cerebral ganglia, pedal ganglia, pleural ganglia, and buccal ganglia, as well as several ganglia associated with the visceral mass. In primitive species of each subclass, the visceral connectives are crossed in a figure-eight pattern (streptoneury) which disrupts the otherwise bilaterally symmetric nervous system (see Fig. 2: *Haliotis* and *Littorina*). This situation is a vestige of the developmental process known as torsion,

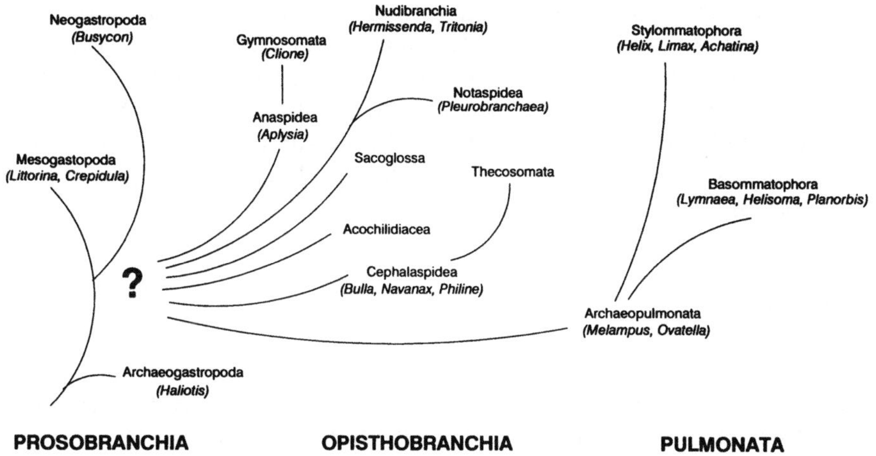

Figure 1. Gastropod taxonomy and speculative relationships (based on Purchon, 1968; Kandel, 1979; and Croll, 1987). Genera pertinent to this chapter are given in italics. Note that the Archaeopulmonata (lower right) are frequently included as primitive members of the Basommatophora.

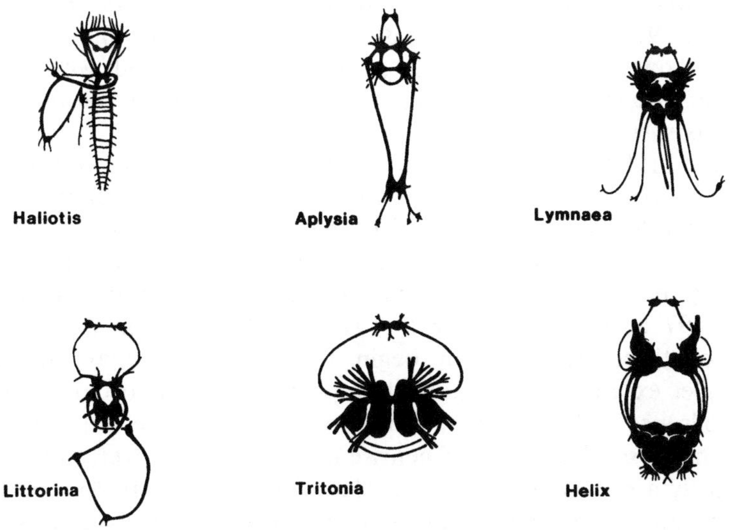

Figure 2. Drawings of the central nervous systems of representative gastropod genera; included are two prosobranchs (*Haliotis* and *Littorina*), two opisthobranchs (*Aplysia* and *Tritonia*), and two pulmonates (*Lymnaea* and *Helix*). These examples are redrawn from the following sources: Bullock and Horridge, 1965 (*Helix*), Kandel, 1979 (*Aplysia, Haliotis, Lymnaea* and *Tritonia*), and Croll, 1987 (*Littorina*).

wherein the visceral mass rotates 180 degrees counter-clockwise to facilitate retraction of the larval foot and velum into the shell as a protective measure (Crofts, 1955; Kandel, 1979). In less primitive species the effects of torison are diminished to varying degrees thus reestablishing symmetry (euthyneury) either by detorsion or an overall shortening of the visceral connectives; the former is common to opisthobranchs (see Fig. 2: *Aplysia*), the latter to pulmonates (see Fig. 2: *Lymnaea*). Less primitive gastropod species also show a general tendency toward cephalization of the nervous system. The distance between the various central ganglia decreases as a result of the shortening of connectives and commissures; this situation may lead to ganglionic fusion (see Fig. 2: *Tritonia* and *Helix*).

Development

Given the extensive use of invertebrates as models of nervous system development, it might be assumed that a wealth of information is available on the neurogenesis of gastropods, yet this is not the case. Only recently have such studies begun to be carried out and these have not included cell lineage experiments. Gastropods, like all molluscs, undergo spiral cleavages (either dextral or sinistral in direction) to give rise to a blastula stage which then undergoes gastrulation to form the ectodermal, mesodermal, and endodermal germ layers (Raven, 1966; Kandel, 1979). Subsequent cell divisions produce the ciliated embryonic stage known as a trochophore, which is maintained within the egg capsule. Organogenesis begins at this time, including development of the nervous system.

Jacob (1984) demonstrated that presumptive neurons of the opisthobranch *Aplysia californica* arise in proliferative zones of the ectoderm and migrate inward to establish the central ganglia. In all species examined, the paired cerebral ganglia are the first to emerge, followed later by the pedal, buccal, pleural, parietal, and visceral ganglia (Kriegstein, 1977a; Marois and Croll, 1992; Goldberg, in press). Differentiation of these neurons can begin soon after terminal mitosis. For example, expression of transmitter phenotypes can occur even before the cells begin their inward migration (McAllister et al., 1983; Goldberg, in press). Initiation of neurite outgrowth, however, appears to be primarily a post-migratory event (Kandel et al., 1980). Among the first neurotransmitters to be expressed by embryonic neurons is serotonin (Goldberg and Kater, 1989; Marois and Croll, 1992). Several studies have suggested that the earliest serotonergic neurons send their axonal projections toward the developing velum, the feeding and locomotory organ of the larva, where they provide modulatory control over the activity of ciliated cells (Marois and Carew, 1990; Diefenbach et al., 1991; Marois and Croll, 1992). In *Helisoma trivolvis*, the earliest serotonergic cells are thought to include the embryonic versions of cerebral ganglion neurons C1, a pair of well-characterized neurons involved in the modulation of adult feeding behavior (Goldberg, in press). We will return to serotonin in the context of both adult feeding and locomotory behaviors.

Temporal development of behaviors

Following the embryonic period the majority of marine prosobranch and opisthobranchs, including *Aplysia californica* (Fig. 3), are hatched to a planktonic veliger phase lasting from days to months before undergoing metamorphosis to become juveniles (Kriegstein, 1977b; Kandel et al., 1980). The veligers depend on the cooordinated ciliary movements of their velar lobes for both locomotion and feeding. Larvae of most freshwater and terrestrial gastropods (including nearly all pulmonates) remain inside their egg capsules, effectively bypassing the free-swimming veliger phase and undergoing an abbreviated metamorphosis (Marois and Carew, 1990). Metamorphosis, which lasts only a few days in species such as *Aplysia*, is characterized by loss of the velum, resulting in a temporary loss of feeding and locomotory behaviors. Velar loss is preceded by recognition of, and settlement onto, an appropriate substrate. Gradually, a shift is made to a crawling form of locomotion based on pedial ciliary activity which is carried through most of the juvenile period before being replaced by muscular based adult behaviors such as pedal wave locomotion or parapodial swimming (Kandel et al., 1980; Marcus et al., 1988). Feeding behavior following velar loss is also gradually changed, first to stereotyped head and radular movements during metamorphosis, and then to the consummatory behavior of adults characterized by coordinated rasping movements, biting, and swallowing. Marois and Croll (1992) have found that a remodeling of axonal projections occurs among identified serotonergic

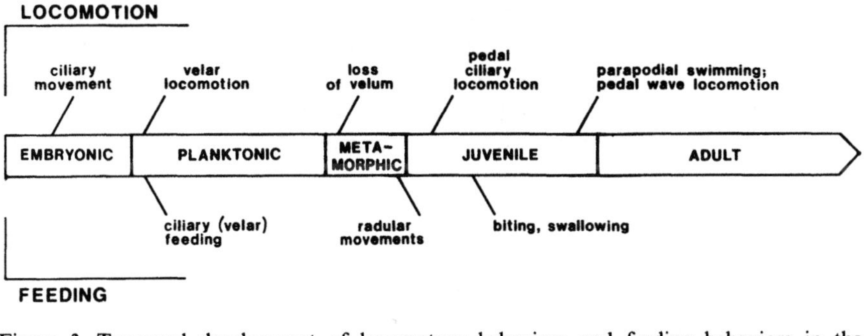

Figure 3. Temporal development of locomotory behaviors and feeding behaviors in the opisthobranch *Aplysia californica*. The *Aplysia* life cycle can be divided into five major phases: embryonic (lasting about 10 days), planktonic (lasting about 30 days), metamorphic (lasting about 3 days), juvenile (lasting about 120 days), and adult. The diagram is based on Kriegstein, 1977a,b; Kandel, 1979; Kandel et al., 1980; and Marcus et al., 1988).

cells of the cerebral ganglia corresponding to the metamorphic period and velar loss. Similar plasticity has been documented in *Haliotis* where metamorphosis is accompanied by loss of serotonin neurons that innervate the velum (Barlow and Truman, 1992). Significant reorganization of the neural circuitry underlying respiration may also take place during metamorphosis and juvenile development of gastropods, but this area has yet to be systematically studied.

Control of rhythmic behaviors by identified neurons

We will compare some of the gastropod neurons known to be involved in feeding, locomotion, and respiration. Rhythmic behaviors have attracted the attention of neurobiologists for about two decades. In terms of the gastropods, feeding has been examined intensively by many groups, largely because of its robust nature, i.e., rhythmic movements and motor output occurs readily in a variety of reduced preparations. Our review of the extensive literature in this area will, alas, do little justice to the field. Knowledge of the neuronal basis of locomotion and respiration is more limited, partly due to the difficulty of eliciting such behaviors in reduced preparations. The neurobiological basis of other rhythmic functions (e.g., heart beat) and non-rhythmic behaviors (e.g., learning) is beyond the scope of this article (but see Salanki, 1973, 1976; Salanki and Rozsa, 1988; Willows, 1985, 1986; Selverston, 1985; Kandel, 1976, 1979; Boer et al., 1987; Jacklet, 1989; Kits et al., 1991).

Pharmacology of rhythmic behaviors

In reviewing the role of serotonin in molluscs, the leech and the planarians, Sakharov (1990) concluded that serotonin "initiates and integrates activities of the systems involved in obtaining food". In part, this concept extends the idea that serotonin is the "invertebrate norepinephrine" (Twarog, 1989). The discussion below regarding feeding and locomotion will highlight the role of serotonin. In the case of respiration, another monoamine transmitter, dopamine, will be highlighted.

Feeding

The gastropods have adapted to an extraordinarily wide range of marine, terrestrial and freshwater habitats. Indeed, "the adaptive radiation of the Gastropoda is principally based on adaptation of the buccal mass and radula to meet the problems faced in feeding in various ways on a wide variety of food substances" (Purchon, 1968). The feeding behavior of various gastropods has received intensive neurobiological attention. Despite the diversity of feeding habits displayed by various pulmonates and opisthobranchs, common rhythmic movements and their underlying neuronal circuitry can be compared between species. Here, we will review the comparative morphology, physiology and pharmacology of some identified effector neurons and a higher order modulatory neuron.

Neurobiologists have studied three principal styles of feeding behavior in gastropods: rasping, browsing, and hunting. The behaviors are reviewed in detail by Audesirk and Audesirk (1985). The raspers (e.g., the pulmonates, *Helix*, *Helisoma*, *Limax*, *Lymnaea*, and *Melampus*) scrape food particles from the substrate by repetitive radula movements that appear monotonous and inflexible (but see below). The browsers, like the opisthobranch, *Aplysia*, use a modified radula to grasp and pull in large pieces of food. Hunting sea slugs (e.g., *Clione*, *Melibe*, *Tritonia* and *Pleurobranchaea*) show the most complex feeding behavior, a distinct "bite-strike" movement first captures the prey, which is then ingested by repetitive swallowing movements. Common to rasping, browsing, and swallowing (of the hunters) are rhythmic protraction retraction cycles of the radula-odontophore complex; herin lies an opportunity for comparative studies. In addition to these muscular activities, some visceral functions have also been examined.

The comparative work discussed below may leave the impression that feeding is inflexible in gastropods. Behavioral studies, principally on *Aplysia* and *Navanax*, however, have clearly shown that feeding is modulated in a complex manner by behavioral state (Audesirk and Audesirk, 1985; Kandel, 1979). Furthermore, recent detailed observations of *Helisoma* have shown a number of distinct behaviors, e.g., biting, regurgitation, and independent swallowing movements. This diversity of behavioral repertoire is reflected as a large number of activity patterns ($n > 14$) that can be recorded from buccal motoneurons (Arnett and Murphy, 1991, 1994; Murphy, 1986; Murphy et al., 1988).

Feeding behavior

A common feature of feeding behaviors of many pulmonates and opisthobranchs are repetitive protraction and retraction movements of the radula. The neuronal basis of these movements is reasonably well understood (Benjamin and Elliot, 1989). The bite strike component of some opisthobranchs is seldom exhibited in reduced preparations, its neuronal circuit is therefore not mapped. However, the same populations of neurons are probably involved in bite-strike and swallowing, and thus this system may provide material for examining the polymorphic network concept of Getting and Dekin (1985) (see also McCrohan, 1988).

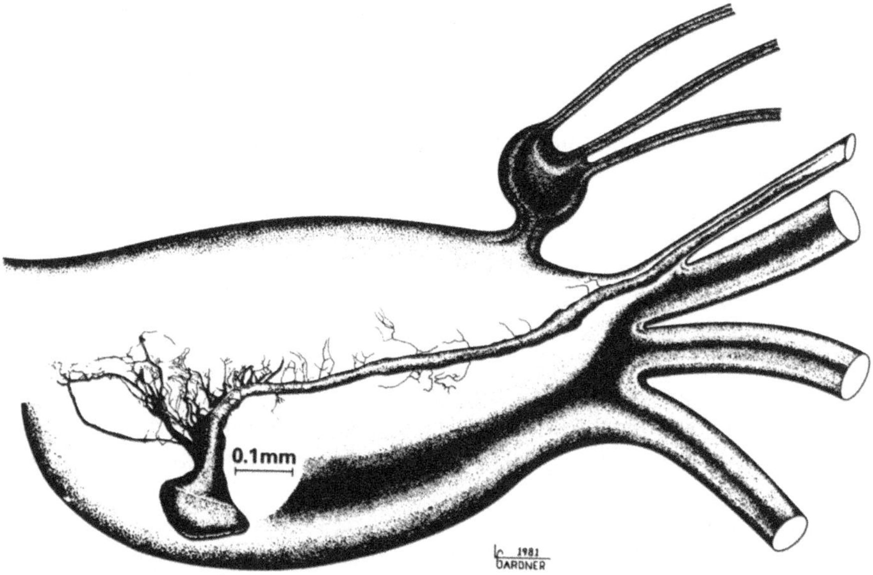

Figure 4. Example of a gastropod neuron. This is an artist's rendering of an HRP stain of a buccal neuron (B5) in the nudibranch *Tritonia diomedea* (see Willows et al., 1988). This neuron is typical of neurons in gastropod ganglia in being monopolar with many axon collaterals ("dendrites") arborizing in the neuropil. For a more extended discussion of neuron morphology in gastropods, see Dorsett (1986). (Reproduced with permission from John Wiley and Sons).

Buccal motoneurons

A typical buccal motoneuron is shown in Figure 4, a schematic which serves as an introduction to gastropod neuronal morphology. Like most ganglionic neurons of gastropods, a single axon arises from the soma (in many this bifurcates immediately) which gives rise to axon collaterals sometimes known as "dendrites" (see discussions by Dorsett, 1986; Berdan et al., 1987). The buccal motoneurons have been mapped in several pulmonate and opisthobranch species (Fig. 5). Comparison of these maps show some striking similarities, for example, the retractor motoneurons tend to form groups of large cells located medially on the posterior dorsal surface. In contrast, the protractor motoneurons are relatively small in size, fewer have been identified, and they tend to be located lateral to the retractor motoneurons. The map of *Philine* differs to some extent in showing a larger number of bigger protractor motoneurons, but the medial, posterior location of retractor motoneurons is still evident. The functional significance of the size difference between protractor and retractor motoneurons is not obvious. It has been postulated that larger neurons innervate a bigger target area (i.e., by analogy with motor units of mammals, see Gillette, 1991); presumably, they also also conduct action potentials faster, but no quantitative data is available to support either of these notions.

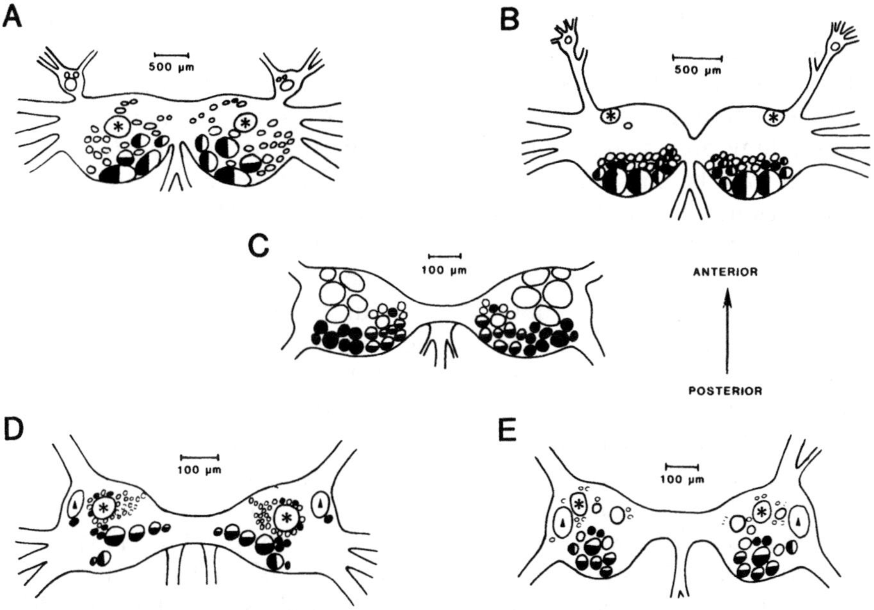

Figure 5. Comparison of the buccal ganglia of some opisthobranchs (A,B,C) and pulmonates (D,E) (see also Dorsett, 1986). Shown are redrawn schematic maps for (A) *Tritonia diomedea* (Willows, 1980; and Bulloch, unpublished data). (B) *Tritonia hombergi* (Bulloch and Dorsett, 1979). (C) *Philine aperta* (Sigger and Dorsett, 1981). (D) *Helisoma trivolvis* (Kater, 1974; Quinlan, 1993) and (E) *Lymnaea stagnalis* (Benjamin and Elliott, 1989). Protractor motoneurons are indicated as fully shaded circles. Retraction occurs in two phases, neurons active in the first phase have left-half shaded circles, neurons active in the second phase have their lower-half shaded. Note that the latter phase is designated as "Flattening" in *T. hombergi* (Bulloch and Dorsett, 1979), "R2" in *Lymnaea* (Benjamin and Rose, 1979) and "S3" in *Helisoma* (Quinlan and Murphy, 1991), and was not reported for *Philine* (Sigger and Dorsett, 1981). Gut motoneurons are labeled with an asterisk, salivary gland effector neurons with an arrowhead.

Other effector neurons

Some of the largest neurons of the buccal ganglia are not motoneurons, but rather effector neurons that innervate the gut and the salivary glands. One or more giant neurons whose axons supply the esophagus are evident in both the buccal ganglia and in the closely associated gastroesophageal ganglia (Fig. 5). A strong case for homology can be made for a large, white buccal neuron known as B11 in *Tritonia diomedea* (Masinovsky et al., 1985), B2 in *Aplysia* (Lloyd et al., 1985) and the ventral white cell in *Pleurobranchaea* (Gillette et al., 1980). Apart from a superficial similarity (re: size, color, gross morphology), these neurons exhibit the same transmitter phenotype, i.e., SCP and acetylcholine in *Tritonia* and *Aplysia* (Lloyd et al., 1985, 1987; Lloyd, Masinovsky and Willows, 1988), control gut motility (Tritonia: Lloyd and Willows, 1988; *Aplysia*: Lloyd, Kupfermann and Weiss, 1988) and modulate the feeding Central Pattern Generator (CPG) (*Pleurobranchaea*: Gillette et al., 1980; *Tritonia*: Willows et al., 1988). It seems reasonable to hypothesize that a homologous neuron in the pulmonates is the giant buccal neuron designated as B5 in *Helisoma* (Kater, 1974), B1 in *Helix* (Altrup, 1987) and B2 in *Lymnaea* (Benjamin and Rose, 1979). These pulmonate neurons send axons to the gut, but their physiology and pharmacology is less well studied. Neuron B5 of *Helisoma* may be cholinergic (Haydon, 1988; Haydon and Zoran, 1989), but does not exhibit SCP_B-like immunoreactivity (Murphy et al., 1985; Murphy and Lukowiak, 1987). The relationship of pharmacology to the function of this gut neuron, i.e., the issue of adaptation during evolution, cannot readily be addressed without more information. A recent study revealed probable homologues of this effector neuron in six additional nudibranchs (Watson and Willows, 1992), but further data on other pulmonates or the prosobranchs are not yet available.

The salivary effector neuron has been examined in diverse gastropod species, e.g., the opisthobranch *Philine* and the pulmonates *Biomphalaria*, *Helisoma*, *Lymnaea*, *Limax*, and *Helix* (Fig. 5) (Prior and Gelperin, 1977; Stewart, 1981; Barber 1983a; Altrup, 1987). This paired buccal neuron (designated as B1 in *Lymnaea*, B2 in *Helix*, and B4 in *Helisoma*) innervates the salivary glands and is presumed to drive secretion, its activity being co-ordinated with that of buccal motoneurons by the feeding Central Pattern Generator (CPG). The case for homology of this neuron across the species listed above, as argued by Barber (1983a), is based on morphology, physiological action, synaptic inputs, and a common putative transmitter (acetlycholine). The peptide SCP_B is a probable co-transmitter in *Helisoma* and *Lymnaea* (Murphy and Lukowiak, 1987; Masinovsky et al., 1988), but not in *Limax* (Prior and Watson, 1988). Curiously, then, there seems to be more variability in the peptide co-transmitter than in primary transmitter in both gut and gland effectors neurons of gastropods.

Modulatory neuron

Higher order modulation of feeding behavior in gastropods is derived from a number of neurons in the cerebral ganglion. Best characterized in this respect is the largest cerebral neuron, a giant serotonergic cell, one of the first true homologues identified in the pulmonates and opisthobranchs (Fig. 6). Originally described in *Helix* as the Metacerebral Giant Cells (Kandel and Tauc, 1966a,b) due to their location in the metacerebral lobes, these cells have been given a variety of names (such as C1, the term which will be used here) in species where the metacerebral lobes cannot be identified. That this neuron is a true homologue in the opisthobranchs *Aplysia*, *Clione*, *Philine*, *Pleurobranchaea*, and *Tritonia*, the basommatophoran pulmonates *Helisoma*, *Limax* and *Helix*, and the stylommatophoran pulmonates *Ariolimax*, *Limax* and *Helix* has been argued extensively elsewhere (Senseman and Gelperin, 1974; Sakharov, 1976; Weiss and Kupferman, 1976; Granzow and Rowell,

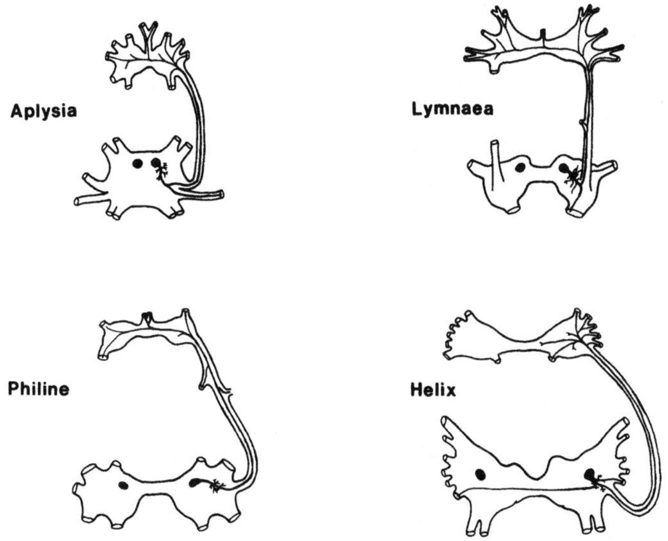

Figure 6. Comparison of the giant serotonergic cerebral neurons C1 in two opisthobranchs (*Aplysia* and *Philine*) and two pulmonates (*Lymnaea* and *Helix*). Each diagram shows the paired cerebral ganglia (lower) and paired buccal ganglia (upper), the somata of both neurons C1, and the main axon projections of the right member of the neuron pair (i.e., RC1). The diagrams are based on the following sources: Senseman and Gelperin, 1974 (*Helix*); Shkolnik and Schwartz, 1980 (*Aplysia*); Goldschmeding et al., 1981 (*Lymnaea*); and Barber, 1983b (*Philine*).

1981; Pentreath et al., 1982; Barber, 1983b; Croll, 1987; McCrohan, 1988). Comparison of morphological, physiological, pharmacological, and behavioural properties of these neurons provide as good a case for true homology as can be generated in the absence of paleontological data. It is interesting to note that, although some aspects of neuron C1 are relatively constant between species (e.g., position of the soma and transmitter phenotype), considerable variation ("plasticity") of axonal branching is evident (Fig. 6). Unfortunately, it is not possible to correlate this variability with feeding behavior or taxonomy. One can conclude that most (but not all) neurons C1 achieve bilateral input to buccal neurons and muscles by means of contralateral projections either in the cerebral and/or buccal ganglia. These differences of adult morphology may reflect a diversity of developmental strategies; ontogenetic studies should provide information in this regard. In fact, recent work has shown that neuron C1 is among the first identifiable serotonin neurons to appear in development of the basommatophoran *Helisoma* (see section on *Development*).

Regarding phylogeny, immunohistochemical studies of the mesogastropod prosobranch, *Littorina*, and the archeogastropod, *Haliotis* (Fig. 1), have suggested that neuron C1 is represented as a small group of serotonin cells (Croll and Lo, 1986; Barlow and Truman, 1992). Why a group of neurons might be replaced by a single giant neuron in more advanced species is worthy of thought. Surely any advantage of simplified connectivity would be offset by the vulnerability of a single cell? (See discussion regarding neuron VD4 below). Nevertheless, it is reasonable to propose the existence of a descending serotonergic cerebral-buccal projection in early gastropod ancestors. The phylogeny of neuron C1, clearly represented in many gastropods, is worthy of further study.

Other modulatory neurons
Both central and peripheral components of feeding circuitry of gastropods are modulated by a large number of other neuroactive agents, e.g., dopamine (Wieland and Gelperin, 1983; Trimble and Barker, 1984), glutamate (Jones et al., 1987; Quinlan and Murphy, 1991), GABA (Arshavsky et al., 1991; Richmond et al., 1991; Richmond et al., 1994), histamine (Chiel et al., 1988) and the peptides, AVT (Richmond et al., 1987), FMRFamide and SCP_B (Murphy et al., 1985, Coates and Bulloch, 1985; Bulloch et al., 1988; Prior and Watson, 1988), buccalin and myomodulin (e.g., Rosen et al., 1989) as well as nitric oxide (Moroz et al., 1993). As the relevant modulatory neurons become identified over the next decade, more opportunities will arise for identification of homologous neurons that regulate feeding behavior.

Locomotion

The locomotion of most metazoan animals involves rhythmic contractions of antagonistic muscles that results in swimming, flying, walking or crawling. In the case of gastropods, the principal means of locomotion are gliding (using cilia on the sole of the foot), crawling (by muscular contractions of the foot), and swimming (by rhythmic body movements) (Trueman, 1983; Audesirk and Audesirk, 1985). For this discussion, we will focus on ciliary locomotion and swimming where comparative data is available.

Ciliary locomotion
Gastropods that employ ciliary locomotion glide smoothly without the muscular locomotory waves seen in, for example, *Helix*, *Limax*, and *Aplysia*. Ciliary locomotion is found in families of some prosobranchs (mesogastropods and neograstropods), and many opisthobranchs and pulmonates (Fig. 1). The pedal cilia are controlled by serotonergic neurons in both the pulmonate *Lymnaea* and the nudibranch *Tritonia diomedea*. In the case of *Lymnaea*, the serotonergic neurons are known as the pedal A-cluster (Syed and Winlow, 1989), a homologous cluster

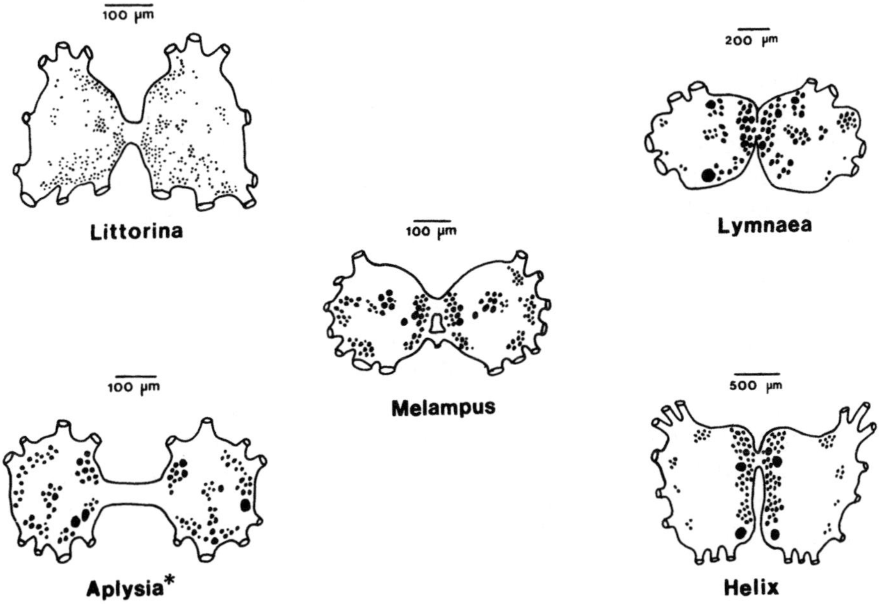

Figure 7. Comparison of serotonin immunoreactive neurons in the paired pedal ganglia of a prosobranch (*Littorina*: Croll and Lo, 1986), an opisthobranch (*Aplysia*: Goldstein et al., 1984), and three pulmonates (*Melampus*: R.L. Ridgway, unpublished data; *Lymnaea*: Kemenes et al., 1989; *Helix*: Hernandi et al., 1989). No distinction between ventral and dorsal cells is made in these drawings. The giant rostral neuron in *Lymnaea* is Left Pedal Dorsal 1 (LPeD1: see Kyriakides et al., 1989), a unique neuron not identified in the other species shown. *Data obtained from juvenile animals.

being evident in *Helisoma* (Syed et al., 1993), *Helix*, *Melampus*, *Aplysia* and *Littorina* (Fig. 7). In the opisthobranchs, a pair of serotonergic neurons has been shown to control the pedal cilia in *Tritonia* (Audesirk, 1978a,b; Audesirk et al., 1979), whereas a candidate group of homologous cells also exists in *Aplysia* (Fig. 7). Taken together, these observations point to the existence of another serotonergic projection in ancestral gastropods, in this case a motor projection from the pedal ganglion.

One mystery that requires thought concerns the role of the "pedal A" serotonergic neurons in animals such as *Helix* and *Aplysia* that locomote via muscular waves. The foot of *Helix* is supplied with serotonin-immunoreactive axons, but their origin(s) are unknown (Sakharov, 1990). It appears that serotonin can stimulate (activate) locomotion in *Aplysia* (Palovcik et al., 1982; Mackey and Carew, 1983), suggesting that the pedal A and or other neurons have central rather than peripheral actions. The concept that serotonin is crucial for locomotion is further explored below in the context of swimming behavior.

Swimming

Swimming behavior is restricted to the opisthobranch subclass of gastropods. The many opisthobranchs that swim do so by *flapping* (using mantle edges, the parapodia, or the metapodia), *undulation* (dorsal-ventral flexions of the whole or parts of the body) or *lateral body bending* (Trueman, 1983; Audesirk and Audesirk, 1985). Our discussion will be restricted to the two genera in which the neuronal basis of swimming has been studied extensively, i.e., the nudibranch *Tritonia* and the gymnosomatid (pteropod), *Clione*. From a comparative point of view, however, the genus *Aplysia* should prove interesting since some species swim (e.g., *A. brasiliana* and *A. fasciata*), whereas the most intensively studied species (*A. californica*) is a purely benthic organism.

Regarding *Tritonia*, escape swimming (like feeding and tuft withdrawal) has been examined in the Pacific species, *T. diomedea* and the Atlantic species, *T. hombergi*. Many homologous neurons can be identified in the central ganglionic ring of these two nudibranchs (Fig. 8)

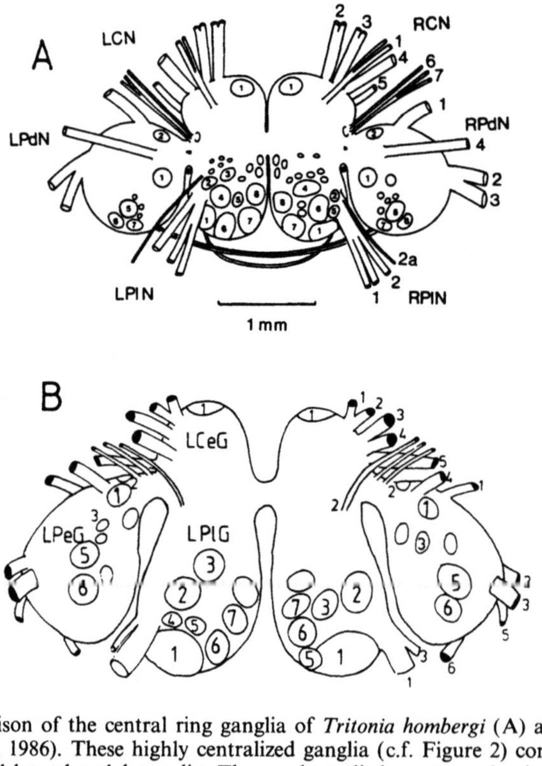

Figure 8. Comparison of the central ring ganglia of *Tritonia hombergi* (A) and *T. diomedea* (B) (from Dorsett, 1986). These highly centralized ganglia (c.f. Figure 2) consist of cerebropleural ganglia and lateral pedal ganglia. The much studied serotonergic giant cerebral cells (C1) are clearly seen, homologous pleural and pedal giant neurons also being indicated. (Abbreviations: Cerebral = C, Ce; Pedal = Pd, Pe; pleural = Pl; N = nerve; G = ganglion. Reproduced with permission from The Company of Biologists (A) and Academic Press (B).

and comparative studies of swimming in the two species has been reported (Willows and Dorsett, 1975; Dorsett, 1976). Interestingly, *T. hombergi* (but not *T. diomedea*) shows rapid habituation to a swimming stimulus, perhaps reflecting a lack of predatory starfish in its environment. To quote Dorsett (1976), "it cannot be established whether *T. hombergi* is in the process of losing the behavior and the neurons are taking on new functions, or has retained a relatively unspecialised behavioral response from the ancestral type, and has lacked the selective pressures necessary for its development".

The neuronal bases of swimming behavior in *Tritonia*, *Clione limacina* (the sea angel) and *Aplysia brasiliana* have some common features (Huang and Satterlie, 1990; McPherson and Blankenship, 1991a,b,c; Arshavsky et al., 1993). Unlike *Tritonia*, which is benthic and swims for short periods to escape predators, *Clione* and *A. brasiliana* are planktonic and use episodes of slow parapodial swimming to maintain the same approximate depth in the water mass. Like *Tritonia*, the swimming motoneurons of *Clione* and *A. brasiliana* are located in the pedal ganglia. Comparison of the efferent neurons for swimming in the three species reveals, at least superficially, differences rather than similarities, e.g., the flexion neurons of *Tritonia* are small groups of non-electrically coupled medial neurons, the swimming motoneurons of *A. brasiliana* are groups of non-electrically coupled central and rostral neurons, whereas the *Clione* efferents are located laterally (close to the wing nerve) and consist of electrically coupled populations. The morphology of the flexion neurons in *Tritonia* is unknown, as is the motoneuron pharmacology of all three species. Interestingly, a medial group of pedal serotonergic neurons modulates swimming in *A. brasiliana* (McPherson and Blankenship, 1991c, see Locomotion section), and serotonin increases locomotor activity in *Clione* (Arshavsky et al., 1993), *Tritonia* (McClellan et al., 1994) and *A. brasiliana* (McPherson and Blankenship, 1991c). In the case of *Clione*, two groups of cerebral interneurons have been identified which can excite and inhibit swimming respectively (Arshavsky et al., 1992). One of these neurons, which may be serotonergic (Arshavsky et al., 1993), is thought to respond to tactile stimulation and initiate escape swimming. Cerebral command neurons also appear to exist in *A. brasiliana* (Blankenship et al., 1993). Once again, these data reinforce the role of serotonin in behavioral arousal of gastropods and point to another descending serotonergic pathway as well as pedal neurons that are involved in a rhythmic behavior (see sections on *Feeding* and *Locomotion*).

In common with most other rhythmic behaviors, swimming is controlled by a central pattern generator (CPG) consisting of a group of synaptically connected interneurons (see reviews by Getting, 1985, 1989; Getting and Dekin, 1985; Bulloch and Syed, 1992; Arshavsky et al., 1993). Throughout the animal kingdom, the key feature of CPG's are two interneurons (or groups of interneurons) which reciprocally inhibit each other and which drive antagonistic pairs of muscles. This model applies to the CPGs for swimming in both *Tritonia* and *Clione*, and also to respiration in *Lymnaea* (see section on *Respiration*). It is not easy to compare homologous neurons between these three CPG's – for example, all are located in different ganglia or groups of ganglia, and little is yet known of their comparative pharmacology. Undoubtedly, these circuits are a result of convergent evolution. Nevertheless, it is notable that the nervous system of many invertebrates and vertebrates has used a common strategy to generate rhythmic motor outputs.

Respiration

The neuronal control of respiration has been studied in a number of opisthobranchs and also the pulmonate, *Lymnaea*. The respiratory apparatus of the opisthobranchs varies from the large gill of *Aplysia* to the rows of branchial tufts of nudibranchs like *Tritonia*, nevertheless, some comparative studies will be reviewed. Both terrestrial and aquatic pulmonates employ lung respiration, and we will conclude this chapter with a brief discussion of relevant neuronal homologues in *Lymnaea* and *Helisoma*.

Gills

One of the first comparative studies of respiratory neurons in gastropods examined the tuft retractor neurons in *T. hombergi* and *T. diomedea* (Fig. 8). This behavior (which appears to be a protective reflex) is controlled by two bilateral pairs of pedal neurons in both species (Pd 1,2 see Fig. 8), whereas the pleural giant cells (Pl 1) cause tuft retraction only in *T. diomedea* (Dorsett, 1974; 1976). These comparative studies have been extended to a variety of other nudibranchs and the slightly more primitive notaspidean, *Pleurobranchaea*, by Dickinson (1979; 1980a). Withdrawl movements of the gill and branchial tufts are controlled by six to eight neurons in all these species, but these neurons are distributed between the cerebral and visceral as well as pedal and pleural ganglia. The giant pleural neurons may well be homologous between all the species examined, but tests for homology will require morphological, biochemical, and molecular comparisons.

The anaspidean *Aplysia* has, of course, served as a cornerstone for behavioral neurobiology. This animal exhibits so-called "respiratory pumping" behavior (contraction and withdrawal of the gill, siphon, and mantle shelf) which is thought to aid respiration and/or clean the mantle (Byrne and Koester, 1978; Byrne, 1983; Croll, 1985). Comparison of four species of *Aplysia* showed that gill motoneurons in the order anaspidea are restricted to the abdominal (parieto-visceral) ganglion

(Dickinson, 1980b) which contrasts to the condition of the two orders (notaspidea and nudibranchia) discussed above. The *Aplysia* gill is protected within the mantle cavity, whereas the gills of *Pleurobranchaea* and the nudibranchs are unprotected. Dickinson (1980b) points out that the circuitry of the latter species may therefore be more adapted to a protective function. Comparison of the behavioral, biophysical, and molecular properties of gill neurons between *Aplysia* and the notaspidean-nudibranch line may therefore provide interesting data regarding evolutionary adaptation.

Lungs

The neural circuit underlying pulmonary respiration in the pond snail *Lymnaea stagnalis* was elucidated recently (Syed et al., 1991; Syed and Winlow, 1991). Although skin respiration occurs in pulmonates, many of these animals breathe largely by opening and closing the lung orifice (the pneumostome), co-ordinated with contraction and expansion of the lung. The motoneurons responsible for these movements in *Lymnaea* are located in the visceral and right parietal ganglia (i.e., the equivalent of the *Aplysia* abdominal ganglion).

The CPG for respiration in *Lymnaea* consists of three interneurons (Syed and Winlow, 1991). That the three cells are sufficient to generate the respiratory rhythm was shown by culture and reformation of a functional three neuron CPG *in vitro* (Syed et al., 1990). The interneuron that controls inspiration (VD4) had not been identified previously in the other basommatophoran (*Helisoma*) we study in Calgary, but we reasoned it would be a small, white cell located near the visceral-right parietal connective (i.e., a mirror image to its position in *Lymnaea*, *Helisoma* being dextral, *Lymnaea* sinistral). This reasoning lead to the identification of neuron VD4 in *Helisoma*, a neuron whose morphology is a remarkable mirror image of its homologue in *Lymnaea* (Fig. 9) (Syed et al., 1993). Although the behavioral function of neuron VD4 in *Helisoma* has not been studied, its size, appearance, morphology, and synaptic targets are very similar to neuron VD4 of *Lymnaea*. However, at the molecular level a surprise arises. Both neurons are peptidergic, but in *Helisoma* a tetrapeptide is predominant (Richmond et al., 1991), whereas VD4 in *Lymnaea* expresses extended heptapeptides and not the tetrapeptide (FMRFamide) (Bright et al., 1993). The significance of alternative splicing in neurons of closely related species is unknown, but points to a new goal of comparative neurobiology, i.e., the understanding of evolution at the molecular level. Parenthetically, neuron VD4 is apparently a unique (unpaired) neuron, its removal from animals prevents respiration (Syed, Ridgway, Bulloch, and Lukowiak, 1992) – single neurons endow vulnerability indeed! (See discussion above regarding neuron C1 in the section on *Feeding*.)

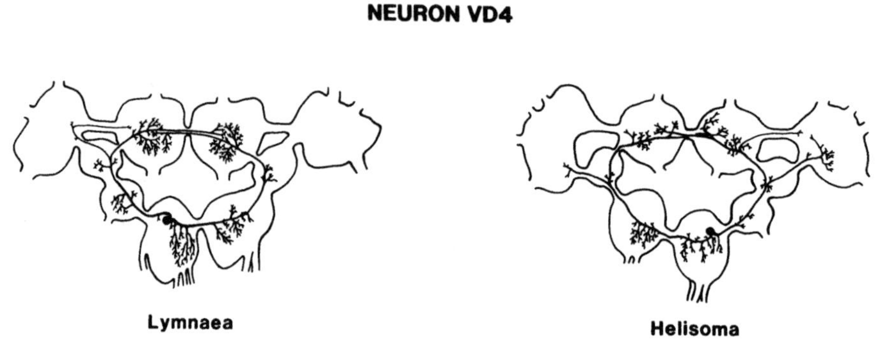

Figure 9. Comparison of neuron VD4 in *Lymnaea stagnalis* and *Helisoma trivolvis*. These schematics are of Lucifer yellow stains (for originals see Syed and Winlow, 1991; Syed et al., 1993). In each case the soma of neuron VD4 is located in the unpaired visceral ganglion, axons and axon collaterals projecting into most of the paired ganglia of the central ring. In these preparations the cerebral commissure has been severed and the cerebral ganglia pinned laterally with their ventral surfaces facing up. All other ganglia are dorsal surface up.

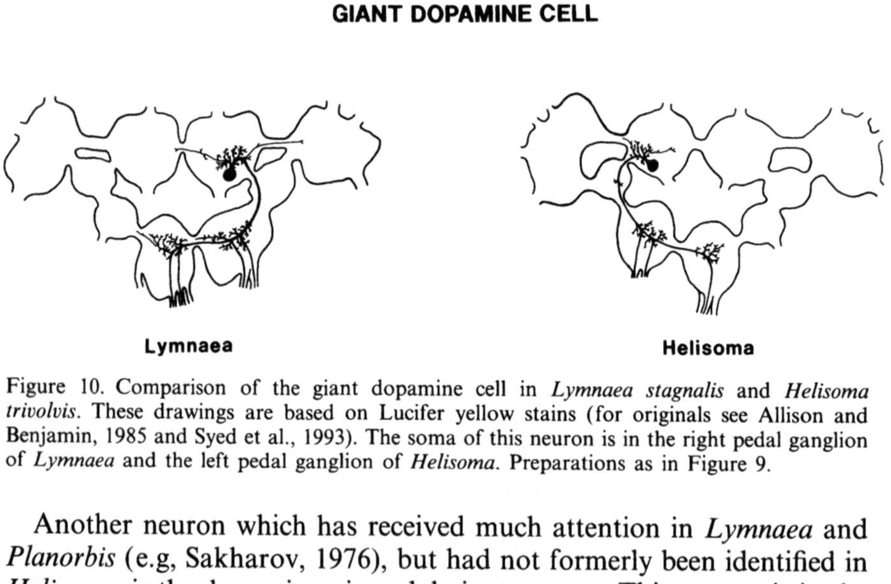

Figure 10. Comparison of the giant dopamine cell in *Lymnaea stagnalis* and *Helisoma trivolvis*. These drawings are based on Lucifer yellow stains (for originals see Allison and Benjamin, 1985 and Syed et al., 1993). The soma of this neuron is in the right pedal ganglion of *Lymnaea* and the left pedal ganglion of *Helisoma*. Preparations as in Figure 9.

Another neuron which has received much attention in *Lymnaea* and *Planorbis* (e.g, Sakharov, 1976), but had not formerly been identified in *Helisoma*, is the dopaminergic pedal giant neuron. This neuron is in the right pedal ganglion of *Lymnaea*. We recently characterized the *Helisoma* homologue by morphological, electrophysiological, and biochemical criteria (Fig. 10) (Syed et al., 1993). The mirror image morphology of the *Lymnaea*/*Helisoma* neurons is again remarkable (see Figs 9 and 10).

The cell culture work mentioned above lead to a new approach for testing neuronal homology. Initiation and maintenance of respiration in *Lymnaea* is the responsibility of giant dopamine cell. Based on size, position, and dopamine content, apparently homologous cells have been located in a variety of pulmonates (Fig. 10) and prosobranchs, but not

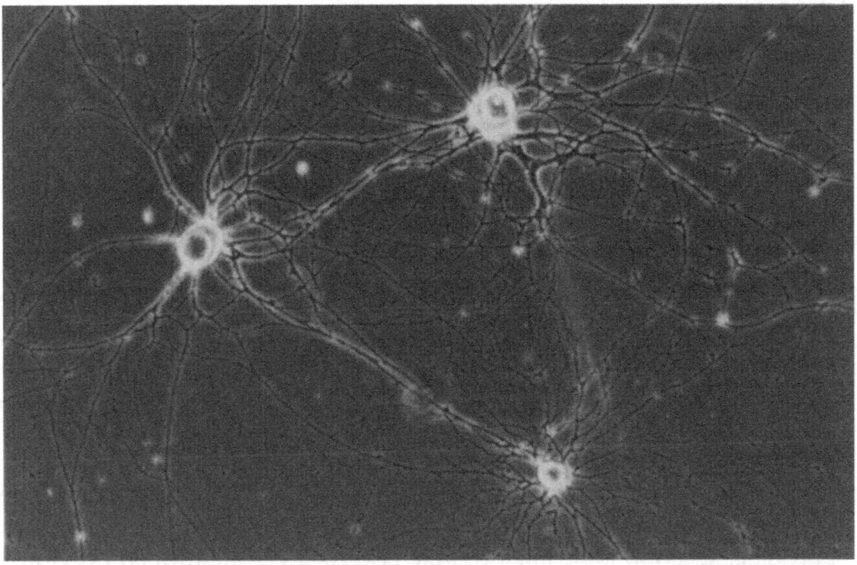

Figure 11. Photomicrograph of identified neurons cultured from both *Helisoma* and *Lymnaea*. Shown are a VJ cell and neuron RPeD1 from *Lymnaea* (upper two cells, left and right respectively) and a VJ cell from *Helisoma* (lower). These neurons were cultured for 2 days in a 1:1 mixture of *Lymnaea/Helisoma* conditioned medium and all show extensive outgrowth (see Syed, Lukowiak and Bulloch, 1992; photograph provided by Dr. Naweed Syed).

in the opisthobranchs (Sakharov, 1976). Using cell culture, however, we showed that the giant dopamine cells of *Lymnaea* and *Helisoma* could form synapses on the post-synaptic neurons of each others' species (Fig. 11) (Syed, Lukowiak and Bulloch, 1992). Not only does this observation strengthen the case for homology of the giant dopamine cells, but it shows that the mechanisms of synapse specificity are conserved between two families of gastropods.

Conclusion

We have briefly reviewed comparative aspects of gastropod neurobiology through an examination of feeding, locomotory, and respiratory behaviors. The diversity of gastropods as a class has facilitated comparative studies that have brought attention both to the adaptive value of these behaviors and to the generality of their underlying neural mechanisms. Although behavioral homologies may appear to be self-evident from such studies, it must be remembered that homology is principally a morphological concept (Croll, 1987). Thus, a description of the behavior based on morphological criteria is required. Kandel (1979) has stated that "the most compelling evidence for homologous behavior

Table 1. Identification of homologous neurons

Minimum criteria of similarity	Acceptable level of variability
(1) Behavioral role	?
(2) Location of soma	Ganglionic displacement can occur.
(3) Major axonal projection	Minor branches.
(4) Transmitter	Alternative splicing of peptides. Peptide co-transmitter may differ or be absent.
(5) Synaptic targets	Number and type of connections?

would be the demonstration that responses are generated by homologous structures...ideally by delineating neural circuits made up of specific homologous cells". This has been the objective of most of the studies we have described in this review, but then the homology question is shifted to the cellular level. To paraphrase an issue raised by David Sandman during the Konstanz symposium: What is the minimum (sufficient) number of similarities that have to be met to consider two neurons as homologues, and how much variability between these neurons is acceptable? Some possible criteria were mentioned previously (e.g., soma location, major axonal projections, synaptic connectivity, biochemical parameters, etc.), but based on the three best gastropod examples discussed in this chapter (neuron C1, the giant dopamine cell, and neuron VD4), we propose the more specific list shown in Table 1. Future studies, especially those involving comparisons of cell lineage and gene expression, should clarify the criteria so that a better definition of cellular homology can be provided.

Acknowledgements
The authors' research was supported by grants from MRC (Canada), NSERC, and the Canadian Network of Centres of Excellence. We thank Drs. O. Breidbach and W. Kutsch for their excellent organization of the Konstanz Symposium, Dr. N. Syed for helpful suggestions on the manuscript, and J. Ma for its preparation.

References

Allison, P. and Benjamin, P.R. (1985) Anatomical studies of central regeneration of an identified molluscan interneuron. *Proc. R. Soc. Lond. B* 226: 135–157.

Altrup, U. (1987) Inputs and outputs of giant neurons B1 and B2 in the buccal ganglia of *Helix pomatia*: an electrophysiological and morphological study. *Brain Res.* 414: 271–284.

Arnett, B.C. and Murphy, A.D. (1991) Correlation of specific neural patterns and buccal muscle contractions with dynamic stages of the feeding cycle of *Helisoma trivolvis*. *Soc. Neurosci. Abstr.* 17: 1391.

Arnett, B.C. and Murphy, A.D. (1994) Neural basis of regurgitation elicited by an emetic in *Helisoma*. *Soc. Neurosci. Abstr.* 20: 22.

Arshavsky, Yu. I., Deliagina, T.G., Orlovsky, G.N., Panchin, Yu. V. and Popova, L.B. (1992) Interneurones mediating the escape reaction of the marine mollusc *Clione limacina*. *J. Exp. Biol.* 164: 307–314.

Arshavsky, Yu. I., Gamkrelidze, G.N., Orlovsky, G.N., Panchin, Yu. V. and Popova, L.B. (1991) Gamma-aminobutyric acid induces feeding behavior in the marine mollusc, *Clione limacina*. *NeuroReport* 2: 169–172.

Arshavsky, Yu. I., Orlovsky, G.N., Panchin, Yu. V., Roberts, A. and Soffe, S.R. (1993) Neuronal control of swimming locomotion: analysis of the pteropod mollusc *Clione* and embryos of the amphibian *Xenopus*. *Trends Neurosci.* 16: 227–233.

Audesirk, G. (1978a) Central neuronal control of cilia in *Tritonia diomedea*. *Nature* 272: 541–543.

Audesirk, G. (1978b) Properties of central motor neurons exciting locomotory cilia in *Tritonia diomedea*. *J. Comp. Physiol.* 128: 259–267.

Audesirk, T. and Audesirk, G. (1985) Behavior of gastropod molluscs. *In:* A.O.D. Willows (ed.): *The Mollusca, Vol.* 8, *Neurobiology and Behavior, Part 1*, Academic Press, Orlando, pp. 1–94.

Audesirk, G., McCaman, R.E. and Willows, A.O.D. (1979) The role of serotonin in the control of pedal ciliary activity by identified neurons in *Tritonia diomedea*. *Comp. Biochem. Physiol.* 62C: 87–91.

Barber, A. (1983a) Nervous control of the salivary glands of the carnivorous mollusc *Philine aperta*. *J. Exp. Biol.* 107: 331–348.

Barber, A. (1983b) Properties of the serotonergic cerebral ganglion neurons of the gastropod mollusc, *Philine aperta*. *Comp. Biochem. Physiol.* 76C: 135–149.

Barlow, L.A. and Truman, J.W. (1992) Patterns of serotonin and SCP immunoreactivity during metamorphosis of the nervous system of the red abalone, *Haliotis refescens*. *J. Neurobiol.* 23: 829–844.

Barnes, R.D. (1980) *Invertebrate Zoology*, Third Edition. Saunders College/Holt, Rinehart and Winston, Philadelphia.

Benjamin, P.R. and Elliott, C.J.H. (1989) Snail feeding oscillator: The central pattern generator and its control by modulatory interneurons. *In:* J.W. Jacklet (ed.): *Neuronal and Cellular Oscillators*. Dekker, Inc., New York.

Benjamin, P.R. and Rose, R.M. (1979) Central generation of bursting in the feeding system of the snail, *Lymnaea stagnalis*. *J. Exp. Biol.* 80: 93–118.

Berdan, R.C., Shivers, R.R. and Bulloch, A.G.M. (1987) Chemical synapses, particle arrays, pseudo-gap junctions and gap junctions of neurons and glia in the buccal ganglion of *Helisoma*. *Synapse* 1: 304–323.

Blankenship, J.E., Laurienti, P.J. and Gamkrelidze, G.N. (1993) Candidate command neurons for *Aplysia* are tentatively identified in cerebral ganglion. *Soc. Neurosci. Abstr.* 19: Part 2, p.1599.

Boer, H.H., Geraerts, W.P.M. and Joosse, J., (1987) *Neurobiology Molluscan Models*. North-Holland Pub. Co., Amsterdam.

Bright, K., Kellett, E., Saunders, S.E., Brierley, M., Burke, J.F. and Benjamin, P.R. (1993) Mutually exclusive expression of alternatively spliced FMRFamide transcripts in identified neuronal systems of the snail *Lymnaea*. *J. Neurosci.* 13: 2719–2729.

Bulloch, A.G.M. and Dorsett, D.A. (1979) The functional morphology and motor innervation of the buccal mass of *Tritonia hombergi*. *J. Exp. Biol.* 79: 7–22.

Bulloch, A.G.M. and Syed, N.I. (1992) Reconstruction of neuronal networks in culture. *Trends Neurosci.* 15: 422–427.

Bulloch, A.G.M., Price, D.A., Murphy, A.D., Lee, T.D. and Bowes, H.N. (1988) FMR-Famide peptides in *Helisoma*: identification and physiological actions at a peripheral synapse. *J. Neurosci.* 8: 3459–3469.

Bullock, T.H. and Horridge, G.A. (1965) *Structure and Function in the Nervous Systems of Invertebrates*, 2 Volumes, Freeman Press, San Francisco.

Byrne, J. (1983) Identification and initial characterization of a cluster of command and pattern-generating neurons underlying respiratory pumping in *Aplysia californica*. *J. Neurophysiol.* 49: 491–508.

Byrne, J. and Koester, J. (1978) Respiratory pumping: neuronal control of a centrally commanded behavior in *Aplysia*. *Brain Res.* 143: 87–105.

Chiel, H.J., Kupfermann, I. and Weiss, K.R. (1988) An identified histaminergic neuron can modulate the outputs of buccal-cerebral interneurons in *Aplysia* via presynaptic inhibition. *J. Neurosci.* 8: 49–63.

Coates, C.J. and Bulloch, A.G.M. (1985) Synaptic plasticity in the molluscan peripheral nervous system: Physiology and role of peptides. *J. Neurosci.* 5: 2677–2684.

Crofts, D.R. (1955) Muscle morphogenesis in primitive gastropods and its relation to torsion. *Proc. Zool. Soc. Lond.* 125: 711–750.

Croll, R.P. (1985) Sensory control of respiratory pumping *Aplysia californica*. *J. Exp. Biol.* 117: 15–27.

Croll, R.P. (1987) Identified neurons and cellular homologues. *In:* M.A. Ali (ed.): *Nervous Systems in Invertebrates*, Plenum Press, New York, pp. 41–58.
Croll, R.P. and Lo, R.Y.S. (1986) Distribution of serotonin-like immunoreactivity in the central nervous system of the periwinkle, *Littorina littorea* (Gastropoda, Prosobranchia, Mesogastropoda). *Biol. Bull.* 171: 426–440.
Dickinson, P.S. (1979) Homologous neurons control movements of diverse gill types in nudibranch molluscs. *J. Comp. Physiol. A* 131: 277–283.
Dickinson, P.S. (1980a) Gill control in the notaspidean *Pleurobranchaea* and possible homologies with nudibranchs. *J. Comp. Physiol. A* 139: 11–16.
Dickinson, P.S. (1980b) Neuronal control of gills in diverse *Aplysia* species: Conservative evolution. *J. Comp. Physiol. A* 139: 17–23.
Diefenbach, T.J., Koehncke, N.K. and Goldberg, J.I. (1991) Characterization and development of rotational behavior in *Helisoma* embryos: Role of endogenous serotonin. *J. Neurobiol.* 22: 922–934.
Dorsett, D.A. (1974) Neuronal homologies and the control of branchial tuft movements in two species of *Tritonia*. *J. Exp. Biol.* 61: 639–654.
Dorsett, D.A. (1976) The evolution of neural mechanisms underlying behavior in *Tritonia*. *In:* P. Spencer Davies, (ed.): *Perspectives in Experimental Biology, Volume* 1, *Zoology*, Pergamon Press, Oxford, pp. 385–397.
Dorsett, D.A. (1986) Brains to cells: The neuroanatomy of selected gastropod species. *In:* A.O.D. Willows (ed.): *The Mollusca, Volume* 9, *Neurobiology and Behavior, Part* 2, Academic Press Inc., Orlando, pp. 101–176.
Getting, P.A. (1985) Neural control of behavior in gastropods. *In:* A.O.D. Willows, (ed.): *The Mollusca, Volume* 8, *Neurobiology and Behavior, Part* 1, Academic Press, Orlando, pp. 269–334.
Getting, P.A. (1989) Emerging properties governing the operation of neural networks. *Ann. Rev. Neurosci.* 12: 185–204.
Getting, P.A. and Dekin, M.S. (1985) *Tritonia* swimming: A model system for integration within rhythmic motor systems. *In:* A.I. Selverston (ed.): *Model Neural Networks and Behavior*, Plenum Press, New York, pp. 3–20.
Gillette, R. (1991) On the significance of neuronal giantism in gastropods. *Biol. Bull.* 180: 234–240.
Gillette, R., Gillette, M.U. and Davis, W.J. (1980) Action-potential broadening and endogenously sustained bursting are substrates of command ability in a feeding neuron of *Pleurobranchaea*. *J. Neurophysiol.* 43: 669–685.
Goldberg, J.I. Neuronal development in embryos of the mollusc, *Helisoma trivolvis*: multiple roles of serotonin. *In:* S. Malholtra (ed.): *Advances in Neural Science, Vol. II*, J.A.I. Press, Greenwich; *in press*.
Goldberg, J.I. and Kater, S.B. (1989) Expression and function of the neurotransmitter serotonin during development of the *Helisoma* nervous system. *Dev. Biol.* 131: 483–495.
Goldschmeding, J.T., Van Duivenboden, Y.A. and Lodder, J.C. (1981) Axonal branching pattern and coupling mechanisms of the cerebral giant neurons in the snail, *Lymnaea stagnalis*. *J. Neurobiol.* 12: 405–424.
Goldstein, R., Kistler, H.B., Jr., Steinbusch, H.W.M. and Schwartz, J.H. (1984) Distribution of serotonin immunoreactivity in juvenile *Aplysia*. *Neurosci.* 11: 535–547.
Granzow, B. and Rowel, C.H.F. (1981) Further observations on the serotonergic cerebral neurones of *Helisoma* (mollusca, gastropoda): The case for homology with the metacerebral giant cells. *J. Exp. Biol.* 90: 283–305.
Haydon, P.G. (1988) The formation of chemical synapses between cell-cultured neuronal somata. *J. Neurosci.* 8: 1032–1038.
Haydon, P.G. and Zoran, M.J. (1989) Formation and modulation of chemical connections: evoked acetylcholine release from growth cones and neurites of specific identified neurons. *Neuron* 2: 1483–1490.
Hernadi, L., Elekes, K. and S-Rozsa, K. (1989) Distribution of serotonin-containing neurons in the central nervous system of the snail *Helix pomatia*. *Cell Tissue Res.* 257: 313–323.
Huang, Z. and Satterlie, R.A. (1990) Neuronal mechanisms underlying behavioral switching in a pteropod mollusc. *J. Comp. Physiol. A* 166: 875–887.
Jacklet, J.W. (1989). *Neuronal and Cellular Oscillators*, Dekker, New York.
Jacob, M. (1984) Neurogenesis in *Aplysia californica* resembles nervous system formation in vertebrates. *J. Neurosci.* 4: 1225–1239.

Jones, P.G., Rosser, S.J. and Bulloch, A.G.M. (1987) Glutamate suppression of feeding and the underlying output of effector neurons in *Helisoma. Brain Res.* 437: 56–68.

Kandel, E.R. (1976) *Cellular Basis of Behavior: An Introduction to Behavioral Neurobiology*, W.H. Freeman and Co., San Francisco.

Kandel, E.R. (1979) *Behavioral Biology of Aplysia*, W.H. Freeman and Co., San Francisco.

Kandel, E.R. and Tauc, L. (1966a) Input organization of two symmetrical giant cells in the snail brain. *J. Physiol. Lond.* 183: 269–286.

Kandel, E.R. and Tauc, L. (1966b) Anomalous rectification in the metacerebral giant cells and its consequences for synaptic transmission. *J. Physiol. Lond.* 183: 287–304.

Kandel, E.R., Kriegstein, A.R. and Schacher, S. (1980) Development of the central nervous system of *Aplysia* in terms of the differentiation of its specific identifiable cells. *Neuroscience* 5: 2033–2063.

Kater, S.B. (1974) Feeding in *Helisoma trivolvis:* The morphological and physiological bases of a fixed action pattern. *Amer. Zool.* 14: 1017–1036.

Kemenes, GY., Elekes, K., Miripi, L. and Benjamin, P.R. (1989) A comparison of four techniques for mapping the distribution of serotonin and serotonin-containing neurons in fixed and living ganglia of the snail *Lymnaea. J. Neurocytol.* 18: 193–208.

Kits, K.S., Boer, H.H. and Joosse, J. (1991) *Molluscan Neurobiology*, North-Holland, Amsterdam.

Kriegstein, A.R. (1977a) Development of the nervous system in *Aplysia californica. Proc. Natl. Acad. Sci. USA* 74: 375–378.

Kriegstein, A.R. (1977b) Stages in the post-hatching development of *Aplysia californica. J. Exp. Biol.* 199: 275–278.

Kyriakides, M.A., McCrohan, C.R., Slade, C.T., Syed, N.I. and Winlow, W. (1989) The morphology and electrophysiology of the neurons of the pedal ganglia of *Lymnaea stagnalis* (L.). Comp. Biochem. *Physiol.* 93A: 861–876.

Lemche, H. (1958) Molluscan phylogeny in the light of *Neopilina. Proc. Int. Congr. Zool., Lond.* 15: 380–381.

Lloyd, P.E. and Willows, A.O.D. (1988) Multiple transmitter neurons in *Tritonia. II.* Control of gut motility. *J. Neurobiol.* 19: 55–67.

Lloyd, P.E., Kupfermann, I. and Weiss, K.R. (1988) Central peptidergic neurons regulate gut motility in *Aplysia. J. Neurophysiol.* 59: 1613–1626.

Lloyd, P.E., Masinovsky, B.P. and Willows, A.O.D. (1988) Multiple transmitter neurons in *Tritonia.* I. Biochemical studies. *J. Neurobiol.* 19: 39–54.

Lloyd, P.E., Frankfurt, M., Stevens, P., Kupfermann, I. and Weiss, K.R. (1987) Biochemical and immunocytochemical localization of the neuropeptides FMRFamide, SCP_A, SCP_B, to neurons involved in the regulation of feeding in *Aplysia. J. Neurosci.* 7: 1123–1132.

Lloyd P.E., Mahon, A.C., Kupfermann, I., Cohen, J.L., Scheller, R.H. and Weiss, K.R. (1985) Biochemical and immunocytological localization of molluscan small cardioactive peptides in the nervous system of *Aplysia californica. J. Neurosci.* 5: 1851–1861.

Mackey, S. and Carew, T.J. (1983) Locomotion in *Aplysia:* Triggering by serotonin and modulation by bag cell extract. *J. Neurosci.* 3: 1469–1477.

Marcus, E.A., Nolen, T.G., Rankin, C.H., Stopfer, M. and Carew, T.J. (1988) Development of behavior and learning in *Aplysia. Experientia* 44: 415–423.

Marois, R. and Carew, R.J. (1990) The gastropod nervous system in metamorphosis. *J. Neurobiol.* 21: 1053–1071.

Marois, R. and Croll, R.P. (1992) Development of serotonin-like immunoreactivity in the embryonic nervous system of the snail *Lymnaea stagnalis. J. Comp. Neurol.* 322: 255–265.

Masinovsky, B., Lloyd, P.E. and Willows, A.O.D. (1985) Morphology of two pairs of identified peptidergic neurons in the buccal ganglia of the mollusc *Tritonia diomedea. J. Neurobiol.* 16: 27–39.

Masinovsky, B., Kempf, S.C., Callaway, J.C. and Willows, A.O.D. (1988) Monoclonal antibodies to the molluscan small cardioactive peptide SCP_B: Immunolabeling of neurons in diverse invertebrates. *J. Comp. Neurol.* 273: 500–512.

McAllister, L.B., Scheller, R.H., Kandel, E.R. and Axel, R. (1983) *In situ* hybridization to study the origin and fate of identified neurons. *Science* 222: 800–808.

McClellan, A.D., Brown, G.D. and Getting, P.A. (1994) Modulation of swimming in *Tritonia*: excitatory and inhibitory effects of serotonin. *J. Comp. Physiol. A* 174: 257–266.

McCrohan, C.R. (1988) Modification of cental pattern generation in invertebrates. *Comp. Biochem. Physiol.* 90A: 17–22.
McPherson, D.R. and Blankenship, J.E. (1991a) Neural control of swimming in *Aplysia brasiliana*. I. Innervation of parapodial muscle by pedal ganglion motoneurons. *J. Neurophysiol.* 66: 1338–1351.
McPherson, D.R. and Blankenship, J.E. (1991b) Neural control of swimming in *Aplysia brasiliana*. II. Organization of pedal motoneurons and parapodial motor fields. *J. Neurophysiol.* 66: 1352–1365.
McPherson, D.R. and Blankenship, J.E. (1991c) Neural control of swimming in *Aplysia brasiliana*. III. Serotonergic modulatory neurons. *J. Neurophysiol.* 66: 1366–1379.
Moroz, L.L., Park, J.-H. and Winlow, W. (1993) Nitric oxide activates buccal motor patterns in *Lymnaea stagnalis*. *NeuroReport* 4: 643–646.
Murphy, A.D. (1986) Selective and independent activation or suppression of subunits of the central pattern generator underlying feeding in *Helisoma*. *Soc. Neurosci. Abstr.* 12: 791.
Murphy, A.D. and Lukowiak, K. (1987). Identified small cardioactive peptide B immunoreactive neurons and patterned motor activity in the buccal ganglia of *Helisoma*. *In:* H.H. Boer, W.P.M. Geraerts and J. Joose (eds): *Neurobiology. Molluscan models*. North Holland Publishing Co., Amsterdam, pp. 124–126.
Murphy, A.D., Lukowiak, K. and Stell, W.K. (1985) Peptidergic modulation of patterned motor activity in identified neurons of *Helisoma*. *Proc. Natl. Acad. Sci. USA* 82: 7140–7144.
Murphy, A.D., Richmond, J.E. and Lukowiak, K. (1988) Differential effects of neuromodulators on patterned motor activity in the buccal ganglia of *Helisoma*. *Soc. Neurosci. Abstr.* 14: 258.
Palovcik, R.A., Basberg, B.A. and Ram, J.L. (1982) Behavioral state changes induced in *Pleurobranchaea* and *Aplysia* by serotonin. *Behav. Neural. Biol.* 35: 383–394.
Pentreath, V.W., Berry, M.S. and Osborne, N.N. (1982). The serotonergic cerebral cells in gastropods. *In:* N.N. Osborne (ed.): *Biology of Serotonergic Transmission*. John Wiley & Sons, New York, pp. 457–512.
Prior, D.J. and Gelperin, A. (1977) Autoactive molluscan neuron: Reflex function and synaptic modulation during feeding in the terrestrial slug, *Limax maximus*. *J. Comp. Physiol.* 114: 217–232.
Prior, D.J. and Watson, W.H., III (1988) The molluscan neuropeptide, SCP_B, increases the responsiveness of the feeding motor program of *Limax maximus*. *J. Neurobiol.* 19: 87–105.
Purchon, R.D. (1968) *The Biology of the Mollusca*. Pergamon Press, Oxford.
Quinlan, E.M. (1993) *The functional organization of the buccal central pattern generator of Helisoma trivolvis: roles of dopamine and glutamate*. Ph.D. Thesis, University of Illinois at Chicago.
Quinlan, E.M. and Murphy, A.D. (1991) Glutamate as a putative neurotransmitter in the buccal central pattern generator of *Helisoma trivolvis*. *J. Neurophysiol.* 66: 1264–1271.
Raven, C.P. (1966). *Morphogenesis: The Analysis of Molluscan Development*, 2nd Edition. Pergamon, New York.
Richmond, J.E., Bulloch, A.G.M., Bauce, L. and Lukowiak, K. (1991) Evidence for the presence, synthesis, immunoreactivity, and uptake of GABA in the nervous system of the snail *Helisoma trivolvis*. *J. Comp. Neurol.* 307: 131–143.
Richmond, J.E., Bulloch, A.G.M., Murphy, A.D. and Lukowiak, K. (1987) The modulation of feeding motor output in *Helisoma* by arginine vasotocin. *Proc. West. Pharmacol. Soc.* 30: 411–413.
Richmond, J.E., Murphy, A.D., Lukowiak, K. and Bulloch, A.G.M. (1994) GABA modulates the buccal motor output of *Helisoma* by two pharmacologically distinct actions. *J. Comp. Physiol. A* 174: 593–600.
Richmond, J.E., Funte, L.R., Smith, W.L., Price, D.A. and Haydon, P.G. (1991) Activation of peptidergic synapse locally modulates postsynaptic calcium influx. *J. Exp. Biol.* 161: 257–271.
Rosen, S.C., Susswein, A.J., Cropper, E.C., Weiss, K.R. and Kupfermann, I. (1989) Selective modulation of spike duration by serotonin and the neuropeptides, FMRFamide, SCP_B, buccalin and myomodulin in different classes of mechanoafferent neurons in the cerebral ganglion of *Aplysia*. *J. Neurosci.* 9: 390–402.
Runnegar, B. and Pojeta, J.,Jr. (1974) Molluscan phylogeny: the paleontological viewpoint. *Science* 186: 311–317.
Sakharov, D.A. (1976) Nerve cell homologies in gastropods. *In:* J. Salanki (ed.): *Neurobiology of Invertebrates. Gastropoda Brain*. Akad. Kiado, Budapest, pp. 27–39.

Sakharov, D.A. (1990) Integrative function of serotonin common to distantly related invertebrate animals. *In:* M. Gustafsson and M. Reuter (eds): *The Early Brain.* Abo Akademi Press, Abo, pp. 73–88.

Salanki, J. (1973) *Neurobiology of Invertebrates: Mechanisms of Rhythm Regulation,* Akademiai Kiado, Budapest.

Salanki, J. (1976) *Neurobiology of Invertebrates: Gastropoda Brain,* Akademiai Kiado, Budapest.

Salanki, J. and Rozsa, K.S. (1988) *Neurobiology of Invertebrates: Transmitters. Modulators and Receptor,* Akademiai Kiado, Budapest.

Selverston, A.I. (1985) *Model Neural Networks and Behavior,* Plenum Press, New York.

Senseman, D.M. and Gelperin, A. (1974) Comparative aspects of the morphology and physiology of a single identifiable neuron in *Helix aspersa, Limax maximus* and *Ariolimax californica. Malac. Rev.* 7: 51–52.

Shkolnik, L.J. and Schwartz, J.H. (1980) Genesis and maturation of serotonergic vesicles in identified giant cerebral neuron of *Aplysia. J. Neurophysiol.* 43: 945–967.

Sigger, J.N. and Dorsett, D.A. (1981) Synaptic actions of oesophageal proprioceptors in the mollusc, *Philine. J. Exp. Biol.* 94: 95–104.

Stewart, W.W. (1981) Lucifer dyes – highly fluorescent dyes for biological tracing. *Nature* 292: 17–21.

Syed, N.I. and Winlow, W. (1989) Morphology and electrophysiology of neurons innervating the ciliated locomotor epithelium in *Lymanaea stagnalis* (L.). *Comp. Biochem. Physiol.* 93A: 633–644.

Syed, N.I. and Winlow, W. (1991) Respiratory behavior in the pond snail *Lymnaea stagnalis.* II. Neural elements of the central pattern generator (CPG). *J. Comp. Physiol. A.* 169: 557–568.

Syed, N.I., Bulloch, A.G.M. and Lukowiak, L. (1990) *In vitro* reconstruction of the respiratory central pattern generator (CPG) of the mollusc *Lymnaea. Science* 250: 282–285.

Syed, N.I., Harrison, D. and Winlow, W. (1991) Respiratory behavior in the pond snail *Lymnaea stagnalis.* I. Behavioral analysis and the identification of motor neurons. *J. Comp. Physiol.* A. 169: 541–555.

Syed, N.I., Lukowiak, K. and Bulloch, A.G.M. (1992) Specific *in vitro* synaptogenesis between identified *Lymnaea* and *Helisoma* neurons. *NeuroReport* 3: 793–796.

Syed, N.I., Ridgway, R.L., Lukowiak, K. and Bulloch, A.G.M. (1992) Transplantation and functional integration of an identified respiratory interneuron in *Lymnaea stagnalis. Neuron* 8: 767–774.

Syed, N.I., Roger, I., Ridgway, R.L. Bauce, L.G., Lukowiak, K. and Bulloch, A.G.M. (1993) Identification, characterization and *in vitro* reconstruction of an interneuronal network of the snail *Helisoma trivolvis. J. Exp. Biol.* 174: 19–44.

Trimble, D.L. and Barker, D.L. (1984) Activation by dopamine of patterned motor output from the buccal ganglia of *Helisoma trivolvis. J. Neurobiol.* 15: 37–48.

Trueman, E.R. (1983). Locomotion in molluscs. *In:* A.S.M. Saleuddin and K.M. Wilbur (eds): *The Mollusca, Volume* 4, *Physiology, Part I.* Academic Press, New York, pp. 155–198.

Twarog, B.M. (1988) Serotonin: History of a discovery. *Comp. Biochem. Physiol.* 91C: 21–24.

Watson, W.H.,III and Willows, A.O.D. (1992) Evidence for homologous peptidergic neurons in the buccal ganglia of diverse nudibranch mollusks. *J. Neurobiol.* 23: 173–186.

Weiss, K.R. and Kupfermann, I. (1976) Homology of the giant serotonergic neurons (metacerebral cells) in *Aplysia* and pulmonate molluscs. *Brain Res.* 117: 33–49.

Wieland, S.J. and Gelperin, A. (1983) Dopamine elicits feeding motor program in *Limax maximus. J. Neurosci.* 3: 1735–1745.

Willows, A.O.D. (1973) Learning in gastropod molluscs. *In:* W.C. Corning and A.O.D. Willows (eds): *Arthropods and Mollusks, Volume* 2. Plenum Press, New York, pp. 187–274.

Willows, A.O.D. (1980) Physiological basis of feeding behavior in *Tritonia diomedea*: II. Neuronal mechanisms. *J. Neurophysiol.* 44: 849–861.

Willows, A.O.D. (1985) *The Mollusca: Volume* 8, *Neurobiology and Behavior, Part 1,* Academic Press, Inc., Orlando.

Willows, A.O.D. (1986) *The Mollusca: Volume* 9, *Neurobiology and Behavior, Part 2,* Academic Press Inc., Orlando.

Willows, A.O.D. and Dorsett, D.A. (1975) Evolution of swimming behavior in *Tritonia* and its neurophysiological correlates. *J. Comp. Physiol.* 100: 117–133.

Willows, A.O.D., Lloyd, P.E. and Masinowsky, B.P. (1988) Multiple transmitter neurons in *Tritonia.* III. Modulation of central pattern generator controlling feeding. *J. Neurobiol.* 19: 69–86.

The Nervous Systems of Invertebrates: An Evolutionary and Comparative Approach
ed. by O. Breidbach & W. Kutsch
© 1995 Birkhäuser Verlag Basel/Switzerland

The cephalopod nervous system: What evolution has made of the molluscan design

B.U. Budelmann

The Marine Biomedical Institute, Department of Otolaryngology, University of Texas Medical Branch, Galveston, Texas 77555-1163, USA

Summary. The cephalopod nervous system is the most complex of any invertebrate nervous system. Although species-specific differences exist, its high level of complexity almost certainly is due to the cephalopods' very active, fast-moving, predatory life style, and their complex behavior and extreme flexibility of response to different environmental situations. Nevertheless, the basic morphological plan of the cephalopod nervous system is still of the ganglionated "molluscan" design. During the course of evolution some of the ganglia became increasingly complex and subdivided into lobes; also, new ganglia were added. This chapter describes the basic morphological plan of the cephalopod nervous system and outlines some species-specific differences that developed as adaptations to different life styles. Special emphasis will be given to those features of the cephalopod nervous system that, within the invertebrates, are unusual or even unique, often rivaling the equivalent parts of the vertebrate nervous system in sophistication; some of these features may characterize higher brain and nervous system function.

Introduction

The molluscan nervous system shows an extreme diversity. It ranges from systems almost as simple as those of flatworms, e.g., in amphineurans (such as *Chiton*) and monoplacophorans (such as *Neopilina*), to systems in cephalopods (such as octopuses, cuttlefishes, and squids) which are as complex as those of lower vertebrates. Despite their complexity, however, the cephalopod nervous systems still show a molluscan design ("Bauplan") (see Bullock and Horridge, 1965).

The basic design that is common to all mollusc nervous systems is a set of (five to) six pairs of ganglia: the cerebral, buccal, pedal, pleural, parietal, and visceral ganglia. The left and right of each pair is linked by a *commissure*, whereas *connectives* run in anterior/posterior direction between the ganglia on one side. Depending upon the mollusc species, this basic design can be modified in various ways.

This paper (1) summarizes the complexity of the cephalopod nervous system and outlines the changes that occurred during the course of evolution, as compared to the basic molluscan design, (2) describes the variability in design of the cephalopod nervous system, (3) highlights the features that are special to the cephalopod nervous system, or are even unique within all the invertebrates, and (4) outlines for what cephalopods use such a complex nervous system.

Gross morphology and basic functions of the cephalopod nervous system

Cephalopods have the largest of all the invertebrate nervous systems; its brain-to-body weight ratio exceeds that of most fishes and reptiles (Packard, 1972). In an adult *Octopus*, the nervous system has an estimated 520 million nerve cells (Young, 1963). As compared to a hatchling, this is a 1000-fold increase in the number of cells and a 5000-fold increase in weight; such dramatic increases are unrivalled for any nervous system (see Packard, 1972). One-third of the nerve cells is in the central part of the nervous system, i.e., in the brain (40 million cells) and in the optic lobes (130 million cells), and two-thirds (350 million cells) are in the periphery (Tab. 1). Most of the peripheral nerve cells are in the nerve cords and ganglia of the arms; this is not surprising if one considers the enormous complexity of the coordinate movements of the *Octopus* arms and suckers, which lack a skeleton and lack joints (see Boyle, 1986).

Table 1. Number of nerve cells in the supra-, peri- and sub-esophageal lobes of the central nervous system of *Octopus vulgaris*. For each lobe, the number refers to the total number of nerve cells of the left and right lobe. Data from Young (1963, 1971)

NUMBER OF NERVE CELLS IN THE CNS OF *OCTOPUS VULGARIS*				
	1	optic		128 940 000
SUPRA	1	vertical		25 066 000
	1	subvertical		810 000
	1	subpedunculate		144 000
	1	subfrontal		5 308 000
	2	frontal superior	(medial/lateral)	1 854 000
	2	frontal inferior	(medial/lateral)	1 085 000
	1	precommissural		78 000
	1	peduncle		142 000
	1	olfactory		136 000
	1	basal anterior		380 000
	1	basal dorsal		1 796 000
	1	basal inter		57 000
	1	basal lateral		127 000
	1	basal median		245 000
	1	buccal superior		150 000
	1	buccal posterior		?
PERI	3	magnocellular	(dorsal/ventral/posterior)	581 000
SUB	1	palliovisceral		108 000
	2	chromatophore	(anterior/posterior)	526 000
	4	vasomotor	(2 × lateral/2 × medial)	1 307 000
	2	funnel	(anterior/posterior)	?
	2	brachial	(pre/post)	341 000
	2	pedal	(anterior/posterior)	207 000
	2	pedal lateral	(anterior/posterior)	36 000
			CNS total	170 000 000

The gross morphology of the cephalopod nervous system, especially of its central parts, varies between the groups. Major differences exist between the nervous system of *Nautilus* and those of decapods (like cuttlefishes and squids) and octopods; but there are also distinct differences between the decapod and the octopod brains. Fortunately, there is a tremendous body of information on the comparative anatomy of cephalopod brains, based on the work of J.Z. Young. This paper is a tribute to his work.

The cephalopod nervous system is divided into a central and a peripheral part (Figs 1 and 2). The central part includes the brain proper and the optic lobes; the large peripheral part includes the nervous system of the body and of the arms.

The brain is arranged around the esophagus. There is a supra-esophageal and a sub-esophageal part, and they are laterally connected by a part that may be considered as peri-esophageal (Fig. 2). Each of these parts is further subdivided into a varying number of lobes. In octopods and decapods, 25 major lobes have been described; however, some are further subdivided, giving a total number of 37 and 38 lobes, respectively (Tab. 1; Young, 1963, 1971). Many of these lobes (the supra-esophageal ones in particular) do not have a single precise function. This is obvious from two facts: (i) almost all brain nerves project to, and/or originate from, more than one lobe, and (ii) almost all lobes, including the sub-esophageal lower motor centers, have two-way connections (Plän, 1987). Obviously, these allow for a high degree of "cross-talking", interaction and fine tuning, and make strictly hierarchical concepts of cephalopod brain function difficult to accept.

It is beyond the aim of this paper to describe the structures and possible functions of all the 38 brain lobes. For comprehensive descriptions and reviews, see Pfefferkorn (1915), Thore (1939), Tompsett (1939), Boycott (1961), Bullock and Horridge (1965), Young (1965a, 1971, 1974, 1976b, 1977a, 1977b, 1979), Wells (1978), Messenger (1979), Boyle (1986), Plän (1987), Mangold (1989), Gillette (1991), and Nixon and Young (in preparation). In brief, the vertical, subvertical, subpedunculate, and precommissural lobes, as well as the superior frontal, inferior frontal, sub-frontal, and part of the posterior buccal lobes, together with the optic lobes, are all involved in the analysis of sensory input, in various types of learning, and in short-term and long-term memory. The various basal lobes (dorsal, median, anterior, lateral, and inter-basal) and the peduncle lobe are higher motor centers; the median basal lobe is involved in the motor control of swimming and respiration, and the lateral basal lobe is involved in the control of the chromatophores and the muscles in the skin. The superior buccal lobe controls the motor activity of feeding.

The magnocellular lobe is the most difficult lobe to understand; it can best be described as an intermediate motor center, i.e., a relay station

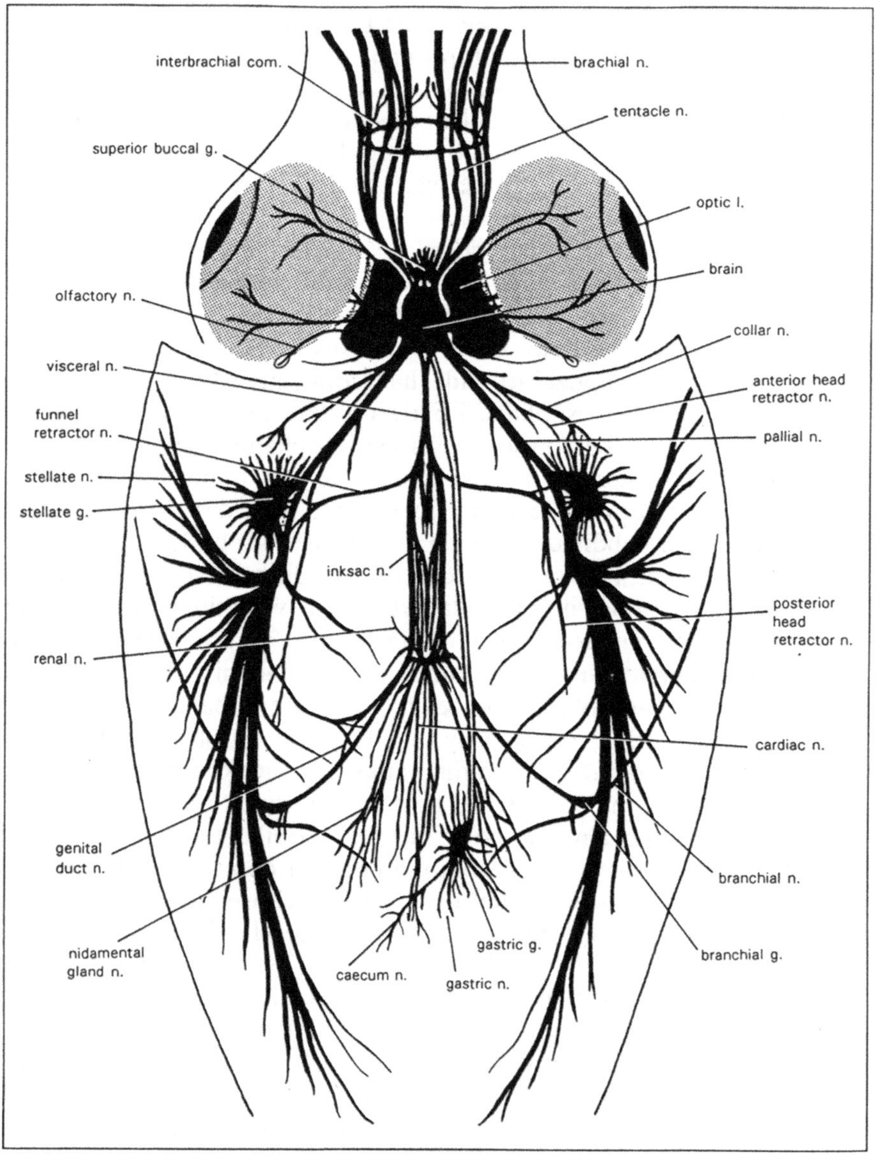

Figure 1. The central and peripheral nervous system of the cuttlefish *Sepia officinalis*. com., commissure; g., ganglion; l., lobe; n., nerve. (Modified from Hillig, 1912.)

between the supra-esophageal higher and the sub-esophageal lower motor centers. The magnocellular lobe is particularly involved in defense and fast escape reactions; the giant fiber system originates there (Young, 1971, 1976b).

The sub-esophageal lobes are all motor lobes. Some are intermediate motor centers, such as the brachial lobes, and some are lower motor centers, such as the chromatophore lobes, the funnel lobe, the vasomotor lobes which control circulation, and the anterior lateral pedal lobe, which is the oculomotor center; some lobes include both intermediate and lower motor centers, such as the pedal and the palliovisceral lobes (Young, 1971, 1976b).

The optic lobes show striking differences in size. They are always the largest lobes of the central nervous system; their volume is up to seven times that of the brain (Fig. 2; Young, 1988). In *Octopus*, each lobe contains 60–90 million nerve cells. The optic lobes are analyzers of the visual input and regulate visual behavior. In addition, they initiate the programs by which the higher motor centers (the basal lobes) produce the various motor activities of the animals. Furthermore, they are centers for simple visual discrimination learning and are involved in the expression of the various chromatic components produced by the chromatophores (Young, 1962, 1971, 1974).

The peripheral nervous system contains twice as many nerve cells (350 million) as the central nervous system (Young, 1963, 1971). All peripheral ganglia are lower motor centers. The largest part of the peripheral nervous system is the five brachial nerve cords; each has one large axial cord, four much smaller peripheral intramuscular cords, and as many brachial ganglia as the arm has suckers; in addition, each sucker has its own small sucker ganglion. Most of the neurons in the brachial and sucker ganglia are (lower) motor neurons that drive the muscles of the suckers, but they also analyze chemosensory and mechanosensory input (Young, 1971).

The most well known of the peripheral ganglia are the stellate ganglia on the inside of the mantle (Fig. 1; Young, 1972). These ganglia are lower motor centers for the movement of the mantle. They house the giant synapse and give rise to the stellate nerves (Young, 1939). Other peripheral ganglia are (see Fig. 1): the inferior buccal and subradula ganglia, which control the buccal and the radula movements (the two inferior buccal ganglia give rise to the "sympathetic nerves"); the large gastric ganglion, which innervates the crop, stomach, cecum, and intestines; the fusiform, cardiac and auricular ganglia, which innervate the vena cava, and the gill and systemic hearts; and the branchial (or gill) ganglia, which innervate the muscles of the gill lamellae (Alexandrowicz, 1960; Young, 1965b, 1967a; Smith and Boyle, 1983).

The giant fiber system is the most well known single part of the cephalopod nervous system (for references, see Young, 1939; Martin, 1977; Martin and Miledi, 1986). It is present in decapods only and is composed of a chain of three giant nerve cells on each side. The first-order giant cell (soma 150 µm in diameter) lies in the ventral magnocellular lobe (Fig. 6B); its axon runs to the palliovisceral lobe where it crosses to the contralateral side, forming a fusion (*Loligo*) or a synapse (*Sepia*) at the chiasma with its contralateral peer. From there the

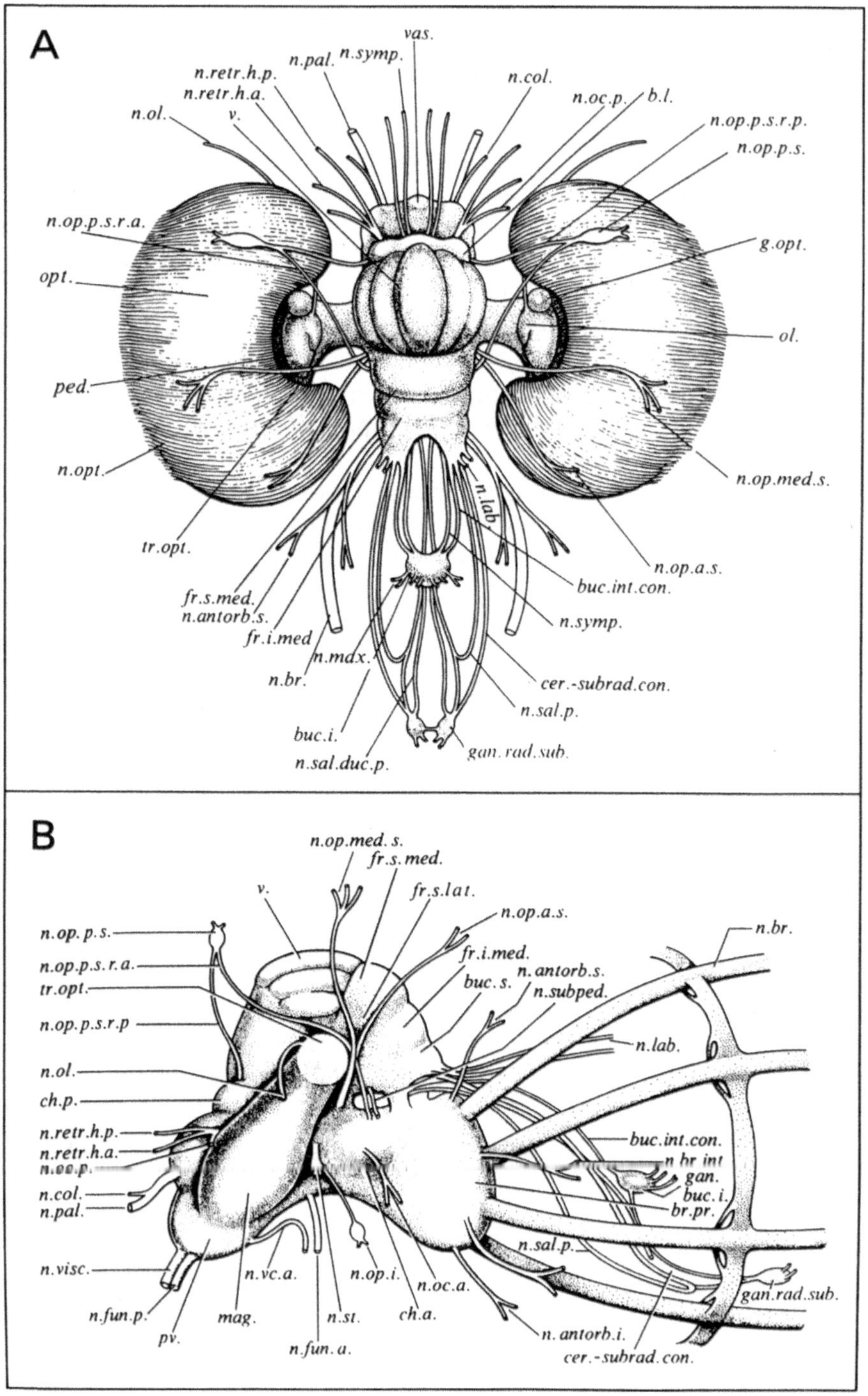
A
vas.
n.pal.
n.symp.
n.retr.h.p.
n.retr.h.a.
n.col.
n.oc.p.
b.l.
n.ol.
v.
n.op.p.s.r.p.
n.op.p.s.
n.op.p.s.r.a.
g.opt.
opt.
ol.
ped.
n.op.med.s.
n.opt.
n.lab
tr.opt.
n.op.a.s.
buc.int.con.
fr.s.med.
n.antorb.s.
n.symp.
fr.i.med
n.mdx.
n.br.
cer.-subrad.con.
n.sal.p.
buc.i.
gan.rad.sub.
n.sal.duc.p.
B
n.op.med.s.
fr.s.med.
v.
fr.s.lat.
n.op.a.s.
n.op.p.s.
n.br.
fr.i.med.
n.op.p.s.r.a.
n.antorb.s.
buc.s.
tr.opt.
n.subped.
n.op.p.s.r.p
n.lab.
n.ol.
ch.p.
n.retr.h.p.
buc.int.con.
n.retr.h.a.
gan.
n.col.
buc.i.
n.pal.
br.pr.
n.sal.p.
n.visc.
n.vc.a.
n.op.i.
n.oc.a.
gan.rad.sub.
n.fun.p.
mag.
n.st.
ch.a.
n.antorb.i.
pv.
n.fun.a.
cer.-subrad.con.

second-order giant cell runs to the stellate ganglion. There, it connects via the giant synapse to the third-order giant cell(s), which drives the mantle musculature; the axons of the third-order giant cells have a diameter of up to 1000 µm. Recent behavioral data on the function of the giant fiber system gave rather surprising results; they showed that the fast escape jetting (at least in restrained squids) is mediated by the small-fiber system (as is the respiratory mantle movement) and the giant axons serve to increase the amplitude of the muscle response (Otis and Gilly, 1990; Gilly et al., 1991).

Structure of cephalopod nerve cells, lobes and ganglia

Most neurons in the cephalopod brain and peripheral ganglia are unipolar neurons. Their somata form the outer perikaryal layer of the lobes and ganglia, whereas their trunks proceed centrally as single fibers (or for some distance together in bundles) and, together with their first-, second-, and higher order collaterals and incoming afferent fibres, form the inner neuropil. The thickness of the perikaryal layer, as compared to the diameter of the neuropil, varies greatly (Young, 1971).

The perikaryal layers are strictly organized when somata of different diameters occur: the largest somata are in the periphery, medium somata are in the middle, and the smallest somata are in the inner layers, close to the neuropil. In some lobes, and in the stellate ganglia, very small microneurons (amacrines) occur in the inner perikaryal layer (e.g., Young, 1971, 1972, 1974).

The pattern of the neuropil varies greatly between the lobes; it is sometimes rather "lobe specific". In lower and intermediate motor centers it is mostly irregular and tangled, whereas in the sensory and memory centers it is more regularly arranged, e.g., in matrices (Young, 1971, 1976b, 1977b, 1979, 1991a). The neuropils of the anterior basal and peduncle lobes are of particular interest because of their system of respectively horizontal (anterior basal lobe) and vertical (peduncle lobe)

Figure 2. Dorsal (A) and lateral (B) view of the central nervous system of *Octopus vulgaris*. Key to abbreviations (Figs 2 and 6): a., anterior; antorb., antorbital; art., artery; b., basal; br., brachial; buc., buccal; c., commissure; ce., cell; cen., central; ceph., cephalic; cer., cerebral; ch., chromatophore; col., collar; con., connective; cort., cortical; cr., crista; d., dorsal; duc., duct; e., extra; fl., fin lobe; fr., frontal; fun., funnel; g., gland; gan., ganglion; gi., giant; h., head; i., inferior; int., inter; l., lobe; lab., labial; lat., lateral; mac., macula; mag., magnocellular; max., maxillary; med., median/medial/middle; mus., muscle; n., nerve; oc., oculomotor; oes., esophagus; ol., olfactory; op., opthalmic; opt., optic; p., posterior; pal ., pallial; pe., pedal; ped., peduncle; per., peri; pil., neuropil; po., post; pr., pre; prec., precommissural; pv., palliovisceral; r., root; rad., radular; retr., retractor; s., superior; sal., salivary; st., statocyst; sth., statolith; sub., sub; subped., subpeduncle; subrad., subradular; subv., subvertical; symp., sympathetic; tr., tract; v., vertical; vas., vasomotor; vc., vena cava; ven., ventral; visc., visceral. (From Young, 1971; reproduced with the permission of Oxford University Press.)

parallel fibers (similar to the situation in the vertebrate cerebellum); both lobes are part of the cephalopod's statocyst-oculomotor reflex (Messenger, 1967, 1979; Young, 1976a; Budelmann and Young, 1984, 1993; Gleadall, 1990).

The organization of the optic lobes is the most elaborate of all the lobes (Fig. 3; Young, 1962, 1971, 1974). It has an outer cortex and a central medulla. The cortex is arranged in three layers: an outer and inner granule cell layer, and between them a plexiform zone; this resembles the organization of the deeper layers of the vertebrate retina. The inverted image from the eye – after a re-inversion in the dorso-ventral chiasma of the optic nerves – is projected rather precisely onto the plexiform layer; there, the first-order neurons meet branches of at least four different types of cells. From there, the second-order neurons run towards the center of the lobe. They form characteristic parallel bundles, which are surrounded by different types of nerve cells (often amacrines) and almost certainly are some kind of feature detectors similar to those in the mammalian visual cortex. Further central in the lobe are rather regularly arranged islands of nerve cells; their axons proceed into the brain via the optic tracts.

Variability in design of the cephalopod nervous system

The design of the cephalopod brain and nervous system shows a variety of forms. This is not surprising since cephalopods inhabit every part of the ocean and each habitat requires a different behavior. Each behavior, in turn, corresponds – together with taxonomical constraints – to different designs of the nervous system, especially of the brain (Young, 1977a; Maddock and Young, 1987).

In cephalopods the brain has been modified in four ways: by a change of position of ganglia, by a fusion of ganglia, by an addition or loss of ganglia, and by a change in the relative volume of ganglia. The differences in volume, as compared to the rest of the brain, can be very large, e.g., almost 50-fold in the optic lobes (15% in *Cirrothauma vs.* 700% in *Grimalditeuthis*), 10-fold in the vertical lobe system (3% in *Cirrothauma vs.* 28% in *Vampyroteuthis*), and 6-fold in the brachial lobes (5% in *Sepioteuthis* vs. 30% in *Architeuthis*) (Maddock and Young, 1987).

There are three main designs of the cephalopod brain: the *Nautilus* brain, the decapod brain and the octopod brain. In that order, they are characterized by an increase in centralization. There are some intermediate forms, such as in *Cirrothauma* (Aldred et al., 1983) and *Vampyroteuthis* (Young, 1977a), and there are forms with very special adaptation, e.g., *Mastigoteuthis*, a squid with long tentacles with minute suckers, has a very large magnocellular lobe (Dilly et al., 1977) and *Pterygioteuthis* has an extra giant fibre system with one large fibre to each arm (Young, 1977a).

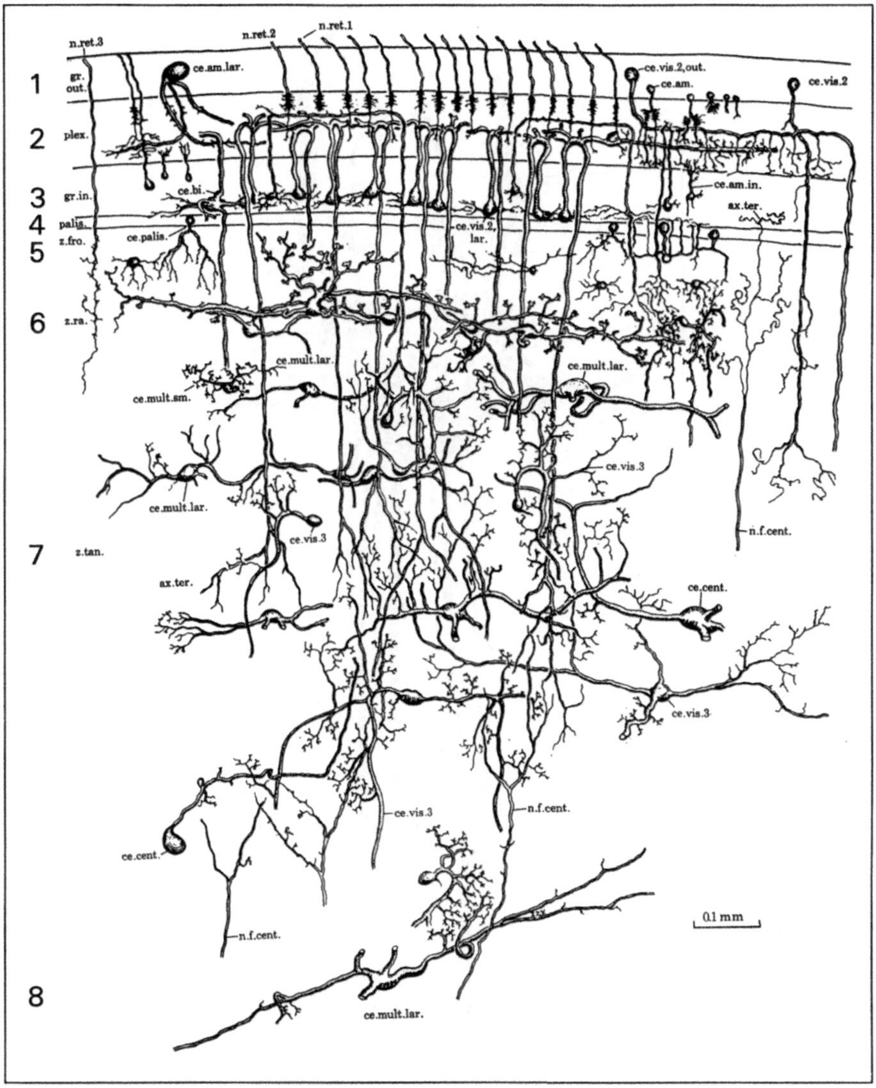

Figure 3. Cellular organization of the optic lobe of a squid (*Loligo*). Drawing of the different cell types seen in a sagittal section. The numbers on the left indicate the following cell layers: 1, outer granule cell layer; 2, plexiform zone; 3, inner granule cell layer; 4, palisade layer; 5, frontier zone of optic lobe; 6, zone of radial columns of medulla; 7, zone of tangential bundles; 8, zone of optic tract bundles in optic lobe. For the abbreviations of the various cell types, see Young (1974). (From Young, 1974; reproduced with permission of The Royal Society.)

The Nautilus nervous system is much simpler than any other cephalopod nervous system (Fig. 4). It lacks the brain structures for an elaborate motor control of the arms and it lacks the higher brain centers for learning and memory. The optic lobes are much smaller; they have

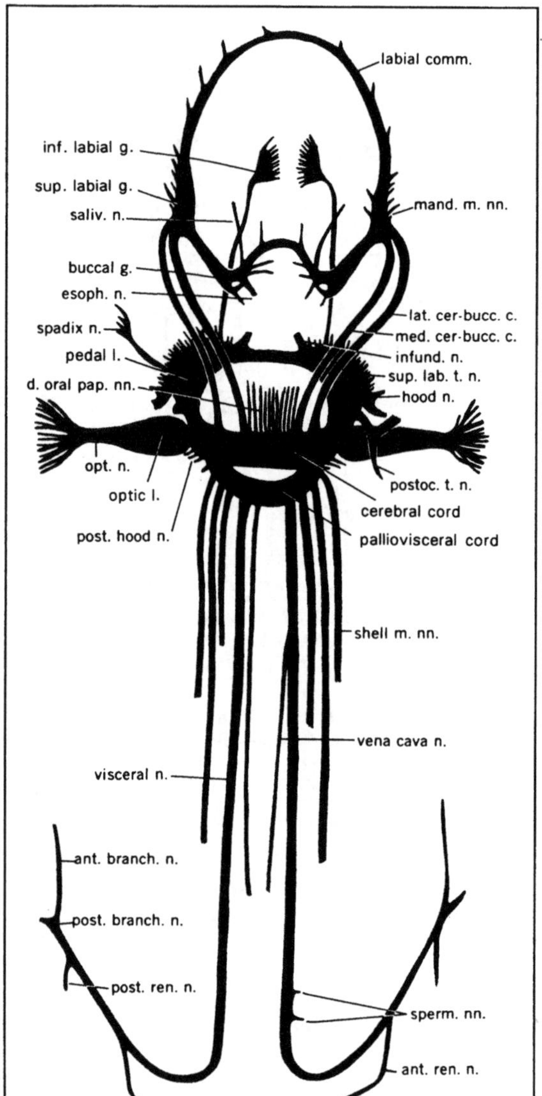

Figure 4. The central peripheral nervous system of *Nautilus pompilius*. Key to abbreviations: ant., anterior; branch., branchial; c, connective; cer.-bucc., cerebro-buccal; comm., commissure; d., dorsal; esoph., esophagus; g., ganglion; inf., inferior; infund., infundibular; l., lobe; lab., labial; lat., lateral; m., muscle; mand., mandibular; med., medial; n./nn., nerve/nerves; opt., optic; pap., papilla; post., posterior; postoc., postocular; ren., renal; saliv., salivary; sup., superior; t., tentacle. (Modified from Griffin, 1900).

no external granular layer. The peripheral system is significantly reduced; it has no giant fibre system and lacks most ganglia, including the stellate ganglia (Griffin, 1900).

The *Nautilus* brain represents a system of three ganglionated nerve cords (although with only little sign of the ganglionated condition): a supra-esophageal cerebral cord, and an anterior and posterior sub-esophageal cord; they are subdivided into 12 lobes (Young, 1965a). The cerebral cord has no clear internal divisions (Fig. 5A). Its anterior-ventral part connects to the buccal ganglia and is involved in feeding. Its dorsal part receives visual and tactile input and may be compared with the frontal and vertical lobe system of the decapod and octopod brains (Fig. 5). The *Nautilus* brain has no distinct vertical, basal, or peduncle lobes. The anterior sub-esophageal cord (or the brachio-pedal cord) lies laterally and is continuous with the cerebral and the posterior sub-esophageal cords. Its origin is unclear; part of it is certainly of pedal origin, but part might be cerebral (Young, 1965a). Its anterior part gives rise to the nerves of the arms, hood, and ocular tentacles; a distinct posterior part is the origin of the funnel nerves. The posterior sub-esophageal cord is the origin of a series of pallial and visceral nerves. The lateral region, where all three nerve cords join, contains the largest cells of the brain. It is an important motor area and may, at least partly, represent the magnocellular lobe of the decapod and the octopod brains (Young, 1965a).

The decapod and octopod nervous systems, in general, are rather similar; however, distinct differences exist and correlate well with the animals' different anatomies and behaviors (Young, 1977a). Overall, the octopod brain is more centralized than the decapod brain, i.e., in the octopod brain the brachial lobe joins the pedal lobes, and the superior buccal lobe is united with the inferior frontal lobe (Fig. 6). In octopods, the brachial and pedal lobes and the inferior frontal lobe system are larger than in decapods (Fig. 6), most certainly because of the octopods' more sophisticated use of their arms and their ability of tactile learning. Decapods, in contrast, have larger optic lobes, larger basal lobes and a larger, but unfolded, vertical lobe; their inferior frontal system is much simplified and they have no supra-brachial commissure. In contrast to octopods, they have a ventral magnocellular commissure. To control the motor activities of their fins, decapods have an additional fin lobe in the posterior sub-esophageal brain, and a peripheral giant fibre system for high-performance swimming by jet propulsion (Young, 1971, 1974, 1976b, 1977b, 1979; Maddock and Young, 1987; Nixon and Young, in preparation).

No major differences occur between the brains of different decapod species, with the exception of the vertical lobe. Its relative volume varies between 6% and 26%; it is generally larger in species that live in shallow water. Between octopod brains some variations occur with respect to the size of the lobes, depending upon in which habitat the animals live, e.g., in shallow water species the inferior frontal and vertical lobes are larger, whereas the optic lobes are often smaller (Maddock and Young,

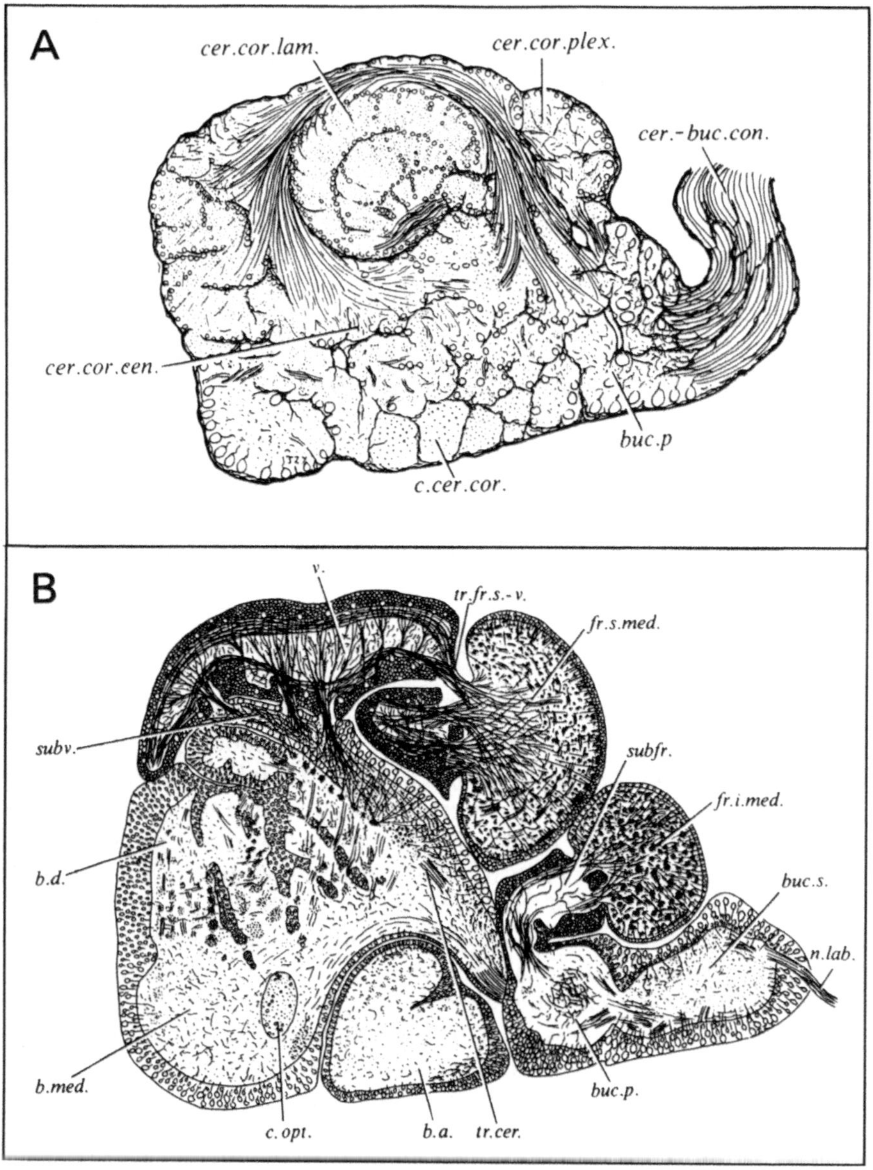

Figure 5. Comparison of the organization of the supra-esophageal part of *Nautilus* and *Octopus* brains. Sagittal section through the cerebral cord of *Nautilus* (A) and through the supra-esophageal lobes of *Octopus* (B). b.a./d./med., anterior/dorsal/median basal lobe; buc.p./s., posterior/superior buccal lobe; c.cer.cor., commissural bundles of cerebral cord; cer.-buc.con., cerebro-buccal connective; cer.cor.cen./lam./plex., central/laminated/plexiform zone of cerebral cord; c.opt., optic commissure; fr.i.med./s.med., median inferior/superior frontal lobe; n.lab., labial nerve; subfr., subfrontal lobe; subv., subvertical lobe; tr.fr.s.-v., superior frontal-vertical lobe tract; v., vertical lobe. (From Young, 1965a, 1971; reproduced with permissions of the The Royal Society and Oxford University Press.)

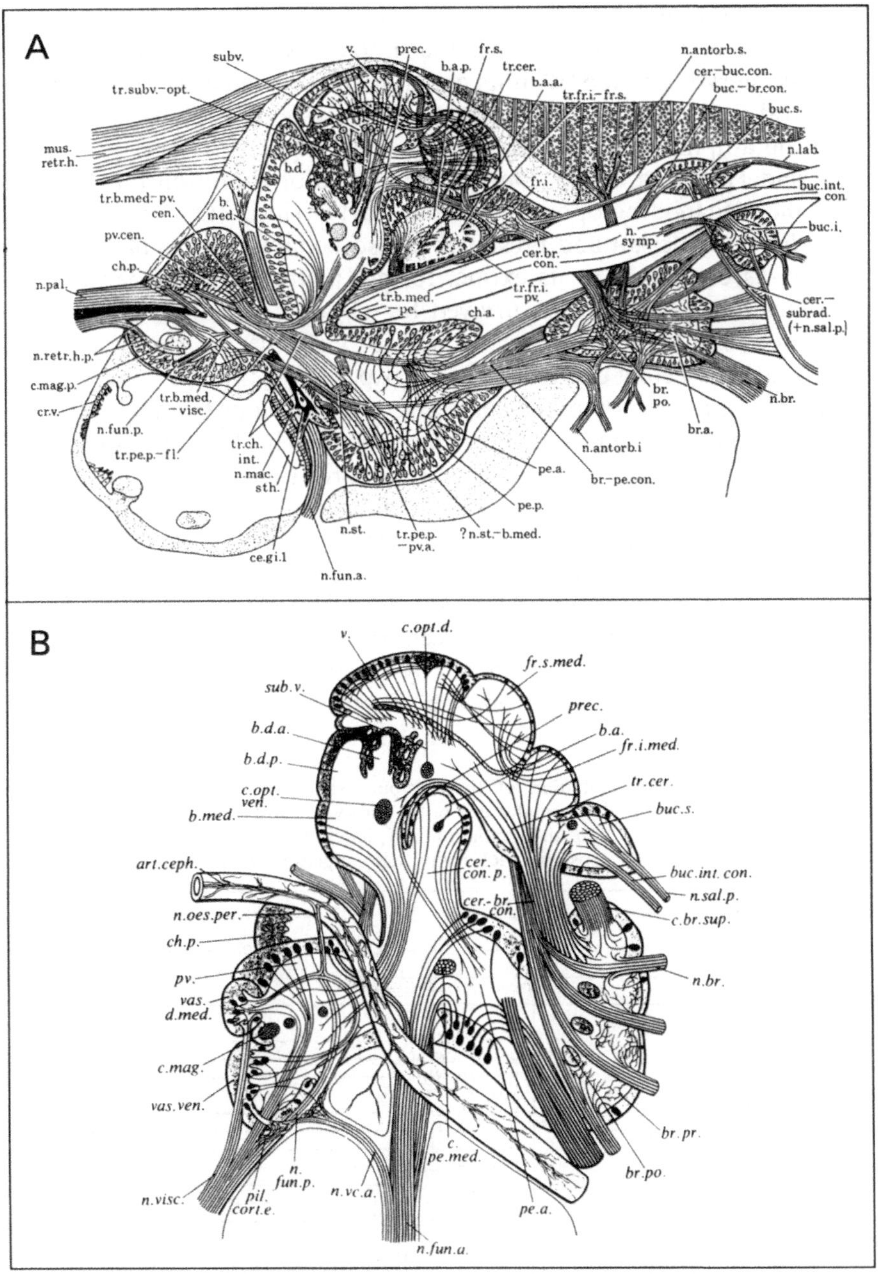

Figure 6. Comparison of the organization and major connections of the brains of a decapod (A; *Loligo*) and an octopod (B; *Octopus*) cephalopod, as seen in sagittal sections. For key to abbreviations, see Fig. 2. (From Young, 1971, 1976b; reproduced with permission of Oxford University Press and The Royal Society.)

1987; Nixon and Young, in preparation). The finned octopods (Cirrata) have an additional fin lobe (Maddock and Young, 1987).

It is still controversial whether the simple (by cephalopod standards) *Nautilus* nervous system with its ganglionated nerve cords represents an ancestral form of the cephalopod nervous system, as some similarities with the amphineuran nervous system may suggest (Thore, 1939; Bullock and Horridge, 1965). During the course of evolution the *Nautilus* and the decapod/octopod lines split early; since then, at least the *Nautilus* body design and presumably also behavior have changed only a little, as compared to its extinct forms (ammonites and belemnites). The fact that *Nautilus* has survived for about 120 million years, however, does not necessarily mean that, for example, its nervous system remained completely unchanged for that period of time (Young, 1965a).

In the *Octopus* embryo, the nervous system develops from five separate sets of ectodermal cells that form the optic, cerebral, pedal and visceral "primordial" ganglia, and the arm nerve cords (Tab. 2; Marquis, 1989). Later in development, the cerebral ganglia fuse and then differentiate into the various supra- and subesophageal lobes, whereas the brachial lobe separates from the pedal lobe; all connectives and commissures are secondarily formed by the ganglia. The magnocellular lobe is thought to be a connective between the supra- and sub-esophageal brain that has later been surrounded by nerve cells; its middle and ventral parts are probably derived from the visceral ganglion (Marquis, 1989).

Table 2. Embryonic differentiation of the nervous system of *Octopus vulgaris*. Roman numericals indicate the developmental stages of the embryo according to Naef (1928); VI-XII, ganglionic stage; XIII-XIX, lobe differentiation; XX, hatchling. ANT, anterior; INF, inferior; INT, inter; LE, left; POST, posterior; RI, right; SUP, superior. (Data from Marquis, 1989.)

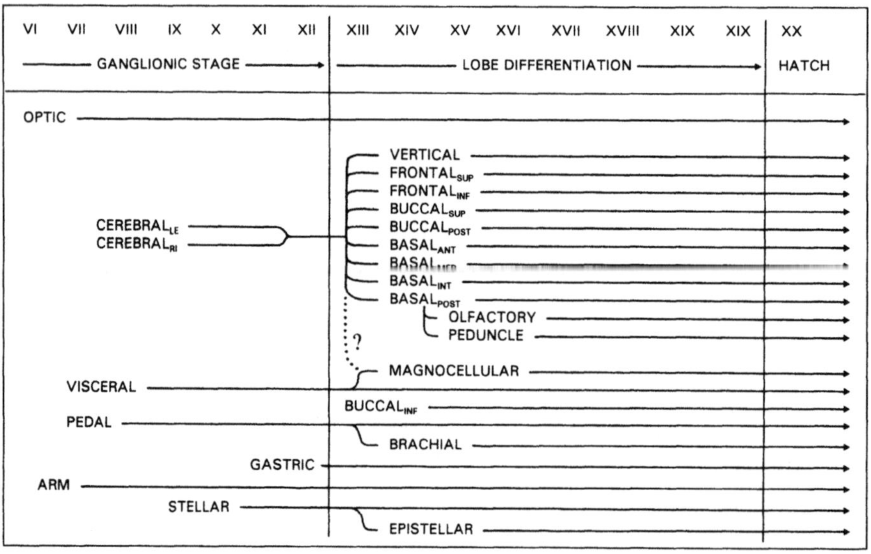

Unfortunately, there is no detailed study of the development of the *Nautilus* nervous system. Thus, further embryological evidence is needed in any attempt to properly homologize the highly specialized parts of the cephalopod brain with the various ganglia of the mollusc, especially the gastropod, nervous system.

Special features of the cephalopod nervous system

There is a number of features of the cephalopod nervous system that is quite unusual by molluscan, invertebrate, or even vertebrate standards, and some of them may characterize higher brain and nervous system function.

High degree of centralization

The nervous system of cephalopods manifests the highest degree of centralization ("cerebralization") for any mollusc (cephalopods > gastropods > bivalves) and, together with insects, the highest for any invertebrate; within cephalopods, the most centralized nervous system is present in octopods (octopods > decapods > nautiloids; Fig. 7). The centralization is acheived by a shortening of the connectives between the ganglia and lobes; shortening of the commissures contributes to the centralization to a much lesser extent. Since invertebrates have unmyelinated axons, a centralization considerably reduces the time for information processing between stimulus reception and behavioral reaction. This, in turn, is an advantageous feature for a predatory life style and complex behavior. In cephalopods, the degree of centralization of the nervous system, in general, corresponds well with the level of complexity of the animal's behavior (Young, 1977a; Nixon and Young, in preparation).

Local interneurons (amacrines or microneurons)

In many of the cephalopod brain lobes there is a large number of very small neurons, with a nucleus diameter of 3–5 μm and very little cytoplasm (Young, 1971, 1977b, 1979). These neurons are local interneurons with fine processes that remain in the outer part of the lobe's neuropil. They are especially abundant in the vertical lobe, where they constitute more than 99% of the nerve cells (25 million), the subfrontal, and the superior and inferior median frontal lobes. They are absent from the chromatophore and fin lobes. Whenever these neurons are present together with larger neurons, they are located in

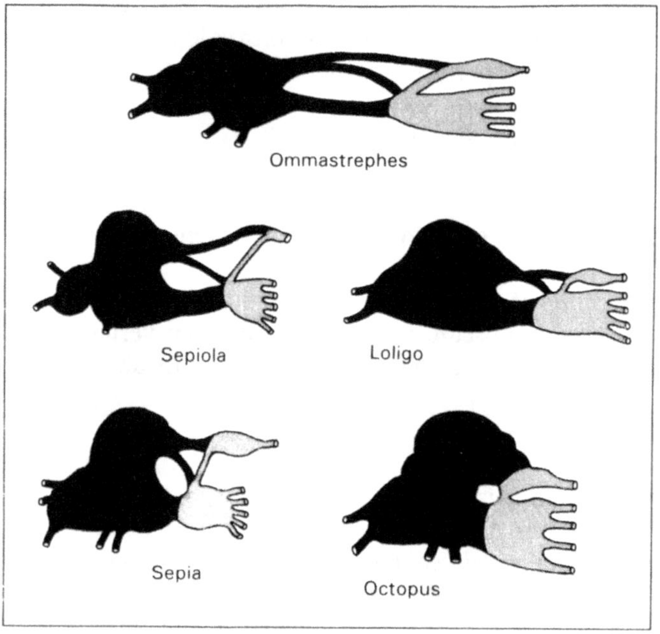

Figure 7. Diagrams of decapod and octopod brains, arranged according to the degree of centralization. The stippled areas outline the superior buccal and brachial lobes. (Modified from Bullock and Horridge, 1965.)

the inner perikaryal layer, close to the neuropil. Unfortunately, almost nothing is known about their function; it is speculated that they are invloved in some kind of lateral inhibition and fine tuning (Young, 1971, 1972).

Absence of somatotopy

In contrast to the situation in the vertebrate and insect brains, there is no obvious somatotopic arrangement of moto- and premoto-neurons in the cephalopod brain (Plän, 1987). The neurons are equally scattered in the relevant brain lobes; for example, motoneurons of the anterior and posterior oculomotor nerves that innervate widely separate muscles of antagonistic function originate (mixed together) throughout the anterior lateral pedal lobe (Budelmann and Young, 1984, 1993), and motoneurons in the posterior chromatophore lobe that innervate neighboring skin areas of the mantle originate from all parts of that lobe (Dubas et al., 1986a, 1986b). The absence of somatotopy perhaps explains why many motoneurons have very long dendritic collaterals; at least, they can provide the neural substrate to coordinate actions of widely separated neurons. In this context, it is also relevant to note that in *Octopus* individual eye muscles can be innervated by up to three eye muscle nerves (Budelmann and Young, 1984).

Blood-brain barrier

Cephalopods have an effective blood-brain barrier, formed by glial cells (Bundgaard and Abbott, 1992). In invertebrates such a barrier is also known in insects, higher arachnids, and decapod crustacea, but is has not been found in molluscs other than cephalopods. Blood-brain barriers are possibly associated with the development of higher integrative functions of a brain, such as pattern vision, fine motor control, and social behavior (Abbot et al., 1986).

Compound field potentials ("EEG")

The compound field potentials of the *Octopus* and *Sepia* brains are essentially different from those of other invertebrates; they are comparable to those of the vertebrate brains (Bullock, 1984; Bullock and Basar, 1988; Bullock and Budelmann, 1991). In some areas they show special event-related potentials, known as "omitted stimulus potentials" (Bullock and Budelmann, 1991). To date, such potentials have not been found in any other invertebrate brain; in humans they are associated with cognitive events.

Efferent innervation of sense organs

The amount of efferent innervation of cephalopod receptor cells is unusually high, even by vertebrate standards. There is an extensive efferent innervation to the retina which comes from the inner granular layer of the cortex of the optic lobe (Young, 1971, 1974); it is involved in the regulation of the size of the receptive fields (Tasaki et al., 1982) and the control of the screening pigment migration (Gleadall et al., 1993). The efferent innervation to the equilibrium receptor organs is particularly extreme: an average of 75% of the fibers in the various statocyst nerves are efferent (Colmers, 1981; Budelmann et al., 1987). They originate from four different areas of the brain and branch considerably at the level of the sensory epithelia; individual receptor cells and neurons can receive more than 15 and 30 efferent endings, respectively (Budelmann and Young, 1984; Budelmann et al., 1987). The majority of the efferents has an inhibitory function and only a few are excitatory (Williamson, 1989); they certainly allow for an extensive modulation and fine tuning of the afferent signals.

Peripheral first-order afferent neurons

The sensory epithelia of the cephalopod equilibrium receptor organs are composed of only secondary sensory cells (maculae), or a mixture of secondary and primary sensory cells (cristae). Some secondary sensory cells are in convergent, others are in divergent, afferent synaptic contact with different types of first-order afferent neurons, which have their cell bodies located in the sensory epithelia very close to the receptor cells (Budelmann and Thies, 1977; Colmers, 1981; Budelmann et al., 1987). Even by vertebrate standards, this is a very unusual configuration for an equilibrium receptor organ; it allows for significant information processing outside the brain, at the level of the receptor epithelia.

Transmitter

A large variety of putative transmitters has been found in the cephalopod brain; within a single sensory organ (such as the statocysts) five or more transmitters can be involved (e.g., Tansey, 1979, 1980; Budelmann and Bonn, 1982; Auerbach and Budelmann, 1986; Kime and Messenger, 1990; Cornwell et al., 1993; Tu and Budelmann, 1994). The biological significance of this multiplicity of transmitters, however, is still obscure. Furthermore, it is very likely that various peptides also play an important, yet unknown, role (Martin and Voigt, 1987).

For what do cephalopods use such a complex nervous system?

Most cephalopods, different from all other aquatic invertebrates, are visually-oriented, very mobile, and fast-moving predators which feed on a variety of organisms. In fact, cuttlefishes and squids are the only invertebrate predators (perhaps with the exception of the swimming crab *Macropipus*; Schäfer, 1954) that regularly chase and feed on adult vertebrates (fishes). Such a lifestyle necessarily requires sophisticated and very effective sense organs and an elaborate nervous system (Budelmann, 1994). Both developed in cephalopods during the course of evolution, most likely in competition with bony fishes (Packard, 1972; Aronson, 1991). Cephalopods are well known for their sophisticated eyes and equilibrium receptor organs; their statocyst-oculomotor system rivals the vertebrate vestibulo-oculomotor system in complexity (Budelmann and Young, 1984, 1993; Budelmann, 1990; Messenger, 1991). Cephalopods evolved neck proprioceptors that participate in the head-to-body coordination (Preuss and Budelmann, 1991, and submitted), and cephalopods evolved a lateral line analogue system that is as sensitive as that of fishes and aquatic amphibians (Budelmann and Bleckmann, 1988; Bleckmann et al., 1991).

Cephalopods have a complex muscle system in their arms and suckers which is capable of well-coordinated and finely tuned movements. Since that system operates without a skeleton and without joints, i.e., without a "simple" antagonistic mode of muscle action, it needs a large number of motoneurons and a sophisticated nervous control (Rowell, 1963, 1966; Kier, 1988; Kier and Smith, 1990; Budelmann and Young, 1985, 1987). In addition, most cephalopods have thousands of chromatophores in their skin which are under direct control of motoneurons in the chromatophore lobes of the brain (Florey, 1969; Dubas et al., 1986a, 1986b). With those chromatophores cuttlefish produce at least 34 different chromatic components (in varying combinations and intensities of expression!) as part of their body patterns (Hanlon and Messenger, 1988).

Finally, cephalopods have a large repertoire of intra- and inter-specific behaviors; the latter show a plasticity that is exceptional and unrivaled among invertebrates (e.g., Moynihan and Rodaniche, 1982; Hanlon and Messenger, in preparation). Cephalopods are also famous for their capability of complex learning, including forms of observational learning, and they are well known for their short- and long-term memories (e.g., Young, 1991a, 1991b; Fiorito and Scotto, 1992).

Concluding remarks

The cephalopod nervous system is certainly the most complex of any of the invertebrate nervous systems and the morphology and function of all its parts are well adapted for the animals' survival in their particular habitats. Specifically its central parts (brain and optic lobes) show distinct grades of expression and complexity between the species that parallel, or allow prediction of, the grade of complexity of the behaviors they control (Maddock and Young, 1987). For example, fin lobes, which are present in all decapods, are present in octopods only in the species of the suborder Cirrata which, different from all other octopods, developed fins; also, the relative volume of the fin lobes correlates quite well with the size of the fins (Maddock and Young, 1987). The deep-sea octopod *Cirrothauma* has very small and simple optic lobes; accordingly, its eyes are small and simple and have no lens (Aldred et al., 1983; Maddock and Young, 1987). In cephalopods that have a large number of chromatophores and repertoire of chromatic components, the chromatophore lobes are either relatively large or house more chromatophore motoneurons per unit of volume.

On the other hand, the extreme complexity of the cephalopod brain and the differentiation of most of its ganglia into several subunits (lobes) makes it often very difficult to identify, or just to discuss, homologies of individual parts of the cephalopod and the mollusc brains; further detailed studies of their origin and development are definitely needed. The complexity of the cephalopod brain, combined with the sophisticated sensory outfit, can certainly also explain why a comparison of the areas of projection and origin of the various brain nerves, at best, are of limited use. For example, in scaphopods, bivalves and gastropods, the afferent fibres of the statocyst nerves project to the cerebral ganglia, whereas in *Nautilus* they project to the magnocellular lobe and in decapods and octopods they project to the anterior and median basal and to the peduncle lobes (which develop from cerebral primordial ganglia), as well as to the magnocellular, pedal, lateral pedal, and palliovisceral lobes. In addition, in decapods and octopods statocyst efferents originate from the pedal, lateral pedal, and magnocellular lobes (Young, 1965a; Budelmann, 1976; Budelmann and Young, 1984; Marquis, 1989; Neumeister and Budelmann, unpublished). With respect to homologies, a similar difficulty arises when one tries to explain why the extraocular eye muscles are innnervated by sub-esophageal motoneurons that are located in the anterior lateral pedal lobe (Budelmann and Young, 1984, 1993).

Despite these difficulties, a study of the cephalopod brain is not just intrinsically interesting. Because of its many analogies with the vertebrate brain, comparative data (similariaties and differences) on the cephalopod and the vertebrate brain in general can contribute to a better understanding of the basics principles of brain morphology, physiology and perhaps pathology (Young, 1967b; Bullock 1986, 1993). Very recently, new powerful techniques became available for the study of the morphology and physiology of the cephalopod brain: tracing of labelled pathways with three-dimensional magnetic resonance microscopy, brain slice recordings, recordings from intact animals with implanted electrodes, and mapping of metabolic brain activity with [^{14}C]deoxyglucose (Bullock and Budelmann, 1991; Williamson and Budelmann, 1991; Novicki et al., 1992; Quast et al., 1992 and in preparation). In addition, some cephalopods can now be maintained, reared, and cultured under laboratory conditions (e.g., Hanlon, 1990) and, therefore, their nervous system is available for studies from the early embryo to the senescent adult. These techniques, together with a tremendous body of morphological and behavioral data on cephalopod brains, finally allow fascinating research on almost every aspect of their development, morphology, and function.

Acknowledgements

The author likes to thank Professor J.Z. Young for many stimulating discussions on various aspects of this paper. The author's participation at the conference was supported by the Marine Medicine Budget of The Marine Biomedical Institue of The University of Texas Medical Branch at Galveston.

References

Abbott, N.J., Lane, N. and Bundgaard M. (1986) The blood-brain interface in invertebrates. *Ann. N.Y. Acad. Sci.* 481: 20–41.

Aldred, R.G., Nixon, M. and Young, J.Z. (1983) *Cirrothauma murrayi* Chun, a finned octopod. *Phil. Trans. R. Soc. Lond. B* 301: 1–54.

Alexandrowicz, J.S. (1960) Innervation of the hearts of *Sepia officincalis. Acta Zool.* 41: 65–100.

Aronson, R.B. (1991) Ecology, paleobiology and evolutionary constraint in the octopus. *Bull. Marine Sci.* 49: 245–255.

Auerbach, B. and Budelmann, B.U. (1986) Evidence for acetylcholine as neurotransmitter in the statocyst of *Octopus vulgaris. Cell Tissue Res.* 243: 429–436.

Bleckmann, H. Budelmann, B.U. and Bullock, T.H. (1991) Peripheral and central nervous responses evoked by small water movements in a cephalopod. *J. Comp. Physiol.* A 168: 247–257.

Boycott, B.B. (1961) The functional organization of the brain of the cuttlefish *Sepia officinalis. Proc. R. Soc. Lond. B* 153: 503–534.

Boyle, P.R. (1986) Neural control of cephalopod behavior. *In:* A.O.D. Willows (ed.): *The Mollusca, Vol. 9, Neurobiology and Behavior, Part 2,* Academic Press, Orlando, pp. 1–99.

Budelmann, B.U. (1976) Equilibrium receptor systems in molluscs. *In:* P.J. MIll (ed.): *Structure and Function of Proprioceptors in the Invertebrates*, Chapman and Hall, London, pp. 529–566.

Budelmann, B.U. (1990) The statocysts of squid. *In:* D. Gilbert, H. Adelman and J. Arnold (eds): *Squid as Experimental Animals.* Plenum Press, New York, pp. 421–439.

Budelmann, B.U. (1994) Cephalopod sense organs, nerves and the brain: adaptations for high performance and life style. *Mar. Behav. Physiol.* 25: 13–33.

Budelmann, B.U. and Bleckmann, H. (1988) A lateral line analogue in cephalopods: water waves generate microphonic potentials in the epidermal head lines of *Sepia* and *Lolliguncula. J. Comp. Physiol.* A 164: 1–5.

Budelmann, B.U. and Bonn, U. (1982) Histochemical evidence for catecholamines as neurotransmitters in the statocyst of *Octopus vulgaris. Cell Tissue Res.* 227: 475–483.

Budelmann, B.U. and Thies, G. (1977) Secondary sensory cells in the gravity receptor system of the statocyst of *Octopus vulgaris. Cell Tissue Res.* 182: 93–98.

Budelmann, B.U. and Young, J.Z. (1984) The statocyst-oculomotor system of *Octopus vulgaris*: eye muscles, eye muscle nerves, statocyst nerves, and the oculomotor centre in the central nervous system. *Phil. Trans. R. Soc. Lond. B* 306: 159–189.

Budelmann, B.U. and Young, J.Z. (1985) Central pathways of the nerves of the arms and mantle of *Octopus. Phil. Trans. R. Soc. Lond. B* 310: 109–122.

Budelmann, B.U. and Young, J.Z. (1987) Brain pathways of the brachial nerves of *Sepia* and *Loligo. Phil. Trans. R. Soc. Lond. B* 315: 345–352.

Budelmann, B.U. and Young, J.Z. (1993) The oculomotor system of decapod cephalopods: eye muscles, eye muscle nerves, and the oculomotor neurons in the central nervous system. *Phil. Trans. R. Soc. Lond. B* 340: 93–125.

Budelmann, B.U. Sachse, M. and Staudigl, M. (1987) The angular acceleration receptor system of *Octopus vulgaris*: morphometry, ultrastructure, and neuronal and synaptic organization. *Phil. Trans. R. Soc. Lond. B* 315: 305–343.

Bullock, T.H. (1984). Ongoing compound field potentials from octopus brain are labile and vertebrate-like. *Electroencephalogr. Clin. Neurophysiol.* 57: 473–483.

Bullock, T.H. (1986) "Simple" model systems need comparative studies: differences are as important as commonalities. *Trends Neurosci.* 9: 470–472.

Bullock, T.H. (1993) How are more complex brains different? *Brain Behav. Evol.* 41: 88–96.

Bullock, T.H. and Basar E. (1988) Comparison of ongoing compound field potentials in the brains of invertebrates and vertebrates. *Brain Res. Rev.* 134: 57–75.

Bullock, T.H. and Budelmann, B.U. (1991) Sensory evoked potentials in unanesthetized unrestrained cuttlefish: a new preparation for brain physiology in cephalopods. *J. Comp. Physiol. A* 168: 141–150.

Bullock, T.H. and Horridge, G.A. (1965) *Structure and Function of the Nervous Systems of Invertebrates.* Freeman, San Francisco and London.

Bundgaard, M. and Abbott, N.J. (1992) Fine structure of the blood-brain interface in the cuttlefish *Sepia officinalis* (Mollusca, Cephalopoda). *J. Neurocytol.* 21: 260–275.

Colmers, W.F. (1981) Afferent synaptic connections between hair cells and the somata of intramacular neurons in the gravity receptor system of the statocyst of *Octopus vulgaris*. *J. Comp. Neurol.* 197: 385–394.
Cornwell, C.J., Messenger, J.B. and Williamson, R. (1993) Distribution of GABA-like immunoreactivity in the octopus brain. *Brain Res.* 621: 353–357.
Dilly, P.N., Nixon, M. and Young, J.Z. (1977) *Mastigoteuthis* – the whip-lash squid. *J. Zool. Lond.* 181: 527–559.
Dubas, F., Leonard, R.B. and Hanlon, R.T. (1986a) Chromatophore motoneurones in the brain of the squid, *Lolliguncula brevis*: an HRP study. *Brain Res.* 374: 21–29.
Dubas, F., Hanlon, R.T., Ferguson, G. and Pinsker, H. (1986b) Localization and stimulation of chromatophore motoneurones in the brain of the squid, *Lolliguncula brevis*. *J. Exp. Biol.* 121: 1–25.
Fiorito, G. and Scotto, P. (1992) Observational learning in *Octopus vulgaris*. *Science* 256: 545–547.
Florey, E. (1969) Ultrastructure and function of cephalopod chromatophores. *Am. Zool.* 9: 429–442.
Gilly, W.F. Hopkins, B. and Mackie, G.O. (1991) Development of giant motor axons and neural control of excape responses in squid embryos and hatchlings. *Biol. Bull.* 180: 209–220.
Gillette, R. (1991) The mollucan nervous system. *In:* C.L. Prosser (ed.): *Neural and Integrative Animal Physiology*, Wiley-Liss, New York, pp. 574–611.
Gleadall, I.G. (1990) Higher motor function in the brain of *Octopus*: the anterior basal lobe and its analogies with the vertebrate basal ganglia. *Ann. Appl. Inf. Sci.* 16: 1–30.
Gleadall, I.G. Ohtsu, K., Gleadall, E. and Tsukahara, Y. (1993) Screening-pigment migration in the *Octopus* retina includes control by dopaminergic efferents. *J. Exp. Biol.* 185: 1–16.
Griffin, L.E. (1900) The anatomy of *Nautilus pompilius*. *Mem. Acad. Sci. Wash.* 8: 103–230.
Hanlon, R.T. (1990) Maintenance, rearing, and culture of teuthoid and sepioid squids. *In:* D.L. Gilbert, W.J. Adelman and J.M. Arnold (eds): *Squid as Experimental Animals*, Plenum Press, New York, pp. 35–62.
Hanlon, R.T. and Messenger, J.B. (1988) Adaptive coloration in young cuttlefish (*Sepia officinalis* L.): The morphology and development of body patterns and their relation to behaviour. *Phil. Trans. R. Soc. Lond. B* 320: 437–487.
Kier, W.M. (1988) The arrangement and function of molluscan muscle. *In:* E.R. Trueman and M.R. Clarke (eds): *The Mollusca, Vol. 11, Form and Function*, Academic Press, San Diego, pp. 211–252.
Kier, W.M. and A.M. Smith (1990) The morphology and mechanics of *Octopus* sucker. *Biol. Bull.* 178: 126–136.
Kime, D.E. and Messenger, J.B. (1990) Monoamines in the cephalopod CNS: an HPLC analysis. *Comp. Biochem. Physiol.* 96C: 49–57.
Maddock, L. and Young, J.Z. (1987) Quantitative differences among the brains of cephalopods. *J. Zool. Lond.* 212: 739–767.
Mangold, K. (1989) Le système nerveux. *In:* P.P. Grassé (ed.): *Traité de Zoologie, Vol.* 5(4), *Céphalopodes*, Masson, Paris, pp. 163–240.
Marquis, F. (1989) Die Embryonalentwicklung des Nervensystems von *Octopus vulgaris* Lam. (Cephalopoda, Octopoda), eine histologische Analyse. *Verhandl. Naturf. Ges. Basel* 99: 23–76.
Martin, R. (1977) The giant nerve fibre system of cephalopods. Recent structural findings. *Symp. Zool. Soc. Lond.* 38: 261–275.
Martin, R. and Miledi, R. (1986) The form and dimensions of the giant synapse of squids. *Phil. Trans. R. Soc. Lond. B* 312: 355–377.
Martin, R. and Voigt, K.H. (1987) The neurosecretory system of the octopus vena cava: A neurohemal organ. *Experientia* 43: 537–543.
Messenger, J.B. (1967) The peduncle lobe: a visuo-motor centre in *Octopus*. *Proc. R. Soc. Lond. B* 167: 225–251.
Messenger, J.B (1979) The nervous system of *Loligo*. IV. The peduncle and olfactory lobes. *Phil. Trans. R. Soc. Lond. B* 285: 275–309.
Messenger, J.B. (1991) Photoreception and vision in molluscs. *In:* J.R. Cronly-Dillon and R.L. Gregory (eds): *Evolution of the Eye and the Visual System*, McMillan, London, pp. 364–397.

Moynihan, M. and Rodaniche, A.F. (1982) *The Behavior and Natural History of the Caribbean Reef Squid Sepioteuthis sepioidea*. Advances in Ethology 25. Paul Parey, Berlin, Hamburg.
Naef, A. (1928) *Die Cephalopoden (Embryologie). Fauna e Flora del Golfo di Napoli.* Monografia 35(I, 2): 1–363. Bardi, Roma and Friedländer, Berlin.
Novicki, A., Messenger, J.B., Budelmann, B.U., Terrell, M.L. and Kadekaro M. (1992) [^{14}C]deoxyglucose labelling of functional activity in the cephalopod central nervous system. *Proc. R. Soc. Lond. B* 249: 77–82.
Otis, T.S. and Gilly, W.F. (1990) Jet-propelled escape in the squid, *Loligo opalescens*: Concerted control by giant and non-giant motor axon pathways. *Proc. Natl. Acad. Sci. USA* 87: 2911–2915.
Packard, A. (1972) Cephalopods and fish: the limits of convergence. *Biol. Rev.* 47: 241–307.
Pfefferkorn, A. (1915) Das Nervensystem der Octopoden. *Z. Wiss. Zool.* 114: 425–531.
Plän, T. (1987) *Functional Neuroanatomy of Sensory-Motor Lobes of the Brain of Octopus vulgaris*. Dissertation, University of Regensburg, Regensburg, Germany.
Preuss, T. and Budelmann, B.U. (1991) A new sense organ in cephalopods: sensory hair cells on the neck of the squid *Lollinguncula brevis*. *Soc. Neurosci. Abstr.* 17: 1403.
Preuss, T. and Budelmann, B.U. (1995) Proprioceptive hair cells on the neck of the squid *Lolliguncula brevis*: a sense organ in cephalopods for the control of head-to-trunk position. *Phil. Trans. R. Soc. Lond. B* (submitted).
Rowell, C.H.F. (1963) Excitatory and inhibitory pathways in the arm of *Octopus*. *J. Exp. Biol.* 40: 257–270.
Rowell, C.H.F. (1966) Activity of interneurons in the arm of *Octopus vulgaris* in response to tactile stimulation. *J. Exp. Biol.* 44: 589–605.
Quast, M.J., Neumeister, H. and Budelmann, B.U. (1992) Tracking cobalt-labelled nerve pathways through an invertebrate brain (*Sepia officinalis*, Cephalopoda) by three-dimensional MR microscopy. *Soc. Magn. Reson. Med. Abstr.* Berlin.
Schäfer, W. (1954) Form und Funktion der Brachyurenschere. *Abhandl. Senckenberg. Naturforsch. Ges.* 489: 1–66.
Smith, P.J.S. and Boyle, P.R. (1983) The cardiac innervation of *Eledone cirrhosa* (Lamarck). (Mollusca: Cephalopoda). *Phil. Trans. R. Soc. Lond. B* 300: 493–511.
Tansey, E.M. (1979) Neurotransmitters in the cephalopod brain. *Comp. Biochem. Physiol.* 64C: 173–182.
Tansey, E.M. (1980) Aminergic fluorescence in the cephalopod brain. *Phil. Trans. R. Soc. Lond. B* 291: 127–145.
Tasaki, K., Tsukahara, Y., Suzuki, H. and Nakaye, T. (1982) Two types of inhibition in the cephalopod retina. *In:* A. Kaneko, N. Tsukahara and K. Uchizono (eds): *Neurotransmitters in the Retina and the Visual Centers*, Biomedical Research Suppl., Tokyo, pp. 41–44.
Thore, S. (1939) Beiträge zur Kenntnis der vergleichenden Anatomie des zentralen Nervensystems der dibranchiaten Cephalopoden. *Pubbl. Staz. Zool. Napoli* 17: 313–506.
Tompsett, D.H. (1939) Sepia. L.M.B.C. *Mem. typ. Br. mar. Pl. Anim.* 32: 1–184.
Tu, Y. and Budelmann, B.U. (1994) The effect of L-glutamate on the afferent resting activity in the cephalopod statocyst. *Brain Res.* 642: 47–58.
Wells, M.J. (1978) *Octopus. Physiology and Behaviour of an Advanced Invertebrate.* Chapman and Hall, London.
Williamson, R. (1989) Electrophysiological evidence for cholinergic and catecholaminergic efferent transmitters in the statocyst of *Octopus*. *Comp. Biochem. Physiol.* 93C: 23–27.
Williamson, R. and Budelmann, B.U. (1991) Convergent inputs to octopus oculomotor neurones deomonstrated in a brain slice preparation. *Neurosci. Lett.* 121: 215–218.
Young, J.Z. (1939) Fused neurons and synaptic contacts in the giant nerve fibres of cephalopods. *Phil. Trans. R. Soc. Lond. B* 229: 465–505.
Young, J.Z. (1962) The optic lobes of *Octopus vulgaris*. *Phil. Trans. R. Soc. Lond. B* 245: 19–58.
Young, J.Z. (1963) The number and sizes of nerve cells in *Octopus*. *Proc. Zool. Soc. Lond.* 140: 229–254.
Young, J.Z. (1965a) The central nervous system of *Nautilus*. *Phil. Trans. R. Soc. Lond. B* 249: 1–25.
Young, J.Z. (1965b) The buccal nervous system of *Octopus*. *Phil. Trans. R. Soc. Lond. B* 249: 27–43.

Young, J.Z. (1967a) The visceral nerves of *Octopus*. *Phil. Trans. R. Soc. Lond. B* 253: 1–22.

Young, J.Z. (1967b) Some comparisons between the nervous systems of cephalopods and mammals. *In:* C.A.G. Wiersma (ed.): *Invertebrate Nervous Systems. Their Significance for Mammalian Neurophysiology*, University of Chicago Press, Chicago and London, pp. 353–362.

Young, J.Z. (1971) *The Anatomy of the Nervous System of Octopus vulgaris*. Clarendon Press, Oxford.

Young, J.Z. (1972) The organization of a cephalopod ganglion. *Phil Trans. R. Soc. Lond. B* 263: 409–429.

Young, J.Z. (1974) The central nervous system of *Loligo*. I. The optic lobe. *Phil. Trans. R. Soc. Lond. B* 267: 263–302.

Young, J.Z. (1976a) The "cerebellum " and the control of eye movements in cephalopods. *Nature* 264: 572–574.

Young, J.Z. (1976b) The nervous system of *Loligo*. II. Suboesophageal centres. *Phil. Trans. R. Soc. Lond. B* 274: 101–167.

Young, J.Z. (1977a) Brain, behaviour and evolution of cephalopods. *Symp. Zool. Soc. Lond.* 38: 377–434.

Young, J.Z. (1977b) The nervous system of *Loligo*. III. Higher motor centres: The basal supraoesophageal lobes. *Phil. Trans. R. Soc. Lond. B* 276: 351–398.

Young, J.Z. (1979) The nervous system of *Loligo*. V. The vertical lobe complex. *Phil. Trans. R. Soc. Lond. B* 285: 311–354.

Young, J.Z. (1988) Evolution of the cephalopod brain. *In:* M.R. Clarke and E.R. Trueman (eds): *The Mollusca, Vol. 12, Paleontology and Neontology*, Academic Press, San Diego, pp. 215–228.

Young, J.Z. (1991a) Computation in the learning system of cephalopods. *Biol. Bull.* 180: 200–208.

Young, J.Z. (1991b) Light has many meanings for cephalopods. *Visual Neurosci.* 7: 1–12.

The Nervous Systems of Invertebrates: An Evolutionary and Comparative Approach
ed. by O. Breidbach & W. Kutsch
© 1995 Birkhäuser Verlag Basel/Switzerland

Common and special features of the nervous system of Onychophora: A comparison with Arthropoda, Annelida and some other invertebrates

F.W. Schürmann

I. Zoologisches Institut, Universität Göttingen, Berliner Straße 28, D-37073 Göttingen, Germany

Summary. An account of the nervous system of the Onychophora, dealing with neuron types, synaptic connectivity, the structural organization of the brain and ventral cord, physiology and behavior is given. Features of the nervous systems of the Onychophora, other arthropods, and worms of different phyla are compared. The onychophoran brain contains the types of condensed neuropils (glomeruli, columns and strata) stated for other arthropods and some polychaetes. The preponderance of globuli cells in the brain and ventral nerve cord, the synaptic neuropils and tracts, mainly made up by tiny nerve fibers, and the nerve-muscle synapse belong to the characters of the onychophoran nervous system.

Introduction

The Onychophora have attracted the attention of zoologists since Sedgwick (1888, 1909) placed this old group with Gondwanaland distrubtion as a living fossil and missing link between annelids and arthropods, a view repeadtedly found in even more recent textbooks. Classification of the Onychophora has remained a matter of debate and subsequent revision will probably continue. The proposed phylogenetic status ranges widely, from incorporating the group among Articulata (Beklemishev, 1969), into the Uniramia (Manton, 1977), or as a basic group of arthropods (Ax, 1984). The recent revival of interest in phylogeny comes from some new studies employing molecular biological techniques, powerful modern tools that can add to the classical approaches comparing structural, physiological, and behavioral characters. Ballard et al. (1992) for example propose from cladistic analysis of 12S ribosomal DNA that the Onychophora should be included into the monophyletic group of the arthropods, relating this group to the myriapods and chelicerata. Analysis and scope of data of this study have been critically inspected (Fortey and Thomas, 1993) and some conclusions questioned. Early cambrian marine fossils of the "armored lobopod" type have been placed near to the extant Onychophora (Ramsköld and Hou Xianguang, 1991).

Phyletic placement depends on morphological, physiological, and biochemical data together. This holds true especially for the nervous system, representing the most sophisticated layout of connected cells and with many parts still poorly understood in biochemical, structural, physiological and developmental terms (Arbas et al., 1991). The nervous system, underlying behavioral adaptations and responding continuously to the constraints of environment, is a major key for evolutionary alterations.

The nervous system of the Onychophora has only been fragmentarily analyzed (for review on morphology see Storch and Ruhberg, 1993).

This chapter deals with the nervous system of Onychophora and lists what is common (i.e., found in some other groups) and what can be proposed as special, (i.e., setting the nervous system apart from others) (for reviews on the nervous system see Horridge, 1965; Schürmann, 1987; on sensory and neurosecretory system, see Paulus, 1979; Tombes, 1979). Different features, from nerve specific molecules, synaptic connectivity to central nervous system layout and behavior are considered. Similarities to other groups, especially platyhelminthes, annelids, and arthropods are discussed (compare Bullock and Horridge, 1965).

Specific molecules of the nervous system

Neuroactive compounds and their receptors are very specific molecules par excellence. Only some investigations on such compounds based on biochemical, pharmacological, and histochemical methods are available (Schürmann, 1987). It is obvious and not surprising that the Onychophora contain the same classical transmitters such as acetylcholine and dopamine, that are so widely spread through the animal kingdom. Immunocytological studies on the occurrence and distribution of transmitters in Onychophora are lacking. Pharmacological studies of Florey and Florey (1965) show insensitivity of dorsal longitudinal muscles to GABA and glutamate, but responses to acetylcholine and potentiation to eserine application. Thus, the nerve-muscle system of onychophorans has as a remarkable feature in common with the annelids, cholinergic connectivity (Golding, 1992), thus differing from insects and crustaceans (Leake and Walker, 1980). In an overview on the evolution of neurotransmitter receptors, onychophorans are incorporated with respect to cholinergic receptors and placed beside the annelids (Venter et al., 1988), but a molecular analysis that would elucidate the refinement of molecules due to evolutionary requirements is still missing.

Cell types and cell connectivity

Neuronal somata

The nervous system of Onychophora contains as basic cell types neurons, glial cells, and tracheocytes (Lane and Campiglia, 1987;

Schürmann, 1987). Three types of neurons are discriminated on the basis of nuclei diameters: giant cells, medium-sized neurons, and small cells of the globuli type (nucleus diameter 5–9 μm; Feodorov, 1926) (Fig. 1). As reported by all authors, globuli cells are so abundant that other types appear as a vey small minority. In Onychophora, the globuli cell type makes up the majority of the neuropil of the brain and ventral nerve cord. The globuli cells with nuclei rich in condensed chromatin are surrounded by a small cytoplasmic rind. The tiny globuli cells of the Onychophora mainly give rise to small axons (diameter < 1 μm) like in other phyla. Globuli cells are encountered in worms (Platyhelminthes: Hadenfeldt, 1929; Rieger et al., 1991: Nemertinea: Hanström, 1928; Annelida: Golding, 1992) and arthropods and molluscs (Bullock and Horridge, 1965; compare Budelmann, Sandeman and Scholtz, Strausfeld et al.; this volume), confined mainly to the brains in these groups. The acquisition of globuli cell concentrations in Platyhelminthes (Notoplana: Hadenfeldt, 1929) and polychaete annelids (Horridge, 1965) is commonly attributed to forms with highly developed brains and behavior (Hanström, 1928). Interestingly, the nemertineans, with a well developed brain with large concentrations of globuli cells, have recently been located within the protostome coelomata (see Audouette and Philippe, 1993)

In insects, massive globuli cell clusters are found in the optic neuropil of the brain (Strausfeld, 1976) and in the mushroom bodies (here named Kenyon cells) of the protocerebrum (Schürmann, 1987). A common feature of globuli cells in insect and crustacean brains appears to be their association with precise geometrical layers or columns of tiny fibers or orderly geometric neuropil arrays with well defined borders.

Clustering of neuron types in Onychophora is not well developed or not reflected in histological preparations, though neuropil compartments are clearly demarcated in the brain.

In this respect, the Onychophora differ very much from annelid and insect brains. The onychophoran nervous system does not rely on a high cytological specialization of neurons, laid down in clusters. In the Onychophora, the prominent cell type of the brain and ventral cord is the globuli cells.

The reason for this striking speciality is not known, but it is tempting to speculate on functional and developmental implications.

If we consider globuli cell masses as an advanced state of evolution, then this peculiarity of the onychophoran nervous system would account for its specialized, non primitive state. Obviously, the occurrence of globuli cell clusters forming special neuropilar compartments is widely correlated with the development of head sense organs (visual and olfactory information processing). Such de novo neuropilar acquisitions of brains assigned to animals forms with an advanced evolutionary state may be simply added or superimposed to the ancient layout of neuropil, instead of increasing and refining it. Interestingly, in some insect groups, the mushroom bodies gain their form by globuli cell division and growth as imagos or may display morphological plasticity (see Schürmann, 1987).

The basic state of globuli cells, as judged from caryological and cytological features, is proposed to be a simple, non-specialized one. Clark (1958) regards the primitive corpora pedunculata globuli neurons

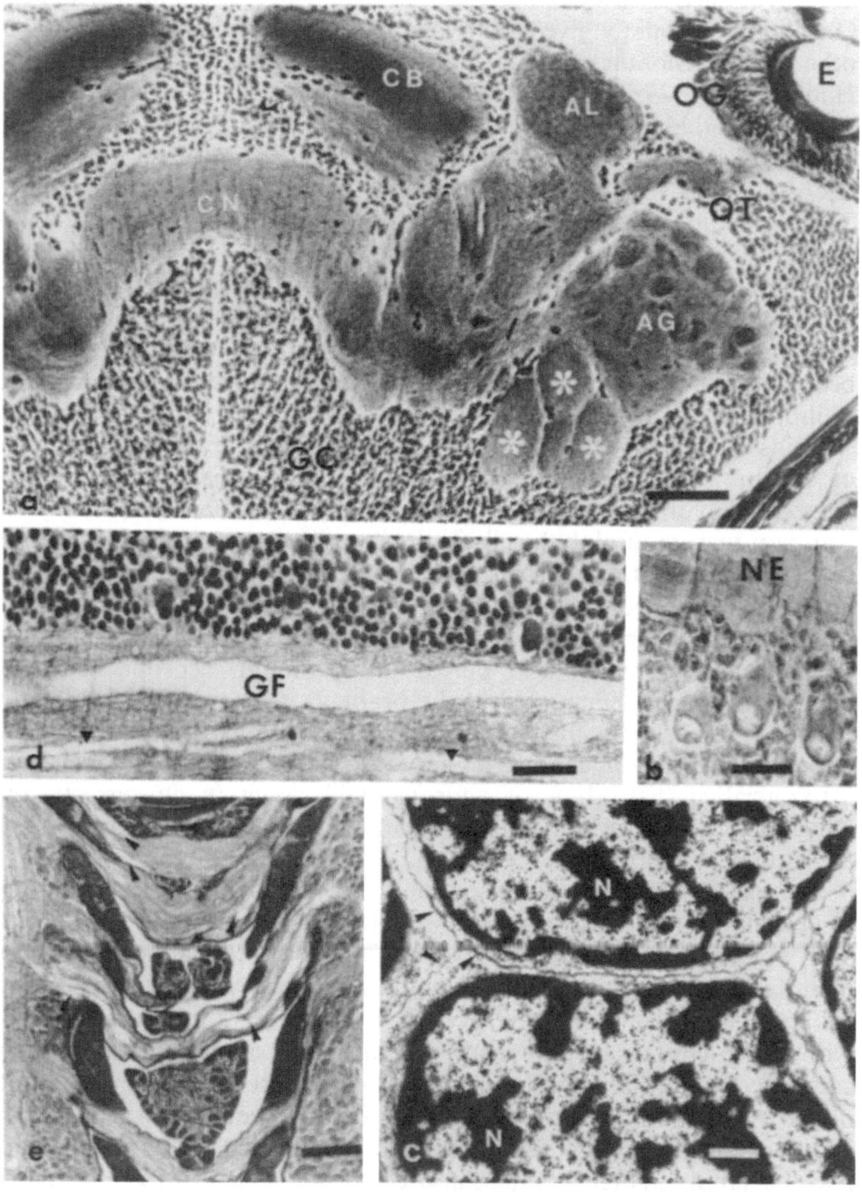

in the polychaete Nephtys as deriving from a plasma-rich nerve cell variety.

The large neurons (among them giant neurons) are found in the brain (Gaffron, 1884; Schürmann, 1987) among globuli cells and appear (partly segmentally repeated) in the ventral cord (Feodorov, 1926) (Fig. 1). They may contain neurofibrils restricted to the soma (Lane and Campiglia, 1987), also observed in annelid giant neurons (Schürmann and Günther, 1973), but there they are found in nerve fibers as well. Ciliary rootlets, found in pericarya of the optic ganglion, have been discussed as a annelid feature (Eakin and Westfall, 1965).

Lack of onychophoran studies showing individually stained neurons forbids a more detailed classification and comparison with other groups.

Synapses

Cell connectivity is regarded here with respect to synapses and non-synaptic membrane specializations. Lack of tight and gap junctions has been stated (Lane, Campiglia, 1987). The absence of a structural basis for extracellularly sealing glial and neuronal compartments makes the nervous system directly accessible to compounds, so that no "blood brain barrier" is established. This is seldom encountered in arthropods, e.g., in crustaceans (Lane and Abbot, 1975).

Glial wrapping of globuli cell somata is widely incomplete, as seen in annelids (Golding, 1992), and, to a smaller extent, in some arthropods (Fahrenbach, 1976; Schürmann, 1987).

Synapses in onychophorans (Figs 2, 3) show the usual characteristics attributed to polarized chemical synapses: pre- and postsynaptic dense membrane appositions, widened synaptic cleft, presynaptic vesicle clusters (Schürmann, 1978a) found in other groups as well (Cobb and Pentreath, 1978; Reuter, 1981). Different forms of clear and dense core vesicles are noticed. Dense core vesicles are apparently accumulated in some neuropilar compartments (e.g., central body) (Fig. 3).

Figure 1. Perikaryal layer and neuropil in brain (a–c) and ventral cord (d, e) of *Peripatoides leuckarti.* (a) Transverse section, showing large globuli cell rind (GC) and neuropil with different areas; antennal glomeruli (AG), antennal lobe (AL), central body (CB), central neuropil (CN), Eye (E), optic ganglion (OG), optic tract (OT), mushroom body stalk parts (asterisks): scale 100 μm. (b) Giant neurons surrounded by globuli cells, neuropil (NE); ventro-caudal brain, dorsal neuropil (NE), scale 100 μm. (c) Globuli cells of the mushroom bodies (N nuclei) with small cytoplasmic rind and with directly adjacent membranes (arrow heads), not separated by glial cells; electron micrograph, scale 0.5 μm. (d) Pericaryal layer with abundant globuli cells and some other larger neurons, giant fiber (GF) and other single axons (triangles) in a tract region; longitudinal section, scale 100 μm. (e) Suboesophageal anterior commissures between the connectives, equipped with large axons (arrowheads) besides tiny fibers: transverse section, scale 100 μm.

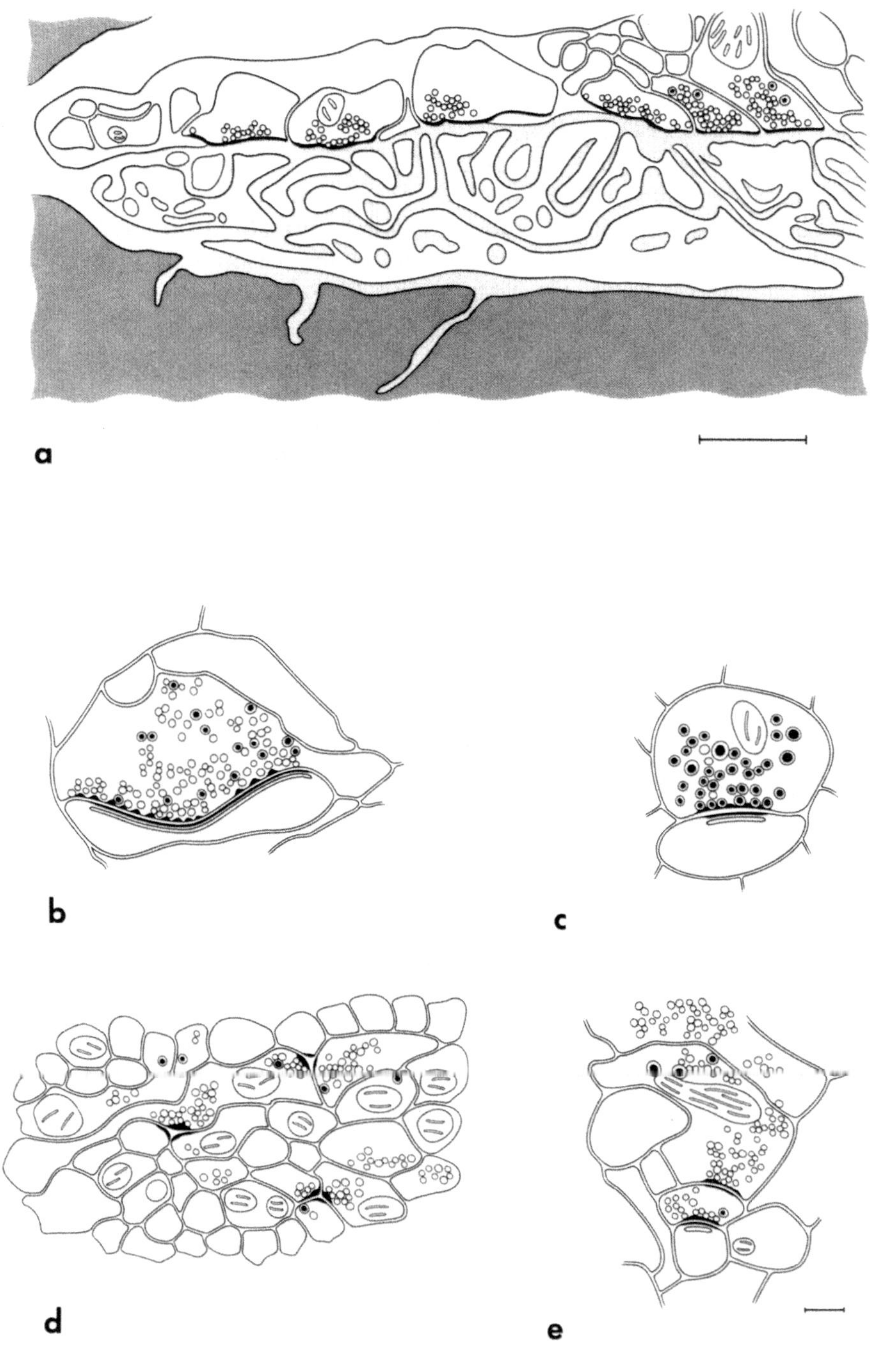
a
b
c
d
e

We have set the onychophoran synapse apart from annelid and insect synapses, mainly according to the extent of membrane appositions (Schürmann, 1987). A greater variety of types is seen for the onychophorans compared to annelids (see Günther and Schürmann, 1973; Golding, 1992). Presynaptic appositions resemble the regular synaptic grids gating the vesicle access and well known from insects and vertebrates (Schürmann, 1980). The onychophoran synapse typically shows a large subsynaptic system of cisternae, resembling the one found at the corresponding sites of fly synapses (Schürmann, 1978a). Comparing these postsynaptic structures of photoreceptor synapses in flies of different phyletic state, Shaw and Meinertzhagen (1986) detected the most elaborate cisternal systems in more recent flies. If diversification of synapses into many types indicates a developed status of a nervous system, we would assign an advanced state to the Onychophora in this respect.

Ultrastructural diversity of synaptic types contributes to the expression and variance of synaptic circuitry patterns (i.e., arrangement of connected fibers). We have noted dyad synapses (not so common as in arthropods), en passant-synapses, convergent and divergent mode of connectivity, serial synapses and complex synaptic configurations as detected for other invertebrate groups (Golding, 1992, Reuter, 1981; Schürmann, 1987).

Complex circuitry is seen in the glomeruli neuropiles, a term first used by Bellonci (1883), comparing olfactory lobes of arthropods and vertebrates. Looking into mushroom body and antennal glomeruli neuropil, we observe some striking similarities in the design of glomeruli.

The convergent mode of wiring of the antennal input in the antennal glomeruli of the Onychophora and in insects is not easily seen in electron micrographs, but is apparently hidden in the mass of tiny fibers of different origin, that represent high integration in a peripheral neuropil (Boeckh and Ernst, 1987).

Figure 2. Synaptic connexions of muscles (a) and of the central nervous system (b–e). (a) Nerve-muscle synapses: six nerve fibers synapses on an end-plate structure, complex subsynaptic foldings (light stippled areas) connect with muscle (dark stippled area); scale 0.2 µm. Scheme, drawn after electron micrograph, from Hoyle and Williams, 1980. (b–e) Synaptic forms in the central nervous system. (b) Synapses with presynaptic bars regularly lined up at the presynaptic membrane, accumulation of clear synaptic vesicles and some dense core vesicles. Note synaptic cleft and large subsynaptic cistern. (c) Synapses with presynaptic dense core vesicles ar the presynaptic membrane, electron dense pre- and subsynaptic appostions and subsynaptic cistern. (d) Synaptic connexions in the mushroom body column; parallel fibers transversely sectioned. Divergent dyad type synapse and two fibers with convergence on a postsynaptic element depicted (e) Serial synaptic connexion of three fibers. Note the subsynaptic cistern at only one synapse. Scale 0.2 µm. All schemes after electron micrographs from Peripatoides.

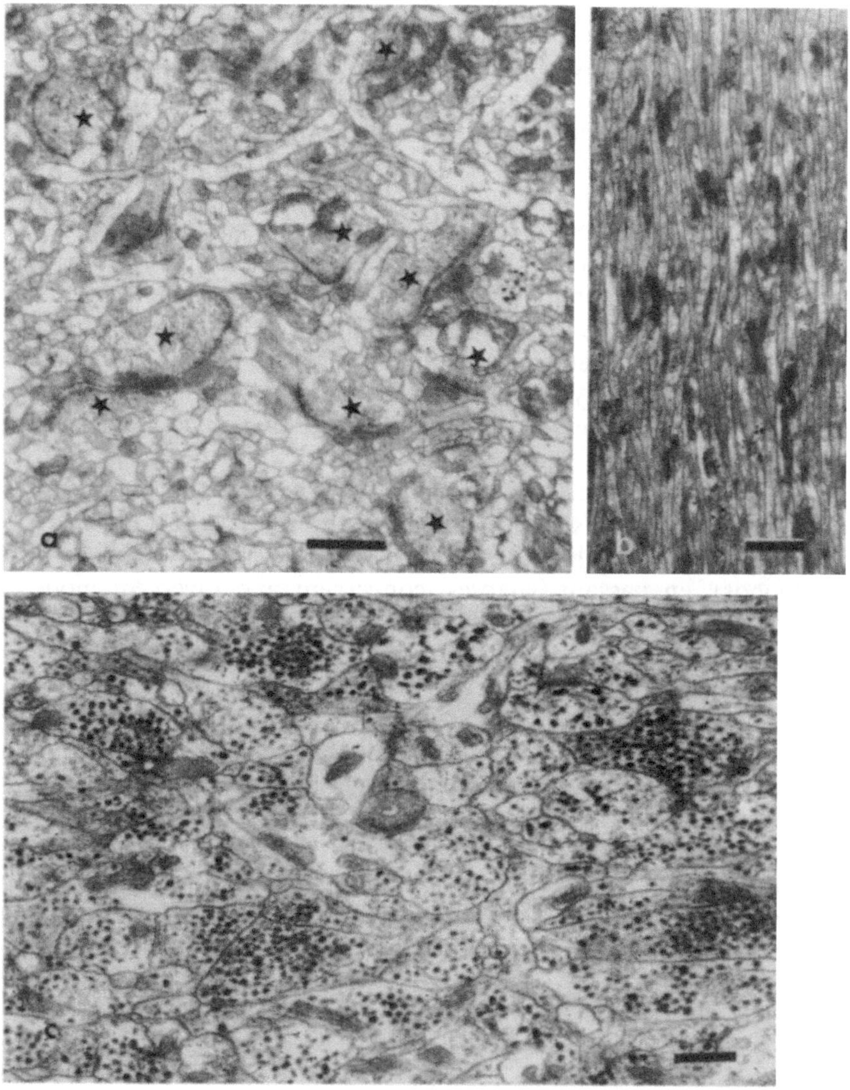

Figure 3. Different forms of glomeruli neuropil in the brain, *Peripatoides leuckarti*, electron micrographs. (a) Upper mushroom body stalk with presynaptic boutons (asterisks), making large synaptic contacts with tiny postsynaptic fibers; scale 1 μm. (b) Small axons arranged in parallel in the stalk column: scale 1 μm. (c) central body neuropil, fibers packed with dense core vesicles. Note paucity of synaptic contacts: scale 0.5 μm.

The similarity of divergent glomeruli, forming patterns in mushroom body subcompartments, and of convergent synapses in distal mushroom body parts in onychophorans and insects is striking (Akert and Steiger, 1967; Schürmann, 1987). However, connections of the divergent boutons with fine dendritic-like fibers in the onychophoran mushroom bodies are made via broad synapses of a non-dyadic type (Fig. 3). The diverse types of glomerular neuropil representing complex condensed circuitry of fine fibers, appear to be an old development (Leise, 1991).

Synapses are found in the neuropil, the somata cortex being devoid of such contacts. As a peculiarity, synaptic connections are found in the compact antennal nerve (Schürmann, 1987) and in heart nerve (Rosenberg and Seifert, 1978). Peripheral synapse location is occasionally seen in other invertebrates (Foelix, 1975; Govind, 1992). Recently, palp nerves of a minute polychaete have been show to contain a synaptic neuropil-like arrangement (Purschke, 1993). These specialized forms live interstitially in littoral sediment and display similarity to onychophoran head and body form.

Nerve-muscle synapses

Nerve-muscle junctions occur in two forms: nerve contacting muscle and, more commonly distributed, muscle arms connected to nerves (Hoyle and Williams, 1980). Synaptic contacts resemble annelid-type junctions (Schürmann, 1978b). The postsynaptic extracellular system penetrates muscles deeply, resembling in this respect vertebrate motor endplates. The multiple onychophoran neuro-muscular synapse (Fig. 2) is considered to be unique (Hoyle and Williams, 1980).

Neuropil organization

The generation of distinct neuropilar compartments, often separated from surrounding fibers masses, is generally assumed to indicate a specialized brain or nerve cord with an advanced state of organization (Hanström, 1928; Leise, 1991) and size difference of such compartments (i.e., areas of condensation of synaptic connectivity) have been intensively investigated for different insect orders (see Rensch, 1959; Schürmann, 1987). Recently, three types of condensed neuropil (modules) have been compared for higher invertebrate and vertebrate representatives (Leise, 1991): spherical neuropils (glomeruli), columns and layers. This general terminology does not seem appropriate for compartments traditionally classified as glomerular neuropil (e.g. mushroom bodies, central body, optic lobes of insects). In insects, the mushroom bodies comprise neuropil textures with glomerular parts (calyces) and columns, and the central body consists of layers with 16 column-like subcompartments.

Brain neuropil areas

In the onychophoran brain (Fig. 4), we find a number of well defined compartments of condensed fiber texture, clearly separated by glial sheaths from surrounding different fibers masses comprising the three types of modules emphasized by Leise (1991), and seen in antennal glomeruli, mushroom body columns, bridge and central body strata. Compartments listed by Holmgren (1916) do not show the clear-cut separation noted for the named ones, but we have to accept them as subdivisions of the brain from electron microscopy and from fiber orientation. We identified at least nine compartments of condensed neuropil in the onychophoran brain (Schürmann, 1987), compared to about seven in the insect brain (optic lobe laminae, central body, bridge, optic tubercles, mushroom bodies, antennal and tritocerebral

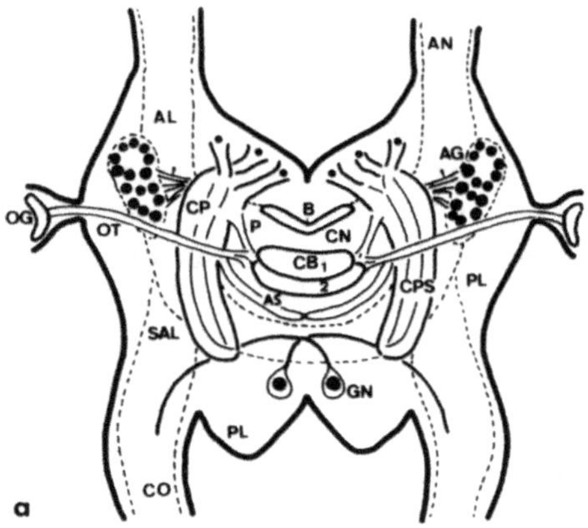

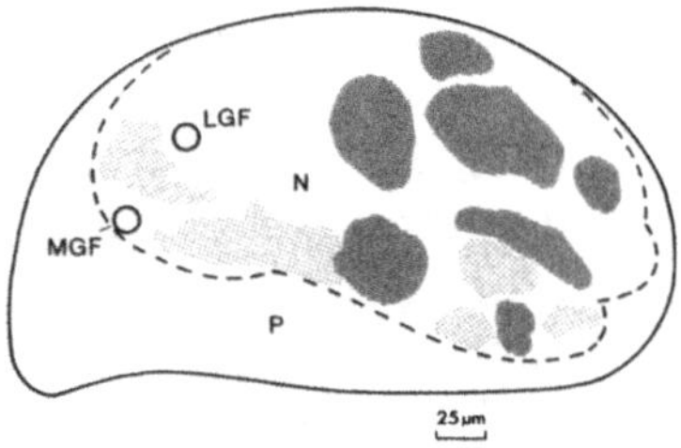

Figure 4. Neuropil compartments in the brain (a) and ventral cord (b): schemes based on histological sections of *Peripatoides leuckarti.* (a) antennal nerve (AN), antennal glomeruli (AG), antennal lobe (AL) and subantennal lobe (SAL), bridge (B), central body (CB) with Lamina anterior (1) and posterior (2) connected by the pedunculus to the mushroom bodies (CP), equipped with stalks (CPS)and accessory stalks (AS); connexions with AG (arrowheads), optic ganglia (OG) are linked to the CB by the optic tracts (OT), central neuropil (CN), perikaryal layer (PL) with giant neurons (GN), connective neuropil (CO) to the ventral cord: dorsal view. (b) Diagram of a transverse section through one lateral nerve cord. The border between the pericaryal layer (P) and neuropil (N) marked by an interrupted line. Dotted areas represent longitudinal bundles of large fibers, medial and lateral giant fibers (MGF, LGF), lined areas mark regions with condensed synaptic neuropil.

lobe glomeruli). In annelids, such compartments have also been found, at least for the advanced polychaetes (Hanström, 1928; Dhainiaut-Courtois and Golding, 1988). Existence and the number and form of such compartments can be taken as a sign for advanced, specialized brains. The brains of Onychophora and diverse arthropods contrast sharply in this respect with those of acoelomate worms and basic polychaetes.

Complex neuropils

Looking for possible homologous structures, we cannot avoid a comparison of antennal glomeruli and the mushroom body system (see Holmgren, 1916; Hanström, 1935; Horridge, 1965).

Antennal glomeruli in Onychophora (about 30 spherical or ovoid bodies of different size in either side hemisphere) are gathered and surround a central mass of fibers (elements of connecting interneurous and sensory fibers) (Fig. 1), forming a compartment similar to the antennal lobe of the insect and the olfactory lobe of the crustacean brain (Boeckh and Ernst, 1987; Sandeman, 1982). In the onychophorans, this neuropil area is situated lateral to proximal parts of the mushroom bodies, so that the connections between these two brain parts are short and do not form the distinct tractus olfactorio-globularis found in insects. In polychaetes, the connection of palpal glomerular neuropil to the mushroom bodies is partly made up by direct input of palpal sensory fibers (Hanström, 1928). In the Onychophora, the rostro-dorsal position and the linkage of antennal glomeruli to the mushroom bodies appears similar to the polychaete situation (see discussion by Feodorov, 1929), but awaits further clarification by single fiber stains.

In a comparative sense, mushroom bodies appear to be compartments considerably variable in size, internal design and position, and unstable in evolution (see Gupta, 1987), although the basic columnar organization, due to the order of gathered local interneurons, seems to be consistently preserved. The position of protocerebral mushroom bodies in the brain differs for insects and onychophorans (dorsally and ventrally, respectively), not properly comparable with the polychaete situation (Horridge, 1965; Orrhage, 1993). As a charactertistic, postional differences of a compartment may be unstable and therefore not a valid reason for excluding homology. The paired mushroom body system is mainly characterized by columnar arrangement of fine parallel fibers for all groups where they have been stated (annelids, insects, onychophora), but with high variance in cell number and columnar design in annelids and arthropods. Recently, a very special, multicolumnar mushroom body system has been described for the crustacean

Hutchinsoniella, a key group for phylogenetic discussion (Elofson and Hessler, 1990). Magnitude, the number of globuli cell fiber channels, and development of subcompartments have to be considered as indicators for the grade of specialization and advanced state of evolution (Hanström, 1928; Schürmann, 1987, Weiss 1981). As stated, the synaptic connectivity of mushroom bodies in insects and onychophorans shows remarkable similarities concerning the overall distribution of synapses in the intrinsic fibers and patterns of connectivtiy (signs for input and output integration). The mushroom bodies of both groups are condensed local circuitry systems (Schürmann, 1987). The mushroom bodies of annelids await a systematic electron microscopic description.

We speculate that the mushroom bodies might have developed from a tract filled with local synaptic circuitry. In the platyhelminthe Notoplana, the outer paired globuli cell masses form distinct column-like fiber tracts, resembling mushroom bodies of polychaetes (Hadenfelt, 1929). In the polychaete Nephtys, the mushroom body globuli cells, varying in number for related species, do not form the typical column-like structure seen in other polychaetes, but contribute to connective fibers (Clark, 1958). An electron microscopical study is needed to define its neuropilar or tract status.

If we accept the often reported trend for the development of sophisti cated mushroom bodies, we have to question the functional implications. Unfortunately, it is not yet possible to assign very many concrete functions to the system to develop a sound overall view valid for the different forms (for review on physiology of best investigated insects see Erber et al., 1987). We tentatively proposed the columnar multichannel structure as an ordered array for time measurement and for slow information processing (Schürmann, 1987), but a detailed explanation of the evolutionary advantage of such a system has not yet been provided.

Distinct compartments are found in the central body and bridge system of most arthropods and Onychophora, with no equivalents in annelids. A common feature of the central body and bridge system is their central position, bridging the hemispheres, and their module structure. Though much progress has been made in understanding their internal architecture in insects (Homberg, 1987), there is a lack of knowledge in connectivity and function. We tentatively suggest that midbrain structures like the central body might have been developed from commissures, adding precise wiring of left-right distributed neurons with partially chiasmatic nature (Williams, 1975). These compartments are packed with peptides and amines in insects (Nässel, 1993) and crustaceans (Schürmann et al., 1991). We predict the same for the Onychophora, as dense core vesicles (storage organelles for these neuroactive compounds) are concentrated here (Fig. 3). The central bodies appear as putative key centers appropriately positioned to serve for modulation of vast parts of the brain.

Tracts

Tracts serve for connecting neuropilar compartments and are characteristically formed by a limited number of fibers with larger diameters (> 1 µm) collected in bundles, to be stained with heavy metals in annelids and arthropods with no major synaptic integration in themselves.

Onychophora apparently differ in this respect, as connections are made mainly by small fibers forming massive bundles, though the ability for layout of tracts with large fibers is present, as seen in the ventral commissures (Fig. 1). In the onychophorans, slow information transfer between compartments is expected to proceed in a slow manner. Even giant fibers work with slow conduction velocities (1.14 ms; Schürmann and Sandeman, 1976). With respect to the predominance of small fibers, onychophorans may resemble acoelomate worms, in not showing well developed fiber tracts with thick axons. A very recent, most detailed study, comparing the design of tracts, commissures, and nerves in annelid brains points out the limitations of the neuroanatomical approach for phylogenetic argument (Orrhage, 1993). More details are needed to consider the evolutionary status of design of neuropil with fine fibers.

Ventral cord

The ventral cord of the onychophorans also shows a mixture of special an common features. The widely separated nerve strands linked by 9–10 commissures (Feodorov, 1926; Pflugfelder, 1968) per neurosomite appear as a ladder-like structure that is different from those of annelids and arthropods. Each neuromere is discerned by the repetition of the two massive leg nerves emerging latero-ventrally, at portions where a slight swelling of the cord comprising accumulated neurons occurs (see Feodorov, 1926). Within the neuromere, rows of somata, mainly of the globuli type (Figs 1,4), are restricted to a medio-ventral position, and not partly intermingled with the neuropil as seen in the longitudinal nerve cord of platyhelminthes (Rieger et al., 1991). The position of the somata rind and the abundance of globuli cells sets the Onchophora apart from annelids and arthropods. We estimate, on the basis of cell counts of cross-section series of small and large neurosomites, that a single, about 1-mm-long neurosomite of Peripatoides spec. contain 10 000–20 000 neurons at one site. The resulting number of 20 000–40 000 neurons per segment contrasts sharply with the 250 neurons of a leech ganglion (Blackshaw, 1981), 800 in the ventral neuromere of the oligochaete Lumbricus (Günter, 1971), and about 2000 in a thoracic and 500 in an abdominal ganglion of an orthopteran insect (Thomas et al., 1984). However, it remains to be demonstrated that all globuli cell perikarya in the ventral cord and brain of the Onychophora form axons, contributing to the neuropil.

In the leech, similarly limited in segments as are onychophorans, the segmental ventral ganglion is involved in several behaviors such as body shortening, swimming, twisting and walking, which are also partly displayed in onychophorans. One substantial difference between the onychophorans and the leeches is the mode of locomotion (walking). We suggest that the high number of neurons in the former, representing augmentation of channels, might account for the unique elaborate muscle and lobopode locomotion apparatus (Hoyle and Williams, 1980), a neuronal attribute invented for organizing locomotion.

The lack of clearly separated ganglia with linking connectives and the overall presence of somata along a nerve strand can be compared with the ventral cord of oligochaetes, and may not represent a primitive layout, but can be regarded as a secondary development like that attributed to the specialized nervous system of Lumbricus. The ventral cord is, as correctly pointed out by Feodorov (1926) not an unsegmented cord ("Markstrang"), but a chain of specialised, well defined neuromeres, reflected by segmental swelling of the cord at leg nerves, repetition of peripheral nerves and commissures with unequal fiber type layout.

The organization of the ventral cord neuropilar fiber masses in Onychophora shares some common features with other articulates: formation of bundles of thicker longitudinal fibers to tracts (with about 130 fibers with diameters >1 μm in each ventral cord) including giant fibers (at least two on each side), and condensation of segmentally repeated synaptic compartments (Fig. 4). The plurisegmental giant fiber system sets the Onychophora apart from acoelomate worms. The layout of neuromeres in onychophorans is unique, concerning position and size of tracts and neuropils. The dorso-ventral structural differences in ventral ganglia, clearly seen in annelids (Günther, 1971) and in arthropods (Leise, 1991), and commonly regarded as a polarization into motor and sensory like neuropils, are not detected for the Onychophora.

The predominance of the globuli cells among somata of other cell types and their restriction to a massive ventral layer in the ventral cord appear as unique features of the Onychophora.

Brain segmentation and innervation

The analysis of head and brain segmentation has always been a principle source for evolutionary comparisons of articulate groups, but has remained a matter of controversy (see Ax, 1984; Willmer, 1990). Three head segments are undoubtedly attributed to the onychophoran head (antennal, feeding claw and oral papillae segment). According to Pflugfelder (1968), who analyzed the development of the head ventral organs, additional segments have to be recognized, a protocerebral one contributing to the brain. Head neuromeres stemming from the ventral organs are fused to form a brain comprising proto-, deuto and tritocerbrum. Unlike insect brains, the onychophoran adult brain does not allow such a subsequent segmentation to be discerned (see Feodorov, 1929; Hanström, 1935). The annelid brain composition of fused neuromers (clarified for the suboesophageal ganglion of oligochaetes; see Spörhase et al., 1987) is still a matter of controversy (see Horridge, 1965). As stated before, deutocerebral antennal glomeruli of the Onychophora are situated lateral to the mushroom bodies very similar to the arrangement of neuropils in polychaete brains, and unlike the situation in insects. In insects, protocerebrum and deutocerebrum are positioned along the neuraxis. As a consequence, insects have developed a tract connecting the protocerebral mushroom bodies with the deutocerebral antennal lobes, consisting of relay interneurons, not seen in polychaete brains (see Hanström, 1928). Fusion of brain parts may reflect an advanced, not a basic status. A postoral suboesophageal ganglion is missing in the Onychophora, but the connectives appear as thickened synaptic neuropils, resembling some polychaete connectives (see Horridge, 1965). The pattern of innervation of the brain (about 15 nerves; see Feodorov, 1929; Horridge, 1965; Pflugfelder, 1968) superficially resembles the annelid situation (Horridge, 1965; Orrhage, 1993), and is different from other arthropod groups. The position of the dorsal antennal nerve in Onychophora resembles that of the antennal, but not the palpal nerve in the polychaete brain. The similarity might be due to analogous constraints exerted by the muscle and sensory system layout. A terminal ganglion is missing in the Onychophora, the two nerve strands fuse dorsally to the alimentary system at the tail end.

The difficulty of comparing the brain and its innervation among onychophorans, arthropods and annelids can be followed in all studies and might reflect the limitation of the classic histological approach for homologizing structures. These difficulties are thoroughly discussed by Feodorov (1926; 1929), who, interestingly, incorporated the myriapods into his considerations. We do not see a compelling reason to homologize brain and ventral cord innervation pattern of Onychophora with other groups.

Physiology and behavior

Studies on physiology and behavior of Onychophora are rare (see Storch and Ruhberg; 1993). The longitudinal giant fiber system is through conducting from head to tail and vice versa, demonstrated by electrophysiological collision experiments (Schürmann and Sandeman, 1976). It can be activated by natural stimuli of head, tail, legs and from all body parts of both sides. It is suitable for left-right coordination which is transmitted by segmental commissures seen from ablation experiments. The system is also involved in fast reactions like annelid-like body shortening and twisting (Fig. 5).

Shortening reflexes typically occur in a variety of annelid species (Zoran and Drewes, 1987), and are organized by slow and fast intersegmental neuron systems. The leech, constituted by a small number of segments (as the Onychophora), represents an animal with a specialized nervous system, lacking fast giant longitudinal axons. Onychophoran walking behaviour has been intensively investigated by Manton (1977), and compared with the locomotion of myriapods and insects.

In fact, the innervation of the muscle system and the leg is so different from insects and other land form arthropods in so many aspects that it has been set apart entirely (Hoyle and del Castillo, 1979; Hoyle and Williams, 1980). Hoyle and Williams considered the simply organized lobopodium as a living model for the formation of an arthropod appendage and emphasized the unparalleled structrue of muscle and nerve muscle junctions. The physiology of muscles and of nerve muscle-junctions differ in general from annelids and arthropods, though they

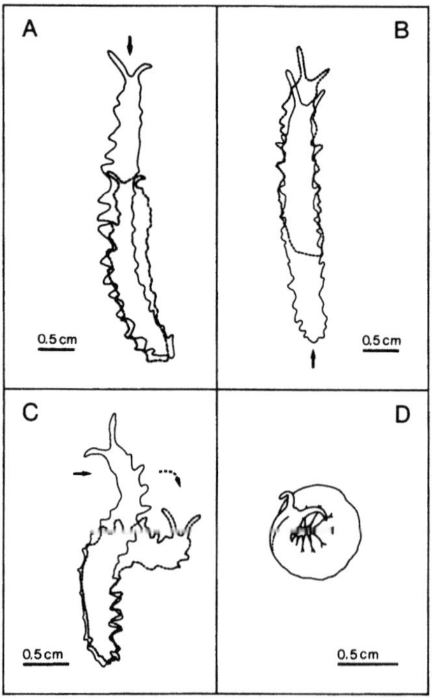

Figure 5. Drawings made from cinephotographs of *Peripatoides leuckarti* to show shortening of the body after touching the head (A) or the tail (B). The animal does not stop walking after the tail is touched. (C) shows the result when the side of the head is touched and (D) the response of a small animal to being pinched.

do share some features (e.g., control of contraction of a muscle fiber by 8–12 motorneurons). No single neuron in the onychophorans exerts strong activation of contraction, which is regulated by the summation of graded potentials. Annelids, onychophorans, and arthropods all show multiple innervation of single muscle fibers. However, Peripatus muscles are reported not to receive fast axons and there is no indication of peripheral inhibition (Hoyle and del Castillo, 1979; del Castillo and Hoyle, 1982).

Conclusions

Onychophorans have unique features in their nervous system, which cannot be ascribed to a primitive or advanced state of its organization in this group. These features occur at different levels in the structure of the nervous system, and can also be found in details of physiology and behavior. The special features of the Onychophora appear rather as the result of the composition of neuronal molecules, organelles, and synapses into circuit layouts and a compartmental neuropil organization, than by a specialization of the cellular elements themselves, which are also found in other invertebrate groups. We propose that globuli cells are a basic "juvenile" cell type that is employed for adding new brain compartments.

Construction of the brain and ventral nerve cord from small, uniform globuli cells is seen as a succesful evolutionary experiment as found in the Onychophora. The brain of the Onychophora could serve as a model system for analyzing the physiological consequences in which globuli cells dominate the neuropils. The similarities in the architecture of the mushroom body system and in antennal glomeruli in the Onychophora and in insects – homologous or analogous in nature – are striking and might indicate a very early development of specialized condensed neuropils in the Cambrian epoque.

Acknowledgments
I thank Mrs. M. Klages and M. Knierim for skillful technical assistance, Mr. B. Baumgarten for the preparation of line drawings and Dr. D. Sandeman, Kensington, University of New South Wales, for critically reading the manuscript.

References

Adouette, A. and Philippe, H. (1993) The major lines of metazaon evolution: Summary of traditional evidence and lessons from ribosomal RNA sequence. *In:* Y. Pichon (ed.): *Comparative Molecular Neurobiology*, Birkhäuser Verlag, Basel, pp. 1–30.

Akert, K. and Steiger, U. (1967) Über Glomeruli im Zentralnervensystem von Vertebraten and Invertebraten. Schweiz. *Arch. Neurol. Neurochir. Pyschiat.* 100: 321–337.

Arbas, E.A., Meinertzhagen, I.A. and Shaw, S.R. (1991) Evolution in nervous systems. *Ann. Rev. Neurosci.* 14: 9–38.

Ax., P. (1984) *Das Phylogenetische System*. Gustav Fischer Verlag, Stuttgart.

Ballard, J.W.O., Olsen, G.J., Faith, D.P., Odgers, W.A., Rowell, D.M. and Atkinson, P.W. (1992) Evidence from 12S ribosomal RNA sequences that onychophorans are modified arthropods. *Science* 258: 1345–1348.

Beklemishev, W.N. (1969) *Principles of Comparative Anatomy of Invertebrates*. Vol. 1. Promorphology. University of Chicago Press.

Bellonci, G. (1883) Sur la structure et les rapports des lobes olfactifs dans les arthropodes supérieur et les vertébrés. *Archiv. ital. Biol.* 3. 191–196.

Blackshaw, S.E. (1981) Sensory cells and motor neurons. *In:* K.F. Muller, J.G. Nicholls and G.S. Stent (eds): *Neurobiology of the Leech*, Cold Spring Harbour, New York, pp. 51–78.

Boeckh, J. and Ernst, K.-D. (1987) Contribution of single unit analysis in insects to an understanding of olfactory function *J. Comp. Physiol.* 161:549–565.

Bullock, T.H. and Horridge, G.A. (1965) *Structure and Function in the Nervous System of Invertebrates*. Vol. II. Freeman & Co, San Francisco.

Clark, R.B. (1958) The micromorphology of the supraoesphageal ganglion of Nephtys. *Zool. Jahrb. Physiol.* 68: 261–296.

Cobb, J.L.S. and Pentreath, V.W. (1978) Comparison of the morphology of synapse in invertebrate and vertebrate nervous system. *Prog. Neurobiol.* 10: 231–252.

Del Castillo, J. and Hoyle, G. (1982) Sontaneous junctional potentials associated with cholinergic transmission in Peripatus muscle. *Comp. Biochem. Physiol.* 73: 451–456.

Dhainaut-Courtois, N. and Golding, D.W. (1988) VI. Nervous system. *In:* W. Weistheide and C.O. Hermans (eds): *The Ultrastructure of Polychaeta*. Microfauna Marina 4. Gustav Fischer Verlag, Stuttgart, pp. 89–110.

Eakin, R.M. and Westfall, J.A. (1965) Fine structure of the eye of Peripatus (Onychophora). *Z. Zellforsch* 68: 278–300.

Elofsson, R. and Hessler, R.R. (1990) Central nervous system of Hutchinsoniella macracantha (Cephalocarida). *J. Crust. Biol.* 10: 423–439.

Erber, J., Homberg, U., and Gronenberg, W. (1987) Functional roles of the mushroom bodies in insects. *In:* A.P. Gupta (ed.): *Arthropod Brain. Its Evolution Development, Structure, and Functions*. John Wiley, New York, pp. 485–512.

Fahrenbach, W.H. (1976) The brain of the horseshoe crab (Limulus polyphemus). I. Neuroglia. *Tiss. Cell* 3: 395–410.

Feodorov, B. (1926) Zur Anatomie des Nervensystems von Peripatus. I. Das neurosomit von Peripatus tholloni Bouv. *Zool. Jahrb. Anat.* 48: 273–310.

Feodorov, B. (1929) Zur Anatomie des Nervensystems von Peripatus. II. Das Nervensystem des vorderen Köperendes und seine Metamerie. *Zool. Jahrb. Anat.* 50: 279–332.

Florey, E. and Florey, E. (1965) Cholinergic neurons in the Onychophora. A comparative study. *Comp. Biochem. Physiol.* 15: 125–136.

Foelix, R.F. (1975) Occurrence of synapses in peripheral sensory nerves of arachnids. *Nature* 254: 146–148.

Fortey, R.A. and Thomas, R.H. (1993) The case of the velvet worm. *Nature* 361: 205–206.

Gaffron, J. (1884) Kurzer Bericht über fortgesetzte Peripatus-Studien. *Zool. Anzeiger* 7: 336–339.

Gardner, C.R. and Walker, R.J. (1982) The role of putative neurotransmitters and and neuromodulators in annelids and related invertebrates. *Prog. Neurobiol.* 18: 81–120.

Golding, D.W. (1992) Polychaeta: Nervous system. *In:* F.W. Harrison and S.L. Gardiner (eds): *Microscopic Anatomy of Invertebrates*. Vol. 7: *Annelida*. Wiley and Liss, New York, pp. 153–179.

Govind, C.K. (1992) Nervous system. *In:* F.W. Harrison and S.L. Gardiner (eds): *Microscopic Anatomy of Invertebrates. Vol. 10: Decapod Crustacea*. Wiley and Liss, New York, pp. 395–438.

Günther, J. (1971) Mikroaantomie des Bauchmarks von Lumbricus terrestris L. (Annelida, Oligochaeta). *Z. Morph. Tiere* 70: 141–182.

Günther, J. and Schürmann, F.-W. (1973) Zur Feinstruktur des dorsalen Rienesfasersystems im Bauchmark des Regenwurms. II. Synaptische Beziehungen der proximalen Riesenfaserkollateralen. *Z. Zellforsch.* 139: 369–396.

Gupta, A.P. (1987) Evolutionary trends in the central and mushroom bodies of the arthropod brain. *In:* A.P. Grupta (ed.): *Arthropod Brain, Its Evolution, Development, Structure, and Functions*. John Wiley, New York, pp. 27–42.

Hadenfeldt, D. (1929) Das Nervensystem von Stylochoplana maculata und Notoplana atomata. *Z. wiss. Zool.* 133: 586–638.

Hanström, B. (1928) *Vergleichende Anatomie des Nervensystems der wirbellosen Tiere*. Springer Verlag, Berlin.

Hanström, B. (1935) Bemerkungen über das Gehirn und die Sinnesorgane der Onychophoren. *Lunds Univ. Arskr. N.F.* 31: 1–37.

Holmgren, N. (1916) Zur vergleichenden Anatomie des Gehirns von Polychaeten, Onychophoren und Xiphosuren. Kung 1. *Svenska Vet. Akad. Handl.* 156: 1–299.

Homberg, U. (1987) Structure and functions of the central complex in insects. *In*: A.P. Gupta (ed.): *Arthropod Brain. Its Evolution, Development, Structure, and Functions.* John Wiley, New York, pp. 347–367.

Horridge, G.A. (1965) Onychophora. *In:* T.H. Bullock and G.A. Horridge (eds): *Structure and Functions in the Nervous System of Invertebrates*, Vol. 1. Freeman, San Francisco, pp. 791–798.

Hoyle, G., and Del Castillo, J. (1979) Neuromuscular transmission in Peripatus. *J. Exp. Biol.* 83: 13–29.

Hoyle, G., and Williams, M. (1980) The musculature of Peripatus and its innervation. *Phil. Trans. Roy. Soc. Ser. B.* 288: 481–510.

Lane, N.J. and Abbott, N.J. (1975) The organization of the nervous system in the crayfish Procambarus clarkii, with emphasis on the blood brain interface. *Cell Tissue Res.* 156: 173–187.

Lane, N.J. and Campiglia, S.S (1987) The lack of a structured blood-brain barrier in the onychophoran Peripatus acacioi. *J. Neurocytol.* 16: 93–104.

Leake, L.D. and Walker, R.J. (1980) *Invertebrate Neuropharmacology.* Blackie & Son, Glasgow.

Leise, E.M. (1991) Evolutionary trends in invertebrate ganglionic structure. *The Neurosciences* 3: 369–377.

Manton, S.M. (1977) *The Arthropoda.* Clarendon Press, Oxford.

Nässel, D.R. (1993) Neuropeptides in the insect brain: a review. *Cell Tiss. Res.* 273: 1–29.

Orrhage, L. (1993) On the microanatomy of the cephalic nervous system of Nereidae (Polychaeta), with the preliminary discussion of some earlier theories on the segmentation of the polychaete brain. *Acta Zool.* 74: 145–172.

Paulus, H.F. (1979) Eye structure and the monophyly of the Arthropoda, pp. 299–383. *In*: A.P. Gupta (ed.): *Arthropod Phylogeny.* Van Nostrand Reinhold Company, New York.

Pflugfelder, O. (1968) *Onychophora*, Gustav Fischer Verlag, Stuttgart.

Purschke, G. (1993) Structure of the prostomial appendages and the central nervous system in the Protodrilida (Polychaeta). *Zoomorph.* 113: 1–20.

Ramsköld, L. and Hou Xianguang (1991) New early cambrain animal and onychophoran affinities of enigmatic metazoans. *Nature* 351: 225–228

Rensch, B. Trends towards progress of brains and sense organs. Cold Spring Harbour Symp. *Quant. Biol.* 24: 291–303.

Reuter, M. (1981) The nervous system of Microstomum lineare (Turbellaria, Macrostomida). II. The ultrastructure of synapses and neurosecretory release sites. *Cell Tissue Res.* 218: 375–387.

Rieger, R.M., Tyler, S., Smith III, J.P.S., Rieger, G.E (1991) Platyhelminthes: Turbellaria. *In*: F.W. Harrison and S.L. Gardiner (eds.): *Microscopic Anatomy of Invertebrates. Vol* 3: *Platyhelminthes and Nemertainea.* Wiley and Liss, New York, pp. 7–140.

Rosenberg, J. and Seifert, G. (1978) Feinstruktur der Innervierung des Dorsalgefäßes von Peripatoides leuckarti (Saenger 1869) (Onychophora, Peripatopsidae). *Zool. Anz.* 201: 21–30.

Sandeman, D.C. (1982) Organization of the central nervous system. *In:* H.L. Atwood and D.C. Sandeman (eds): *The Biology of Crustacea, Vol.* 3: *Neurobiology*: *Structure and Function.* Academic Press, New York, pp. 1–61.

Schürmann, F.-W. (1978a) A note on the structure of synapses in the ventral cord of the onychophoran Peripatoides leuckarti. *Cell Tissue Res.* 186: 527–534.

Schürmann, F.-W. (1978b) On the ultrastructure of nerve-muscle contacts in Onychophora. *Experientia* 34: 774.

Schürmann, F.-W. (1980) Methods for special staining of synaptic sites. *In:* N.J. Strausfeld and T.A. Miller (eds): *Neuroanatomical Methods. Experimental Entomology*, Vol 1 A, Springer, New York, pp. 241–262.

Schürmann, F.-W. (1987) The architecture of the mushroom bodies and related neuropils in the insect brain. *In:* A.P. Gupta (ed.): *Arthropod Brain: Its Evolution, Structure and Functions*. John Wiley, New York, pp. 231–264.

Schürmann, F.-W. (1987) Histology and ultrastructure of the onychophoran brain. *In:* A.P. Gupta (ed.): *Arthropod Brain: Its Evolution, Structure and Functions*. John Wiley, New York, pp. 159–180.

Schürmann, F.-W. and Günther, J. (1973) Zur Feinstruktur des dorsalen Riesenfasersystems im Bauchmark des Regenwurms. I. Die Somata der Riesenfasern. *Z. Zellforsch.* 139: 351–368.

Schürmann, F.W. and Sandeman, D.C. (1976) Giant fibres in the ventral nerve cord of Peripatoides leuckarti (Onychophora). *Naturwiss.* 63: 580.

Schürmann, F.W., Sandeman, R. and Sandeman, D. (1991) Dense core vesicles and non-synaptic exocytosis in the central body of the crayfish brain. *Cell Tissue Res.* 265: 493–501.

Sedgwick, A. (1888) The development of the cape species of Peripatus. Part 4. Q. Jl. *Microsc. Sci.* 18: 373–398.

Sedgwick, A. (1909) *A Student's Textbook of Zoology*. Allen, London.

Shaw, S.R., and Meinertzhagen, I.A. (1986) Evolutionary progression at synaptic connections made by identified homologous neurones. *Proc. Natl. Acad. Sci. USA* 83: 7961–7965.

Spörhase-Eichmann, U., Gras, H. and Schürmann, F.-W. (1987) Patterns of serotonin-immunoreactive neurons in the central nervous system of the earthworm Lumbircus terrestris L. II. Rostral and caudal ganglia. *Cell Tissue Res.* 249: 625–632

Storch, V. and Ruhberg, H. (1993) Onychophora. *In:* F.W. Harrison and S.L. Gardiner (eds): *Microscopic Anatomy of Invertebrates. Vol. 12: Onychophora Chilopoda, and Lesser Protostomata.* Wiley and Liss, New York, pp. 11–56.

Strausfeld, N.J. (1976) *Atlas of an Insect Brain.* Springer Verlag, Berlin.

Thomas, J.B., Bastiani, M.J., Bate, M. and Goodman, C.S. (1984) From grasshopper to Drosophila: a common plan for neuronal development. *Nature* 310: 203–207.

Tombes, A.S. (1979) Comparison of arthropod neuroendocrine structures and their evolutionary significance. *In:* A.P. Gupta (ed.): *Arthropod Phylogeny.* Van Nostrand Reinhold Company, New York, pp. 645–667.

Venter, J.C., Di Porzio, U., Robinson, D.A., Shreeeve, S.M., Lai, J., Kerlavage, Fracek, S.P. Lentes, K.-U. and Fraser, C.M. (1988) Evolution of neurotransmitter receptor systems. *Progr. Neurobiol.* 30: 105–169.

Weiss, M.J. (1981) Structural patterns in the corpora pedunculata of Orthoptera: A reduced silver analysis. *J. Comp. Neurol.* 203: 515–553.

Williams, J.L.D. (1975) Anatomical studies of the insect central nervous system: A ground-plan of the midbrain and an introduction to the central complex in the locust Schistocerca gragaria (Orthoptera). *J. Zool.* 167: 67–86.

Willmer, P. (1990) *Invertebrate Relationships. Patterns in Animal Evolution.* Cambridge University Press, Cambridge.

Zoran, M.J. and Drewes, C.D. (1987) Rapid escape reflexes in aquatic oligochaetes: variations in design and function of evolutionary conserved giant fiber systems. *J. Comp. Physiol.* 161: 729–738.

The Nervous Systems of Invertebrates: An Evolutionary and Comparative Approach
ed. by O. Breidbach & W. Kutsch
© 1995 Birkhäuser Verlag Basel/Switzerland

Comparative aspects of the chelicerate nervous systems

R. Wegerhoff[1] and O. Breidbach[2]

[1]*Biochemisches Institut I, Olshausenstr. 40, D-24098 Kiel, Germany*
[2]*Institut für Angewandte Zoologie, An der Immenburg 1, D-53121 Bonn, Germany and Fakultät und Institut für Mathematik, Universitätsstr. 150, NA 5-26, D-44801 Bochum, Germany*

Summary. Main features of the neuroanatomical organization of recent Chelicerata are described. Thereby, it is possible to characterize a general scheme of the chelicerate brain that includes neuropil regions as the "central body" and the "corpora pedunculata". Application of immunohistochemistry allows a description of identified neurons associated with these neuropil regions. Such neurons have been characterized as being homologous in different arachnid species. The chapter discusses whether these neurons form part of a ground plan of the arachnid brain.

Introduction

Within the arthropod phylum, chelicerate nervous systems attracted only minor interest during recent decades. Thus, most comparative work dates back to the classical studies of Holmgren (1916) and Hanström (1919, 1921, 1923, 1926, 1928, 1935). For many of the chelicerate groups, their description provide the only data availabe at present. Likewise, more detailed comparative neuroembryological studies are missing. Apart from the work of Legendre (1958a,b), Pross (1966), and Weygold (1975), most studies date back to the end of the 19th century and the first decades of this century (Packard, 1893; Patten, 1896, 1912; Lambert, 1909; Johansson, 1933, Iwanoff, 1933; for review see Anderson, 1973). Accordingly, up to now, molecular biological techniques have not been employed to the developing chelicerate nervous systems.

Thus, in comparative neurobiology these arthropods seems to be somehow neglected, in spite of their evolutionary success which is likewise reflected in their importance for terrestrial ecosystems (Nentwig, 1987; Foelix, 1992) and in some of the most elaborate behavioral adaptations of arthropods, the web-building behaviors of spiders (Peters, 1969; Shear, 1986).

History and systematics

The Chelicerata are a monophylum within the Arthropoda. They were named according to their first appendages, the chelicerae. Their body plan is characterized by a subdivision into two tagmata, the prosoma, where all appendages are concentrated, and an opistosoma. First fossil forms were found dating to the pre-cambriam era. Fossils of the Xiphosuran, the most "primitive" recent Chelicerata, were found in the Cambrian fossil layers (Hammen, 1989). During the Paleozoic era, a subgroup of chelicerata, the Eurypterida, formed the largest arthropods known (Bergström, 1979). This group gave rise to the scorpions which became terrestrial in the late Silurian (Hammen, 1989). Non-scorpionid Arachnida, like Acarida, developed in Early Devonian times or earlier (Bergstöm, 1979). All recent groups of Chelicerata were definitely present in the carboniferous era (Hammen, 1989).

With regard to specifications of the mouthparts and the body segmentation patterns, the most recent classification of Chelicerata (Hammen, 1989) subdivides the subphylum Chelicerata into two superclasses, the Rostrostomata, which include among others the Araneida and Mylioso-mata, consisting of Opilionida, Xiphosura, and Scorpionida (Fig. 1). However, the common systematics of Remane, which we adopted in this review, separates the Xiphosura from the group of Pantopoda and Arachnida (Tab. 1) (Remane, 1956).

External morphologies

The central nervous system of the oldest living Chelicerata, the Xiphosura, consists of a brain, defined as the region anterior to the circumesophageal connective, the circumesophageal connective, four free ganglia and a terminal ganglion consisting of three fused neuromers (Fig. 1; Patten and Redenbaugh, 1900; Scharrer, 1941). Within the adult Arachnida there is a tendency to fuse the free abdominal ganglia to the subesophageal nerve mass; Scorpionida possess seven free abdominal ganglia and a subesophageal nerve mass formed by nine neuromers. Solifugida and Uropygida possess one free abdominal ganglion, and in most other Arachnida, all ventral nerve cord ganglia are fused to form a condensed subesophageal ganglion (Babu, 1985).

Peripheral nervous system

A common feature of arachnids is the occurrence of peripheral synapses of afferent neurons (Foelix, 1985). The organization of these presumptive peripheral integration units has been studied in detail in the leg nerves of Arachnida (Foelix, 1975). These nerves consist of afferent fibers arising from the leg receptors, and of efferent fibers innervating the leg muscles. Within the nerves, the axons of sensory neurons form axo-axonal synapses with afferent fibers. On the receptor level, not only axon-axonal, but even

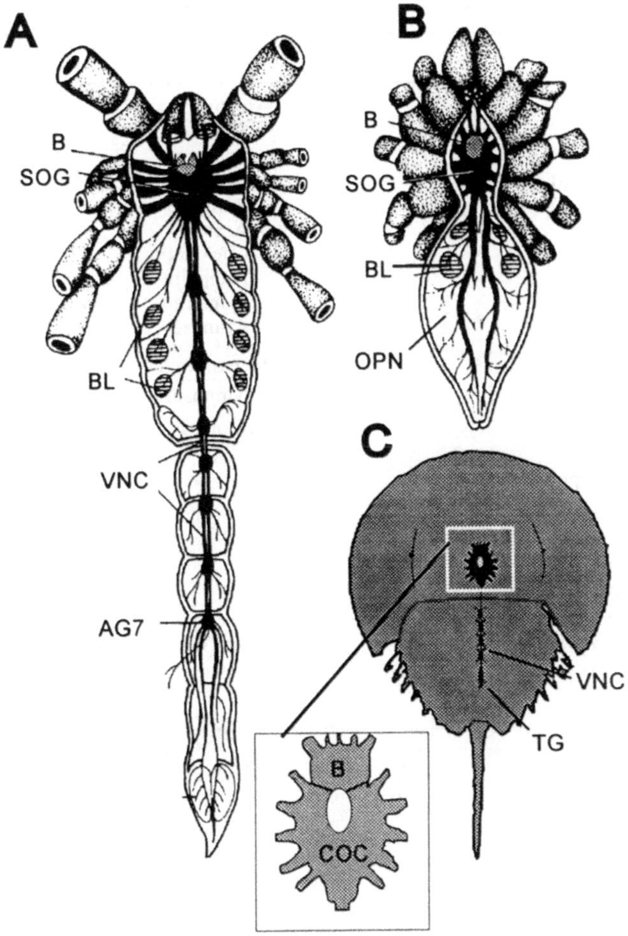

Figure 1. (A–C) Topography of the central nervous system of Chelicerata. A) Scorpion (*Heterometrus*), the ventral nerve cord consists of seven free abdominal ganglia and a fused subesophageal ganglion (SOG). B) Spider (*Poecilotheria*), note all abdominal ganglia are fused to compose the subesophageal nerve mass (SOG). C) Xiphosura (*Limulus*), the ventral nerve cord consists of three fused ganglia composing the terminal ganglion (TG) and of four free abdominal ganglia; the anterior region of the central nervous system consists (insert) of the circumesophageal connectives (COC) and of the brain (B). AG7: 7th abdominal ganglion, BL: book lungs, VNC: ventral nerve cord. (A,B after Babu, 1985).

axo-dendritic, axo-somatic and axo-glial contacts are present. In *Limulus*, such receptor-associated interneuronal contacts form complex neuronal plexi (Hays and Barber, 1982). These are associated with chemosensory sensilla and are composed of collaterals from afferent fibers (Fig. 2).

Electrophysiological recordings of the giant fiber system of the Amblypygids, the whip spiders, gave evidence for an intergrative function of peripheral interneuronal contacts (Igelmund and Wendler, 1991b).

Table 1. Systematics of Chelicerata according to van der Hammen (1989) and Remane (1956)

Chelicerata	Chelicerata
Rostrostomata	Arachnida
Neosternata	Scorpiones
Cyptognomae	Solifugae
Anactinotrichida	Opiliones
Ricinulei	Ricinulei
Arachnida	Acari
Uropygi	Chelonethi
Amblypygi	Pedipalpi
Araneida	Palpigradi
Coxisternata	Araneae
Epimerata	Merostomata
Palpigradi	Xiphosura
Actinotrichida	
Apatellata	Pantopoda
Solifugae	
Pseudoscorpionida	
Myliosomata	
Xiphosura	
Scorpionida	
Opilionida	
After: van der Hammen (1989).	After: Remane (1956).

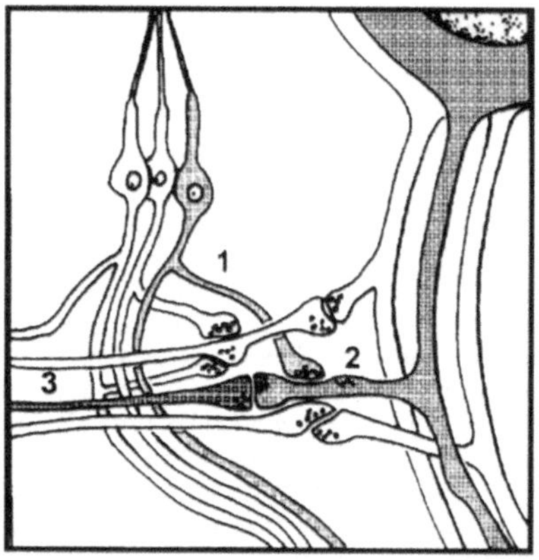

Figure 2. Schematic drawing of the neural plexus in *Limulus*. Synapses are formed by axon collaterals of sensory neurons serving chemoreceptive sensilla (1–3); see text for details. (After Hayes and Barber, 1982).

The giant fiber system is found in the first pair of extremely elongated legs, the "whips", which are functionally analogous to an insect antenna (Foelix, 1985). The somata of the giant neurons are situated peripherally and their dendrites are innervated by en passant-synapses of mechanosensitive bristles (Foelix and Troyer, 1980). Ultrastructural reconstructions show that about 90% of the afferent synapses of the giant neurons are found on their axons (diameter 20 μm). Activation of a single mechanosensitive bristle is sufficient to elicit the giant neuron. Igelmund and Wendler (1991a,b) showed that the giant fiber system could not be accounted for as an escape system, but rather those authors discussed a role in mating and intraspecific aggressive behavior.

Neuroembryology

The basic development of the nervous system is alike throughout the Chelicerata. The ventral nerve cord develops from ventro-lateral embryonic ectoderm that thickens bilaterally, to form a pair of segmental ganglion rudiments (Anderson 1973). In more detail, this has been described for the Araneae *Cupiennius salei.* There, the ventro-lateral embryonic ectodem forms individual segmental hemiganglia that later fuse to produce the subesophageal ganglion complex. Neurogenesis starts though proliferative invagination of cells from multiple germinal points in the ectoderm. No large neuroblasts were found (Chaboud et al., 1990). This type of neuronal development seems to be a common feature of the ventral nerve cord of Arachnida (Weygold, 1985). Development of the prochelicerate ganglion, which includes the optic lobes, is more complicated.

At least gross morphologically, brain development in the arachnids seems to be remarkably conserved. Anderson (1973) reports on common features in the details of brain formations in spiders, uropygids and opilionids. In general, the arachnid brain is formed by an invagination of the apical ectodem, the chephalic fold. The inner walls of this fold thicken, merge and give rise to the brain with the optic masses, to which also the walls of the lateral vesicle contribute. The anlagen of the optic masses of the lateral eyes derive from those lateral vesicles (Weygold, 1985; Fig. 3).

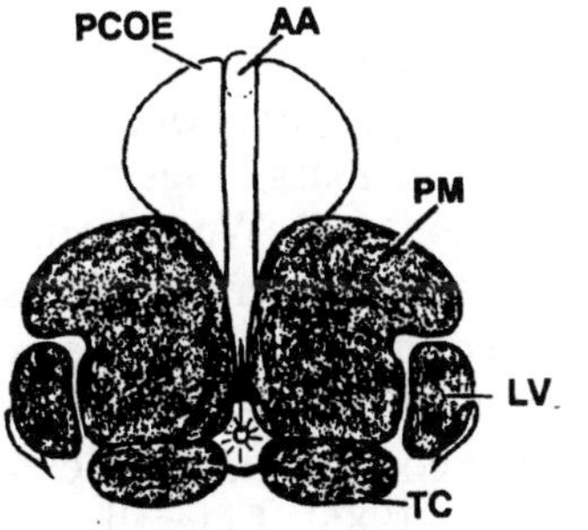

Figure 3. Schematic drawing of the brain anlage of the arachnid *Tarantula marginemaculata.* The brain rudiment of the spider does not show a clear division into lobes. It consists of a large protocerebral mass (PM), the two chelicerae ganglia which are regarded as the tritocerebrum (TC) and of the lateral vesicles (LV) which form the optic masses of the lateral eyes. AA: anterior aorta, PCOE: precheliceral coelom. (After Weygold, 1975).

The arachnid brain is separated into a tritocerebrum and a protocerebrum. The latter is developed from two coelom cavities, but morphologically no clear division into lobes is discernable. As no molecular markers were tested up to now, the neuromeral character of the arachnid brain anlage is not clarified. The segmental organization of the protocerebrum, thus, remains doubtful, whereas the so-called tritocerebum represents the cheliceral ganglia. A deutocerebrum is lacking or might be totally fused to the protocerebrum (Weygold, 1975, 1985).

One of the most prominent neuropils of the later arachnid brain is the so-called 'central body'. It is formed by two *Anlagen*, found at the anterior-most part of each developing brain hemisphere. With the invagination of the brain *Anlage*, it rotates first to an apical position and then shifts to a posterior dorsal position (Fig. 4; Legendre, 1958a,b; Pross, 1966). A comparable rotation is performed by the *Anlagen* of other brain neuropil areas. Thus, the optic lobes move to an apical position, where they remain in opilionids and mites; whereas in arachnids with a foreward flexed opistosoma, they are moved on to an anterior postition (Pross, 1966).

Subesophageal ganglion

Data on the neuronal organization of the ventral nerve cord ganglia are present for Scorpionida (Root, 1985), Opilionida and Araneae. But only for the Opilionida and Araneae is a neuroanatomical framework outlined in such detail as to allow a comparative analysis of their neuropil organizations (Babu, 1965, 1985; Babu and Barth, 1984, 1989; Babu et al., 1985; Breidbach and Wegerhoff, 1993).

Gross morphology

The subesophageal ganglion of the adult spiders and opilionids is formed by one paired pedipalpal ganglion, four pairs of leg ganglia, and a condensed abdominal nerve mass (Figs 5A, 6A,B; Babu, 1965; Babu and Barth, 1984; Breidbach and Wegerhoff, 1993). Cell bodies are concentrated on the ventral and lateral surfaces of the ganglionic mass (Babu, 1965, 1969). The condensed neuropils are connected by longitudinal tracts and commissures that show a reptitive organization. The longitudinal tracts and the commissures are concentrated in the median parts of the ganglia (Babu, 1985). Laterally, in the pedipalpal and leg ganglia, two distinct main neuropil areas are present, the dorsal motoric and the ventral sensoric neuropil. (Figs 5C, 6D-F).

The repetitive, homonomous pattern of the ventral nerve cord ganglia is even obvious on the level of identified neurons. CCAP-immunoreactive

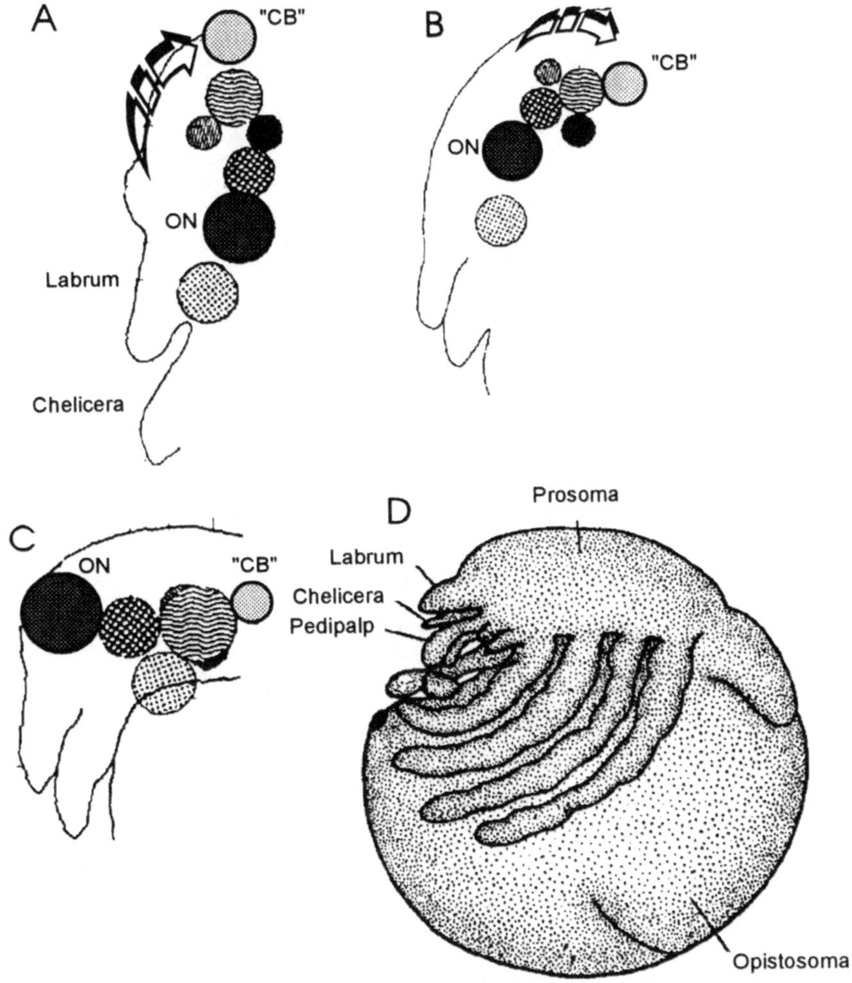

Figure 4. (A–D) Schematic drawing of the rotation of neuropil anlagen during embryonic development of the aranean brain. (A–C) Position of the different neuropil *Anlagen* as indicated; "CB": "central body", ON: optic neuropil: the "central body" shifts from an anterior to a dorsal position (A, arrow) and than ato a posterio-dorsal position (B, arrow; C). The optic neuropil anlage shifts from a median to a frontal position. D) Lateral view of a entelegynid spider embryo (Latrodectus mactans). (A–C after Pross, 1966; D after Rempel, 1957).

neurons were found in a serial repetitive pattern in each pedipalpal and leg ganglion (Fig. 6C). Descending histamine immunoreactive neurons of *C. salei* possess repetitively organized side projections with extended arborizations in the motor and sensory areas of all neuromers (Schmid and Dunker, 1993).

The subesophageal ganglion of the spider *C. salei* comprises 11 pairs of longitudinal tracts. There are six major, presumably motoric, tracts:

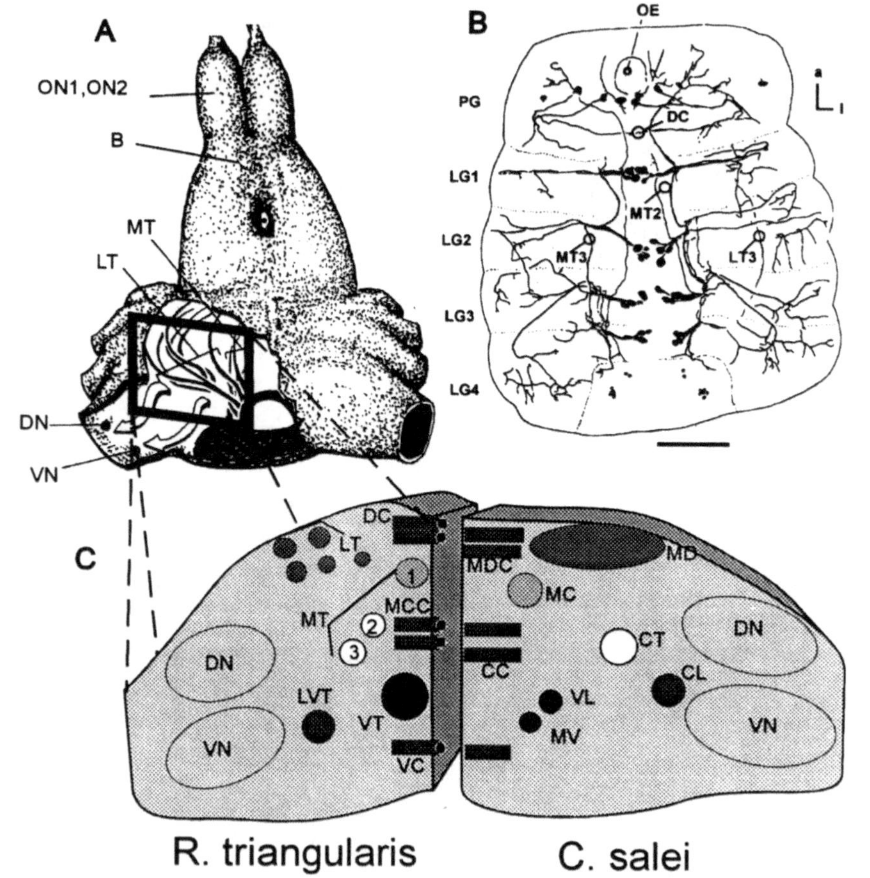

Figure 5. (A–C) Anatomy of the subesophageal nerve mass of Araneae and Opilionida. A) Caudal view of the central nervous system of the Opilionida (*Rilaena triangularis*). The longitudinal tracts in the subesophageal nerve mass are outlined schematically; the box indicates the position of C. B) Serotonin immunopositive neurons in the subesophageal nerve mass of *Rilaena triangularis*, ventral view, camera lucida drawing, scale bar 100 μm. In all leg ganglia (LG1-LG4) serotonin immunoreactive neurons show similar projection patterns. C) Scheme of the tract and commissural organization of arachnid leg ganglia of *Rilaena triangularis* (left side) and *Cupiennius salei* (right side). Longitudinal tracts and commissures are outlined, topologically corresponding tracts are shown in the same gray scale. In both species five commissures were found: a: anterior, B: brain, CC: central commissure, CL: centro–lateral tract, CT: central tract, DC: dorsal commissure, DN: dorsal neuropil (motoric), LG1–4: leg ganglia 1–4, LT: lateral tracts, LVT: latero-ventral tract, MCC: mid-central commissure, MD: mid-dorsal tracts, MT: median tracts (1–3), ON1, ON2: optic neuropil 1,2, VC: ventral commissure, VL: ventro-lateral tract, VN: ventral neuropil (sensoric), VT: ventral tract. (A,B,C (*Rilaena*), after Breidbach and Wegerhoff 1993, C (*Cupiennius*) after Babu et al. 1985, Babu and Barth, 1989).

the mid-ventral, the ventral lateral, the centro-lateral, the central, the midcentral, and the mid-dorsal tracts (Babu and Barth, 1984, 1989; Babu, 1985). The other mid-centrally located five pairs of longitudinal tracts are interpreted as sensory tracts (Fig. 6E,F; Babu et al., 1985;

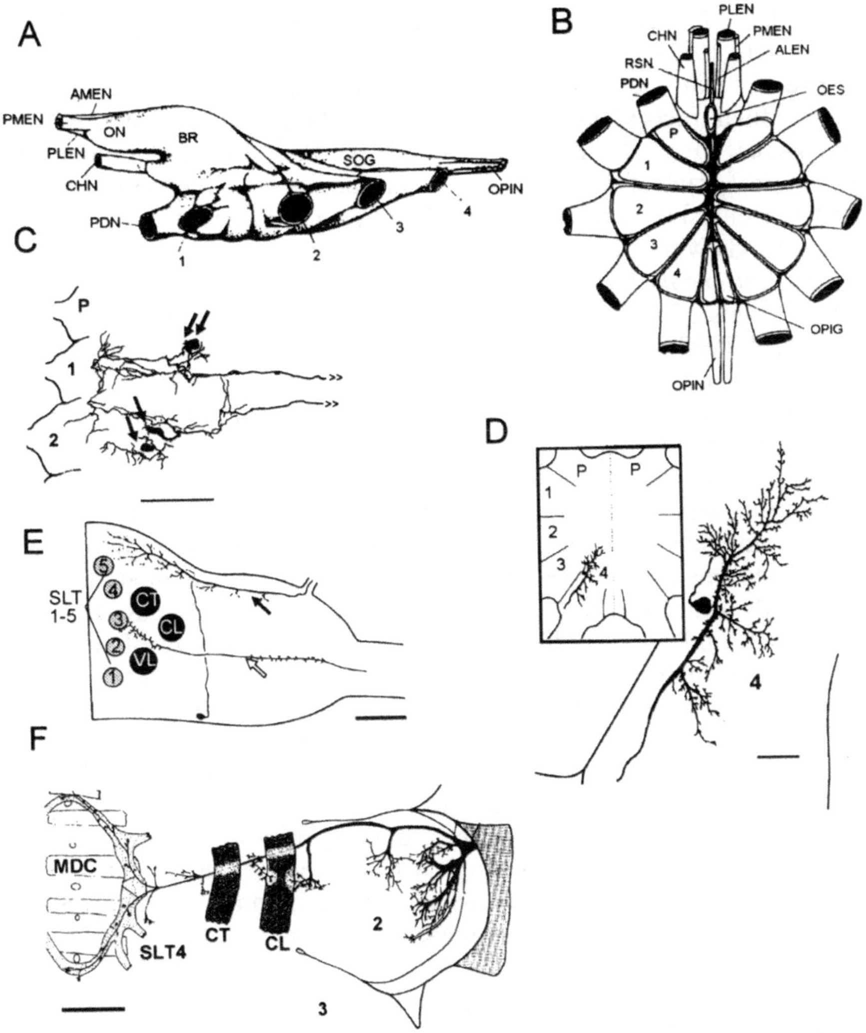

Figure 6. (A-F) Anatomy of the subesophageal nerve mass of (*Cupiennius salei*); A, B) Morphology of the central nervous system. A) lateral view; B) ventral view. C) CCAP immunopositive neurons (arrows) in the first and second leg ganglia (1, 2), note repetitive organization; scale bar 200 µm. D) excitatory motor neuron for muscle c2; cell soma is located in the ventral cell rind and gives rise to vertically ascending process that arborizes in the dorsal neuropil of the 4th leg ganglion; the box indicates the position of the neuron in the ventral nerve mass; scale bar 100 µm. E) Schematic drawing of the vertical organization of ventral sensory-hair input (open arrow) and motor neurons (black arrow); scale bar 200 µm. F) Branching pattern of a dorsal root axon from lyriformic organ; scale bar 200 µm. 1–4: leg ganglia 1–4, ALEN: anterio lateral eye nerve, AMEN: anterior median eye nerve, BR: brain, CHN: cheliceral nerve, CL: centro-lateral tract, CT: central tract, MDC: mid-dorsal commissure, OES: esophagus, OPIC: opisthosomal ganglion, OPIN: opisthosomal nerve, ON: optic neuropil, P: pedipalpal ganglia, PDN: pedipalpal nerve, PLEN: posterior lateral eye nerve, PMEN: posterior median eye nerve, RSN: rostral nerve, SLT1–5: sensory longitudinal tracts 1–5, SOG: subesophageal nerve mass (A,B after Babu and Barth, 1984: D, E after Milde and Seyfarth 1988: F after Babu and Barth 1989).

Babu and Barth, 1989). The organization is likewise found in the Pholcidae *Pholcus phalanagoides* and *P. opilionoides* (Wegerhoff and Breidbach, 1989).

The six pairs of motoric tracts of the spider correspond to 10 longitudinal tracts identified in the hemiganglion of the opilionid *Rilaena triangularis* (Fig. 5C; Breidbach and Wegerhoff, 1993). Topologically, the ventral tract of *R. triangularis* corresponds with the ventral lateral and midventral tract of *C. salei*: the lateral ventral tract of *R. triangularis* corresponds with the central lateral tract of *C. salei*; the median tracts 2,3 of *R. triangularis* correspond with the central tract in *C. salei*, and the lateral tracts of *R. triangularis* correspond with the mid dorsal tract of *C. salei*. Likewise, the architectures of the commissures are equivalent in the spider and in the opilionid. In both arachnid groups, a dorsal, a mid-dorsal, a central, a mid-central, and a ventral commissure are present (Babu and Barth, 1989; Wegerhoff and Breidbach, 1989; Breidbach and Wegerhoff 1993). Thus, the principal organization of the subesophageal nerve mass is structurally similar in these different arachnids. Comparative analyses of the distributions of serotonin immunoreactive neurons and CCAP-like immunoreactive neurons in spiders and opilionids demonstrated (Fig. 5B; Seyfarth et al., 1990; Breidbach, 1992; Breidbach and Wegerhoff, 1993) that even the topology of identified neurons is similar in these two groups. The identified neurons are situated in equivalent positions; they share not only structural but even immunohistochemical characteristics. Furthermore, only slight variations in their shapes are evident within these different arachnid species. Thus, they can be regarded as homologous neurons (Kutsch and Breidbach, 1994). So far, not only the principal gross structural organizations of this neuropil, but even details of their histological characteristics up to the level of identified neurons are conserved, showing a common *Bauplan* of the subesophageal nerve masses of these arachnids.

Supraesophageal ganglion

Gross morphology

As the descriptions of the segmentation patterns of the arachnid brain neuropil is steeped in controversy, a comparative analysis can only point out structural specifications of neuropil regions as the commissural and tract organizations of this neuropil. The principal morphology of the nervous systems and the pattern of neuro-embryogenesis is highly conserved in Chelicerata. Thus, a comparative neuroanatomy within this phyllum is possible, but correlations of the chelicerate brain structures with the supraesophageal ganglion groups may be crucial (see Breidbach, this volume).

Prominent sensory input into the brain comes via the optic nerves, which terminate in the optic masses (Hanström, 1928). Since in Chelicerata, number and position of the eye differ, the distribution and structure of the appropriate sensory neuropils varies for the different species (Hanström, 1928; Strausfeld and Barth, 1993).

Hitherto, comparative anatomical characterizations of the chelicerate brain relied on the studies of Hanström, (1919, 1921, 1923, 1926, 1928, 1936; compare with: Bullock and Horridge, 1966; Gupta, 1987). Hanström suggested homologies of the chelicerate neuropils with their name-sakes in insects, thus describing a chelicerate "corpora pedunculata," a "bridge," and a "central body" (Hanström, 1928). However, these neuropil areas are no true homologues between these two different arthropod group (see Breidbach, this volume).

The adult Xiphosuran brain is a large, nearly spherical body with a flat or slightly concave dorsal surface (Fig. 7A; Fahrenbach and Chamberlain, 1987). In the anterior dorsal region, the brain is innervated by the optic nerves from the lateral, median and ventral eyes. Except for the optic masses and the "central body," the entire surface of the brain is formed by the lobes of the "corpora pedunculata" (Fahrenbach, 1977). They extend through the whole ventral hemisphere of the Xiphosuran brain. Anteriorily and laterally, these even extend upward to the dorsal surface. They are composed of approximately 5×10^7 globule cells each (Fahrenbach, 1977). Their structural characteristics, small somata, fine axonal projections, and restricted branching patterns are analogous to onychophoran globule cells (see Schürmann, this volume) or to mandibulate globule cells like the insect Kenyon cells which are found in the insect mushroom bodies. The Xiphosuran "corpora pedunculata" get indirect input from chemosensory sensilla of the legs and the opisthosoma (via the posterior lateral neuropil), presumably from the visual system and from peripheral mechanoreceptors (Chamberlain, 1978). FMRFamide like immunoreactivity but no presence of catecholamines was found in this neuropil (O'Connor et al., 1982; Watson et al., 1984).

The "central body" of the Xiphosura is found superficially on the dorso-central part of the brain. Its lateral parts are anterior-medially compressed due to the massive development of the "corpora pedunculata," and form a horseshoe-shaped mass of neuropil (Fahrenbach and Chamberlain, 1987). This is subdivided into two distinct lobes, a dorsal and a ventral one, and is dorsally overlaid by a compressed layer of cell bodies. These send short, fine-dimensioned fibers into the upper "central body" lobe.

The main sensory input into this neuropil is due to the optic tracts, by which secondary optic interneurons innervate the "central body" neuropil (Hanström 1926) and to the ocellar ganglia. From these, primary optic interneurons innervate the "central body" (Fahrenbach and Chamberlain, 1987).

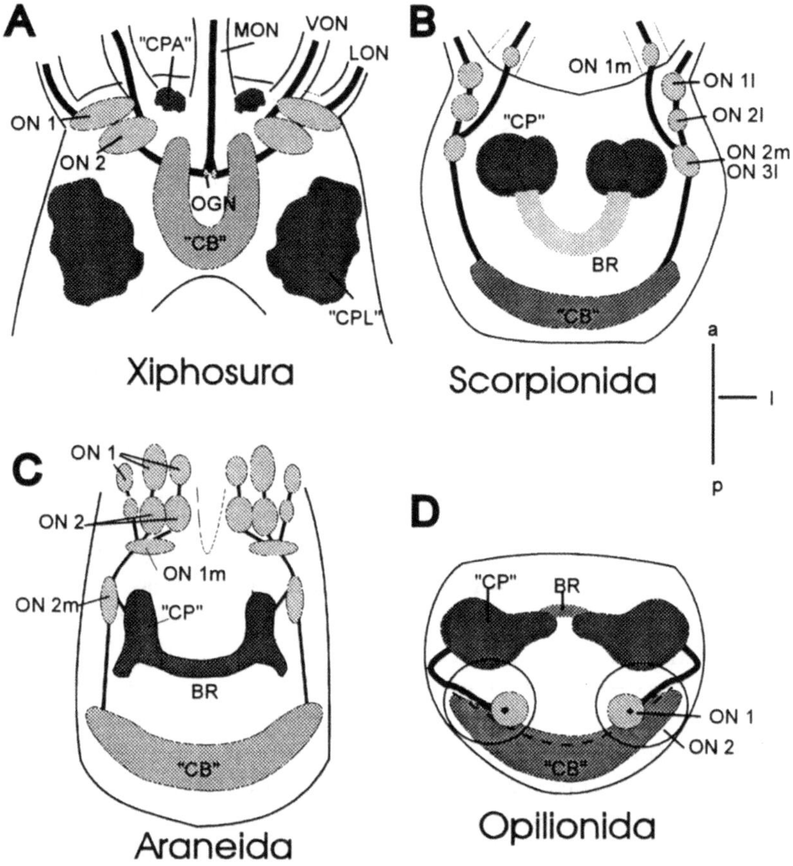

Figure 7. (A–D) Schemes of Chelicerata brains, dorsal view, outlining the main neuropil areas as the "central body" ("CB") and the "corpora pedunculata" ("CP") and their connections to the optic neuropils (ON). a: anterior, BR: bridge, "CPA": anterior lobe of the "corpora pedunculata", "CPL": lateral lobe of the "corpora pedunculata", 1: lateral, LON: lateral optic nerve, MON: median optic nerve, p: posterior, OGN: optic ganglion neuropil, ON 1, 2, 31: first to third optic neuropil of the lateral eyes, ON 1, 2 m: first and second optic neuropil of the median eyes, VON: ventral optic nerve. (A after Fahrenbach and Chamberlain, 1987; B after Hanström, 1923; C after Strausfeld et al., 1993).

By immunohistochemistry and glyoxylic-acid-histofluorescence, presence of catecholamine, serotonin (Chamberlain et al., 1983), substance P (Chamberlain and Engbretson, 1982) and proctolin (Wyse, 1983) was demonstrated for the "central body" neuropil. FMRFamide-immunostaining labels two distinct horizontal strata of immunopositive fibers, an anterior and a posterior one, in the central body of *Limulus polyphemus* (Wegerhoff, unpublished results).

Most data on the organizations of the brains of Scorpionida, Solifugida, Uropygyda, and Amblypygida date back to the classical studies of Hanström and show no significant differences to the principal brain organizations of spiders (Fig. 7; for review, see Babu, 1985).

In the Arcarine synganglion, brain structures are highly condensed. Thus by classical neuroanatomy, brain neuropil areas are difficult to recognize. According to Ioffe (1963), the tick *Dermacentor pictus* seems to have reduced its "central body". Ioffe attributed this to reductions of the eyes and the optic masses in ticks. She proposed, that the dorsal glomeruli in this species correspond to the "corpora pedunculata" of arachnids (for review, see Binnington, 1987).

Detailed information about the neuroanatomy of the arachnid brains is only availabe for spider and harvestman.

The supraesophageal ganglion of the harvestman *R. triangularis* consists of two cheliceral ganglia and the brain. It is characterized by a prominent input via the optic nerves that terminate in the optic masses. The havestman only possess two median eyes. Accordingly, the optic masses consist of two pairs of optic lobes, the 1st and the 2nd optic lobes. Caudally to the 2nd optic lobe, the first unpaired neuropil region, the so-called central body (superior associative neuropil, Breidbach and Wegerhoff, 1993) is situated. This neuropil is lined with a pair of laterally situated associative neuropil areas, the so-called corpora pedunculata. These are connected by the so-called bridge neuropil (Fig. 7D). Ventral to these, a second unpaired neuropil region, the inferior bilateral associative neuropil is situated. Five main commissures were characterized in the supraesophageal ganglion of *R. triangularis*, including an optic commissure that runs ventrally adjacent to the "central body" neuropil (Fig. 9). The neuropil areas are connected by prominent tracts. Sensory input from the optic masses is directed to both the "central body" neuropil and the "corpora pedunculata".

The "corpora pedunculata" of the spider *C. salei* (Babu and Barth, 1984; Babu, 1985) show topological characteristics (position in the ganglion, principal architecture, coupling to pathways in the brain, immunohistochemistry) similar to those of the "corpora pedunculata" of *R. triangularis* (Breidbach and Wegerhoff, 1993).

The "corpora pedunculata" of the spider *C. salei* serve as an optic neuropil for the secondary eyes (Strausfeld and Barth, 1993). *C. salei* is equipped with one pair of principal eyes and three paris of secondary eyes. Principal and secondary eyes serve two different sets of optic neuropils in the brain (Figs 7C, 8; Strausfeld and Barth, 1993). The 2nd optic lobes of the secondary eyes are connected homotopically via T-cells to the "corpora pedunculata". There is no evidence that the "corpora pedunculata" of *C. salei* are supplied by afferents other than those arising from the 2nd optic neuropil of the secondary eyes (Strausfeld and Barth, 1993). Thus, in this species, the "corpora pedunculata" seem to serve as a third order neuropil in the visual pathway from the secondary eyes.

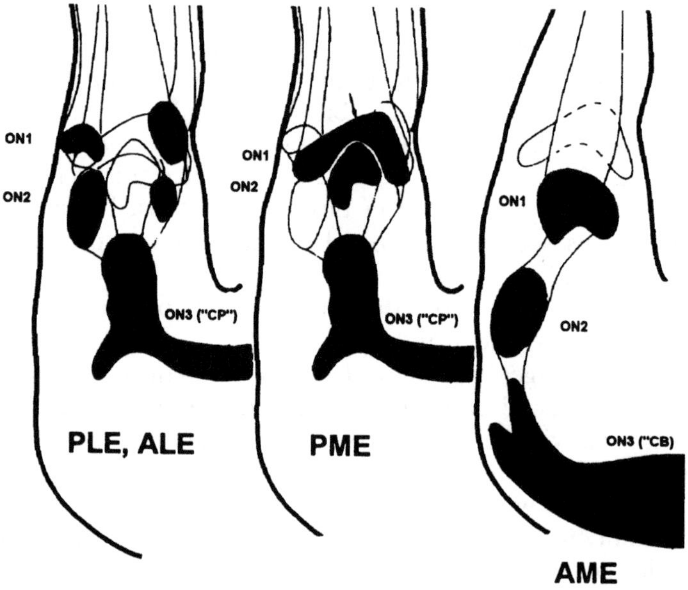

Figure 8. Two visual systems in the brain of *Cupiennius salei.* The "corpora pedunculata" ("CP") receives optic information from the secondary eyes (posterior and anterior lateral eyes [PLE, ALE) and the posterior median eyes (PME)] and serves as their third order optic neuropil (ON3), the principal eyes (anterio median eyes, AME) supply the "central body" ("CB"), which is their third order optic neuropil. (After Strausfeld and Barth, 1993).

According to the analysis of Strausfeld et al. (1993), the optic pathway of the principal eyes includes the "central body" as a third-order optic neuropil. Sensory input into this neuropil, however, is not exclusively from the 2nd-order optic neuropil of this pathway, but is even due to neurons arising from the midbrain. Furthermore, Weltzien (1988) reported on connections between the 2nd-order neuropil of the secondary eyes with the "central body" optic neuropil. These data support the interpretation of Bullock and Horridge (1966) which regarded the "central bodies" of arachnids as a visual center. Strausfeld et al. (1993) described that the "central body" is mainly supplied by tangential and small-field (columnar) inputs by the 2nd-order neuropil in the pathway from the principal eyes. Thus, they interpret the neuroarchitecture of the "central body" as reminiscent of complex retinotopic neuropils of Insecta and Crustacea.

The aranean "central body" consists of two sausage like lobes that extend horizontally in the dorsal posterior protocerebrum. These lobes are subdivided into many distinct strata (Weltzien, 1988). Apically, they are covered by a dense layer of small neuronal somata which give rise to small descending fibers, which penetrate the "central body" neuropil, and pass rostrally (Strausfeld et al., 1993). Thus, they constitute a columnar organization of the "central body" neuropil. This pattern is similarily found in opilionids, whose "central body" is likewise formed by two horizontal lobes that are subdivided into many strata (Breidbach and Wegerhoff, 1993). Already by comparison of the behavior of opilionids, where no capacity for the formation of webs is developed, it was clear that the arachnid "central body" neuropil does not serve exclusively for coordination of web building behavior. But even for spiders, no evidence has been found to connect diversifications in the organizations of the "central body" neuropil with web building behavior (Hanström, 1926). In a volumetric analysis of the morphologies of the "central bodies" of different spiders that reconsidered the volumetric measurements of Hanström (1928), Weltzien and Barth (1991) could demonstrate that, even in spiders, the "central body" has no specific role in web-building. This interpretation is supported by analysis of the development of this neuropil. In both web-building spiders and those that do not build webs, a "central body" neuropil is already clearly recognizable in spiderlings which have not even left the egg sac (Weltzien and Barth, 1991).

Homologous neurons

As in the insect and crustacean brain, immunohistochemistry allows the identification and morphological characterization of identified neurons in arachnids (for reviews see Nässel, 1993; Kutsch and Breidbach, 1994). Immunohistochemical studies show structurally similar sets of immunoreactive neurons in spiders and in opilionids. Thus, Seyfarth et al. (1990) described serotonin immunoreactive neurons of the nervous system of *C. salei* in locations corresponding to those described for an opilionid. Seyfarth et al. (1990) report on a prominent group of serotonin immunopositive neurons with terminations in the optic lobes. According to the position of their somata, axonal pathways and principal architectures, these neurons correspond to serotonin-immunoreactive neurones described for the harvestman *R. triangularis* (Fig. 9; Breidbach and Wegerhoff, 1993). Thus, a comparative analysis of pathway characteristics of serotonin immunoreactive neurons provides evidence for structural homologies in the brain of Araneae and Opiliones.

Further hints for the conservation of structural pecularities of neuropil regions in arachnids can be shown by immunostaining with antisera against neuropeptides (see also Agricola and Bräunig, this volume). Thus, the pattern of FMRFamide-like immunoreactive sites in the "central body" of various spiders and opilionids is structurally similar. FMRFamide-immunohistochemistry in both, spider and opilionid "central bodies," detects two strata of immunoreactivity: In the anterior lobe of the "central body," two layers were stained that show a high amount of immunoreactivity. In the posterior lobe of the "central body" a bright

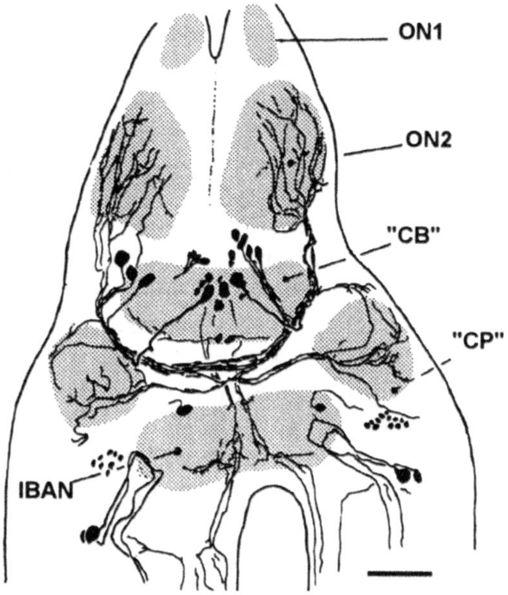

Figure 9. Serotonin-immunoreactive neurons in the brain of the harvestman *Rilaena triangularis*, whole-mount preparation, caudal view; see text for details; scale bar 100 µm; "CB": "central body", "CP": "corpora pedunculata" IBAN: inferior bilateral associative neuropil, ON 1, 2: first and second optic neuropil. (After Breidbach and Wegerhoff, 1993).

band of immunoreactivity was discerned (Fig. 10). This patterns was found in Theraphosidae (*Brachypelma albopilosa*), in Agelenidae (*Tegenaria atrica*), in Araneidae (*Zygella x–notata*), and in the opilionid *R. triangularis* (Phalangiidae) (Wegerhoff, unpublished observations).

Immunostaining with an anti CCAP (crustacean cardioactive peptides) antibody even allows the characterization of identified homologous neurons in different arachnid species (Breidbach et al., 1994). CCAP immunoreactivity labels 14 clusters of immunoreactive neurons in the protocerebrum of the spider *Tegenaria atrica* and the opilionid *R. triangularis*. In both species, these immunopositive cells possess the same number of neurons and share structural and topological characteristics. Two sets of bilateral symmetrical neurons, associated with the optic lobes and the arachnid "central body" were analyzed in more detail, comparing the opilionid *R. triangularis*, the bird spider *Brachypelma albopilsoa*, the wandering spider *C. salei* and the Araneae *Meta segmentata* and *Nephila clavipes*. Thereby, 16 neurons were identified as homologous neurons in the opilionid and the spider brain. The morphologies of these homologous neurons are conserved, showing similar peculiarities of their axonal pathways and their arborization patterns (Fig. 11). Thus, these phylogenetically persisting neurons seem to present part of the principal *Bauplan* of the arachnid brain.

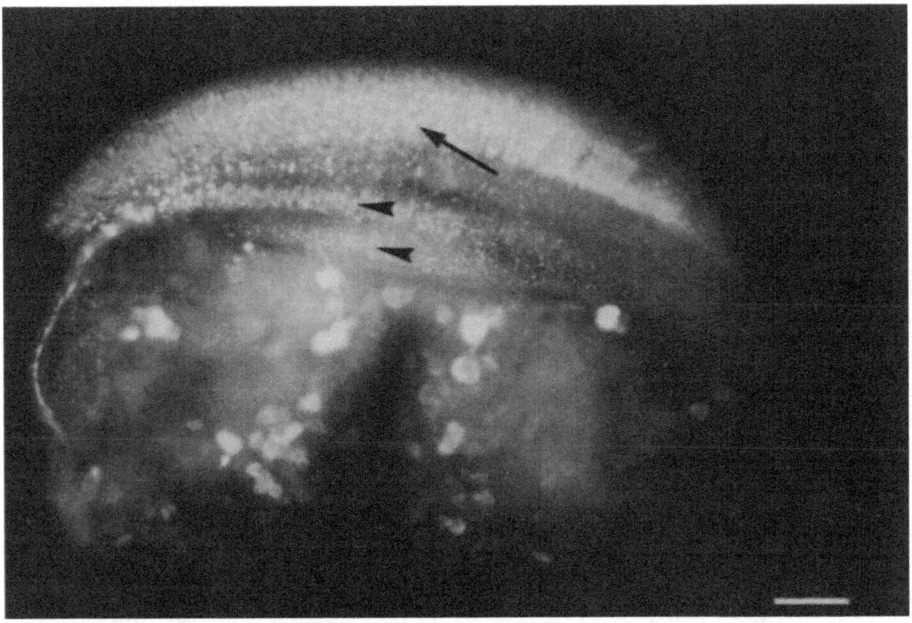

Figure 10. FMRFamide-like immunostaining in the "central body" of the spider *Tegenaria atrica* , horizontal Vibratome section, Rhodamin-conjugated secondary antibody, two distinct anterior layers (arrowheads) and a bright band in the posterior lobe (arrow) are labeled. Scale bar 50 µm.

Conclusion

Both in the brain and in the ventral nerve cord, identification of homologous neurons, as the CCAP or serotonin-immunoreactive neurons, allow to describe a scaffold of common cellular characteristics in the nervous systems of different Arachnida. These identified neurons enable to identify homologous neuropil areas and tract-substructures in different species. According to our view, in two species, an area – or even a nerve – which is supplied in a similar fashion by structurally conserved, homologous neuron is defined as a homologous structure.

The high degree of structural persistences in the nervous systems of arachnids, which is obvious even at the level of identified neurons, shows that the structural organizations of the protocerebrum of the Arachnida and presumably even of the Chelicerata in general are remarkably conserved. Likewise, the subesophageal nerve mass at least of opilionids and spiders is identical in its principal structural features. Thus, it seems possible to define a common scheme for the *Bauplan* of the arachnid nervous systems. The few data on the neuroanatomy and neuroembryonology of Xiphosura and scorpions, fit into such a general

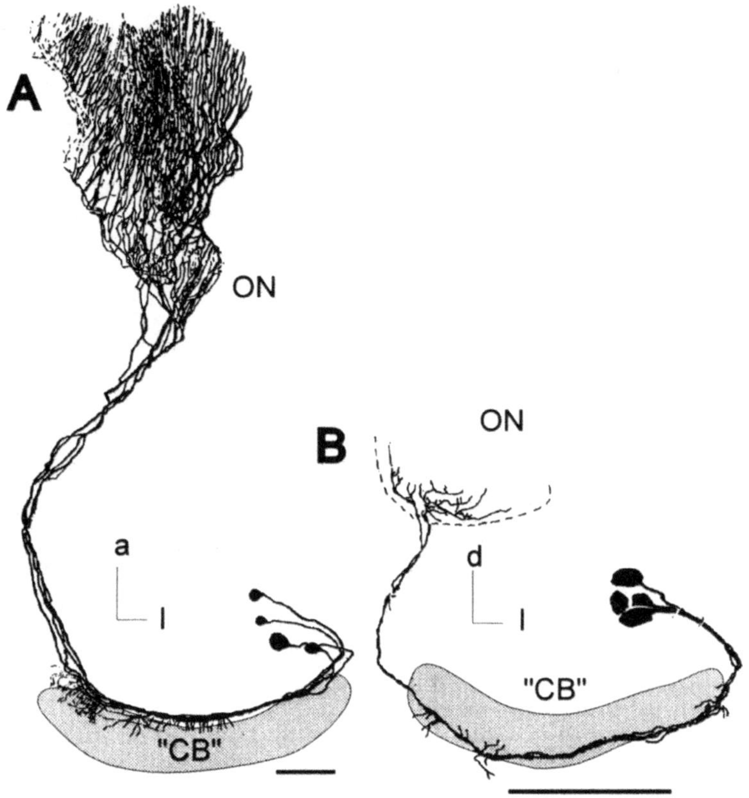

Figure 11. A,B: Homologous CCAP-immunoreactive neurons in the brain of the spider *Cupiennius salei* (A) and the harvestman *Rilaena triangularis* (B). Neurons of the dorso-frontal lateral cluster, showing the same number of somata, similar axonal pathway (contralateral projection ventrally to the "central body" ("CB")) and similar aborizations (within the "CB" and the contralateral optic neuropil (ON)), scale bar 100 μm. a: anterior, d: dorsal, l: lateral.

scheme and may allow to sketch a general *Bauplan* for the brain of Chelicerata.

Acknowledgments
Research was supported by the DFG and the Landesregierung NRW.

References

Anderson, D.T. (1973) *Embryology and Phylogeny in Annelids and Arthropods*. Pergamon Press, Oxford.

Babu, K.S. (1965) Anatomy of the central nervous systems of arachnids. *Zool. Jb. Anat.* 82: 1–154.

Babu, K.S. (1969) Certain histological and anatomical features of the central nervous system of a large Indian spider *Poecilotheria*. *Am. Zool.* 9: 113–119.

Babu, K.S. (1975) Postembryonic development of the central nervous system of the spider *Argiope aurantia* (Lucas). *J. Morphol.* 146: 325–342.

Babu, K.S. (1985) Patterns of arrangement and connectivity in the central nervous system of arachnids. *In:* F.G. Barth: *Neurobiology of Arachnids*. Springer Verlag, Berlin, pp. 3–19.

Babu, K.S. and Barth, F.G. (1984) Neuronanatomy of the central nervous system of the wandering spider, *Cupiennius salei* Keys. *Zoomorphology* 104: 325–342.

Babu, K.S. and Barth, F.G. (1989) Central nervous projection of mechanoreceptors in the spider *Cupiennius salei* Keys. *Cell Tissue Res.* 258: 69–82.

Babu, K.S., Barth, F.G. and Strausfeld, N.J. (1985) Intersegmental sensory tracts and contralateral motor neurons in the leg ganglia of the spider *Cupiennius salei* Keys. *Cell Tissue Res.* 241: 53–57

Bergström, J. (1979) Morphology of fossil arthropods as a guide to phyolgentic relationships. *In:* A.P. Gupta (ed.): *Arthropod Phylogeny*. Van Norstrand Reinhold, New York, pp. 3–56.

Binnington, K.C. (1987) Histology and ultrastructure of the acarine synganglion. *In*: A.P. Gupta (ed.): *Arthropod Brain*. J. Wiley & Sons, New York, pp. 95–110.

Breidbach, O. (1990) Constant topological organization of the Coleopteran metamorphosing nervous system, analysis of persistent elements in the nervous system of *Tenebrio molitor*. *J. Neurobiol.* 21: 990–1001.

Breidbach, O. (1992) Ist das Arthropoden-Hirn zweimal enstanden? *Natur und Museum* 122: 301–310,

Breidbach, O. and Dircksen, H. (1991) Crustacean cardioactive peptide-immunoreactive neurons in the ventral nerve cord and the brain of the meal beetle *Tenebrio molitor* during postembryonic development. *Cell Tissue Res.* 265: 129–144.

Breidbach, O. and Wegerhoff, R. (1993) Neuroanatomy of the central nervous system of the harvestman, Rilaena traingularis (HERBST 1799) (Arachnida; Opiliones) – principal organization, GABA-like and Serotonin-immunohistochemistry. *Zool. Anz.* 230: 55–81.

Breidbach, O., Dircksen, H. and Wegerhoff, R. (1994) Common bauplan of peptidergic neurons in the arachnid brain – CCAP (Crustacean cardioactive peptide) – immunoreactive neurons in the protocerebrum of different arachnid species. *Cell Tissue Res.; in press.*

Bullick, T.H. and Horridge, G.A. (1965) *Structure and Function in the Nervous System of Invertebrates*. Vol. 2, Freeman, San Francisco.

Chaboud, F., Seyfarth, E.A. and Reichert, H. (1990) Neuronal development in the spider nervous system. *In:* N. Elsner and G. Roth (eds): *Brain, Perception, Cognition*. Thieme Verlag, Stuttgart, p. 368.

Chamberlain SC. (1978) *Neuroanatomy of the Visual Affernts in Limulus polyphemus*. Ph.D. Thesis and Special Report. ISR-S17, Syracuse, New York.

Chamberlain, SC. and Egbertson, G.A. (1982) Neuropeptide immunoreactivity in *Limulus*. I. Substance P-like immunoreactivity in the lateral eye and protocerebrum. *J. Comp. Neurol.* 208: 304–315.

Chamberlain, SC. and Wyse, G. (1986) An Atlas of the brain of the horseshoe crab *Limulus polyphemus*. *J. Morpol* 187: 363–386.

Chamberlain, SC., Battelle, B.A. and Wyse, G.A. (1983) Localization of serotonin-like immunoreactivity in the *Limulus* protocerebrum. *Soc. Neurosci. Abstr.* 9: 76.

Fahrenbach, W.H. (1977) The brain of the horseshoe crab (*Limulus polyphemus*) II. Architecture of the corpora pedunculata. *Tissue Cell* 9: 157–166.

Fahrenbach, W.H. and Chamberlain, S.C. (1987) The brain of the horseshoe crab, *Limulus polyphemus*. *In:* A.P. Gupta (ed.): *Arthropod Brain*. J. Wiley & Sons, New York, pp. 63–94.

Foelix, R.F. (1985) Sensory nerves and peripheral synapses. *In:* F.G. Barth (ed.): *Neurobiology of Arachnids*. Springer Verlag, Berlin, pp. 189–200.

Foelix, R.F. (1975) Occurrence of synapses in peripheral sensory nerves of arachnids. *Nature* 254: 146–148.

Foelix, R.F. (1992) *Biologie der Spinnen*. Thieme Verlag, Stuttgart-New York.

Foelix, R.F. and Troyer, D. (1980) Giant neurons and associated synapses in the peripheral nervous system of whip spiders. *J. Neurocytol.* 9: 517–535.

Gupta, A.P. (1987) Evolutionary trends in the central and mushroom bodies of the arthropod brain. A dilemna. *In:* A.P. Gupta (ed.): *Arthropod Brain*. J. Wiley & Sons, New York, pp. 27–44.

Hammen, L. Van der (1989) *An Introduction in Comparative Arachnology*. SPB Academic Press, Den Haag.

Hanström, B. (1919) *Zur Kenntnis des zentralen Nervensystems der Arachnoiden und Pantopoden*. Inaug. Diss. Lund AB, Skanska.

Hanström, B. (1921) Über die Histologie und vergleichende Anatomie der Sehganglien und der Globuli der Araneen. *Kungl. Svenska. Vetensk. Acad. Handl.* 61: 1–39.

Hanström, B. (1923) Further notes on the central nervous system of arachnids scorpions, phalangids and trap-door spiders. *J. Comp. Neurol.* 35: 249-272.

Hanström, B. (1926) Untersuchungen über die relative Größe der Gehirnzentren verschiedener Arthropoden unter Berücksichtigung der Lebensweise. *Z. Mikr. Anat. Forsch.* 7: 135–190.

Hanström, B. (1928) *Vergleichende Anatomie des Nervensystems der wirbellosen Tiere*. Springer Verlag, Berlin.

Hanström, B. (1935) Fortgesetzte Untersuchungen über das Araneengehirn. *Zool. Jb. Anat.* 569: 455–478.

Hayes, W.F. and Barber, S.B. (1982) Peripheral synapses in *Limulus* chemoreceptors. *Comp. Biochem. Physiol.* 72: 287–293

Holmgren, E. (1916) Zur vergleichenden Anatomie des Gehirns der Polychäten, Onychophora, Xiphosurne, Arachniden, Crustaceen, Myriapoden und Insekten, *Kung. Svenska. Venten. Handl.* 56: 1–303.

Igelmund, P. and Wendler, G. (1991a) The giant fibre system in the forelegs (whips) of the whip spider *Heterophrynus elaphus* Pocock (Arachnida: Amblypygi). *J. Comp. Physiol.* A 168: 63–73.

Igelmund, P. and Wendler, G. (1991b) Morphology and physiology of peripheral giant interneurons in the forelegs (whips) of the whip spider *Heterophrynus elaphus* Pocock (Arachnida: Amblypygi). *J. Comp. Physiol.* A 168: 75–83

Ioffe, I.D. (1963) Structure of the brain of *Dermacentor pictus* Herm. (Chelicerata, Acarina) *Zool. Zh.* 42: 1472–1484.

Iwanoff, P.P. (1933) Die embryonale Entwicklung von *Limulus moluccanus*. *Zool. Jb. Anat. und Ont. d. Tiere* 56: 163–348.

Johansson, G. (1933) Beiträge zur Kenntnis der Morphologie und Entwicklung des Gehirns von *Limulus polyphemus*. *Acta Zool.* 14: 1–100.

Kutsch, W. and Breidbach, O. (1994) Homologous structures in the nervous systems of Arthropda. *Adv. Insect. Physiol.* 24: 1–113.

Lambert, A.E. (1909) History of the procephalic lobes of *Epeira cinera*. A study in arachnid embryology. *J. Morphol.* 20: 412–459.

Legendre, R. (1958a) Contribution a l'étude du développment embryonnaire des araignées. *Bullk. Soc. Zool. France* 83: 60–75.

Legendre, R. (1958b) Contribution a l'étude du systéme nerveaux des aranédes. *Ann. Bio.* 34: 139–223.

Milde, J.J. and Seyfarth, E.A. (1988) Tactile hairs and leg reflexes in wandering spiders: physiological and antomical correlates of reflex activity in the leg ganglia. *J. Comp. Physiol.* A 162: 623–631.

Nässel, D.R. (1993) Neuropeptides in the insect brain: a review. *Cell Tissue Res.* 273: 1–29.

Nentwig, W. (1987) *Ecophysiology of Spiders*. Springer Verlag, Berlin.

O'Connor, E.F., Watson, W.H. and Gordon, A.W. (1981) Identification and localization of catecholamines in the nervous system of *Limulus polyphemus*. *J. Neurobiol.* 13: 49–60.

Packard, A.S. (1893) Further studies on the brain of *Limulus polyphemus* with notes on its embryonology. *Mem. Nat. Acad. Sci.* 6: 289–331.

Patten, W. (1896) Variations in the development of *Limulus polyphemus*. *J. Morphol.* 12.

Patten, W. and Redenbaugh, W.A. (1899) Studies of *Limulus*. II The nervous system of *Limulus polyphemus*, with observations upon the general anatomy. *J. Morphol.* 16: 91–200.

Peters, H.M. (1969) Maturing and coordination of web-building activity. *Am. Zoologist* 9: 223–227.

Pross, A. (1966) Untersuchung zur Entwicklungsgeschichte der Araneae (*Pardosa hortesis* (Thorell)) unter besonderer Berücksichtigung des vorderen Prosomaabschnittes. *Z. Morph. Ökol. Tiere* 58: 38–108.

Remane, A. (1956) *Die Grundlagen des natürlichen Systems, der vergleichen-den Anatomie und der Phylogenetik*. Akademische Verlagsgesellschaft, Greest and Portig, Leipzig.

Rempel, J.G. (1957) On the embryology of the black widow spider, *Latrodectus mactans*. *Can. J. Zool.* 35: 35–74.

Root, T.M. (1985) Central and peripheral organization of scorpion locomotion. *In:* F.G. Barth (ed.): *Neurobiology of Arachnids*. Springer Verlag, Berlin, pp. 337–347.

Scharrer, B. (1941) Neurosecretion IV. Localization of neurosecretory cells in the central nervous system of *Limulus. Biol Bull. – Woods Hole* 81: 96–104.

Schmid, A. and Duncker, M. (1993) Histamine immunoreactivity in the central nervous system of the spider *Cupiennius salei. Cell Tissue Res.* 273: 533–545.

Seyfarth, E.A., Hammer, K. and Grünert, K. (1990) Serotonin-like immunoreactivity in the CNS of spiders. *In:* N. Elsner and G. Roth (eds): *Brain, Perception, Cognition.* Thieme Verlag, Stuttgart, p. 321.

Strausfeld, N.J. and Barth, F.G. (1993) Two visual systems in one brain: Neuropils serving the secondary eyes of the spider *Cupiennius salei. J. Comp. Neurol.* 328: 43–62.

Strausfeld, N.J., Weltzien, P. and Barth, F.G. (1993) Two visual systems in one brain: Neuropils serving the principal eyes of the spider *Cupiennius salei. J. Comp. Neurol* 328: 63–75.

Sherar, W.A. (1985) *Spiders. Webs, Behaviour and Evolution.* Stanford University Press, Stanford.

Watson, W.H., Groome, J.R., Chronwall, B.M., Bishop, J. and O'Donohue T.L. (1984) Presence and distribution of immunoreactive and bioactive FMRF amide-like peptides in the nervous system of the horseshoe crab. *Limulus polyphemus. Peptides* 5: 585–592.

Wegerhoff, R. and Breidbach, O. (1989) Anatomie des Ventralganglions der Zitterspinnen (Pholcidae). *Verh. D. Zool. Ges.* 83: 472–473.

Weltzien, P. (1988) *Vergleichende Neuroanatomie des Spinnengehirnes unter besonderer Berücksichtigung des Zentralkörpers.* PhD. Thesis, Frankfurt.

Weltzien, P. and Barth, F.G. (1991) Volumetric measurements do not demonstrate that the spider brain "central body" has a special role in web building. *J. Morphol.* 207: 1–8.

Weygold, P. (1975) Untersuchungen zur Embryologie und Morphologie der Geißelspinne *Tarantula marginemaculata* C.L. Koch (Arachnida, Amblypygi, tarantulidae). *Zoomorph.* 82: 137–199.

Weygold, P. (1985) Ontogeny of the arachnid central nervous system. *In:* F.G. Barth (ed.): Neurobiology of Arachnids. Springer Verlag, Berlin, pp. 20–37.

Weyse, G. (1983) Serotonin-, proctolin-, and bombesin-like immunoreactivity: Histochemical localization in *Limulus* nervous system. *Soc. Neurosci. Abstr.* 9: 75.

The Nervous Systems of Invertebrates: An Evolutionary and Comparative Approach
ed. by O. Breidbach & W. Kutsch
© 1995 Birkhäuser Verlag Basel/Switzerland

Conservation *versus* change in early axonogenesis in arthropod embryos: A comparison between myriapods, crustaceans and insects

P.M. Whitington

Department of Zoology, University of New England, Armidale, NSW 2351, Australia

Summary. Studies in the embryos of two evolutionary distant winged insects, the grasshopper and the fruitfly, have revealed a remarkable conservatism in the pattern of early axonogenesis in the CNS, and it has been suggested that all arthropod nervous systems are constructed using the same embryonic plan. We have tested this idea by examining early axonogenesis in the CNS of several groups of arthropods. The spatio-temporal pattern of axon growth from central neurons in two crustaceans, the woodlouse (Isopoda) and the freshwater crayfish (Decapoda), is not identical to that seen in insects. However, we have observed many similarities which encourage us to believe that the insects and the crustaceans share a common, ancestral developmental program for the initial construction of central axon pathways. The pattern of early axonogenesis in the apterygotic insect, the silverfish (Thysanura) is very similar to that seen in the winged insects. However, differences are seen in the pathways taken by apparently homologous neurons and in the relative timing of axonogenesis. These differences are comparable in extent to those seen between the two crustaceans and the winged insects, providing further support for the proposition that those two groups share a common *Bauplan* for CNS development. Substantial differences exist between the Insecta and a myriapod, the centipede, in the pattern of early axonogenesis. In the centipede embryo, the first axons from neurons in the segmental neuromeres arise when there are many more central neurons than in the insects and the pioneering neurons are widely separated from each other. In addition, the pattern of neurons involved in the early phase of axonogenesis bears no obvious similarities to that seen in the insects. These data may indicate that the Myriapoda are more distantly related to the Insecta than are the Crustacea.

Embryos, nervous systems and evolution

Major advances in our understanding of the regulation of animal development have given fresh impetus to the resolution of some classical problems in evolutionary biology. In this article, I will review our knowledge of the comparative pattern of early axonogenesis in the CNS of various groups of arthropods. These studies have the potential to shed light on two fundamental questions of an evolutionary nature:

What are the evolutionary relationships between the major arthropod groups? Are arthropods monophyletic or polyphyletic? Soon after the name Arthropoda was adopted in the middle part of the last century, debate began in earnest on this question (see Tiegs and Manton, 1958, for an account of the early history of this subject) and has continued to the present day (Shear, 1992). In the early part of this century, evolutionary biologists rigorously applied Haeckel's Biogenetic Law in an attempt to reconstruct phylogenetic relationships from embryological data. While it has long been recognized that the developmental sequence of an organism does not strictly recapitulate its evolutionary history, embryological affinity continues to provide one of the most useful characters in phylogenetic studies. Classical embryologists mainly used features at the tissue grade of organization for this purpose – we can now extend this analysis to the cellular, molecular and genetic levels. Nervous systems provide a rich source of material for such comparisons because of the wide variety of cell types present and the complex structure of any individual cell. The relative conservatism of nervous system structure, compared to, say, epidermal tissue, also offers advantages for reconstructing relationships between distantly related groups.

How have developmental processes been modified during evolution to yield the wide diversity of nervous systems seen in adult arthropods? As Wolpert (1992) has so elegantly put it, "In a sense all evolution of multicellular animals is the brilliant result of altering developmental programmes". In the first instance, we would like to know whether a common, conserved programme for early neural development exists across all of the major arthropod groups or whether these groups show a fundamentally different mode of neural development from the outset. The next step will be to determine how programs for nervous system development have changed during arthropod evolution. Changes in the properties of individual neurons and in the wiring of the nervous system could, in principle, result from a variety of means; changes in neurogenesis or neuron elimination; changes in the fate assigned to a particular neuron; changes in the pathfinding ability of an axon; or changes in the capacity of neurites to establish and maintain synaptic connections with other neurons. Since the nervous systems of ancient arthropod embryos are not available for examination, the only method we have for addressing this question is to compare neural development in a variety of extant arthropod species. Clearly, to reconstruct changes, the nervous systems of the organisms under examination must share enough similarities that one can confidently homologize the neural elements. To deduce the direction of change, one must also have a sound knowledge of the phylogenetic relationship between the organisms in question. In practice, this might mean drawing comparisons between relatively closely related groups (e.g., within a class). Ultimately, one would like to have an understanding of the genetic and molecular basis for the developmental changes that take place during evolution. Such knowledge will indicate the constraints that developmental processes have placed on the final morphology of the nervous system and, in turn, the behavioral outcomes that are available for natural selection to act upon.

The contribution of the insects

The development of the nervous system in one group of arthropods, the Insecta, has been the subject of intense research activity over the last 10–15 years. Our understanding of the cellular, molecular and genetic basis for phenomena such as neuron determination and axon pathfinding in the insects is now more detailed than for any other animal group, including the vertebrates. Indeed, much of the conceptual framework of modern developmental neurobiology derives from studies in the insects.

The greatest advances in our understanding of neural development in the insects have been made in two main areas: genetic and cellular mechanisms for neuron determination in the CNS (Campos-Ortega and Jan, 1991; Doe, 1992) and PNS (Jan and Jan, 1990; Ghysen et al., 1993); and cellular, molecular and genetic mechanisms for axon pathfinding and synaptogenesis (Harrelson and Goodman, 1988; Nose et al., 1992; Keshishian and Chiba, 1993; Whitington, 1993).

There are two main reasons for the analytical power of the insect nervous system: (a) by working with insects, one has the ability to monitor the development of single, identified neurons – this affords a great degree of precision both to descriptive and experimental studies conducted at a cellular level; (b) one particular insect, the fruitfly *Drosophila melanogaster* offers the opportunity of using genetic techniques to analyse the cellular basis for nervous system development and for identifying and characterizing genes regulating these processes.

These recent insect studies have employed a relatively small number of species in a few orders of pterygotic (winged) insects – Orthoptera, Blattodea, Lepidoptera and Diptera – and the bulk of our knowledge rests upon studies in but two insects – the grasshopper and the fruitfly. While the other Arthropoda – the apterygotic insects, the Crustacea, the Myriapoda and the Chelicerata – also possess individually identifiable neurons, and therefore lend themselves to the same detailed cellular studies as have been conducted in insects, little is known about embryonic neural development in these groups at this level of resolution.

Recent studies have begun to examine axonogenesis at the level of individually identified neurons in these other arthropod groups. I shall review work which has been carried out on the pattern of early axonogenesis in the central nervous system of five arthropod embryos: the crustaceans *Porcellio scaber* (woodlouse), *Procambarus clarkii* and *Cherax destructor* (freshwater crayfish), the primitive, wingless insect *Ctenolepisma longicaudata* (silverfish) and the myriapod *Ethmostigmus rubripes* (centipede).

Early axonogenesis in the CNS of winged insects

Gross structure of the insect CNS

The insect CNS consists of a brain and a chain of ganglia, which lie posterior to the oesophagus and comprise the ventral nerve cord. There are three gnathal ganglia (mandibular, maxillary and labial), three thoracic ganglia (pro-, meso- and metathoracic) and between eight to 11 abdominal ganglia. In the adult, the ganglia may show various degrees of fusion, but all arise embryonically as separate neuromeres. Adjacent ganglia are connected by a pair of longitudinal connectives and each neuromere possesses a pair of transverse commissures and a number of peripheral nerves. The latter comprise: (a) an intersegmental nerve, which has two roots, one arising from the neuromere immediately anterior to the nerve (anterior root), the other from the neuromere in the same segment as the nerve (posterior root); and (b) a variable number of segmental nerves (Fig. 1). A similar pattern of early axonogenesis is seen in all segmental neuromeres, although some intersegmental differences exist for particular neurons (see MP3 progeny below).

Early axonogenesis in the grasshopper embryo

The embryonic development of the insect nervous system has been described in many classical papers, dating back to the turn of the century (reviewed in Anderson, 1973). However, Bate was the first to apply the modern concept of identified neurons to the study of insect embryogenesis, choosing as his subject, the grasshopper embryo. Following on from Bate's pioneering descriptions of neurogenesis in the CNS (Bate, 1976a) and axonogenesis in the limb bud (Bate, 1976b) and CNS (Bate and Grunewald, 1981), Goodman and co-workers published a series of papers which documented in exquisite detail the pattern of early axonogenesis in the CNS of this insect embryo (Bastiani et al., 1986; Bastiani and Goodman, 1986; du Lac et al., 1986; Goodman et al., 1981; Raper et al., 1983; Taghert et al., 1982). I shall describe in detail the establishment of the major axon pathways in the CNS of the grasshopper embryo – the longitudinal connectives, the transverse commissures, the intersegmental nerve and the segmental nerve – as a basis for comparison with other insects and arthropods.

Establishment of longitudinal connectives

The longitudinal connectives are pioneered on each side of the body by axons from a quartet of segmentally repeated neurons, MP1, dMP2, vMP2 and pCC (Fig. 2). The two MP1 neurons arise by the division of

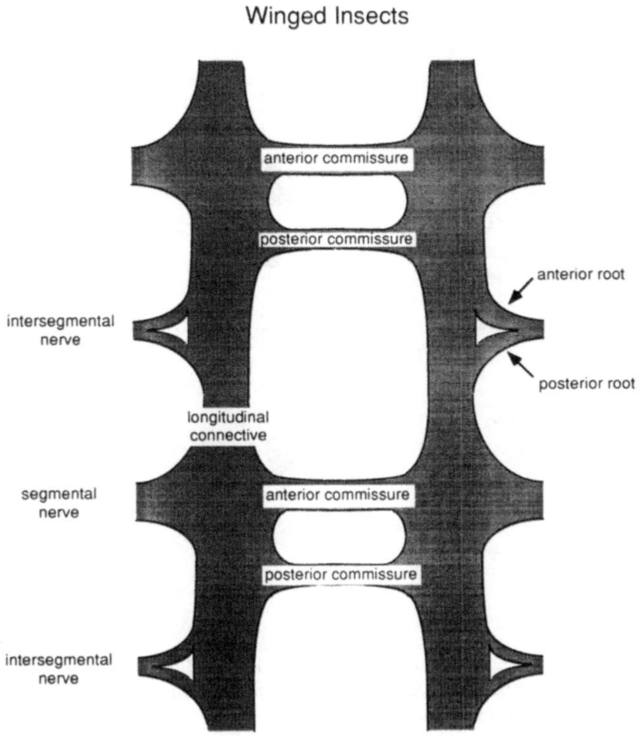

Figure 1. Gross morphology of ventral ganglia in a generalized winged insect.

a single midline precursor cell, MP1, while the MP2 neurons arise by the division of a pair of MP2 precursors which lie on either side of the midline (Goodman et al., 1980). There are six MP precursors in the grasshopper embryo, MP1–MP6. The axonal development of the other MP progeny is described below. The MP1, dMP2, vMP2 and pCC neurons send an axon dorsally to the basal lamina, which covers the developing CNS. The MP1, dMP2 and vMP2 axons then grow laterally for a short distance, bifurcate and initially run both anteriorly and posteriorly along the future position of the connectives: one of the branches is quickly retracted and the branch running in the opposite direction continues to extend along the nerve cord (Bastiani et al., 1986). The dMP2 and MP1 axons grow out at the same time and extend in a posterior direction, fasciculating with each other and associating closely with the soma of the pCC neuron. The pCC axon grows around the lateral edge of the soma of its sibling aCC. It then extends towards, contacts and grows anteriorly along posteriorly coursing dMP2/MP1 axons: it pauses at the level of the dMP2/MP1 somata and continues its anterior growth when the dMP2/MP1 axons from the next most anterior segment arrive in its vicinity. The vMP2 axon grows anteriorly in a

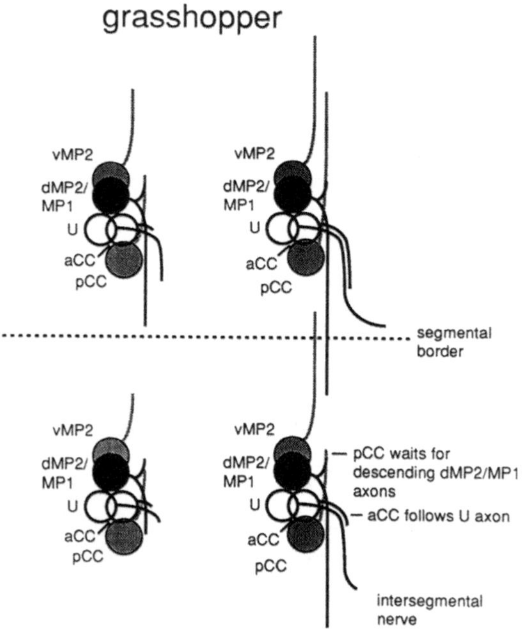

Figure 2. Neurons involved in the establishment of the longitudinal connectives and intersegmental nerve in the grasshopper embryo. The relative times at which axons are sent out from each neuron and the trajectories taken are indicated.

separate fascicle to the dMP2/MP1/pCC axons, associating with the medio-lateral surface of an identified glial cell. The connection between adjacent neuromeres is therefore established by the combined anterior growth of the vMP2 axon and the posterior growth of the dMP2/MP1 pair.

Establishment of intersegmental nerve

The anterior root of the intersegmental nerve in the grasshopper is pioneered by a pair of axons, from the U1 and U2 neurons, which extend laterally somewhat later than the vMP2/dMP2/MP1 axons (Bastiani and Goodman, 1986), (Fig. 2). The U axons cross the MP1/dMP2 fascicle and grow posteriorly several microns lateral to the latter. When the U axons reach the segment border, they diverge laterally from the connective towards an identified glial cell, the segment boundary cell (SBC) to pioneer the anterior root of the intersegmental nerve.The aCC neuron sends out a growth cone at about the same time as its sibling pCC, but waits until the U axons have grown past, before turning posteriorly down the connective, fasiculating with the U axons. Like the U axons, the aCC axon shows a high affinity for the SBC, as it turns laterally into the intersegmental nerve. Ablation experiments attest to the importance of the U axons in guiding the posterior growth of the aCC axon down the connective (du Lac et al., 1986) and of the SBC in guiding the lateral growth of the aCC and U axons into the intersegmental nerve.

Establishment of anterior commissure

The development of the anterior commissure has not been followed in detail. However, Goodman et al., (1981) report that this commissure is pioneered by unidentified neurons which lie lateral to the connectives. An early axon in the anterior commissure is that of the MP4 neuron, which extends anteriorly from the medially located soma, pioneering the median fibre tract. This axon bifurcates just before the anterior commissure and the processes extend laterally in the anterior commissure (Fig. 3). Later growing axons of the DUM neurons follow the median fibre tract pioneered by the MP4 neuron.

Establishment of posterior commissure

The establishment of the posterior commissure in the grasshopper embryo has recently been described by Myers and Bastiani (1993). This axon pathway is pioneered by the Q1 neuron, which lies just lateral to the pCC neuron. The Q1 neuron initially extends an axon anteriorly along the longitudinal MP1/dMP2 fascicle, at about the same time as the pCC neuron begins axonogenesis. It contacts the aCC and pCC somata shortly after outgrowth, at which time it turns medially toward the MP1 soma, and grows further medially, crossing the midline while extending along the axon of its contralateral homologue (Fig. 3).

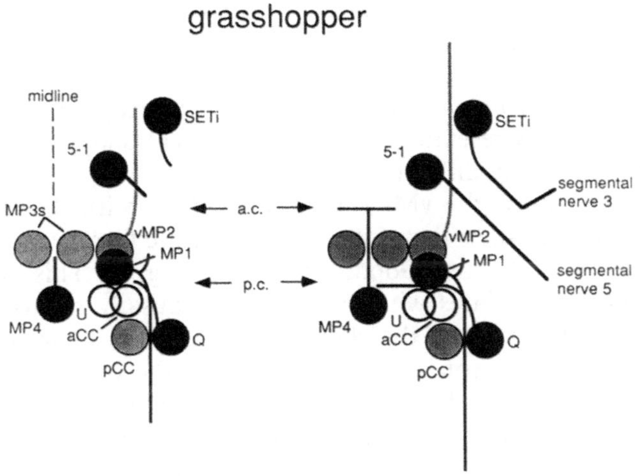

Figure 3. Neurons involved in the establishment of the anterior and posterior commissures (a.c. and p.c.) and segmental nerves in thoracic ganglia of the grasshopper embryo.

Establishment of segmental nerves
The segmental nerves in the grasshopper embryo are pioneered by a set of axons which grow dorsally to the basal lamina and then turn laterally in a parallel bundle (Whitington, 1989). At the edge of the ganglion these axons diverge to form the various nerve roots. The pioneer of nerve root 3 in thoracic segments is the identified leg motoneuron SETi, while nerve root 5 is pioneered by a neuron, 5–1, which lies just medial to the connectives and anterior to the anterior commissure (Fig. 3). These axons grow out in quick succession, beginning axonogenesis just after the pioneers for the connectives.

Axonogenesis by other MP neurons
The remaining MP neurons, MP3, MP5, and MP6 begin axonogenesis somewhat later than the MP1 and MP2 progeny. One of the MP3 progeny develops into a distinctively-shaped neuron, the H-cell, so-called because it extends axons laterally into the posterior commissure, each of which subsequently branches anteriorly and posteriorly along the connectives (Kuwada and Goodman, 1985). This pattern of axon growth, seen in the thoracic segments, is modified in abdominal segments (Bate et al., 1981). The sibling of the H-cell sends a single axon anteriorly along the connective. Axon growth from the MP5 and MP6 progeny has not been described in detail, but MP6's axon is reported to bifurcate in the posterior commissure (Goodman et al., 1981).

Early axonogenesis in the Drosophila *embryo*

The only other insect in which early axonogenesis has been described in comparable detail to the grasshopper embryo is the *Drosophila* embryo (Jacobs and Goodman, 1989; Klämbt et al., 1991; Goodman et al., 1984; Sink and Whitington, 1991a,b).

The establishment of the major axon pathways in the CNS of Drosophila closely resembles that seen in the grasshopper, but with the following interesting differences.

Firstly, in the fly, the vMP2 axon fasciculates with the pCC, dMP2 and MP1 axons, rather than following a separate fascicle, as seen in the grasshopper (Jacobs and Goodman, 1989), (Fig. 4). This difference in axon affinities correlates with, and may be causally related to, differences in the expression of the glycoprotein, fasciclin II, which is one of a class of putative axon guidance molecules, the fasciclins, which are expressed on subsets of axon pathways during specific phases of embryogenesis (Grenningloh et al., 1990). The fly vMP2 neuron expresses fasciclin II in the region where it fasciculates with the pCC axon, whereas the grasshopper vMP2 neuron does not express fasciclin II (Grenningloh et al., 1991).

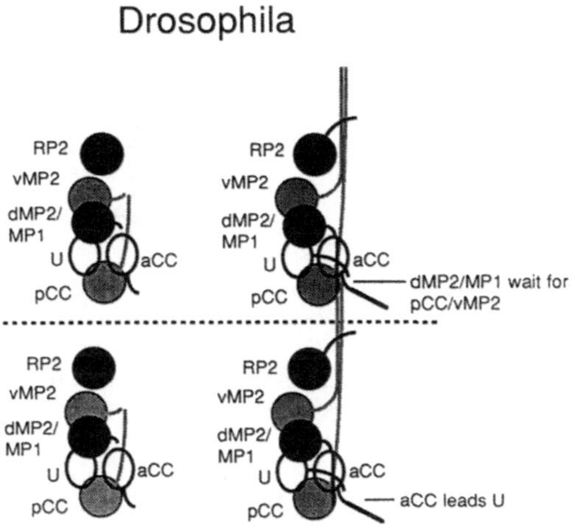

Figure 4. Neurons involved in the establishment of the longitudinal connectives and intersegmental nerve in the Drosophila embryo.

Secondly, the fly MP1/dMP2 axons extend posteriorly beyond the aCC soma in their own segment only after the anteriorly projecting pCC/vMP2 fascicle from the next segment reaches them (Fig. 4) (Jacobs and Goodman, 1989). Hence, the intersegmental connection is established solely by anteriorly projecting axons. In the grasshopper, the posterior advance of the MP1/dMP2 axons is independent of the anterior growth of the vMP2 axon and the intersegmental connection is therefore established by both anteriorly and posteriorly projecting axons. Bossing and Technau (personal communication) have recently shown that the fly MP1 neuron sends out an anterior axon collateral after it has formed its posterior branch.

Thirdly, the fly aCC axon pioneers the intersegmental nerve, whereas in the grasshopper it is preceded by the U axons (Fig. 4), (Jacobs and Goodman, 1989). It has been suggested (Jacobs and Goodman, 1989) that this is a consequence of the relative proximity of the SBC, the apparent axon guidance cue, to the aCC axon: in the grasshopper, the aCC axon is not within filopodial reach of the SBC at the point where it turns into the intersegmental nerve, and must therefore follow the U axons in order to reach this cell. In the fly, on the other hand, the SBC is within filopodial reach of the aCC neuron, before it turns away from the longitudinal connective. This explanation begs the question of the cue(s) used by the U axons in the grasshopper to navigate from the connective into the route of the intersegmental nerve.

Since the neurons that pioneer the anterior commissure have not been identified in either the grasshopper or the fly, and the pioneers of the posterior commissure are only known definitively in the grasshopper, it is difficult to draw precise comparisons between these species with

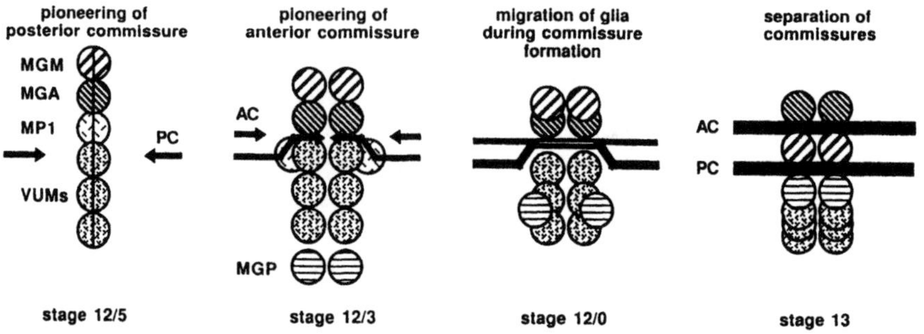

Figure 5. Relationship of the anterior and posterior commissures in the Drosophila embryo at early stages of their formation (from Klämbt et al., 1991).

respect to commissure formation. Jacobs and Goodman (1989) claim that the RP1/3 neurons, motoneurons which project contralaterally in the anterior commissutre and out the intersegmental nerve, are amongst the earliest axons into the anterior commissure in the fly. However, Sink and Whitington (1991) find that the anterior commissure is already relatively well-developed when RP1 and RP3 begin axonogenesis. The RP1/2 neurons have been reported to exist in the grasshopper (Doe et al., 1986) but no details are available concerning the spatio-temporal sequence of their axon growth. A similarity exists between the fly and the grasshopper in that the neuron that pioneers the median fibre tract, (MP4 in the grasshopper and the anterior-most VUM cell in the fly, its likely homologue) bifurcates into the anterior commissure shortly after the latter forms (Fig. 5) (cf., Klämbt et al., 1991 and Goodman et al., 1981). However, the relationship between the anterior and posterior commissures clearly differs between these insects. In the grasshopper, the anterior commissure is separated from the posterior commissure from the earliest stages of its formation (Myers and Bastiani, 1993). In contrast, in the fly, the anterior and posterior commissures are initially in contact at the midline and become separated later as a result of the posterior migration of the MGM glia (Fig. 5) (Klämbt et al., 1991). This difference in axon association may be a trivial consequence of the relatively closer packing of cells in the developing *Drosophila* CNS, compared to the grasshopper. It would help to resolve the issue if the pioneer for the posterior commissure in *Drosophila* were known. Klämbt et al. (1991) report that the pioneering axon arises from a position two cell diameters lateral to the midline, which would place the neuron at an approximately equivalent position to neuron Q in the grasshopper.

Studies in the *Drosophila* embryo have identified the pioneer of the posterior root of the intersegmental nerve – neuron RP2 (Jacobs and Goodman, 1989). This axon extends somewhat later than the pioneers of the conectives, and heads in an antero-lateral direction towards the aCC/U fascicle to find the posterior root of the intersegmental nerve (Fig. 4). The RP2 neuron has been identified in the grasshopper embryo but its spatio-temporal pattern of axon growth has not been described.

Whether the *Drosophila* embryo possesses homologues to the grasshopper neurons MP3, 5 and 6 is presently unclear. *Drosophila* has been claimed to lack a homologue to the MP3 progeny (Thomas et al., 1984). However, Bossing and Technau (personal communication) have recently identified a pair of neurons in *Drosophila* which occupy a similar position to the MP3 progeny and which show a similar axonal morphology to these grasshopper neurons, although the putative homologue to the H-cell sib sends axon branches into both connectives, rather than into only one.

Early axonogenesis in other winged insects

It has been claimed that the pattern of early axonogenesis in the CNS of a range of winged insects, including the moth *Manduca*, the fly *Calliphora*, the bug *Rhodnius* and the cockroach *Periplaneta* is fundamentally similar to that seen in the grasshopper and *Drosophila* (Thomas et al., 1984). However, with the exception of a figure showing axon morphologies of presumed homologues to the aCC, pCC, vMP2, dMP2 and MP1 neurons in the moth, no details of the pattern of axonogenesis were presented for these other winged insects. Wang and Denburg (1992) have shown that the cockroach embryo possesses a homologue of the grasshopper MP4 neuron, which pioneers the median fibre tract. Figure 1 of this paper also shows a pair of neurons in the corresponding position to the vMP2/dMP2/MP1 set in the grasshopper. Both of these cells send axons anteriorly along the connective and are claimed to pioneer this tract. If these cells are homologous to two of the vMP2/dMP2/MP1 neurons, the cockroach would appear to differ from the grasshopper and *Drosophila*, since only one of these neurons (vMP2) projects anteriorly in the latter two insects.

Conclusions

While some differences are found, there is clearly a remarkable degree of conservatism in the pattern of early axonogenesis in the CNS of at least two winged insects, the fly and the grasshopper. Both insects possess a set of at least nine identified neurons – aCC, pCC, RP2, MP1, dMP2, vMP2, RP1, Us and MP4 – which are involved in the initial establishment of the grid of longitudinal and transverse axon bundles making up each segmental neuromere. The pathways taken by the axons of corresponding neurons is virtually identical in the two species, and there are only minor changes in the order of axon outgrowth. Since the grasshopper is a relatively primitive winged insect, while *Drosophila* is advanced, this result implies an early and subsequent conservative evolution of the developmental programme for CNS development in the winged insects. Evolutionary modifications to this programme have involved changes in the association of axons and minor changes in the relative timing of axon outgrowth and axonal morphology (e.g. an additional branch in the fly MP1 neuron compared to its grasshopper homologue).

Early axonogenesis in the CNS of crustaceans

The question then arises as to when, during arthropod evolution, the programme for early CNS development present in the winged insects arose. In the same paper in which they demonstrated strong conservatism between the grasshopper and fly embryos in CNS development, Thomas et al. (1984) claimed that the same set of central neurons is also present in the embryo of the crayfish *Procambarus*. On this basis they concluded that "all arthropod nervous systems seem to be constructed using the same embryonic plan". Unfortunately, no figures of axon morphologies of crayfish neurons were shown in this report. This finding has been widely cited in subsequent studies. For example, it was taken (together with a reported strong similarity between insect and crayfish embryos in the pattern of neural expression of the segmentation gene *engrailed*) as evidence that these two groups "share a relatively recent common ancestor" (Patel et al., 1989), showing that the finding has considerable phylogenetic, as well as developmental significance. Doe (1992) repeats the claim that "early CNS development is virtually identical among arthropods", while Arbas et al. (1991) state that "the cellular pattern of early neurogenesis generalizes widely between the progenitors of the ventral nerve cord in insects and crustaceans".

Given the far-reaching implications of this claim for strong conservatism in the pattern of early axon outgrowth between the insects and the crustaceans, we decided to re-address the issue by examining axonogenesis in the CNS of two crustaceans: the woodlouse (Isopoda) and the freshwater crayfish (Decapoda) (Whitington et al., 1993).

Early axonogenesis in the CNS of the woodlouse and crayfish

The mature embryonic ventral nerve cord in both the woodlouse and the crayfish is very similar to that seen in the winged insect at a gross morphological level. Pairs of longitudinal connectives join the segmental ganglia, each of which has a single anterior and posterior commissure and paired segmental and intersegmental nerves (the latter with anterior and posterior roots) (Fig. 6A, B). In addition, both the woodlouse and the crayfish possess an unpaired medial nerve which runs along the whole length of the ventral nerve cord, linking adjacent ganglia. This median nerve appears to be pioneered by a neuron in the manidibular segment, which sends a process posteriorly well before the appearance of other axons in the CNS (Fig. 7).

Somata in the CNS can be clearly visualized in both the woodlouse and crayfish embryo under Nomarski optics. Within each species, the arrangement of these somata is relatively constant between embryos of the same age and, up to late stages of embryogenesis, individual somata can be reliably identified from soma position alone. Figures 8A, B show the arrangement of these somata in the dorsal-most layers of the CNS at 50% of development in the woodlouse and crayfish embryos, respectively, shortly after the first axons appear in the CNS. We concentrated on the set of neurons that lie in dorsal positions between the connectives at the earliest stages of axonogenesis, since this corresponds to the region in which the conserved insect neurons lie. Neurons were impaled with microelectrodes and injected with Lucifer Yellow (LY) to determine their axon morphology. Embryos were subsequently incubated with anti-LY antibody and HRP-conjugated secondary antibody to deposit a stable DAB reaction product in the injected neurons. We have restricted our observations to the thoracic ganglia in both species. No significant variation in axon morphology between segments was observed for any of the neurons under study.

We have used similarities in some position with respect to surrounding cells, and in axon morphology (see below) as criteria for homology and assigned a common alphabetical label to provisionally homologous neurons in the two species.

Establishment of longitudinal connectives

The first axons to appear in the CNS of these embryonic crustaceans lie in the future position of the longitudinal connectives, but they have a different origin in the two species. In the crayfish, the very first axons originate from neurons located in the brain. These axons arrive at segmental ganglia, at around the time that neurons in those ganglia are beginning axonogenesis. The latter axons fasciculate with the brain-derived axons. In the woodlouse, the pioneering axons in the longitudinal connectives arise from segmental neurons (see below) and axons from the brain descend later in embryogenesis. This may represent an example of heterochronic change in development, associated with the relative precocity of head segment formation in the crayfish compared to the woodlouse (Scholtz, 1992).

Discounting the apparent precocious posterior growth of axons from neurons in the brain in the crayfish embryo, the longitudinal connectives are pioneered by a similar set of neurons in both crustacean species

(Fig. 9A, B). In the woodlouse, neuron M, which lies just posterior to the future posterior commissure, grows out in advance of all other axons. A single, broad, lammelipodial process extends from the antero-lateral region of the soma and advances between the region of contact of the neighbouring somata D and E. This process enwraps the lateral and medial faces of cell E and it subsequently thins into an axon, which grows anteriorly, eventually extending to the brain. The posteriorly directed process on the lateral side of cell E is eventually retracted but persists for at least the next 10% of development. The advancing anterior axon of the M neuron in the next most posterior segment contacts this posterior process and, in this way, the connection between adjacent ganglia is established. Posteriorly directed axons from other neurons do not develop until after this connection has been made. Later in development (from around 60% onwards), the M neuron extends collateral processes out the segmental nerve in every segment through which the main axon courses.

Axon M is followed closely by those of D and K. Each of these axons grows from the antero-lateral side of the soma and advances anteriorly (Fig. 9A). Axons M, D and K run along the medial border of the

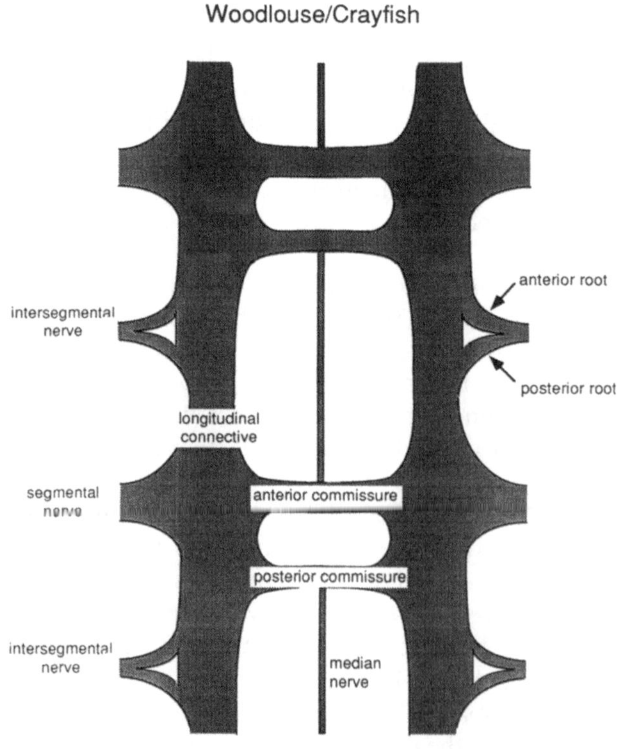

Figure 6A.

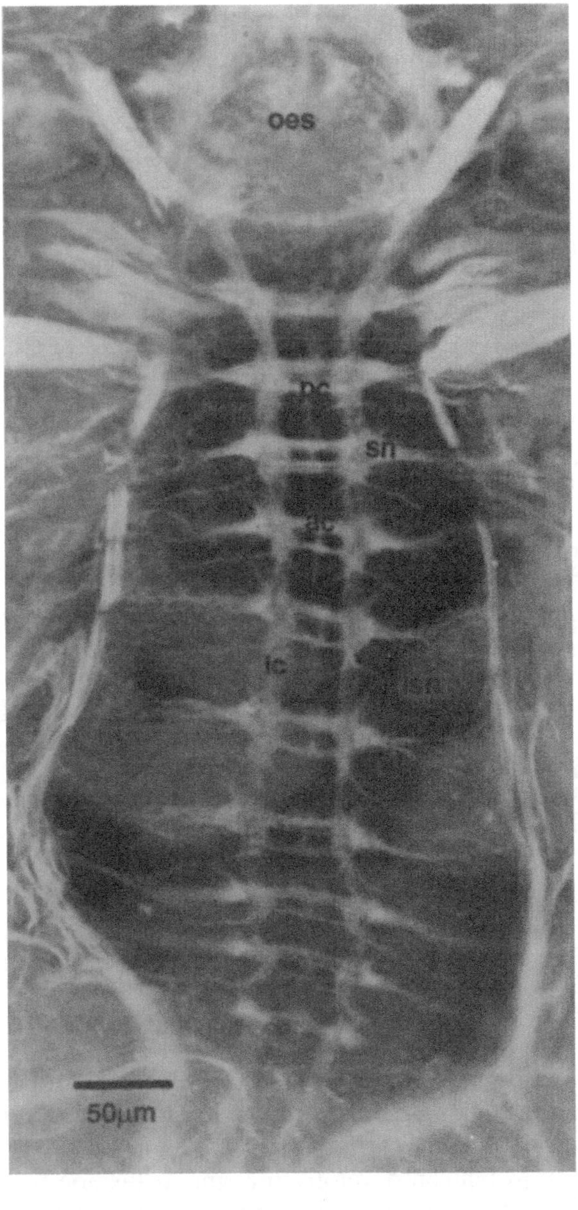

Figure 6B. Gross morphology of ventral ganglia in woodlouse and crayfish embryos. (A) Schematic diagram. (B) Photomicrograph of a rhodamine-phalloidin stained preparation of a crayfish ganglion (from Whitington et al., 1993), showing the ventral nerve cord with anterior and posterior commissures (ac, pc), intersegmental nerve (isn) and segmental nerve (sn) and longitudinal connectives (lc), and oesophagus (oes).

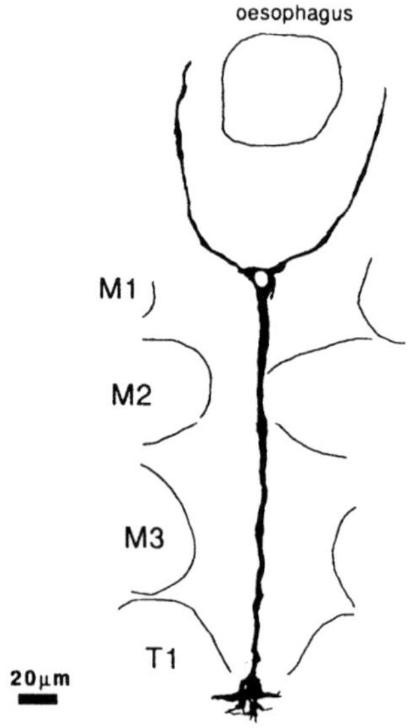

Figure 7. A large neuron in the mandibular ganglion pioneers the median nerve in the crayfish embryo. Camera lucida drawing of an LY fill.

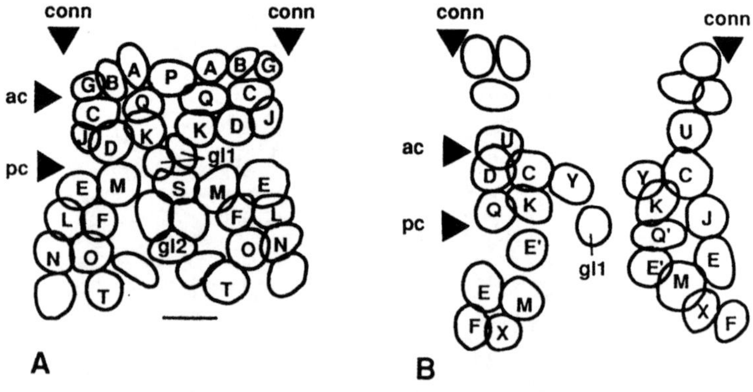

Figure 8. Arrangement of dorsal somata lying between the connectives in the woodlouse (A) and crayfish (B) embryos at 50% of embryogenesis. Scale bar = 10 μm. Arrows indicate the future positions of the anterior and posterior commissures (ac, pc) and the longitudinal connectives (conn). Scale bar = 10 μm.

connective and fasciculate with each other. Both D and K continue to advance anteriorly along the connective, D extending as far as the brain by 75% of development.

The crayfish M neuron develops in a similar fashion to its woodlouse equivalent, except that its soma lies in a more posterior position and as a consequence, its axon grows around the anterior, rather than the posterior side of neuron E (Fig. 9B). There is also a minor difference in relative timing of outgrowth; the crayfish axon M is accompanied by axons D and K, rather than being a clear leader of these neurons. A

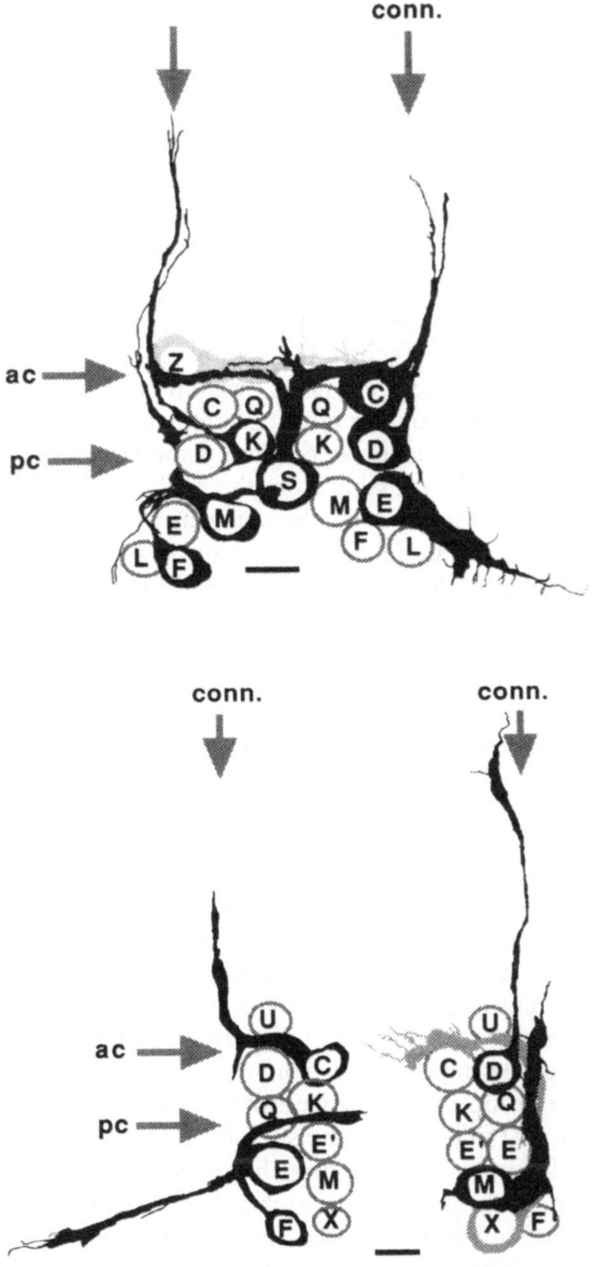

A

B

Figure 9. Early differentiating dorsal neurons in the woodlouse (A) and crayfish (B) embryos. Drawings were made in different embryos between 50 and 55% of embryogenesis and superimposed to show their approximate spatial relationships. These figures do not accurately reflect the relative timing of outgrowth of these axons. Camera lucida drawings made from LY fills. Scale bar = 10 μm.

further species difference is seen later in development: the crayfish M neuron does not form the efferent collateral branches present in the post 60% woodlouse embryo.

In no case has a dorsal neuron been filled at early stages of axonogenesis in either of these crustaceans that sends an axon in a posterior direction down the nerve cord.

Establishment of intersegmental nerve

The anterior root of the intersegmental nerve is pioneered by neuron E. In both species, axon E grows out very early, at around the same time as cell M (Fig. 9). It extends a broad lamellipodium in a postero-lateral direction, which subsequently thins into an axon which continues to extend into the periphery along its original course.

The posterior root of the intersegmental nerve is pioneered in both species by neuron C. In the woodlouse, axon C is sent out just after M and grows anteriorly, crossing from the medial to the lateral side of the connective as it progresses up the connective (Fig. 9A). When axon C reaches axon E in the next most anterior segment, it follows this axon out the intersegmental nerve. Collateral branches are often, but not invariably, found on the axon of C where it crosses other axons in the connective, or turns into the periphery. The crayfish C neuron shows a similar behaviour, except that it grows out at around the same time as M and from the outset, follows a more lateral trajectory in the connective to the pioneering axons M, D and K (Fig. 9B).

Establishment of commissures

Anterior commissure. A major difference appears to exist between the woodlouse and the crayfish in the manner of formation of the anterior commissure. In the woodlouse, the lateral portion of this commissure is pioneered by the medial growth of an axon from neuron Z, which lies just ventral and lateral to the future position of the commissure. It meets the laterally growing axon from cell S, thereby completing the anterior commissural pathway (Fig. 9A). It then advances across the midline, fasciculating with the axon of its contralateral homologue. Cell S is one of a group of midline neurons. It sends a single axon anteriorly along the dorsal midline. At the site of the future anterior commissure, the axon bifurcates and each branch grows laterally, meeting the medially directed axon of cell Z on each side (Fig. 9A). When the S axons meet the longitudinal connective pathway, they grow anteriorly. S and Z are amongst the first axons to arise in the woodlouse.

The crayfish possesses a neuron with a similar axon morphology to the woodlouse S neuron, but it is not the pioneer of the anterior commissure, as its axon grows out relatively late in embryogenesis. Axon X, which grows up the ipsilateral longitudinal connective from a posteriorly located soma, enters the anterior commissure at an early stage of its development (Fig. 9B) and may be the pioneer of this axon tract. To date, we have not identified a crayfish equivalent to the woodlouse Z neuron, which pioneers the lateral part of the anterior commissure.

Posterior commissure. The formation of the posterior commissure has much more in common in the crayfish and woodlouse. In both species, it appears to be pioneered by neuron F, which is located just posterior to E. This cell extends an axon around the lateral and anterior faces of cell E (Fig. 9A, B) before growing directly towards the midline, and turning anteriorly when it contacts the connective. The posterior commissure forms somewhat later than the anterior commissure, between 50 and 55% of development.

Other early differentiating dorsal neurons

We have examined the axon morphology of all of the other dorsal neurons located between the connectives in the woodlouse and crayfish and their morphologies are shown in Figures 10A, B. All of these neurons send out axons somewhat later than the cells discussed above, but still differentiate relatively early. Provisional homologues in the two crustaceans include neurons A, N, and J. Judged on its soma position and the initial course of its axon, neuron Q in the woodlouse is homologous to neuron Q in the crayfish, although its final axon morphology is actually more similar to the crayfish neuron Q′, which lies in a more ventral position.

Neurons that are present in the woodlouse, but that have no apparent counterpart in the crayfish, or vice-versa, include V, Y, R, I, H, L, E′ and T, as well as most of the cells that lie anterior to the anterior commissure.

A comparison of early axonogenesis in the woodlouse and crayfish

The pattern of early axonogenesis in the woodlouse shows many similarities to that seen in the crayfish embryo: the longitudinal connectives, the posterior commissure, the intersegmental nerve and the median nerve are all established by a similar pattern of axon growth from neurons in similar positions. The overall similarity in the axon morphologies of early differentiating neurons in the two species argues persuasively for a common evolutionary program for the formation of the major axon pathways in the CNS and many neurons (such as A, C, D, E, F, J, K, M, O) can be confidently homologized on the basis of soma position, axon morphology and relative timing of axonogenesis. Given the relatively distant relationship between these two crustaceans this result points to a relatively conservative plane for early neurogenesis within the crustacean subclass Eumalocostraca.

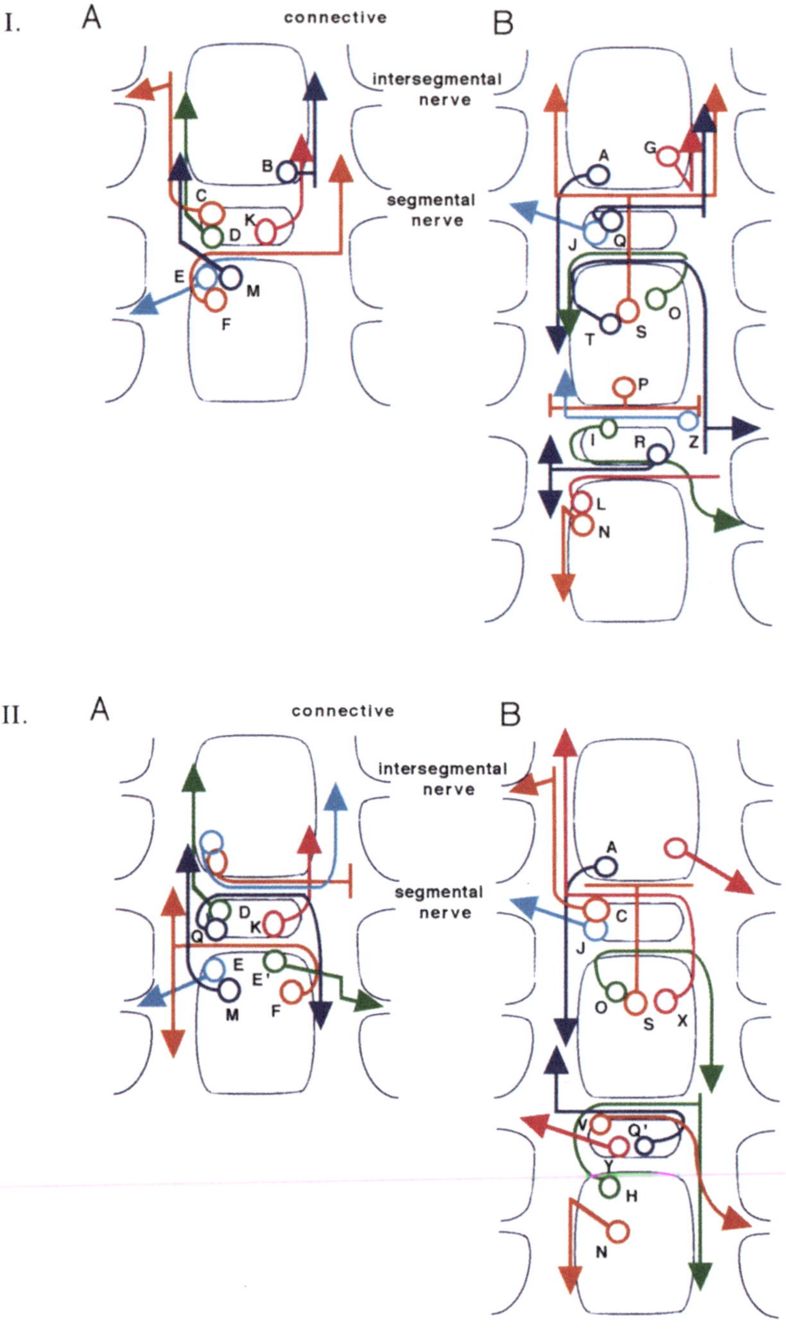

Figure 10. Axon morphologies of dorsal neurons in the woodlouse (I) and crayfish (II) embryos. To minimise overlap, a different subset of the segmentally iterated neurons studied is shown for each of three segments for both species. The cells shown in A lie dorsal to those in B.

On the other hand, some significant differences are seen between these two crustaceans. Firstly, some neurons that are clearly homologous, based on their spatio-temporal pattern of axon outgrowth, occupy slightly different positions with respect to neighbouring neurons. For example, neuron C in the woodlouse lies in an extreme dorsal position, directly adjacent to the longitudinal connectives, while its crayfish equivalent lies in a more ventral and medial position. This difference in soma location results in a different initial course for the axon of this neuron in the two species, but once the axon enters the connective it follows a virtually identical trajectory to the other species. This means that one should not seek to apply strict equivalence of soma position as a criterion for neuron homology between two species.

Secondly, some obviously homologous neurons show differences in axon morphology, such as additional branches. Examples are provided by neurons M, E and F: neuron M in the woodlouse sends collateral branches from its longitudinal axon into peripheral nerves, whereas its crayfish equivalent lacks such branches; the woodlouse E neuron develops an additional contralaterally projecting neurite which the crayfish neuron lacks; the crayfish F neuron projects both anteriorly and posteriorly in the contralateral connective, whereas its woodlouse homologue only forms an anterior axon.

Thirdly, there is a number of differences in the relative timing of axon outgrowth. For example, neuron M in the woodlouse is a clear leader of all other segmental neurons whereas it grows out at about the same time as several other neurons in the crayfish.

Fourthly, some neurons in one species do not appear to have equivalents in the other species. Whether this results from a drastic alteration in axon morphology of homologous neurons or from a different pattern of neuron generation is unlikely to be resolved by studying axon morphology alone. In such cases, we must turn our attention to other characters, such as cell lineage or expression of homologous genes to identify homologues.

A comparison of early axonogenesis in the crustaceans and winged insects

Our observations in the woodlouse and crayfish embryos (Whitington et al., 1993) do not support the claim by Thomas et al., (1984) that the pattern of early axonogenesis in the CNS is identical in the crustaceans and insects. If we restrict our attention to the set of nine neurons shown to be common to the grasshopper and fly – aCC, pCC, RP2, MP1, dMP2, vMP2, RP1, Us and MP4 – the following differences have been observed between the insects and the two crustaceans examined: (a) the absence in either crustacean embryo of neurons with an MP1 or dMP2-like morphology, i.e. early outgrowing axon running posteriorly down the longitudinal connective; (b) the absence in the woodlouse of an apparent homologue to the RP1 neuron, although the V neuron of the crayfish does possess a similar morphology to the RP1 neuron. Furthermore, cells with axon morphologies like the crustacean neurons N, Q, H, J, O, X, Y, I, T and R are not represented in published accounts of early axonogenesis in the CNS of insect embryos, although insect equivalents to these neurons may exist in more ventral regions of the ganglion.

Clearly then, the insect and crustacean patterns are not identical. But does a comparison between these two arthropod groups support the idea that they share a common ancestral program for early axonogenesis? For this purpose, attention should be restricted to primitive characters in each group. If, on this basis, we exclude relatively late differentiating neurons (those crustacean neurons that begin axonogenesis after 60%), and neurons that differ between different crustacean species or between different insect species, we are left with the following differences when comparing the woodlouse and the crayfish with the winged insects (i.e., the grasshopper and the fly): (a) crustaceans lack neurons with a pattern of axon growth like insect dMP2 and MP1 neurons (pioneers in the longitudinal connectives with posteriorly projecting axons); (b) the pioneer of the posterior commissure in crustaceans, F, is located medial to the connectives, whereas the insect pioneers for the posterior commissure are found lateral to the connectives; (c) the early differentiating crustacean neurons F, Q and J are not represented in the winged insects.

Weighed against these differences, there exist a number of similarities: (a) The crustacean neuron M, like the insect neuron pCC which lies in a similar position, sends an axon anteriorly, pioneering the longitudinal connectives. The initial behavior of M's axon as it grows around the soma of neuron E is strikingly similar to that shown by pCC's axon, as it grows around the soma of aCC; (b) The crustacean neurons E and E′, like the insect neurons aCC and U which lie in equivalent positions, pioneer the intersegmental nerve; (c) The crustacean neurons D and K, like the insect neuron vMP2, send an axon ipsilaterally and anteriorly up the connective. Like vMP2, D and K fasciculate with another axon (M in crustaceans, pCC in insects) to pioneer the intersegmental link in the connectives. In both cases these pioneering axons grow along the medial edge of the future connective; (d) Neuron S in the woodlouse, like the insect neuron MP4, sends an axon anteriorly along the midline, pioneering the median fibre tract and bifurcates at the anterior commissure; (e) The crustacean neuron C, like the insect neuron RP2, sends an axon out of the ipsilateral intersegmental nerve into the next most anterior segment and pioneers the posterior root of that nerve.

We feel that the extent of similarity in axon growth between the insects and the crustaceans supports the hypothesis that these two

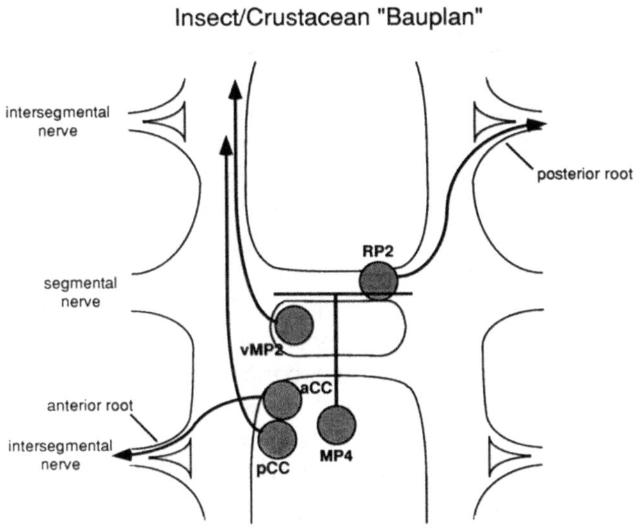

Figure 11. The insect/crustacean "Bauplan" – common features of early axonogenesis in the winged insects and crustaceans.

groups of arthropods share a common ancestral program for early axonogenesis and that the neurons identified above as being similar in the insects and crustaceans are homologous. However, sufficient differences exist between the species examined to suggest that comparisons of the pattern of axon growth alone will not resolve the issue. Evidence that the putative homologues share other characters, such as a common lineage or expression of homologous genes, would provide further support for this hypothesis.

If indeed the insects and crustaceans do share a common plan for early axonogenesis, the aspects of this plan that appear to have been conserved during evolution are: (a) the establishment of the anterior section of the longitudinal connectives; (b) the establishment of the anterior and posterior roots of the intersegmental nerve; and (c) the establishment of the median fibre tract (Fig. 11).

If it can be established with confidence that crustacean and insect embryos share a common plan for early axonogenesis in the CNS, we can begin to analyze the developmental mechanisms by which changes that exist between insects and crustaceans have arisen.

Early axonogenesis in the CNS of apterygotic insects

Strong conservatism exists in the pattern of early axonogenesis when a relatively primitive winged insect is compared with an advanced winged insect, whereas some differences are revealed when those insects are compared with two crustaceans. Just how conservative, then, is the insect pattern? To address this question we examined the pattern of early axonogenesis in a more primitive insect, the thysanuran *Ctenolepisma longicaudata* (silverfish). While there is some controversy as to how closely related the hexapods Protura, Collembola and Diplura are to the pterygotic insects, it is widely accepted that the two entognathous hexapod orders, the Archaeognatha and Thysanura, are monophyletic with the latter (Kristensen, 1991). It is therefore highly likely that the silverfish shares a common program for early axonogenesis with the winged insects. On this basis, neurons with similar axon morphologies in corresponding positions in the two groups can be classed as homologues with some confidence and one can therefore begin to dissect the developmental basis for any differences that exist between these species. In addition, an examination of the extent of such differences should help us to evaluate the significance of the differences observed between the insects and the crustaceans.

General organization of the embryonic silverfish ganglion

The organization of major axon pathways in the embryonic silverfish CNS shows some differences to that found in the winged insects. In particular, four separate commissures are found in the CNS, rather than two. The presence of four commissures has also been reported in the Archaeognatha (Larink, 1969). Otherwise the arrangement of axon tracts conforms to the general winged insect pattern, with paired longitudinal connectives, an intersegmental nerve with posterior and anterior roots and a broad segmental nerve (Fig. 12). Dil backfills on the intersegmental nerve of silverfish embryos helped to clarify the relationship between the four commissures in this insect with the pair of commissures found in winged insects. Motoneurons in the ganglion anterior to the filled nerve were found between the second and third commissures. In the winged insects the corresponding motoneurons lie between the anterior and posterior commissures, showing that the anterior pair of commissures in the silverfish is equivalent to the anterior commissure of winged insects, while the posterior pair is equivalent to the posterior commisure of winged insects.

The arrangement of dorsal somata that lie between the connectives in the silverfish embryo at an early stage of axonogenesis is shown in Figure 13. We have filled all of these cells from the earliest stages of axonogenesis. Based on their positions and the pattern of axon outgrowth, several of these cells are clearly homologues of identified neurons in the winged insects and have been named accordingly. (No similarity or homology is implied when crustacean neurons share a common letter with a silverfish neuron). In addition to the neuronal

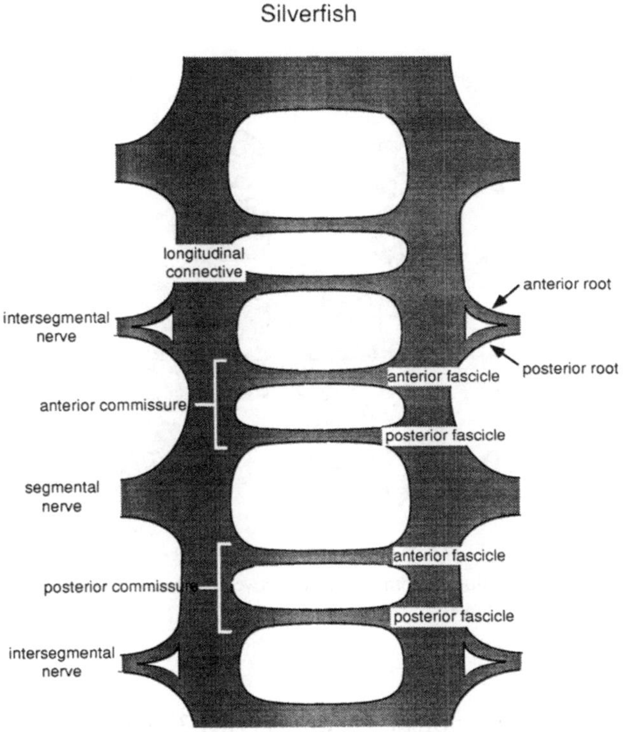

Figure 12. Gross morphology of ganglia in the silverfish embryo. Each neuromere possesses an anterior and a posterior commissure, each of which consists of an anterior and a posterior fascicle.

somata shown, a number of glial cells is found in an extreme dorsal position. These cells, which are arranged in a variable pattern, are flattened dorsoventrally and before axonogenesis begins, have sent out processes which cover most of the dorsal side of the ganglion.

Establishment of longitudinal connectives

The first axons to appear in the CNS of the silverfish embryo arise from the neuron pair vMP2 and dMP2 (Figs 14, 15). These two cells consistently lie in close contact. Their axons grow out together at around 20% of embryogenesis, and initially run in a lateral direction between cells RP2 and I. They then turn anteriorly, often sending a branch in a posterior direction. The anteriorly coursing axons fasciculate closely with each other as they advance to the next segment. They contact the posterior branch of their homologues in the next segment, thereby establishing the interganglionic connection and shortly after, the posterior branch of the axon is retracted. Somewhat later (approx. 23%) the vMP2/dMP2 axon pair is joined by the pCC axon, which fasciculates with the vMP2/dMP2 pair as it grows anteriorly. Around

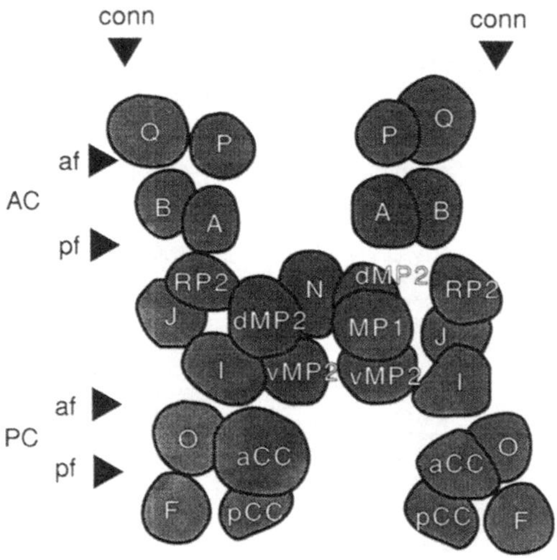

Figure 13. Arrangement of dorsal somata lying between the connectives in the silverfish embryo at 20% of embryogenesis.

24%, MP1, which lies dorsal to dMP2/vMP2, begins axonogenesis. Like the MP2 neurons, the MP1 axon usually bifurcates when it confronts the longitudinal connective, but unlike the former, advances in a posterior, rather than an anterior direction (Fig. 14).

Establishment of intersegmental nerve

aCC is an early contributor to, if not the pioneer of, the anterior root of the intersegmental nerve, growing in a postero-lateral trajectory across the connective at around 23%, before turning laterally into the periphery (Fig. 14). The posterior root of the intersegmental nerve is pioneered by RP2, which sends out an axon around 25%, in an antero-lateral direction. The axon crosses the connective and advances to meet the anterior root of the intersegmental nerve.

Establishment of commissures

The anterior fascicle of the posterior commissure is pioneered by neuron I, which sends out an axon in a medial direction at around 21% (Figs 14, 15). This axon meets and fasciculates with its contralateral homologue as it crosses the midline. On reaching the contralateral commissure it turns anteriorly and finally grows laterally out of the segmental nerve. Neuron F is an early contributor to, if not the pioneer of, the posterior fascicle of the posterior commissure. Its axon grows laterally

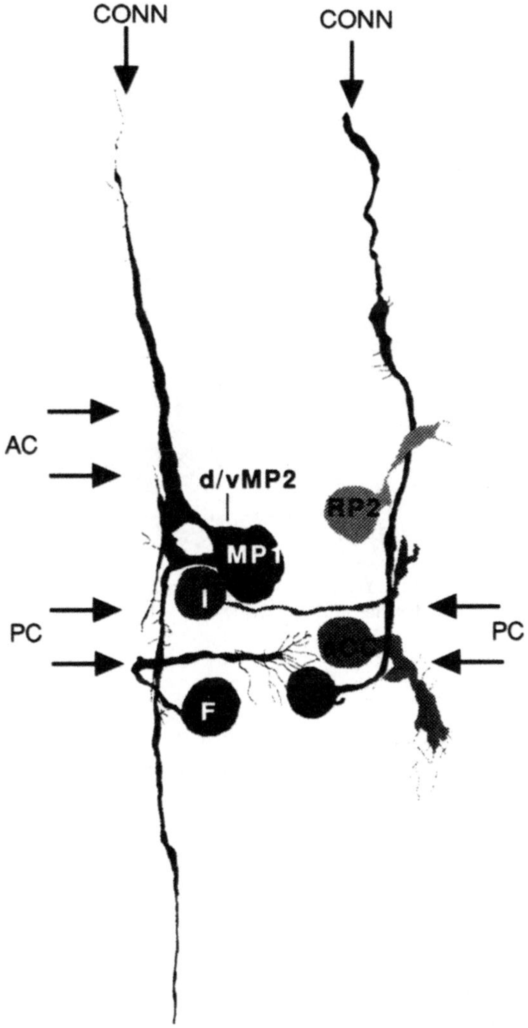

Figure 14. Early differentiating dorsal neurons in the silverfish embryo. Drawings were made in different embryos between 20 and 25% of embryogenesis and superimposed to show their approximate spatial relationships. These figures do not accurately reflect the relative timing of outgrowth of these axons. Camera lucida drawings made from LY fills.

before making a sharp medial turn, grows across the midline and, on reaching the connective, turns anteriorly.

We have not yet identified the pioneers of the fascicles of the anterior commissure. Neither an MP4 homologue nor an early differentiating medial fibre tract is conspicuous. Neuron N, a large, apparently unpaired midline cell between the anterior and posterior commissures, sends an axon along the midline at a relatively late stage ($\sim$25%). Its

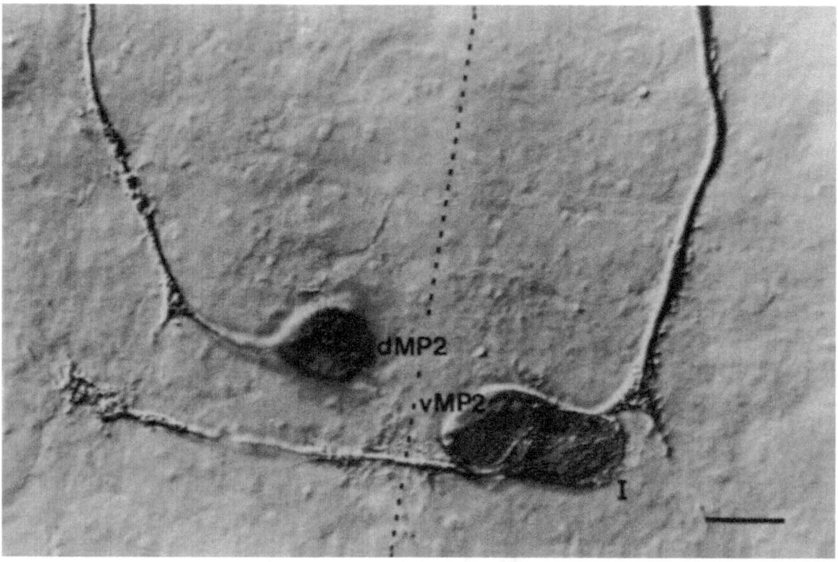

Figure 15. Photomicrograph of LY-filled silverfish neurons dMP2, vMP2 and I. Anterior is up and the dotted line indicates the midline. Scale bar = 10 μm.

axon bifurcates at the anterior commissure and one branch turns posteriorly into the connective and finally projects into the posterior commissure. This may be a homologue to one of the MP3 progeny.

Other early differentiating axons
The axon morphologies of other dorsal neurons lying between the connectives are shown in Figure 16.

A comparison of early axonogenesis in the silverfish, the winged insects and the crustaceans

The pattern of early axonogenesis in the silverfish embryo shows many similarities to that found in the winged insects. The silverfish neurons aCC, pCC, RP2, RP3, MP1 and vMP2 show a very similar axon morphology to their namesakes in the winged insects. Given that the Thysanura and the pterygotic insects are almost certainly monophyletic, it seems very likely that these similarities represent homologies.

On the other hand, there are some significant differences between these two groups of insects:

a. The apparent silverfish homologue to the winged insect neuron dMP2 (based on its close relationship with vMP2 which shows an identical pattern of axonogenesis in the two groups) sends an axon anteriorly,

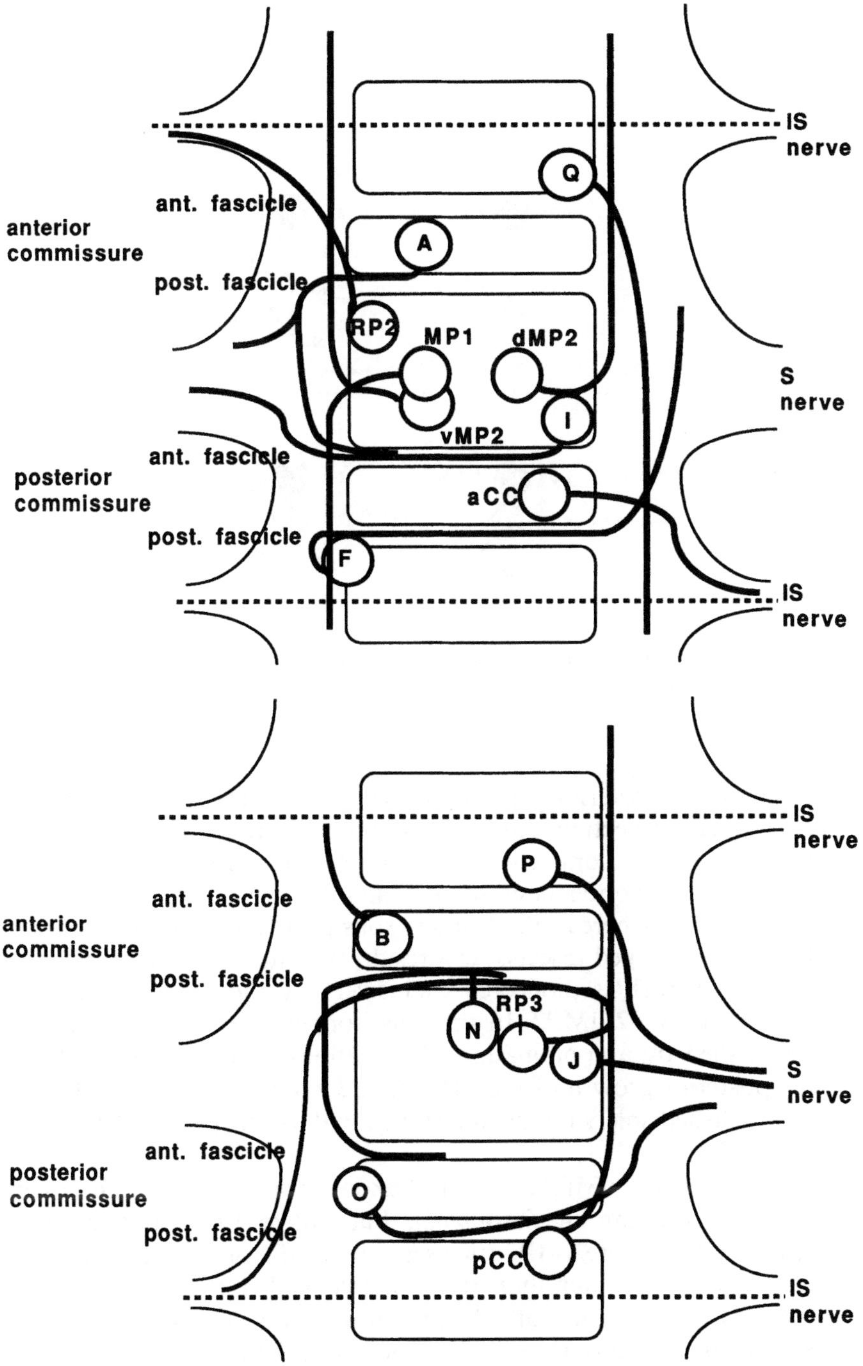

Figure 16. Axon morphologies of dorsal neurons in the silverfish embryo. Neurons shown in the top figure lie dorsal to those shown in the bottom. Dotted lines indicate the segmental borders.

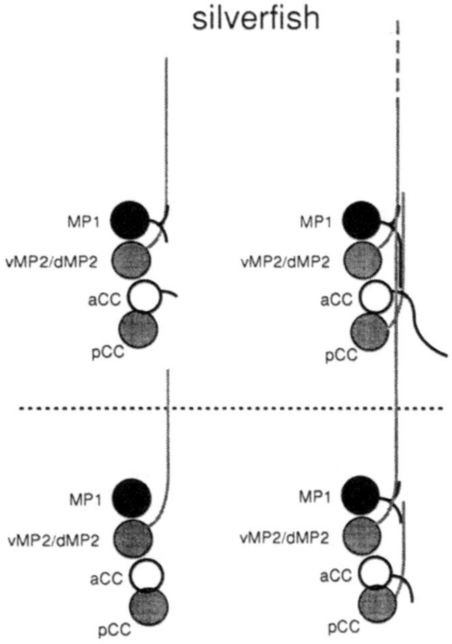

Figure 17. Neurons involved in the establishment of the longitudinal connectives and intersegmental nerve in the silverfish embryo.

rather than posteriorly as in the winged insects. MP1 is the only posteriorly projecting axon from the dorsal group of neurons which lies between the connectives and commissures (Fig. 17).

b. MP1 begins axonogenesis some time after the vMP2/dMP2 pair in the silverfish, whereas all of these axons grow out at around the same time in the winged insects. As a result, the intersegmental connection in the silverfish is pioneered exclusively by the anteriorly growing axon pair vMP2/dMP2. In the grasshopper, the same connection is established by a combination of the anterior growth of vMP2 and the posterior growth of the MP1/dMP2 axon pair, while in the fly it is established solely by the anterior growth of the pCC/vMP2 fascicle (Fig. 17).
c. The pCC, MP1, vMP2 and dMP2 axons grow in a common fascicle, like the situation in *Drosophila*, but unlike the situation in the grasshopper, where vMP2 runs separate to the other axons (Fig. 17).
d. aCC grows out later than the pioneering vMP2/dMP2 pair in the silverfish, whereas in both the grasshopper and the fly, it extends at around the same time as these axons (Fig. 17).
e. RP2 grows out much later than the pioneers of the longitudinal connectives in the silverfish, whereas in the insects it begins axonogenesis shortly after vMP2 and MP1.

f. There is no apparent equivalent in the winged insects to the silverfish neuron I, which pioneers the anterior fascicle of the posterior commissure.

In summary, three classes of changes appear to have taken place during the divergence of the pterygotic from the apterygotic insects:

a. Changes in the relative timing of axon outgrowth from homologous neurons;
b. Changes in the pattern of axon fasciculation; and
c. Changes in axon trajectories, particularly in the polarity of axon growth along the connectives.

These changes mirror similar, although less pronounced changes seen during the evolution of the winged insects. For example, the MP1 neuron in *Drosophila* projects both posteriorly and anteriorly along the connective, while its grasshopper homologue only forms a posterior branch (Bossing and Technau, personal communication), an example of a change in the polarity of axon growth.

Given the close evolutionary relationship between the winged insects and the Thysanura, it should be possible to dissect the developmental basis for these differences. For example, do the silverfish and winged insect vMP2/dMP2 neurons share a common lineage? Does the silverfish dMP2 neuron more closely resemble a *Drosophila* vMP2 neuron in its expression of putative axon guidance molecules? The dMP2 neuron in both the grasshopper and silverfish, like the vMP2 neuron, explores both anterior and posterior directions along the connectives during its initial growth. This may indicate that a relatively minor change in the pathfinding ability of the dMP2 neuron was required to transform it into a posteriorly, rather than an anteriorly projecting neuron.

In addition, a closer analysis of the substrate followed by the dMP2 axon in the silverfish and a comparison with its grasshopper homologue may reveal why the dMP2 and vMP2 axons grow with opposite polarity in one species and in the same direction in the other.

The differences observed between the silverfish and the winged insects are particularly interesting when placed in the context of the differences that exist between the winged insects and the crustaceans. Some of the characters possessed by both crustacean embryos that are not possessed by the winged insects, are present in the silverfish. These include: two anteriorly-directed pioneer axons for the longitudinal connectives arising from dorsal neurons in the central rectangle bounded by the commissures and connectives; and apparent equivalents to neurons F and J, which among the first axons to project into the posterior commissure and segmental nerve, respectively. If these neurons can be considered to be a primitive feature for insects, the case for a common ancestral program for early axonogenesis in the crustaceans and the insects gains further support.

This program would include the following features:
(a) the establishment of the longitudinal connectives by one or two anteriorly projecting axons from cells in the central rectangle; (b) the establishment of the anterior root of the intersegmental nerve by a neuron just posterior to the posterior commissure and medial to the connectives; and (c) the establishment of the posterior root of the intersegmental nerve by a neuron located near the intersection of the anterior commissure and the connectives (see Fig. 11).

Early axonogenesis in the CNS of myriapods

Thus far, we have dealt with the pattern of early axonogenesis in two of the major arthropod groups, the insects and crustaceans. What of the other groups, the Myriapoda, which are generally held to be closely related to the line which gave rise to the insects, and the Chelicerata, which are thought to be much more distantly related? To date, no investigations of axon growth have been carried out at the level of individual neurons in the Chelicerata. However, some information at this level of analysis is available for a representative of one myriapod group, the Chilopoda – the animal in question being the scolopendromorph centipede *Ethmostigmus rubripes* (Whitington et al., 1991).

Axonogenesis in the CNS of the centipede embryo begins at a stage when the left and right sides of the embryo have been separated by the intervening growth of a monolayer of extra-embryonic ectoderm. The first indication of axonogenesis is a bundle of about 10–15 axons visible under Nomarski optics located on the dorsal surface of the hemiganglionic primordia, running along the antero-posterior axis of the embryo. The bundle is lateral to the medial border of the ganglion and appears to be only one axon thick in the dorso-ventral axis. Intracellular LY fills of individual axons in this bundle show that they are thin (2–4 μm), and run parallel to the antero-posterior axis with few lateral arborizations (althoug filopodia extend laterally from the axon at some points). These axons do not appear to arise from segmental neurons, but from neurons located in the brain. As development proceeds these longitudinal axons extend further posteriorly and the bundle becomes wider and thicker and quickly extends to the posterior end of the embryo.We have termed this axon bundle the primary tract.

Axons originating from neurons in the segmental ganglia do not appear until some time later, at which stage the hemiganglia are still widely separated. Due to the relatively large number of cells present in the ganglia at the early stages of axonogenesis, we have not been able to construct maps of the positions of all somata in dorsal positions as we have done in the crustacean and insect embryos. However, certain somata are noticeably larger than surrounding cells and dye fills of these somata show that they have axons which run either parallel with or at right angles to the antero-posterior axis. In all cases their axons run directly dorsal to the ganglion surface where they expand into a growth cone with a tuft of filopodia. The antero-posteriorly running axons fasciculate closely with the primary tract, which therefore represents the forerunner of the longitudinal connectives.

We have filled several such segmental neurons and they are shown in schematic form in Figure 18. Neuron A lies close to the posterior border of the segment, at the same dorsoventral level as the longitudinal axon bundle and near its medial border and is easily recognizable from its large size at that stage. It appears to be one of the earliest segmental neurons to begin axonogenesis as the axon extends anteriorly for half the length of the segment when the axons of other enlarging neurons have only just reached the dorsal surface of the ganglion. The axon continues to extend in an anterior position. Other early differentiating neurons to send axons into the longitudinal connectives include, B, C, D, E, F and G. These axons run either anteriorly, posteriorly or in both directions along the cord.

A group of neurons with medially directed axons is found just posterior to the anterior segmental border and lateral to the primary tract. We can distinguish at least two types of neurons in this group based on their axon morphology (Fig. 18). One type, represented by neuron H, and consisting of at least two different neurons, is located in a deep ventral position within the ganglion. The axons of these neurons initially course in a medial or postero-medial direction and then turn posteriorly upon entering the primary tract. The other type, represented by neuron I, is located in a more dorsal position, and sends axons medially to cross the primary tract. These axons extend across to the contralateral hemiganglion, reaching or just traversing the contralateral primary tract at the time of fusion of the two hemiganglia. Their growth cones show extensive filopodial branching within this contralateral primary tract.

Nomarski optic observations of unstained embryos reveal the first signs of transverse axons crossing the ipsilateral primary tract at a stage when longitudinally coursing axons from segmental neurons have extended for some distance along the tract.

A comparison of axon growth in the centipede and the insects

The pioneering central axons in centipedes clearly have a different origin to those in the insects and crayfish, although, as we have seen, a similar

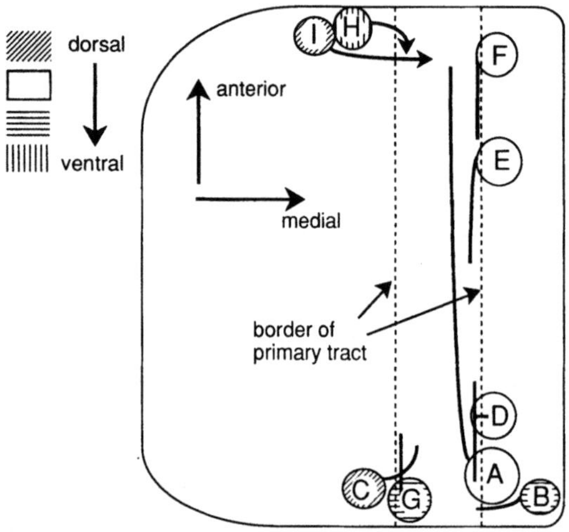

Figure 18. Diagramatic summary of soma positions and axonal morphology of early differentiating segmental neurons in the centipede embryo. The position of the somata in the dorso-ventral axis is indicated in the legend. The lines running from the somata represent the course taken by the axon while their length represents the relative degree of axon growth attained by the neuron.

difference exists between the two crustaceans, the woodlouse and the crayfish. The question as to whether the second wave of axonogenesis in the centipede embryo, from segementally repeated neurons, is similar to the insect/crayfish pattern, remains open at this time. The patterns in the two groups appear to bear few similarities, although as we could not identify all early differentiating neurons in the dorsal layers, we cannot be sure that our sampling method has not missed a population of cells that corresponds to the insect pioneering neurons. Nonetheless, two clear and major differences exist between the centipede and the insects in relation to early axonogenesis in the CNS: (a) there are many more cells in the CNS of the centipede embryo at the earliest stages of axonogenesis than in insects; and (b) the neurons contributing to the early growth of the connectives and commissures are much more widely separated in the centipede than in the insects. Observations by Tiegs (1940; 1947) on the structure of central ganglia at the time of appearance of the first central fibre tracts in the pauropod *Pauropus sylvaticus* and the symphylan *Hanseniella agilis* suggest that in these myriapods, as in the centipede, axonogenesis begins in the central ganglia when the latter contain a large population of neurons.

Patterns of early axonogenesis and arthropod phylogeny

The patterns of early axonogenesis in the various arthropods described above promise to shed light on two issues in arthropod phylogeny: the relationship between (a) the Insecta and the Crustacea, and (b) the Insecta and the Myriapoda.

The phylogenetic relationship between the crustaceans and the insects has been hotly debated for many years. One theory, based largely on differences in embryonic development (Anderson, 1973), as well as appendage structure and function (Manton, 1977), holds that the Insecta and Crustacea are polyphyletic, their most recent common ancestor being a flatworm-like animal (Fig. 19A). A number of workers has criticised the arguments advanced by proponents of this theory on the grounds that they do not follow strict cladistic methodology (Kristensen, 1991). Recently, Kukalova-Peck (1992) presented evidence that the uniramous insect limb, which appears to be fundamentally different in structure to the biramous crustacean limb, arose from an ancestral limb which possessed accessory outer branches. Indeed, she proposes that this ancestral insect appendage, together with the biramous appendages of crustaceans, trilobites and chelicerates arose by reduction

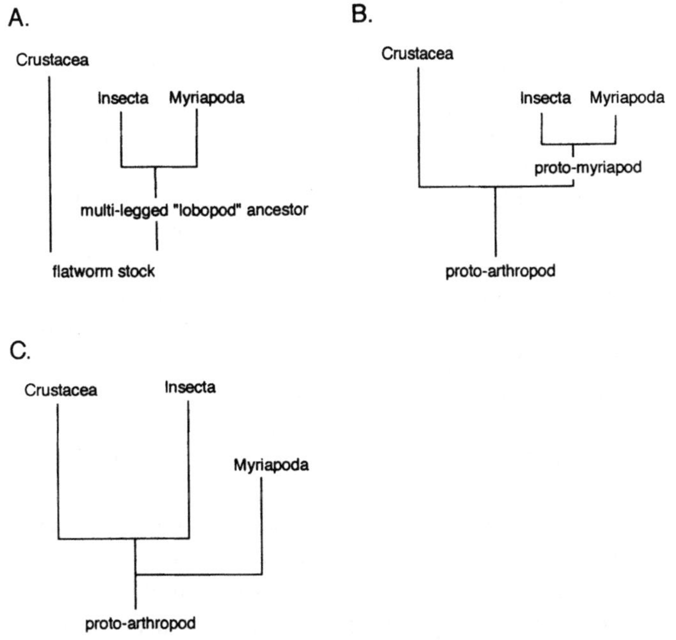

Figure 19. Three views of the evolutionary relationships between the Insecta, Crustacea and Myriapoda. (A) proposes that the Crustacea and the Insecta/Myriapoda lines arose independently from non-arthropod (flatworm?) ancestors. (B) proposes that the Crustacea and Insecta/Myriapoda share a common arthropod ancestor. (C) differs from A and B in postulating an early origin of the Myriapoda from near the base of the arthropod evolutionary tree.

from an ancestral polyramous leg. If true, this finding removes one of the major difficulties of the monophyletic theory of arthropod evolution.

Another view holds that the common ancestor of the Insecta and Crustacea was already a proto-arthropod and that these arthropods are therefore monophyletic (Fig. 19B). The main support for this view comes from the remarkable similarity of several characters common to both of these groups, particularly the compound eye.

If our conclusion that the insects and crustaceans share a common plan for early axonogenesis in the CNS is correct, the argument for a monophyletic relationship between them is strengthened, assuming that the pattern of early axonogenesis is a synapomorphic, rather than a plesiomorphic character. An examination of early axonogenesis in annelids would help to answer the latter question. While a number of studies have been carried out on axon growth in leech embryos (e.g., Kuwada and Kramer, 1983; Kuwada, 1984; Gao and Macagno, 1987), a thorough description of the establishment of central axon pathways is not yet at hand.

While controversy still surrounds the relationship between the insects and crustaceans, there has been near universal agreement for several decades that insects arose from a proto-myriapod stock (Fig. 19A, B). Synapomorphies linking the Insecta and the Myriapoda include the absence of second antennae, the presence of malphighian tubules, absence of a palp on the mandible, presence of tentorial arms, absence of pretarsal levator muscles and presence of water-absorbing limb-base vesicles (exsertile vesicles) (Kristensen, 1991). The presence of tracheae in insects and myriapods, on the other hand, is now considered to represent an apomorphic trait.

It is therefore highly surprising that early axonogenesis in the centipede and, as far as can be determined from older studies, other myriapods, shows few similarities to that seen in the Insecta. The differences in this respect are especially striking when one considers the close similarities between the insects and crustaceans in the establishment of central axonal pathways.

Two possibilities can be entertained: (a) the current view that insects and myriapods are derived from common proto-myriapod stock is correct and for some unexplained reason, the program for early axonogenesis in the myriapods has undergone considerable modification subsequent to their divergence from the insects or (b) myriapods are more distantly related to the insects than the latter are to the crustaceans.

While the latter possibility would seem unlikely in the face of the large number of apparently derived characters shared by the insects and myriapods, a recent study of 12S rRNA sequences in a range of arthropods (Ballard et al., 1992) suggests that myriapods may represent an early offshoot of the main arthropod line (Fig. 19C). Further studies of early axonogenesis in the myriapods and determination of DNA sequences of homologous insect/myriapod genes may help to resolve this issue.

Conclusions

The study of comparative arthropod neural development is at an exciting stage. Techniques that have successfully revealed the pattern and mechanisms of axonogenesis in the insects at the level of individually identified neurons are being applied to a range of other arthropods. The resulting data promise to shed new light on fundamental and long-standing questions in evolutionary and developmental biology. These studies will need to be extended to include a wider range of arthropod groups, and a greater number of representative species. Descriptions of axonogenesis should be supplemented by studies of neurogenesis and the expression of homologous gene products at the level of identified neurons. With this background, we may hope one day to understand how genetic change during evolution has generated the array of nervous system structure and behaviour seen in the arthropods.

References

Anderson, D.T. (1973) *Embryology and Phylogeny in Annelids and Arthropods.* Pergamon Press, Oxford.

Arbas, E.A., Meinertzhagen, I.A. and Shaw, S.R. (1991) Evolution in nervous systems. *Ann. Rev. Neurosci.* 14: 9–38.

Ballard, J.W.O., Olsen, G.J., Faith, D.P., Odgers, W.A., Rowell, D.M. and Atkinson, P.W. (1992) Evidence from 12S ribosomal RNA sequences that onychophorans are modified arthropods. *Science* 258: 1345–1348.

Bastiani, M.J. and Goodman, C.S. (1986) Guidance of neuronal growth cones in the grasshopper embryo. III. recognition of specific glial pathways. *J. Neurosci* 6: 3542–3551.

Bastiani, M.J., du Lac, S. and Goodman, C.S. (1986) Guidance of neuronal growth cones in the grasshopper embryo. I. recognition of a specific axonal pathway by the pCC neuron. *J. Neurosci.* 6: 3518–3531.

Bate, C.M. (1976a) Embryogenesis of an insect nervous system: I. A map of the thoracic and abdominal neuroblasts in *Locusta migratoria. J. Embryol. Exp. Morph.* 35: 107–123.

Bate, C.M. (1976b) Pioneer neurones in an insect embryo. *Nature* 260: 54–56.

Bate, C.M. and Grunewald, E.B. (1981) Embryogenesis of an insect nervous system II: a second class of neuron precursor cell and the origin of the intersegmental connectives. *J. Embryol. Exp. Morph.* 61: 317–330.

Bate, M., Goodman, C.S. and Spitzer, N.C. (1981). Embryonic development of identified neurons: segment-specific differences in the H cell homologues. *J. Neurosci.* 1: 103–106.

Campos-Ortega, J.A. and Jan, Y.N. (1991) Genetic and molecular bases of neurogenesis in *Drosophila melanogaster. Ann. Rev. Neurosci.* 14: 399–420.

Doe, C.Q. (1992) The generation of neuronal diversity in the *Drosophila* embryonic central nervous system. *In: Determinants of Neuronal Identity.* Academic Press, New York, pp. 119–154.

Doe, C.Q., Bastiani, M.J. and Goodman, C.S. (1986) Guidance of neuronal growth cones in the grasshopper embryo. IV. temporal delay experiments. *J. Neurosci.* 6: 3552–3563.

du Lac, S., Bastiani, M.J. and Goodman, C.S. (1986) Guidance of neuronal growth cones in the grasshopper embryo. II. recognition of a specific axonal pathway by the aCC neuron. *J. Neurosci.* 6: 3532–3541.

Gao, W.-Q. and Macagno, E.R. (1987) Extension and retraction of axonal projections by some developing neurons in the leech depends upon the existence of neighbouring homologues. I. The HA cells. *J. Neurobiol.* 18: 43–59.
Ghysen, A., Dambly-Chaudière, Jan, L.Y. and Jan, Y.-N. (1993) Cell interactions and gene interactions in peripheral neurogenesis. *Genes and Development* 7: 723–733.
Goodman, C.S., Bate, M. and Spitzer, N.C. (1981) Embryonic development of identified neurons: origin and transformation of the H cell. *J. Neurosci.* 1: 94–102.
Goodman, C.S., Bastiani, M.J., Doe, C.Q., du Lac, S., Helfand, S.L., Kuwanda, J.Y. and Thomas, J.B. (1984) Cell recognition during neuronal development. *Science* 225: 1271–1279.
Grenningloh, G., Rehm, E.J. and Goodman, C.S. (1991) Genetic analysis of growth cone guidance in Drosophila – fasciclin II functions as a neuronal recognition molecule. *Cell* 67: 45–57.
Grenningloh, G., Bieber, A.J., Rehm, E.J., Snow, P.M., Traquina, Z.R., Hortsch. M., Patel, N.H. and Goodman, C.S. (1990) Molecular genetics of neuronal recognition in Drosophila – evolution and function of immunoglobulin superfamily cell adhesion molecules. *Cold Spring Harb. Symp. Quanti. Biol.* 55: 327–340.
Harrelson, A.L. and Goodman, C.S. (1988) Growth cone guidance in insects: fasciclin II is a member of the immunoglobulin superfamily. *Science* 242: 700–707.
Jacobs, J.R. and Goodman, C.S. (1989) Embryonic development of axon pathways in the *Drosophila* CNS. 2. behavior of pioneer growth cones. *J. Neurosci.* 9: 2412–2422.
Jan, Y. N. and Jan, L.Y. (1990) Genes required for specifying cell fates in Drosophila embryonic sensory nervous system. *Trends Neurosci.* 13: 493–498.
Keshishian, H. and Chiba, A. (1993) Neuromuscular development in *Drosophila*: insights from single neurons and single genes. *Trends Neurosci.* 16: 278–283.
Klämbt, C., Jacobs, J.R. and Goodman, C.S. (1991) The midline of the Drosophila central nervous system – a model for the genetic analysis of cell fate, cell migration, and growth cone guidance. *Cell* 64: 801–815.
Kristensen, N.P. (1991) Phylogeny of extant arthropods. *In:* I. Naumann (ed.): *The Insects of Australia.* Vol. 1, second edition, Melbourne University Press, Victoria, pp. 125–140.
Kukalova-Peck, J. (1992) The 'Uniramia' do not exist: the ground-plan of the Pterygota as revealed by Permian Diaphanopterodea from Russia (Insecta: Paleodictyopteroidea). *Can. J. Zool.* 70: 236–255.
Kuwada, J.W. (1984) Normal and abnormal development of an identified leech motor neuron. *J. Embryol. Exp. Morph.* 79: 125–137.
Kuwada, J.Y. and Goodman, C.S. (1985) Neuronal determination during embryonic development of the grasshopper nervous system. *Dev. Biol.* 110: 114–126.
Kuwada, J.Y. and Kramer, A.P. (1983) Embryonic development of the leech nervous system: primary axon outgrowth of identified neurons. *J. Neurosci.* 10: 2098–2111.
Larink, O. (1969) Zur entwicklungsgeschichte von *Petrobius brevistylis* (Thysanura, Insecta). *Helgoländer wiss. Meeresunters.* 19: 111–155.
Manton, S.M. (1977) *The Arthropoda: Habits, Functional Morphology and Evolution.* Clarendon Press, Oxford.
Myers, P.Z. and Bastiani, M.J. (1993) Cell–cell interactions during the migration of an identified commissural growth cone in the embryonic grasshopper. *J. Neurosci.* 13: 115–126.
Nose, A., Van Vactor, D. and Goodman, C.S. (1992) Development of neuromuscular specifity in *Drosophila. Cold Spring Harb. Symp. Quant. Biol.* 57: 441–449.
Patel, N.H., Poole, S.J., Coleman, K.G., Goodman, C.S., Ellis, M.C., Martinblanco, E. and Kornberg, T.B. (1989) Expression of engrailed proteins in arthropods, annelids, and chordates. *Cell* 58: 955–968.
Raper, J.A., Bastiani, M.J. and Goodman, C.S. (1983) Pathfinding by neuronal growth cones in grasshopper embryos: II selective fasciculation onto specific axonal pathways. *J. Neurosci.* 3: 31–41.
Scholtz, G. (1992) Cell lineage studies in the crayfish *Cherax destructor* (Crustacea, Decapoda) – germ band formation, segementation, and early neurogenesis. *Roux's Archiv. Dev. Biol.* 202: 36–48.
Shear, W.A. (1992) Insect fossils – end of the uniramia taxon. *Nature* 359: 477–478.
Sink, H. and Whitington, P.M. (1991a) Location and connectivity of abdominal motoneurons in the embryo and larva of *Drosophila melanogaster. J. Neurobiol.* 22: 298–311.

Sink, H. and Whitington, P.M. (1991b) Pathfinding in the central nervous system and periphery by identified embryonic Drosophila motor axons. *Development* 112: 307–316.

Taghert, P.H., Bastiani, M.J., Ho, R.K. and Goodman, C.S. (1982) Guidance of pioneer growth cones: filopodial contacts and coupling revealed with an antibody to lucifer yellow. *Dev. Biol.* 94: 391–399.

Thomas, J.B., Bastiani, M.J., Bate, M. and Goodman, C.S. (1984) From grasshopper to *Drosophila*: a common plan for neuronal development. *Nature* 310: 203–207.

Tiegs, O.W. (1940) The embryology and affinities of the Symphyla, based on a study of *Hanseniella agilis*. *Q. Jl. microsc. Sci.* 82: 1–225.

Tiegs. O.W. (1947) The development and affinities of the Pauropoda, based on a study of *Pauropus sylvaticus*. *Q. Jl. microsc. Sci.* 88: 165–267.

Tiegs, O.W. and Manton, S.M. (1958) The evolution of the Arthropoda. *Biol. Rev.* 33: 255–337.

Wang, L.S. and Denburg, J.L. (1992) A role for proteoglycans in the guidance of a subset of pioneer axons in cultured embryos of the cockroach. *Neuron* 8: 701–714.

Whitington, P.M. (1989) The early development of motor axon pathways in the locust embryo – the establishment of the segmental nerves in the thoracic ganglia. *Development* 105: 715–721.

Whitington, P.M. (1993) Axon guidance factors in invertebrate development. *Pharmacology and Therapeutics* 58: 263–299.

Whitington, P.M., Leach, D. and Sandeman, R. (1993) Evolutionary change in neural development within the arthropods: axonogenesis in the embryos of two crustaceans. *Development* 118: 449–461.

Whitington, P.M., Meier, T. and King, P. (1991) Segmentation, neurogenesis and formation of early axonal pathways in the centipede, *Ethmostigmus rubripes* (Brandt). *Roux's Archiv. Dev. Biol.* 199: 349–363.

Wolpert, L. (1992) Gastrulation and the evolution of development. *Development* Suppl. 7–13.

The Nervous Systems of Invertebrates: An Evolutionary and Comparative Approach
ed. by O. Breidbach & W. Kutsch
© 1995 Birkhäuser Verlag Basel/Switzerland

Homologous structures, exemplified by motoneurones of Mandibulata

W. Kutsch and R. Heckmann

Fakultät für Biologie, Universität Konstanz, D-78434 Konstanz, Germany

Summary. Studies of identified neurones at a distinct place within a species are required to understand how far variability or plasticity exert an influence on an individual structure. Primarily, such a study is independent of whether different profiles are attributed to genetic or epigenetic programs. Finally, the basic structure of the element in question should be constructed. This crucial structure, then, can be used for several approaches, which are based on *the concept of the identified neurone*: (1) Ontogenetical studies: addressing the question of when during ontogeny is this neurone present in its final (adult) form? (2) Homonomy ("serial homology"): addressing the question as to how far this element is reiterated along a segmented body. And, how far will the same genetic program be modified depending on the function of different segments? (3) Homology (in its classical sense): addressing the question, how far can a specific element be characterized in different species, thereby considering the different systematic levels of the animal kingdom? Is there a basic "Bauplan" of the nervous system that is valid for the various groups of Insecta, Tracheata, Mandibulata, Arthropoda?

We will demonstrate that the structure of identified motoneurones can be employed to answer several of the questions raised above. In several instances, the profiles of motoneurones can be used to identify apparently homonomous (serial homologous) or homologous muscles. For Insecta, many similarities are present probably reflecting features of the basic nervous "Bauplan" of a primitive ancestor. Although there are still several similarities with Insecta apparent, conformance becomes less with Chilopoda, and is even more reduced in Crustacea.

Definition of homology

From the beginning onwards, including the pre-Darwinian time, the term "homologous" was used in a comparative sense. Owen (1843) defined a homologue as "the same organ in different animals, under every variety of the form and function." However, with such a definition the problem of the underlying cause was not addressed. This definition merely replaced the word "homologous" with "same". In view of some preciseness, Owen (1848) applied two criteria for a description of "sameness": relative position and connectivity.

The causation of the existence of homologous structures became clear after adopting the "Theory of Evolution" (Darwin, 1859). "Homology of a feature in different supra-individual units of living nature is due to the former existence of the feature in a common stem species" (Ax, 1989). This proclamation emphasizes the notion that a claim for true homology can be applied only when comparing a specific structure in different animals (or plants). This classical definition points to a "common precursor", the common ancestor; in a genetical view, this idea implies, that homologous structures develop starting off from genomically common sequences (Dumont and Robertson, 1986). By investigating and comparing the level of equivalence in different animals their grade of relateness may be deciphered. Furthermore, by studying animals at different phylogenetic levels it should become possible to construct a more general "Bauplan," representing the basic organization of a hypothetical common ancestor (Wagner, 1989).

In pursuit of this idea it becomes clear that claims for "sameness" or "similarity" are of basic importance. However, which concept can be applied to distinguish genuine homology from analogy? For the latter, similarity of structures are likewise apparent, but they result from the need of common function and are not dependent on common inheritance. It must be emphasized that homology does not necessarily mean similarity; homologous structures descending from common precursors may very well be quite different (Mayr, 1969). Simpson (1967) (compare also Dobzhansky et al., 1977) propounds two criteria, which may assist in a decision for possible homology:

1. Minuteness of resemblance: With respect to any given feature exact likeness suggests homology.
2. Multiplicity of similarities: The greater the number of equivalent features evident among any two organisms the more likely it is that homology, caused by common ancestry, is responsible for the similarities of any one feature.

With such statements it becomes obvious that pertinent studies must rely on empirical data; a decision on possible homology must often be based upon probabilistic inferences.

Remane (1956) compiled a catalog of three criteria which may define possible homology:

1. Criterion of position: Homology may be inferred for a structure in an equivalent position within a comparable framework.
2. Criterion of specific quality: Similar structures may be homologized when they match in several specific features irrespective of position within the organism. Trustworthiness increases in parallel to the grade of complexity and equivalence of the compared structures.
3. Criterion of continuity: Dissimilar and differently placed structures may be homologous as long as there are intermediate stages satisfying one or both of the aforementioned criteria. Such intermediate stages may be due to ontogenetic processes or may be genuine systematic interjacent stages.

At the time when this canon was accumulated, mainly morphological features were considered. However, modern approaches of cell and molecular biology as well as physiological criteria can easily be incorporated into such a scheme, too (see later).

The concept of identified neurones

General

It becomes apparent that a claim for homology basically relies on anatomical-histological criteria. With respect to the diversity of the nervous system (whether within one species or comparing different ones) it often becomes difficult to define the relevant structures. We, therefore, propose to revert to the *concept of the identified neurone*. This indicates that we compare structures at the level of individual unique cells; especially, we will consider the individual neural elements. In the search for possible homology a comparative approach requires certain guide lines. We have compiled a catalog which mainly is based on anatomical-histological criteria, but it also includes other assays including recent molecular biological methods (see also Kutsch and Breidbach, 1994).

1. Position of a certain cell body depending on its relation to other somata or ganglionic structures.
2. Soma size based on an absolute or relative measurement.
3. Track of the appropriate neuropilar segment(s) and axon(s) with respect to the ganglionic framework.
4. Arrangement of the general dendritic field, number and position of major dendrites, characters concerning the neuropile penetration.
5. Cell ramification along peripheral targets (esp. applicable for MNs).
6. Determination of the progeny of individual neuroblasts.
7. Specification of various physiological (esp. electrophysiological) criteria.
8. Characterization of transmitter or neuromodulator content and release.
9. Expression of specific genes, especially during certain phases of ontogeny.

This canon, of course, is valid for any neural element, whether motoneurone, interneurone, or sensory neurone. This chapter will mainly be devoted to accumulating some knowledge on identified motoneurones (MNs) in Mandibulata. Other studies will apply similar criteria when dealing with sensory or interneurones (Boyan and Williams; Meier and Reichert, this volume).

Specific

The current concept to employ the identified neurone in a comparative approach depends largely on the reliability to discriminate an individual cell and to correlate its features with those of a presumed equivalent in another context. Therefore, this approach mainly relies on a determination of "sameness" or close "similarity."

In the following, we will elaborate this strategy by: (1) presenting examples for a specific segment, (2) demonstrating the structural modification of an apparently reiterated element along a segmented body, (3) finally, comparing possible candidates of homologous neurones in different species. The two first steps will elucidate criteria of a neural element developing from an identical genetic disposition. The last step, however, will show elements based on some changes in genetic disposition of the whole organism. Thereby the major aspect in the search for homology is addressed, i.e., the variation of a precursor element developed by a common ancestor. Most examples will be drawn from our studies with locusts, which are thought to be relatively "primitive" insects, close to the fundamental organization of the early insects (Hennig, 1969; Sharov, 1966). Additional information will be presented from more advanced insects, such as a beetle. Finally, several cases will be compiled employing different levels, taxa, of the Mandibulata.

Variability of identified neurones

Of paramount importance for the concept of the identified neurone is a knowledge of the grade of possible variability of the cell in question. If it turns out that enormous variability occurs, even at the light-microscopical level, then it becomes difficult to envisage a stable nervous system including definite neuronal circuits.

Goodman et al. (1979) studied the arborization of an identified interneurone (IN) in *Locusta*, the movement sensitive DCMD. They favored an identical genetic background and exploited the possibility to raise clones of isogenic locusts. It turned out that the configuration of this IN was not very homogeneous but that various major branches were missing in a large percentage of preparations. Interestingly enough, Satterlie (1985) demonstrated that this variability was considerably reduced when employing animals caught in the wild. Apparently, long inbreeding in cultures may result in an increased variability of neural components. This result should be drawn upon when studying and implementing the importance of genetic vs. epigenetic factors controlling the structuring of neural elements.

We have investigated possible variability by employing identified MNs. A comparison of such neurones does not indicate extraordinary differences among different preparations (Fig. 1). It is doubtless that differences in the soma position occur with respect to other ganglionic structures and the course of the neuropilar segment. However, the track of the neuropilar segment, the relative position and size of the bilaterally arranged dendritic fields, as well as the position and route of the

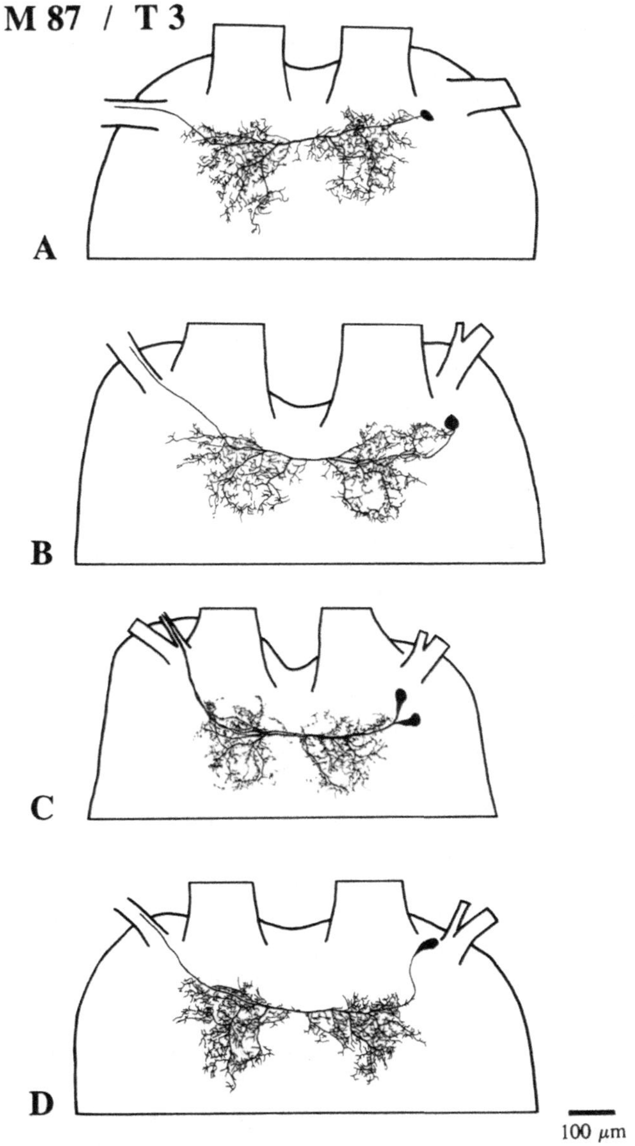

Figure 1. Comparison of four preparations in which one (A, B, D) or two (C) MNs are shown which supply a ventral mesothoracic longitudinal muscle (see Fig. 5A) in *Schistocerca gregaria*. While there is some variation of the soma's position, the track of the neuropilar segment and the structure of both bilateral dendritic fields are rather congruent. The mesothoracic muscle is innervated by MNs placed in the metathoracic ganglion (*T3*) of which only the anterior part is shown (G. Steffens, unpublished).

major dendrites remain rather invariant. Dendritic areas may be substituted by intrusion from neighboring major dendrites (H. Schneider, personal communication). Altogether, the dendritic branches of each individual cell fulfill the criteria of a constant and rather invariant penetration of the neuropile. We, therefore, presume that at least at the light-microscopical level a considerable variability of the MNs does not exist. Therefore, criteria such as general (as opposed to strictly invariant) position of soma, track of the neuropilar segment, size and position of the dendritic fields(s), may be used to identify and designate individual MNs.

Ontogeny of identified MNs

Studies of identified neurones should also consider the possibility that the structure and physiology of an element in question might change features throughout ontogeny (this conforms to the third postulate, raised by Remane, see above). Reverting to the principle of Haeckel, "the biogenetic law of recapitulation" (in the present context we will not discuss its original meaning in detail), one should be aware that putative homologous organs and individual cells of different species may change their characteristics throughout ontogeny within a different time domain. Considering a possible diversion of ontogenetic processes such a situation might impede the search for homology. Therefore, ontogenetic studies must supplement the normal approach which mostly relies on a comparison of adult stages. The early expression of the cell in question needs to be investigated. An early ontogenetic manifestation might especially be of relevance when comparing animals of a more "primitive" status with those associated with a more "advanced" level. Furthermore, such studies might give a hint as to the importance of specific cell structures. With respect to the development of neurones the following points may be raised: When during embryogenesis (or even later) does a specific neurone appear? Does it change its position? When and how does the axon grow? When and how are certain major dendrites formed? When are the specific dendritic fields elaborated? In the study of a specific neuro-muscular system the innervation pattern might assist in a search for possible adequate structures. Peripheral branching of specific nerves is rather fixed and persists throughout ontogeny, whether in Hemimetabola (Wittig, 1955) or in Holometabola passing metamorphosis (Tsujimura, 1988; Yack and Fullard, 1990; Urbach et al., 1994).

Hemimetabola

There are only a few examples in which the development of identified MNs in Hemimetabola, especially in locusts, has been described (Myers and Ball, 1987; Myers et al., 1990; Whitington and Seifert, 1981; Steffens and Kutsch, 1992). We, particularly, have chosen an ensemble of MNs which serve the dorsal longitudinal muscles (DLMs). With respect to proposition of homology this specific neuro-muscular system proved to be of great advantage because the DLMs are present throughout the whole group of Mandibulata. The whole complement of the locust's DLMs consists of one large and two small DLMs of which the latter ones partly or fully degenerate with adulthood (Wiesend, 1957; see also Breidbach and Kutsch, 1990; Urbach et al., 1994). From early work on locusts (Neville, 1963) it was known that the large DLM is innervated by at least 5 MNs. It was also known that the large DLM consists of five muscle bundles; four of the relevant MNs are present in the anterior and one is seen in the posterior ganglion.

For adult locusts (*Locusta* or *Schistocerca*), a Co- or Ni-backfill (combined with an intensification after Timm's; for complete procedure, see Kutsch and Schneider, 1987) of the relevant nerve renders two groups of nerve cells distributed throughout two ganglia (Fig. 2) (Breidbach and Kutsch, 1990; Kutsch and Breidbach, 1994; Kutsch and Heckmann, in preparation). When studying the mesothoracic DLMs, then several cells ($n = 11$; see Tab. 1) are visible in the next anterior (=prothoracic) ganglion, while five cells are present in the segment specific (=mesothoracic) ganglion. For the anterior group the MNs are distributed in the caudal, ipsilateral or medial region, with some dendrites penetrating the contralateral side. The posterior group comprises five cells. Two contralateral somata (one large, one small) are placed in the rostral region; there is another contralateral one which barely shows any other structure beside that of the soma; furthermore, a median posterior cell and a small median cell are present. The larger median cell represents a typical dorsal unpaired median cell (DUM; see Hoyle, 1978); the small medium cell was termed "central cell," due to its position (Breidbach and Kutsch, 1990). Both contralateral cells display two dendritic fields, a small contralateral and a large ipsilateral field (the terminology contra- vs. ipsilateral is used with respect to the segmental location of the innervated muscle).

This arrangement is well established for the adult stage. Locusts pass five larval instars before reaching the imaginal stage. The DLMs are present from the first larval instar onwards (Wiesend, 1957; Bernays, 1972). Recent studies demonstrate that these muscles have been developed (i.e., they have reached their typical adult configuration) at an embryonic stage of about 70% (Xie et al., 1992).

For the appropriate MNs it is obvious that the whole ensemble exists throughout all preimaginal stages (Fig. 3C). Based on light-microscopy studies, no consistent change, except of the general increase in size, is seen throughout postlarval development. The only consistent difference, compared to the adult stage, is observed when measuring the diameter of both contralateral somata; while in the adult the somata differ by a factor of almost 3 (73 vs. 23 μm), the difference for a first larval instar is only marginal (16 vs. 13 μm). Also, for a 95% embryo the whole neural set is present. The dendritic fields are almost identical to the adult ensemble (Fig. 3B). For a 65% embryo all MNs can be visualized (Fig. 3A). However, it becomes clear that at this stage of embryogenesis the dendritic fields are not yet elaborated as they typically are at later

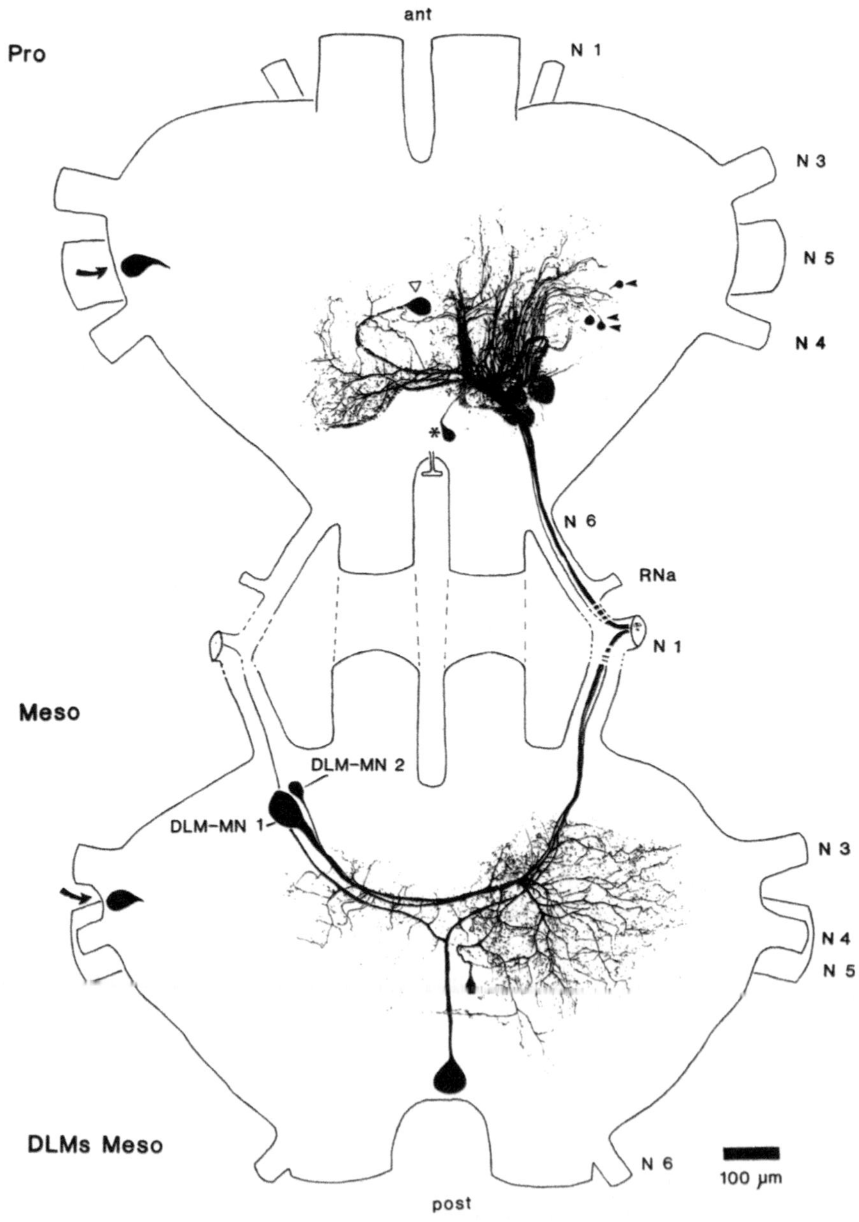

stages. For this ensemble it is obvious that the appropriate MNs are present at a stage equivalent to about two-thirds of embryogenesis, while the dendritic fields reach their final structure throughout the last third of embryogenesis. Differences in soma sizes are not yet apparent. Studies of earlier embryonic stages demonstrate (G. Steffens, unpublished) that the contralateral MNs are born at about 30–33% of embryogenesis. The typical growth of the axons towards the ganglionic midline is seen at a stage of about 35%. With 40–42% the axons leave the ganglion. With 50% the axons have reached the peripheral target structure, the DLM; at this stage only a few longitudinally-arranged side branches within the ganglion are apparent. With 55% a conspicuous growth of the ipsilateral dendritic field is detectable, while on the contralateral side several shorter side branches develop. Ultimately, the smaller contralateral and larger ipsilateral field will be constituted. It should be recalled that such a precise description of the early ontogeny certainly is of interest for studies concerning developmental processes *per se*. However, with respect to the studies concerning possible homology, they might be prerequisite to claim for similarities when employing the nervous system of other, more "primitive" insects or other Mandibulata (see general remark in section on "Ontogeny of identified MNs").

Holometabola

As an example of a more "advanced" insect, we have chosen the beetle *Zophobas* (Breidbach and Kutsch, 1990). The Coleoptera belong to the group of Holometabola which are characterized by a pronounced rearrangement of body structures throughout metamorphosis (Pipa, 1978; Gilbert and Frieden, 1981; Sehnal, 1985).

For the beetle segmental DLMs exist, too. This muscle ensemble is present throughout the larval and adult stages without considerable changes of morphological gross structures (Lee, 1964; Kalogianni et al., 1989); even individual motor units can be proven for both stages (Urbach et al., 1994). For this muscle ensemble, staining shows two groups of MNs distributed throughout two ganglia (Breidbach and

---

Figure 2. Camera lucida drawing of the MNs supplying the DLMs in the mesothorax of an adult *Locusta migratoria*. The MNs are split up into two groups, positioned in the anterior (prothoracic) and posterior (mesothoracic) ganglion. Several of the different cell types as well as single neurones can easily be identified. *DLM-MN1*, *DLM-MN2* indicate the large and small contralateral MN; *arrowheads* indicate the three small cells which supply the small DLM3; the *asterisk* is the posterior dorsal medium ("podo") cell; the *triangle* is the soma of the "loop cell"; the *curved arrow* shows the apparent homonomous contralateral somata from which axons exit via N1, 3, 6 (not documented here). In the posterior ganglion the small central cell (cc) and the large dorsal unpaired median cell (DUMDL) are also shown (Kutsch and Heckmann, unpublished).

Table 1. Identified MN-types serving the DLMs in different segments of *Schistocerca*; adult and embryonal stage

	Anterior neurone group								Posterior neurone group					
Muscle(s)	Ganglion	lMN	mMN	sMN	loop	C1, 3, 6	podo	Sum	Ganglion	contra	C1, 3, 6	DUM	cc	Sum
DLMs Pro Adult (M 49 & 56)	SOG	3	1	3	1	?	1	9	Pro	2	?	1	1	4
DLMs Meso Adult	PRO	4	1	3	(1)	(1)	(1)	11	MESO	2	(1)	1	1	5
DLMs Meso Embryo 95%	PRO	4	1	3	1	1	1	11	MESO	2	1	1	1	5
DLMs Meta Adult	MESO	4	1	3	(1)	(1)	(1)	11	META	2	(1)	1	1	5
DLMs Meta Embryo 65%	MESO	4	1	5	1	1	2	14	META	2	1	1	1	5
DLMs As 7 Adult	A 6	3	1		1	1	1	7	A 7	2	1	2	1	6

The MNs are characterized according to several criteria: The neurones are split into a group positioned in the next anterior ganglion and a group positioned in the (posterior) ganglion associated with the muscles. *lMN*, large motoneurones; *mMN*, medium sized motoneurones; *sMN*, small motoneurones; *loop*, neurone with a contralaterally looping neuropilar segment; *contra*, neurones with a contralateral soma position; *podo*, soma in a posterior-dorsal position; *DUM*, dorsal unpaired median neurone; *cc*, central cell; *C1, 3, 6*, apparent homonomous neurones in both groups located contralaterally, their axons split and exit via nerves N1, 3, 6 (thorax) or N1, 2 and posterior connective (abdomen). Ganglion: *SOG*, suboesophageal; *PRO*, *MESO*, *META* pro-, meso-, metathoracic; *A6*, *A7* sixth and seventh abdominal. The typical number of cells for each type is given. *?*, this neurone type has not been detected, though it might exist; *(1)*, a single cell seen in only a few preparations. For the prothorax, two muscles (nomenclature after Snodgrass, 1929) have been selected. (Kutsch and Heckmann, in preparation.)

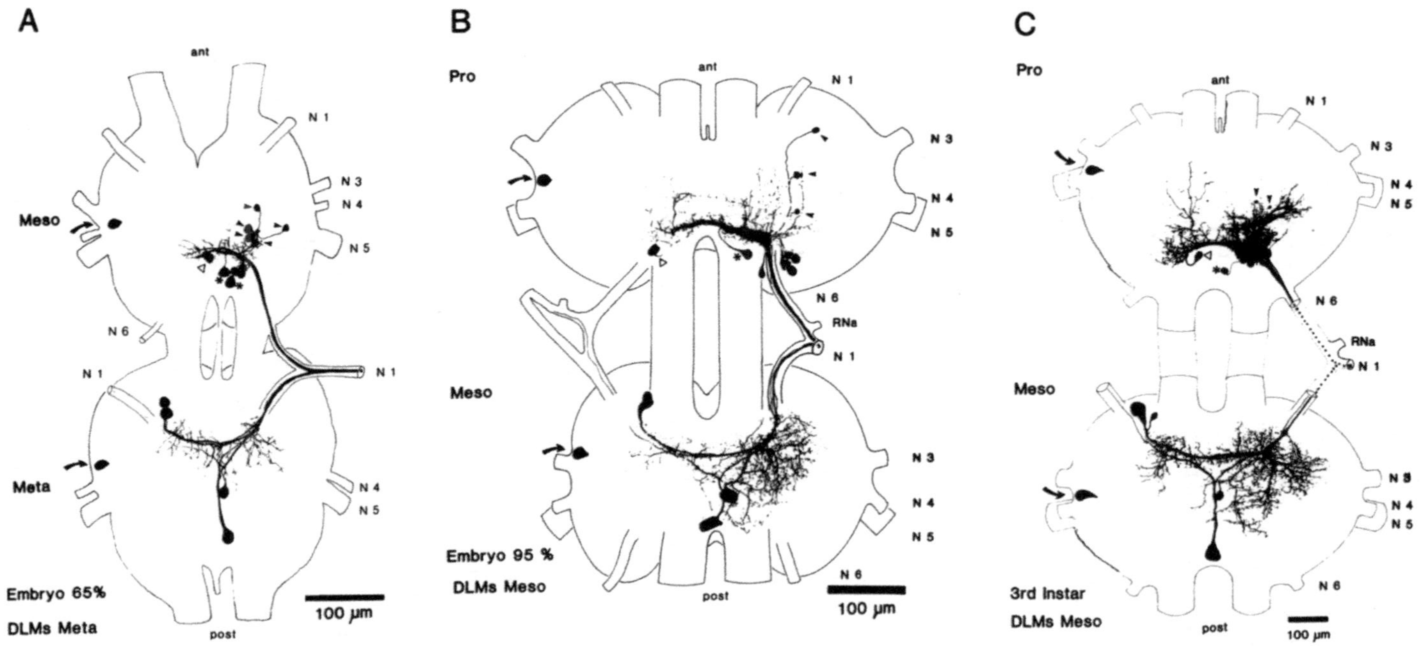

Figure 3. Representation of the MNs serving the thoracic DLMs throughout ontogeny of a locust. For the 65% embryo (A) all cells are present, although the dendritic fields are still scarce. For the 95% embryo (B), also the dendritic fields resemble those of an imago (see Fig. 2). Additionally, for the cell and dendritic representation a stain of a third larval instar (C) is shown. Indication of several specific cell types: *arrowheads* small cells located ipsilaterally in the anterior ganglion; *asterisk* is the soma of a median cell in the posterior dorsal region ("podo"); *triangle* is the "loop cell"; *curved arrows* are homonomous contralaterally neurones by which axons leave the ganglion via N1, 3, 6 (not documented) (Kutsch and Heckmann, unpublished).

Kutsch, 1990; for more information concerning a possible homology with the locust ensemble, see below). One of the somata is situated contralaterally. It shows a rather limited variability of its dendritic fields (see Fig. 8 in Breidbach and Kutsch, 1990). The MNs are present in the larval stage. The structure of the identified contralateral MN is rather similar when the larval and adult stages are compared. This is of special importance since it demonstrates that the process of metamorphosis which separates the larval and imaginal instar by a radical transformation of other structures does not seriously impair the structure of certain identified MNs (a similar issue can be raised for different neuro-muscular systems of a variety of Holometabola; for summary see Breidbach and Kutsch, 1990).

Structure of homonomous neurones

Segmentation results in a linear reiteration of a great number of organs and structures. It is conceivable that reiteration of segments is one solution to the problem to reach a more complex organization of the whole organism and, thereby, to increase its efficiency (Kästner, 1965; Leise, 1990). It is obvious that any difference in the expression of such reiterated structures evolves irrespectively of an identical genomic background. Possible structural and physiological differences, therefore, depend on the position along the longitudinal axis. Ultimately, they will result in different segmental function. Appropriate comparative studies might indicate the importance of epigenetic influences on the expression of specific structures. How far can the same genetic program be influenced to fulfill specific requirements of the various segments? Owen (1848) was already aware of the similarity of organs along a segmented animal. For this similarity he adopted the term "serial (general) homology," whereas for a congruence observed between different animals he chose the expression "special homology". In later studies it was favored to restrict the term "homology" exclusively to the case of interspecific comparison. Therefore, Bronn (1858; see also Kluge, 1977; Ax, 1989) suggested to replace the aprocryphal expression "serial homology" by the term "homonomy". Taking into account the classical meaning of "homology" which may give some clues to understand evolutionary traits, Bronn's suggestion should be accepted to avoid confusion regarding the typical interspecific "homology" compared to the phenomenon of similarity dependent on intraspecific serial reiteration ("serial homology").

Hemimetabola

The notion of segmentally reiterated muscles in Orthoptera belonged to the first examples for which homonomy was claimed (e.g. Voss, 1905;

Snodgrass, 1929, 1935). It is not unsound to infer that homonomous muscles are served by homonomous MNs. Therefore, one might choose a reiterated specific muscle system to understand how far a neuro-muscular ensemble *in toto* changes depending on its segmental position. Such a study, ultimately resulting in the question of whether function determines the structures of identified neurones, has already been addressed towards leg-MNs of *Schistocerca* (Wilson and Hoyle, 1978; Wilson, 1979). All three leg pairs are almost equally used during walking, but the hindlegs are also employed for jumping. Homonomous MNs serving the appropriate leg muscles are situated at a similar position within each segmental ganglion. Morphological structures of the apparent homonomous MNs as well as their neighboring neurones are similar, irrespective of leg function.

The set of DLMs is present along the body of a locust, from the prothorax to the abdominal segments. Due to the very close matching of both pterothoracic segments which bear the wings (meso- and metathorax), it might not be surprising that for both segments not only an identical set of MNs is found, but also that their dendritic fields rather match. The DLMs of the prothorax are not composed to form such a uniform set as is characteristic for the pterothorax. Several longitudinal muscles are present which can be distinguished due to separate points of insertion. Even though for this whole set a similar number of MNs is apparent (Tab. 1), their distribution and structure of dendritic fields is equivalent to those of the other thoracic segments. The major dendritic fields show some modification, but it becomes apparent that the main features are prevailing throughout all segments.

Along each abdominal segment a set of DLMs is present and expresses a general uniformity. With respect to the thoracic system some differences are apparent, especially regarding the smaller DLMs. Even though, the typical split into two groups of MNs, their number and soma position are equivalent to those of the thorax (Tab. 1). Certain major features of the dendritic fields elaborated for the thorax can also be traced for the abdominal segments (Fig. 4).

To generalize such a homonomy-concept other muscle sets need to be employed. In view of the knowledge accumulated for the DLMs, we have chosen a parallel system consisting of the ventral longitudinal muscles. Such a muscle system is present in all segments of *Schistocerca*; however, homonomy of each muscle is not always obvious. A judgment relying only on the origin and insertion points often leaves the question open of whether for specific muscles homonomous partners can be claimed along the thoracic and abdominal segments. Emanating from stains of the MNs there are convincing cases of equivalence, which due to a multiplicity of similarity seem to accomplish the criterion of homology or – to be more precise, and respecting the situation of reiteration – of homonomy (Fig. 5). We, therefore, tend to reverse the

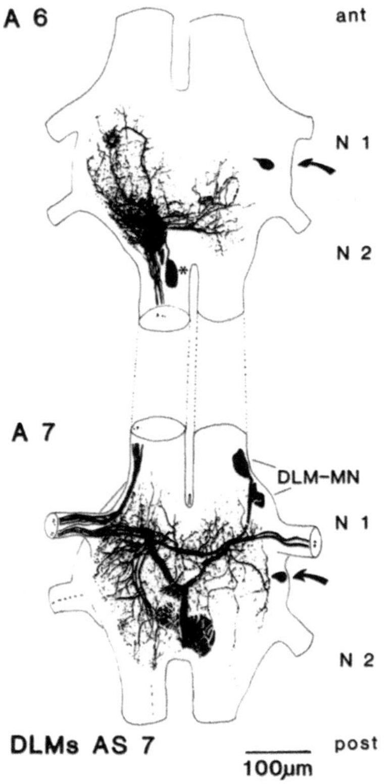

Figure 4. Drawing of the MNs supplying the DLMs of the seventh abdominal segment of an adult *Schistocerca*. The two sets of MNs split up into an anterior group and a posterior group, such as is typical for the thoracic segments. Similar cell types can also be identified (compare with Fig. 2). In the posterior ganglion there are two DUM cells with a large soma and an extensive branching pattern. Indication of specific cell types: *DLM-MN* both contralateral MNs, typical for the abdomen is the almost equal size; *asterisk* is the "podo"-cell; the *curved arrow* is the contralateral cell with the axons exiting via N1, 2 and caudad via the ipsilateral connective (Kutsch and Heckmann, unpublished).

normal approach with respect to muscle homology. Usually, muscles are said to be homonomous (or homologous, whenever appropriate) due to their position and insertion points (Mickoleit, 1961). We suggest also to elucidate the MNs organization and, judging from the grade of equivalence, to claim possible muscle homonomy or homology.

Holometabola

An equivalent approach was carried out for a holometabolous insect, the beetle *Zophobas*. Those MNs which initially were associated with only one of the thoracic DLMs were also detectable in other segments (Breidbach and Kutsch, 1990). Apparently, there is a fixed set of MNs

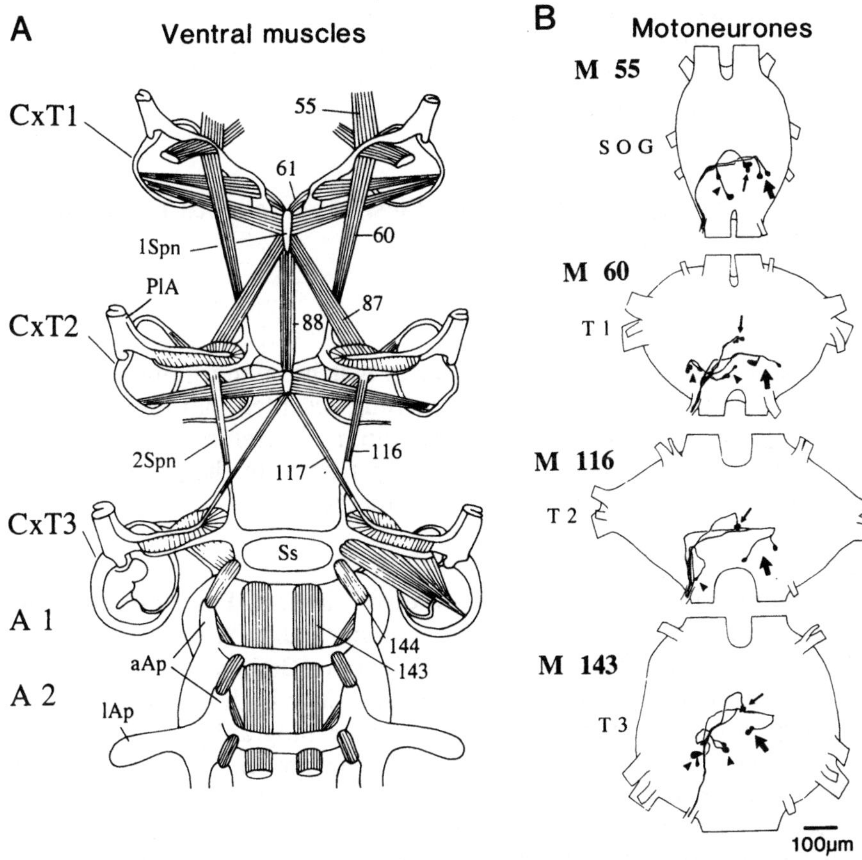

Figure 5. (A) Sketch of the ventral longitudinal muscles of a locust. (B) Demonstration of the MNs which serve a single ventral muscle, in a series from the prothorax to the first abdominal segment. For the MNs only the axons, neuropilar segments, and somata positions are drawn. With respect to the muscle's position the MNs are placed in the next anterior ganglion. *CxT1-CxT3* coxal rim of the first to third legs; *A2, A3* first and second abdominal segments; *SOG* suboesophageal ganglion; *T1–T3* pro-metathoracic ganglion. Indication of specific neural features: *arrowheads* pointing to the ipsilateral soma group; *thin arrows* pointing to the central soma group, additionally characterized by a rostrally oriented neuropilar segment; *thick arrows* pointing to the contralateral soma group, in which the cells often exhibit a looping of the neuropilar/neurite segment. The combination of all these criteria for all the employed segmental muscles suggests a homonomy of the neuro-muscular ensemble ((A) modified after Snodgrass (1929, 1935); (B) G. Steffens, unpublished).

serving the DLMs, which is reiterated with all its specific structures from the thorax to the abdomen. At least for one segmental nerve a scheme concerning the position and targets of its side branches can be compiled. This scheme is valid all along the different segments (Urbach et al., 1994). This matching of branches is congruent, even comprising the level of individual muscle units.

An interesting case of segmental reiteration is known for lepidopteran larvae. Equivalent muscles can be shown in proleg-bearing and non-proleg-bearing segments (Hinton, 1955). Weeks and Ernst-Utzschneider (1989) demonstrate that obvious leg muscles are present in both segment types and that several of the MNs can be identified as apparent homonomous ("putative homologs").

Classical homology: Structure of identified MNs, followed throughout the phylogenetic system of Mandibulata

Insecta

The taxon Insecta represents a group in which an enormous variety of different forms have evolved. It is generally accepted that their origin is monophyletic (Hennig, 1969; Ross, 1965). Therefore, a search for homology within the nervous system might point to the prototype of a nervous system developed by an early ancestor.

Orthoptera

Orthoptera belong to the more "primitive" insects. For closely related locusts, such as *Locusta* and *Schistocerca*, particular MNs can be shown which exhibit no obvious major differences (Breidbach and Kutsch, 1990). This of course may not be surprising since both locust species are not only closely related but share a similar biography and display a similar behavioral repertoire. It needs to be proven whether different behavioral repertoires can be associated with changes in the neural structure. Thereby, to avoid other uncertainties the aspect of close relationship should be respected. Wilson et al. (1982) studied an apparent homologous leg-MN in gryllid and acridid insects. Notwithstanding the jumping specialization of the acridid-leg, eminent differences of the MN-structure are not apparent considering both insect groups, except for a denser dendritic mesh in acridids.

The chosen preliminaries (close relationship, different behavioral repertoires) are especially respected when exploring the order Phasmida. In this insect group a wide selection of wing profiles exists, from winged to non-winged species; there is a wide range of flight ability, from excellent to non-flight (for summary see Kutsch and Kittmann, 1991). Since flight capability is one of the most conspicuous behaviors developed by the insects, we started a comparative approach in Phasmida (Kutsch and Kittmann, 1991; Kittmann and Kutsch, unpublished). For all studied Phasmida two large MNs can be associated with one of the major flight muscles, the subalar muscle (a depressor of the wing) of both winged segments or the equivalent in non-winged species. The somata are placed ipsilaterally and exhibit a large ipsilateral dendritic field. (Occasionally, a small cell is detected, too, albeit with a rather limited branching.) The basic structure of the neurones seems not to

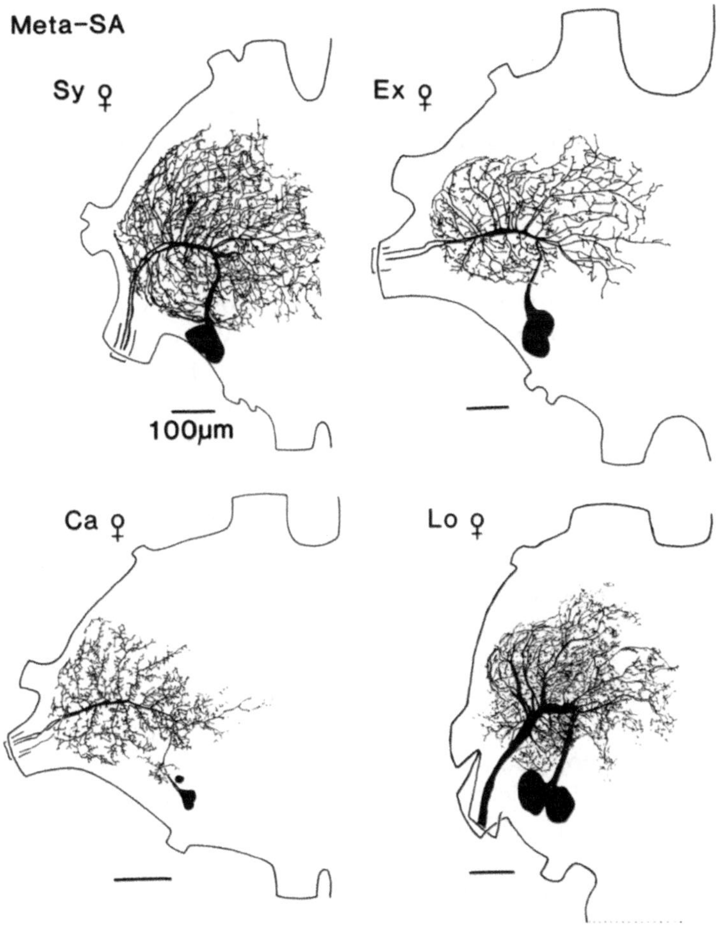

Figure 6. Comparison of the MNs serving a large pleural muscle (metathoracic subalar muscle) in different members of the Phasmida. *Sipyloidea sipylus* (*Sy*) flies very well with its large hindwings; *Extatosoma tiaratum* (*Ex*) has both wing pairs reduced to small wing scales; this species is unable to fly freely but moves the vestiges rhythmically in a wind stream; *Carausius morosus* (*Ca*) is a non-winged species. For comparison the apparent homologous MNs in *Locusta migratoria* (*Lo*) are shown. It becomes apparent that for all species similar neural feature can be shown. However, it also becomes obvious that the "good" fliers (*Sy*, *Lo*) exhibit a denser dendritic field, and that for the non-winged animal (*Ca*) the anterior median neuropile is not penetrated (Kittmann and Kutsch, unpublished).

depend on flight ability (Fig. 6). There appears, however, to exist some correlation when considering the grade of fine dendritic branching, which becomes less towards the non-winged *Carausius*. Furthermore, there occurs some reduction in the anterior median dendritic field in the non-fliers. When a comparison is drawn considering less closely related excellent fliers, such as the locusts (Fig. 6, Lo), then the general equivalence of the MNs of locusts and phasmids is impressive, too.

With respect to weak fliers in Phasmida (Fig. 6, Ex), the equivalence with locusts is manifested even when including the anterior median dendritic field. Altogether, the multitudinous similarities argue for homologous MNs seen in both insect groups and substantiate the claim for homologous muscles (Marquardt, 1940). This result indicates that a complete divergence of the lifestyles does not, automatically, result in a dramatic difference of motoneural structure.

Hemimetabola versus *Holometabola*

Mainly considering the mode of metamorphosis, the pterygote Insecta are separated into Hemi- and Holometabola. With respect to the chosen DLM-ensemble, it is quite obvious that for both groups a similar number and distribution of the characteristic two MN-sets is present (Breidbach and Kutsch, 1990). Furthermore, the dendritic fields are rather equivalent. This is especially apparent when single MNs are compared for both pterygote groups (Breidbach and Kutsch, 1990; Fig. 7). With respect to the large contralateral MN there are several equivalencies (position of soma, a small contralateral paired with a large ipsilateral dendritic field, relative extension of both fields, course of the major dendrites). This multiplicity points to a true case of homology involving far-related insect groups. The distinctness of the DLM-system has stimulated numerous studies throughout the Pterygota. Irrespective of the staining procedure, it becomes apparent that the number of MNs involved, their position and general structure are rather similar, allowing individual identification (Tab. 2). All this points to a common neuro-muscular ensemble in Pterygota.

Taylor and Truman (1974) studied MNs of the abdominal ganglia in *Manduca*. The similarity of several MN-groups of this holometabolous insect with those of some hemimetabolous insects allowed those authors to assume that there might be "a basic plan for the arrangement of MNs" in insect segmental ganglia.

Apterygote Insecta

It is generally accepted that the pterygote insects stem from apterygote ancestors (Hennig, 1969). In the search for a generalization of MN structures such primary apterygote insects also need to be employed; they might even be of eminent importance in the pursuit to elucidate the archaic insect nervous system. In the silverfish, *Lepisma saccharina*, a set of DLMs is present, too (Barlet, 1953, 1954; Birket–Smith, 1974). This segmental muscle ensemble is supplied by a set of MNs, which are distributed throughout two ganglia (Heckmann and Kutsch, 1990). The somata are found at positions similar to those observed for the aforementioned insects. There are, in general, similar dendritic fields (see, for instance, large ipsilateral vs. smaller contralateral) (Fig. 8). However, the absolute grade of matching becomes reduced compared to any of

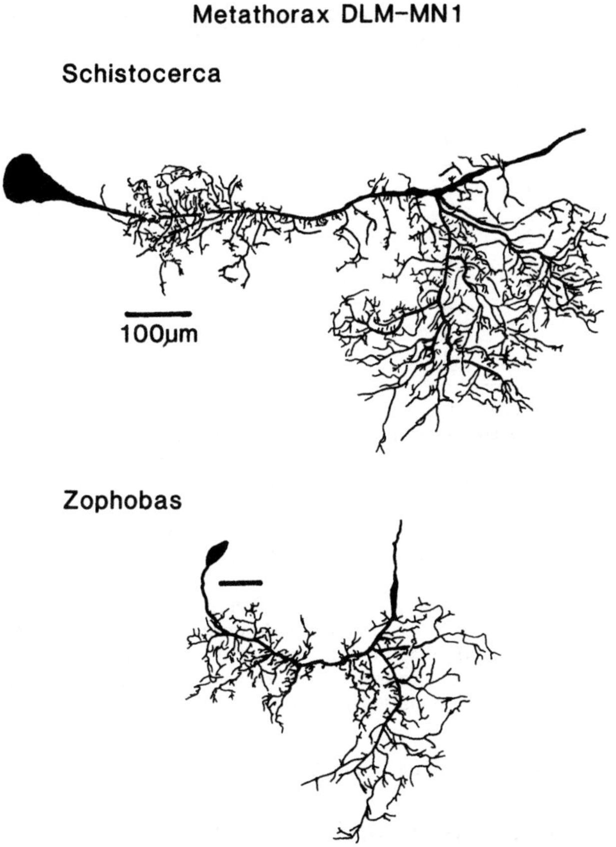

Figure 7. Comparison of an identified metathoracic MN (*DLM-MN1*) in a hemi- and a holometabolous insect. The similarity of the cells (contralateral position of the soma, difference of ipsi- and contralateral dendritic fields, structuring of the ipsilateral field mainly stemming from a bifurcation of a large primary dendrite) is striking, suggesting a homology. Locust: *Schistocerca gregaria* adult; beetle: *Zophobas morio* last larval stage (compiled from Breidbach and Kutsch, 1990).

the other tested insect species. Althogether, at least for this neuro-muscular system, an equivalence is obvious and points to the concept of homology (Heckmann and Kutsch, 1990). Apparently, the appropriate MNs belong to a very ancient neural system that expresses a crucial conservatism.

Chilopoda versus *Insecta; the Tracheata*

The Chilopoda, together with the Diplopoda, comprise the group Myriapoda. They represent a sister group to the Insecta; both are combined in the taxon Tracheata (Gupta, 1979). The Myriapoda are multisegmented animals which exhibit a more or less homogeneous

Table 2. Comparison of type and number of MNs supplying the unisegmental DLMs of different insect orders with the innervation of a Chilopoda and Crustacea

Systematic unit	split	iaMN	cpMN	DUM	loop	cc	C1, 3, 6
Lepidoptera[1]	+	7	2	+	(+)	+	(+)
Diptera[2]	+	4	1	+	?	?	?
Coleoptera[3]	+	7	2	+	1	1	?
Heteroptera[4]	+	7	1	+	?	?	?
Blattaria[5]	+	6	2	+	+	+	(+)
Saltatoria	+	8	2	+	1	1	+
Odonata[6]	+	4	1	?	?	?	?
Zygentoma	+	6	2	+	+	+	–
Chilopoda	+	9	4	–	–	–	–
Crustacea[7]	+	4	3	–	?	?	–

The MNs have been classified according to common features. *split*, neurones are distributed throughout two ganglia (next anterior and muscle segment associated); *iaMN*, ipsilateral somata, anterior neural set; *cpMN*, contralateral somata, posterior neural set; *DUM*, neurones with DUM-like morphology; *loop*, soma located contralaterally, close to the midline, neurite and neuropilar segment loops contralaterally before entering the ipsilateral nerve; *cc*, central cell; *C1, 3, 6*, neurone with soma in contralateral position, axon splits and exists via nerves N1, 3, 6. *+*, at least one cell present; *–*, no cell present; *?*, cell might be present, but has not been reported; *(+)*, some indication for presence of this cell type. (1) *Bombyx*, Tsujimura (1988, 1989); (2) *Drosophila* Ikeda and Koenig (1988); (3) *Zophobas*, Breidbach and Kutsch (1990); (4) *Dysdercus*, Davis (1977); (5) *Periplaneta*, Davis (1983); (6) *Hemianax* and *Aeschna*, Simmons (1977); (7) *Procambarus*, Mittenthal and Wine (1978). Occasionally, some additional information from other authors has been included; information from own work: *Schistocerca* (Saltatoria), *Lepisma* (Zygentoma), *Lithobius* (Chilopoda). (Heckmann and Kutsch, in preparation.)

segmentation. Such a status is supposed to represent a primitive character (Kästner, 1963).

For Chilopoda, a system of DLMs is present, too. The appropriate muscles are segmentally organized, incorporating uni- to multi- (up to three) segmental arrangements (Rilling, 1960, 1968). For *Lithobius forficatus*, an advanced member of the Chilopoda (Kästner, 1963), stains of the equivalent nerve result in a large number of MNs. Interestingly, these MNs are split into two ganglia, an anterior and a posterior one, whereby the latter group belongs to the segment of the investigated muscle (Heckmann and Kutsch, 1990) (Fig. 9). In the anterior ganglion the dendritic field is strictly ipsilateral, while in the posterior ganglion the few cells exhibit a dendritic field predominantly extending into the contralateral side.

Although there are many such similarities compared to the well-developed insect scheme, there are also certain dissimilarities. The following should be noted: the number of MNs is different; for the Chilopoda neither a contralateral field for the anterior group nor a DUM-cell is present, the dendritic fields of the contralateral MNs do not match for the two groups. However, an interesting observation should be mentioned. The contralateral group of MNs belonging to the segment-specific ganglion show several longitudinal tracts with only a sparse additional branching (Fig. 9, *G3*). A similar neural branching pattern of

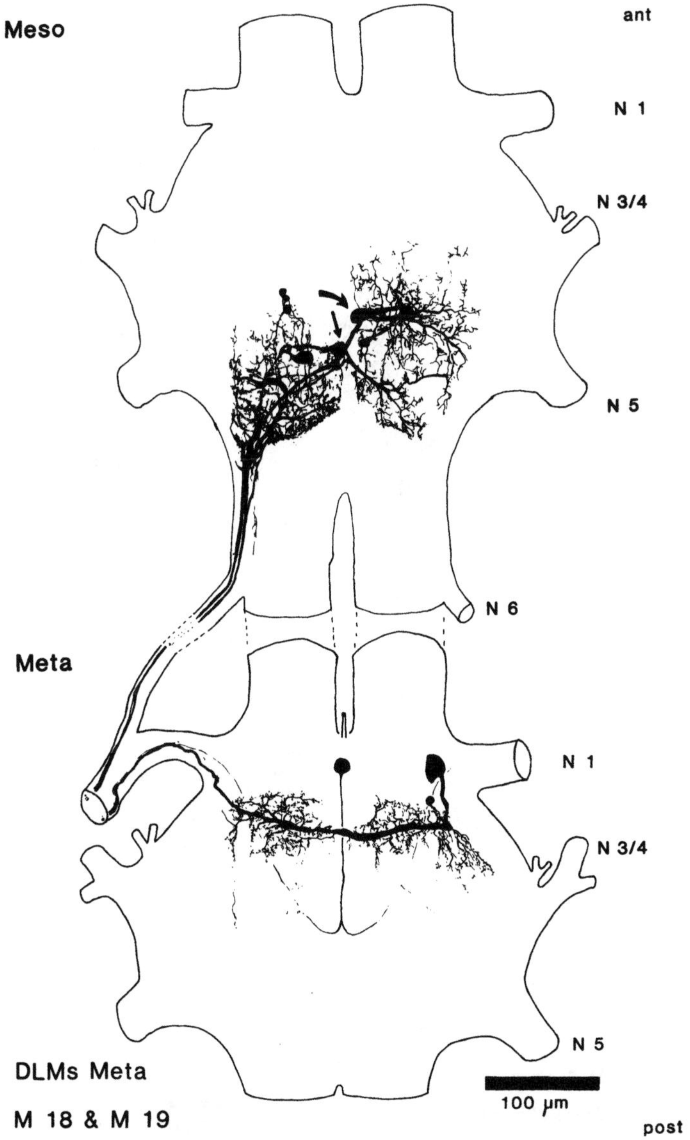

Figure 8. Set of MNs supplying the metathoracic DLMs (M18, M19) of the apterygote *Lepisma saccharina*. The split of neurones, originating from an anterior (*Meso*) and posterior (*Meta*) ganglion, and several characteristic cell types repeat features as elaborated for the pterygote insects. *Straight arrow*, indication of the soma of a midline cell; *curved arrow*, contralaterally looping cell; in the posterior ganglia a large and a small contralateral MN with a bilateral dendritic field are present, as well as a DUM cell with an anterior cell position (Heckmann and Kutsch, unpublished).

a DLM-MN is seen when staining embryos (c. 50% of embryogenesis) of *Schistocerca* (G. Steffens, unpublished). Whether this is a true case favoring the "recapitulation argument" needs to be studies. Presently, a

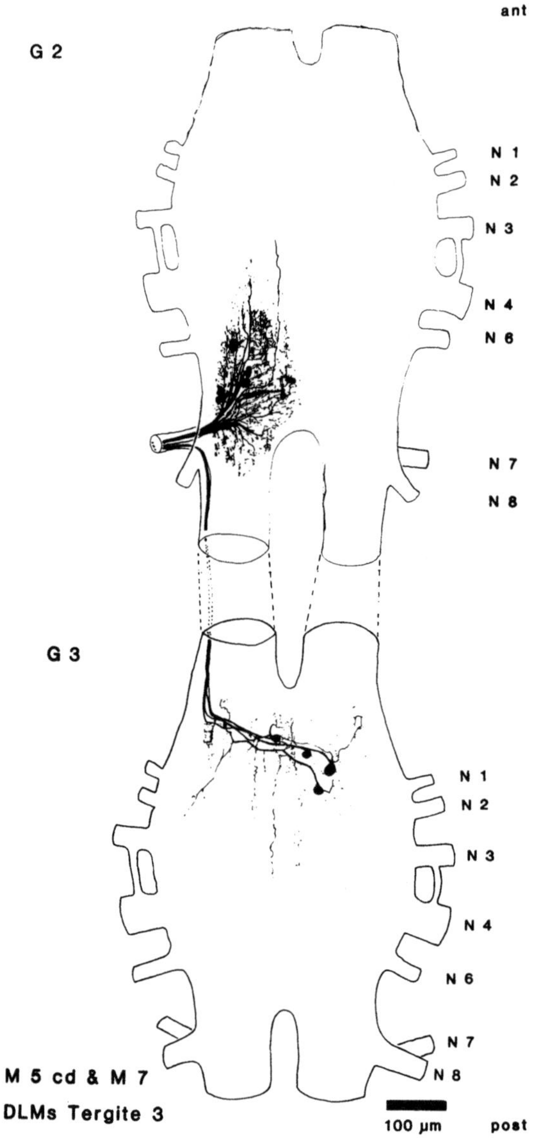

Figure 9. Set of MNs supplying the group of unisegmental DLMs in the centipede *Lithobius forficatus*. The MNs are split into an anterior and a posterior set (ganglion *G2*, *G3*). For the posterior set the somata are seen in a slightly contralateral position; their dendritic field is restricted to several longitudinally oriented major side branches (Heckmann and Kutsch, unpublished).

claim of sound homology, depending on an accumulation of multitudinous, specific similarities, still becomes difficult to support for these MNs of the two tracheate sister groups. In general, however, we are inclined to accept some sort of homology, even when remote, that

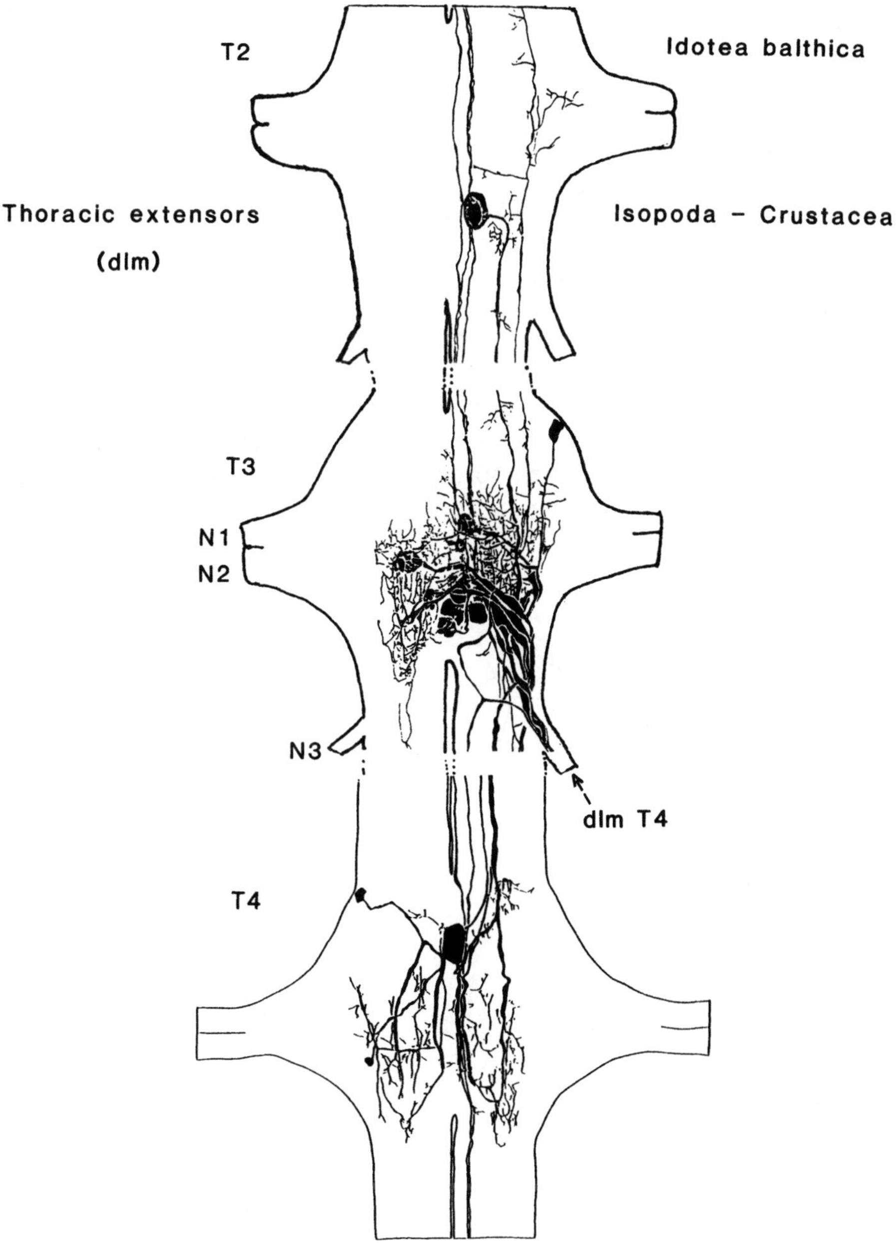

Figure 10. Set of MNs innervating the dorsal extensors in the fourth segment of the thorax of a Crustacea (*Idotea balthica*). The split of MNs in an anterior and posterior ganglion (*T3*, *T4*) is apparent. Note the small anterior contralateral cell in the posterior ganglion; the somata of the MNs in the anterior ganglion are almost exclusively ipsilateral with several contralateral dendrites (H. Schneider, unpublished).

suggests a common neural "Bauplan" for Tracheata (Heckmann and Kutsch, 1990).

Crustacea

Among the Mandibulata the Crustacea represent the group most distant from the Insecta. With respect to body organization the Crustacea probably constitute the group in which the body groundplan has been most varied (Barnes, 1980). For an initial step to study the nervous system in a comparative approach (comparison with the other upper-mentioned Mandibulata), we have chosen a member of the marine Isopoda (Malacostraca; "higher" Crustacea). This group exhibits a rather homogeneous segmentation. Among the muscle system, there is also a segmental DLM. The appropriate MNs are distributed throughout two ganglia, comprising an anterior ganglion and a ganglion adja-

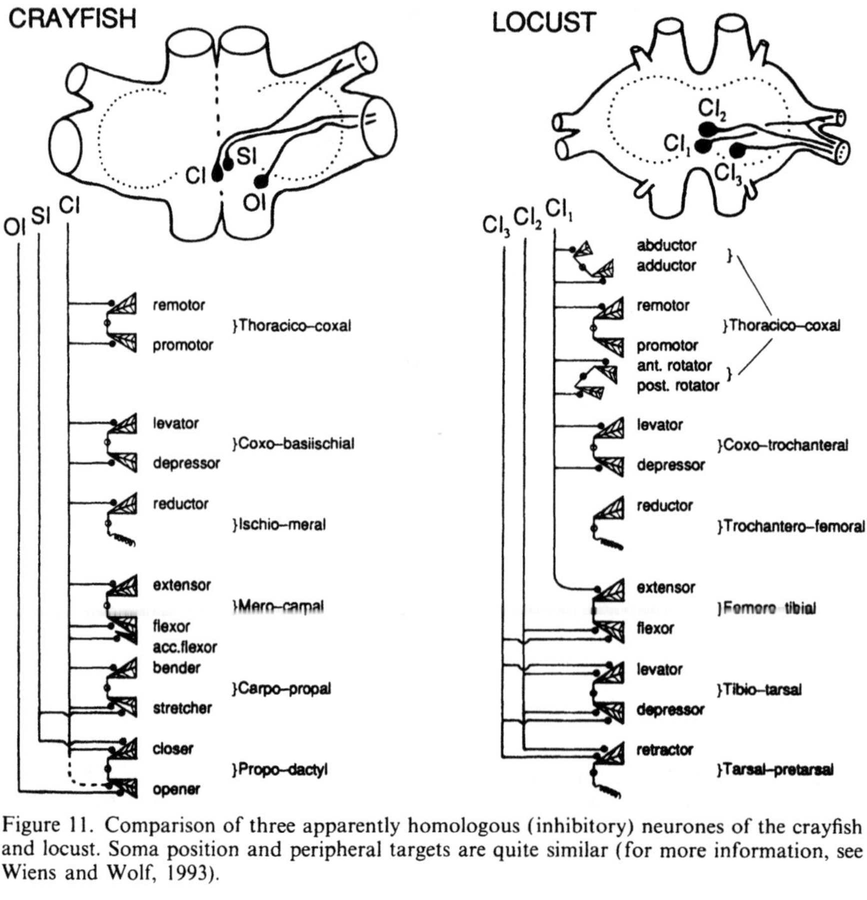

Figure 11. Comparison of three apparently homologous (inhibitory) neurones of the crayfish and locust. Soma position and peripheral targets are quite similar (for more information, see Wiens and Wolf, 1993).

cent to the muscle's segment (Fig. 10). An additional cell is visible in the second anterior ganglion. For the segment-specific ganglion, among others, a contralateral soma is present. For the anterior ganglion a prominent ipsilateral dendritic field is observed. However, considering other details, there are too many dissimilarities to claim a sound homology with a general scheme as elaborated for the insects. Even though the general similarity of Crustacea and Insecta is intriguing, direct comparisons of MNs for both groups are scarce. Mittenthal and Wine (1978) suppose that homologous MNs exist for decapod Crustacea and Insecta. Recently, a quite convincing example was given in which motoneural homology of Crustacea and Insecta species can be assumed. Wiens and Wolf (1993) present a case of MN-matching for

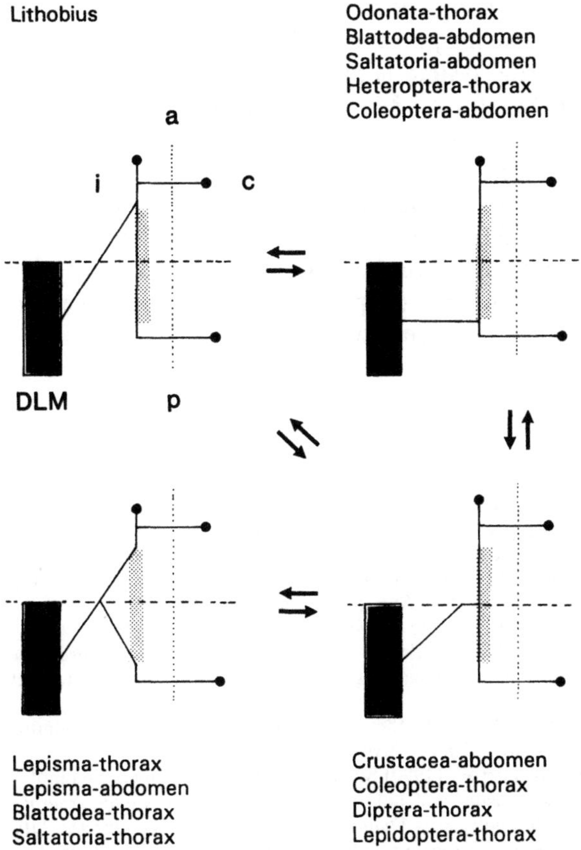

Figure 12. Schematic representation of the innervation pattern of the DLMs found in different taxa of the Mandibulata. Arrows mark the possibility to transform one pattern into another, employing only a simple step of axons' unification. Even in the same animal different patterns may be realized; see, for instance, Saltatoria-thorax vs. – abdomen. *i*, ipsilateral; *c*, contralateral hemiganglion; *dotted line*, ganglionic midline; *dashed line*, segmental border; *dotted area*, ipsilateral connective; *a*, anterior, *p*, posterior ganglion (Heckmann and Kutsch, unpublished).

crayfish and locust (Fig. 11). They have chosen the peculiar inhibitory neurones serving the different muscles that follow a specific branching mode. In both groups three such neurones per segment are present. The position of the somata as well as the modality to serve the peripheral target, i.e., the different muscles, are, unexpectedly, similar. It is assumed that for each locust inhibitory neurone a homologue exists in the crayfish ganglion.

Conclusions

We have especially employed the system of DLMs and their innervation using a comparative approach to search for similarities in Mandibulata. It becomes apparent that there are multitudinous similarities which, due to the relevant catalog, point towards homology of this motor system throughout this arthropod group. If this can be accepted, then one may postulate that this system is part of a basic neuro-muscular "Bauplan" which is common for all Mandibulata (Fig. 12). Given such a homology, one may infer that this particular system has been developed by a common ancestor. Whether such conservative structures are the rule or the exception needs further studies. Maybe by adding future knowledge on other, preferably even on all neuro-muscular systems, we might be able to assemble a (theoretical) compendium constituting the original ensemble with which the (hypothetical) common ancestor of Insecta or even Mandibulata was endowed.

References

Ax, P. (1989) Homologie in der Biologie – ein Relationsbegriff im Vergleich von Arten. *Zool. Beitr. N.F.* 32: 487–496.

Barlet, J. (1953) Morphologie du thorax de *Lepisma saccharina* L. (Aptérygote Thysanoure). II. Musculature 1. Bull. *Ann. Soc. Entom. Belg.* 89: 214–236.

Barlet, J. (1954) Morphologie du thorax de *Lepisma saccharina* L. (Aptérygote Thysanoure). II. Musculature 2. Bull. *Ann. Soc. Entom. Belg.* 90: 299–321.

Barnes, R.D. (1980) *Invertebrate Zoology*. Saunders, Philadelphia.

Bernays, E.A. (1972) The muscles of newly hatched *Schistocerca gregaria* larvae and their possible functions in hatching, digging and ecdysial movements (Insecta: Acrididae). *J. Zool.* 166: 141–158.

Birket-Smith, S.J.R. (1974) On the abdominal morphology of Thysanura (Archaeognatha and Thysanura s. str.). *Entom. Scand.*, Suppl. 6: 1–67.

Breidbach O. and Kutsch W. (1990) Structural homology of identified motoneurones in larval and adult stages of hemi- and holometabolous insects. *J. Comp. Neurol.* 297: 392–409.

Bronn, H. (1858) *Morphologische Studien über die Gestaltungsgesetze der Naturkörper überhaupt und der organischen insbesondere.* C.F. Winter'sche Verlagshandlung, Leipzig, Heidelberg.

Darwin, C. (1859) *The Origin of Species by Means of Natural Selection, or the Preservation of Favoured Races in the Struggle for Life.* Murray, London.

Davis, N.T. (1977) Motor neurons of the indirect flight muscles of *Dysdercus fulvoniger*. *Ann. Entom. Soc. Am.* 70: 377–386.

Davis, N.T. (1983) Serial homologies of the motor neurons of the dorsal intersegmental muscles of the cockroach, *Periplaneta americana* (L.). *J. Morph.* 176: 197–210.

Dobzhansky, T., Ayala, F., Stebbins, G.L. and Valentine, J.W. (1977) *Evolution*. Freeman, San Francisco.

Dumont, J.P.C. and Robertson, R.M. (1986) Neuronal circuits: an evolutionary perspective. *Science* 233: 849–853.

Gilbert, L.I. and Frieden, E. (1981) *Metamorphosis – A Problem in Developmental Biology*. Plenum Press, New York.

Goodman, C.S., Pearson, K.G. and Heitler, W.J. (1979) Variability of identified neurons in grasshoppers. *Comp. Biochem. Physiol.* 64 A: 455–462.

Gupta, A.P. (1979) *Arthropod Phylogeny*. v. Nostrand Reinhold, New York.

Heckmann, R. and Kutsch, W. (1990) Common neural 'Bauplan' in Tracheata? Innervation of the dorsal longitudinal muscles. *In:* N. Elsner and G. Roth (eds): *Brain-Perception-Recognition*, Thieme Verlag, Stuttgart, p. 39.

Hennig, W. (1969) *Die Stammesgeschichte der Insekten*. Kramer, Frankfurt.

Hinton, H.E. (1955) On the structure, function, and distribution of the prolegs of the Panorpoidea with a criticism of the Berlese-Imms theory. *Trans. R. Entom. Soc. Lond.* 106: 455–540.

Hoyle, G. (1978) The dorsal, unpaired, median neurons in the locust metathoracic ganglion. *J. Neurobiol.* 9: 43–57.

Ikeda, K. and Koenig, J.H. (1988) Morphological identification of the motor neurons innervating the dorsal longitudinal flight muscle of *Drosophila melanogaster*. *J. Comp. Neurol.* 273: 436–444.

Kalogianni, E., Consoulas, C. and Theophilidis, G. (1989) Anatomy and innervation of the abdominal segmental muscles in the larval and adult *Tenebrio molitor* (Coleoptera). *J. Morph.* 202: 271–279.

Kästner, A. (1963) Tracheata. *In: Lehrbuch der Speziellen Zoologie, Teil I: Wirbellose*. Gustav Fischer Verlag, Jena.

Kästner, A. (1965) Articulata. *In: Lehrbuch der Speziellen Zoologie. Vol. I, Wirbellose*. Gustav Fischer Verlag, Stuttgart.

Kluge, A.G. (1977) *Chordate Structure and Function*. Second edition, Macmillan, New York, London.

Kutsch, W. and Breidbach, O. (1994) Homologous structures in the nervous systems of Arthropoda. *Adv. Insect Physiol.* 24: 1–113.

Kutsch, W. and Kittmann, R. (1991) Flight motor pattern in flying and non-flying Phasmida. *J. Comp. Physiol. A* 168: 483–490.

Kutsch, W. and Schneider, H. (1987) Histological characterization of neurones innervating functionally different muscles of *Locusta*. *J. Comp. Neurol.* 261: 515–528.

Lee, W.-Y. (1964) *A study of the development of the musculature from the larva to the adult in Tenebrio molitor L. (Tenebrionidae, Coleoptera)*. Ph.D. thesis, Univ. Minnesota.

Leise, E.M. (1990) Modular construction of nervous systems: a basic principle of design for invertebrates and vertebrates. *Brain Res. Rev.* 15: 1–23.

Marquardt, F. (1940) Beiträge zur Anatomie der Muskulatur und der peripheren Nerven von *Carausius (Dixippus) mososus*. *Br. Zool. Jb. Anat.* 66: 63–128.

Mayr, E. (1969) *Principles of Systematic Zoology*. McGraw-Hill, New York.

Mickoleit, G. (1961) Zur Thoraxmorphologie der Thysanoptera. *Zool. Jb. Anat.* 79: 1–92.

Mittenthal, J.E. and Wine, J.J. (1978) Segmental homology and variation in flexor motoneurons of the crayfish abdomen. *J. Comp. Neurol.* 177: 311–334.

Myers, C.M. and Ball, E. (1987) Comparative development of the extensor and flexor tibiae muscles in the leg of the locust, *Locusta migratoria*. *Development* 101: 351–361.

Myers, C.M., Whitington, P.M. and Ball, E.E. (1990) Embryonic development of the innervation of the locust extensor tibiae muscle by identified neurons: formation and elimination of inappropriate axon branches. *Dev. Biol.* 137: 194–206.

Neville, A.C. (1963) Motor unit distribution of the dorsal longitudinal flight muscles in locusts. *J. Exp. Biol.* 40: 123–136.

Owen, R. (1843) *Lectures on the Comparative Anatomy and Physiology of the Invertebrate Animals*. Longman, Brown, Green, Longmans, London.

Owen, R. (1848) *On the Archetype and Homologies of the Vertebrate Skeleton*. J. v. Voorst, London.

Pipa, R.L. (1978) Patterns of neuronal reorganization during the postembryonic development of insects. *Int. Rev. Cytol.*, Suppl. 7: 403–438.

Remane, A. (1956) *Die Grundlagen des natürlichen Systems der vergleichenden Anatomie und der Phylogenetik*. Akademische Verlagsgesellschaft, Geest and Portig, Leipzig.

Rilling, G. (1960) Zur Anatomie des braunen Steinläufers *Lithobius forficatus* L. (Chilopoda). Skelettmuskelsystem, peripheres Nervensystem und Sinnesorgane des Rumpfes. *Zool. Jb. Anat.* 78: 39–128.
Rilling, G. (1968) *Lithobius forficatus.* Grosses Zool. Prakt. 13b. Gustav Fischer Verlag, Stuttgart.
Ross, H.H. (1965) *A Textbook of Entomology*, Third edition, Wiley, New York.
Satterlie, R.A. (1985) Structural variability of an identified interneurone in locusts from a wild population. *J. Exp. Biol.* 114: 691–695.
Sehnal F. (1985) Morphology of insect development. *Ann. Rev. Entomol.* 30: 89–109.
Sharov, A.G. (1966) *Basic Arthropodan Stock with Special Reference to Insecta.* Pergamon Press, Oxford.
Simmons, P. (1977) The neuronal control of dragonfly flight. I. Anatomy. *J. Exp. Biol.* 71: 123–140.
Simpson, G.G. (1967) *Principles of Animal Taxonomy.* Columbia Univ. Press, New York.
Snodgrass, R.E. (1929) The thoracic mechanism of a grasshopper, and its antecedents. *Smithson. Misc. Coll.* 82: 1–112.
Snodgrass, R.E. (1935) The abdominal mechanisms of a grasshopper. *Smithson. Misc. Coll.* 94: 1–89.
Steffens, G.R. and Kutsch, W. (1992) Embryonic development of identified motor neurones in the locust. *In:* N. Elsner and D.W. Richter (eds): *Rhythmogenesis in Neurons and Networks*, Thieme Verlag, Stuttgart, p. 630.
Taylor, H.M. and Truman, J.W. (1974) Metamorphosis of the abdominal ganglia of the tobacco hornworm, *Manduca sexta. J. Comp. Physiol.* 90: 367–388.
Tsujimura, H. (1988) Metamorphosis of wing motor system in the silk moth, *Bombyx mori* L. (Lepidoptera: Bombycidae): anatomy of the sensory and motor neurons that innervate larval mesothoracic dorsal musculature, stretch receptors, and epidermis. *Int. J. Insect Morph. Embryol.* 17: 367–380.
Tsujimura, H. (1989) Metamorphosis of wing motor system in the silk moth, *Bombyx mori*: origin of wing motor neurons. *Dev. Growth Differ.* 31: 331–339.
Urbach, R., Breidbach, O. and Kutsch, W. (1994) Comparative anatomy of muscle sets in larval and adult stages of *Zophobas morio* (Coleoptera, Tenebrionidae). Zoomorph. 114: 47–57.
Voss, F. (1905) Über den Thorax von *Gryllus domesticus*, Mit besonderer Berücksichtigung des Flügelgelenks und dessen Bewegung. II. Die Muskulatur. *Z. wiss. Zool.* 78: 355–521.
Wagner, G.P. (1989) The biological homology concept. *Annu. Rev. Ecol. Syst.* 20: 51–69.
Weeks, J.C. and Ernst-Utzschneider, K. (1989) Respecification of larval proleg motoneurons during metamorphosis of the tobacco hornworm, *Manduca sexta*: segmental dependence and hormonal regulation. *J. Neurobiol.* 20: 569–592.
Whitington, P.M. and Seifert, E. (1981) Identified neurons in an insect embryo: the pattern of neurons innervating the metathoracic leg of the locust. *J. Comp. Neurol.* 200: 203–212.
Wiens, T.J. and Wolf, H. (1993) The inhibitory motoneurons of crayfish thoracic limbs: identification, structures, and homology with insect common inhibitors. *J. Comp. Neurol.* 336: 261–278.
Wiesend P. (1957) Die postembryonale Entwicklung der Thoraxmuskulatur bei einigen Heuschrecken mit besonderer Berücksichtigung der Flugmuskeln. *Z. Morph. Ökol. Tiere* 46: 529–570.
Wilson, J.A. (1979) The structure and function of serially homologous leg motor neurons in the locust. I. Anatomy. *J. Neurobiol.* 10: 41–65.
Wilson, J. A. and Hoyle, G. (1978) Serially homologous neurones as concomitants of functional specialisation. *Nature* 274: 377–378.
Wilson, J.A., Phillips, C.E., Adams, M.E. and Huber, F. (1982) Structural comparison of a homologous neuron in gryllid and acridid insects. *J. Comp. Neurol.* 13: 459–467.
Wittig, G. (1955) Untersuchungen am Thorax von *Perla abdominalis* Burm. (Larve und Imago). *Zool. Jb. Anat.* 74: 491–570.
Xie, F., Meier, T. and Reichert, H. (1992) Embryonic development of muscle patterns in the body wall of the grasshopper. *W. Roux's Arch. Dev. Biol.* 201: 301–311.
Yack, I.E. and Fullard, J.H. (1990) The mechanoreceptive origin of insect tympanal organs: a comparative study of similar nerves in tympanate and atympanate moths. *J. Comp. Neurol.* 300: 523–534.

The Nervous Systems of Invertebrates: An Evolutionary and Comparative Approach
ed. by O. Breidbach & W. Kutsch
© 1995 Birkhäuser Verlag Basel/Switzerland

Developmental mechanisms, homology and evolution of the insect peripheral nervous system

T. Meier[1] and H. Reichert[2]

[1]*Department of Physiology, University of Colorado Health Sciences Center 4200 East Ninth Ave., Denver CO 80262, USA*
[2]*Laboratory of Neurobiology, Department of Zoology, University of Basel, Switzerland*

Summary. Segmentally homologous neurogenesis and neuronal differentiation processes characterize the formation of the insect peripheral nervous system. These processes have been analyzed at the level of identified neurons and nerve branches in the grasshopper. During early embryogenesis a simple peripheral nerve scaffolding is established in each body segment. This scaffolding consists of serially reiterated segmental nerves, which are pioneered by afferent neurons situated in limb bud homologs, and of serially reiterated intersegmental nerves, which are pioneered by homologous body wall afferents. Subsequently, identified sets of serially homologous sensory neurons differentiate in a stereotyped spatiotemporal pattern in each segment and project their axons onto these nerves. Data on serial homology in the development of the peripheral nervous system, which were obtained on the basis of comparative structural and developmental data in the grasshopper, are supported by a large body of developmental and genetic data obtained using *Drosophila* wildtype and mutants. Indeed, a comparison of the results obtained in the grasshopper with those obtained from *Drosophila* shows that virtually the same pattern of peripheral nerves and major sensory structures is formed in both species. These findings indicate that the construction of the peripheral nervous system in extremely divergent modern insects relies on conserved developmental mechanisms that evolved in ancestral insects over 300 million years ago.

Introduction

The insect nervous system is a complex structure. Understanding the way this nervous system develops and deducing the possible mechanisms by which it evolved are daunting challenges. The most obvious reason for this is the fact that the total number of neurons in insect nervous systems is very large. In orthopteran insects for instance, the brain and associated optic lobes can contain over one million neurons; numerous further nerve cells are found in the ventral nerve cord and the peripheral nervous system (Farrel and Kuhlenbeck, 1964). These neurons and sensory cells are interconnected into highly sophisticated circuits; analyzing the cellular and molecular mechanisms responsible for correctly interconnecting them is certainly a difficult task (for review see Arbas et al., 1991). However, the number of neurons that needs to be interconnected in different neuronal circuits in the developing insect nervous system is considerably smaller than the total number of nerve cells present. This is because there is an enormous degree of modularized and repetitive neuronal circuitry in the insect nervous system. It is important to realize that this repetitive modularization of a relatively small number of neuronal types greatly reduces the size of the developmental and the evolutionary problem involved in constructing the circuits of the insect nervous system (Dumont and Robertson, 1986; Reichert, 1994). Modular circuitry is the hallmark of the segmental nervous system of insects, where it is based at least in part on serial homology (Robertson et al., 1982; Wilson et al., 1982; Pearson et al., 1985). The ontogenetic and phylogenetic mechanisms which give rise to the serially homologous nervous structures in the segmentally organized peripheral nervous system of insects are the focus of this review.

How does serially homologous circuitry develop? Are the developmental mechanisms involved in its construction evolutionarily conserved among different insect species? These questions can be addressed by investigating the developmental origin of the segmentally organized peripheral nervous systems of selected insect species (Boyan and Ball, 1993). In some of these species evidence is accumulating that serially repeated pathfinding mechanisms and serially repeated molecular guidance cues are involved in guiding the axons of homologous sensory neurons in the body segments and appendages into their target regions. Furthermore, comparative developmental studies in these model systems indicate that aspects of the underlying basic neurodevelopmental program are evolutionarily highly conserved. Specifically, a comparison of cell biological data obtained in the grasshopper *Schistocerca* with molecular genetic data obtained in the fly *Drosophila* shows that the pattern of peripheral innervation and peripheral sensory structures is comparable to a high level of resolution in both species (Meier et al., 1991). This indicates that the construction of the peripheral nervous system in extremely divergent modern insects relies on highly conserved cellular and molecular mechanisms that evolved in ancestral insects over 300 million years ago.

The biological homology concept

The concept of homology is generally used for the reconstruction of phylogenetic history. The empirical basis of all homology concepts is the recognition of conservative morphological features that are used to identify other structures as "the same" in different organisms, defining phylogenetic homology, or in different regions of the same body, thereby defining serial homology (Wagner, 1989). Among three homology concepts (Wagner, 1989), the "Biological Homology Concept" seems to be the most suitable to investigate the insect peripheral nervous system (PNS) with respect to its evolution. According to this concept's mechanistic rather than genealogical basis (Roth, 1984) the analysis of developmental mechanisms is essential for the definition of homology. According to Wagner (1989), one-to-one mapping of single characters in two species is only meaningful if each of the two elements compared is developmentally individualized. In order to investigate homology of insect peripheral nervous systems, this would imply that clear-cut homology only can be stated if the structures of interest (nerves, sensory neurons) can be traced during development.

In this review, a comparative approach has been chosen to uncover serial and phylogenetic homology of developmentally individualized elements of the peripheral nervous system in two distantly related insect species (Kristensen, 1981), the dipteran *Drosophila* and the orthopteran *Schistocerca*, following the "Biological Homology Concept". It is noteworthy in this respect that phylogenetic homologous characters need not share common pathways of ontogenetic development. As will be shown in this review, the embryonic PNS of *Drosophila* and *Schistocerca* contains homologous elements despite the drastically different way of germ band development: In the fly embryo segments differentiate almost synchronously, producing a so-called long germ-band, whereas in the grasshopper embryo segments are added from a terminal growth zone, generating a short germ-band (Sander, 1988).

Serial homology in the development of appendage innervation

Insects are arthropods. As this term implies, insects have jointed appendages, and typically a pair of appendages are located on each body segment. The different appendages are often greatly modified depending on their segment of origin. Thus, typical walking legs are found only on the three thoracic segments. In the head, the three gnathal appendages are modified for feeding and function as mouthparts. In the pregenital abdominal segments, appendage-like structures may be transiently expressed during development but are no longer found on the adult animal. How is the innervation of all of these appendages formed during embryonic development? Are serially repeated developmental mechanisms involved in constructing the innervation of all of these external structures?

An examination of mature adult innervation does not necessarily answer these questions since a great deal of fusion, shift of position or fragmentation of nervous structures can occur during embryonic and postembryonic development (Matsuda, 1976). From an evolutionary standpoint, serial homology is a reflection of the homonymous segmentation of ancient forms in which comparable structures were present in most body segments (Dobzhansky et al., 1977; Raff and Kaufmann, 1983). If the fundamental aspects of the developmental mechanisms for constructing different appendages have been conserved in evolution, basic similarities in the way in which very different appendage types in insects are built may still be evident during embryogenesis.

An attractive preparation for studying the way in which peripheral innervation of body appendages develops, is the embryonic grasshopper. In this animal, the embryonic neurons are large, individually identifiable and experimentally accessible. Due to the pioneering work of a number of investigators in the last decade (Bentley and Keshishian, 1982; Ho and Goodman, 1982; Keshishian and Bentley, 1983a,b; Caudy and Bentley, 1986; Whitington, 1989), a great deal is now known about the way in which the innervation of the prototypic thoracic legs is formed (see Fig. 1; thoracic leg). Briefly summarized, at about 30% of embryonic development, a cell within the epithelium at the anterior tip of the embryonic limb bud enlarges and divides, giving rise to a pair of pioneer neurons, called Ti1 neurons. These cells extend filopodia distally and proximally and shortly thereafter an axonal growth cone emerges from the proximal side of the cells. The growing axon navigates along the inner surface of the limb bud epithelium. During this process, it follows a series of preaxogenesis guidepost neurons spaced along its pathway to the CNS. Some of these guidepost neurons, the Cx1 neurons and the neurons Fe1 and Tr1, become the earliest follower neurons of the Ti1 axon pathway; by fasciculation with the Ti1 pathway they also reach the CNS (Keshishian and Bentley, 1983a: Lakes-Harlan and

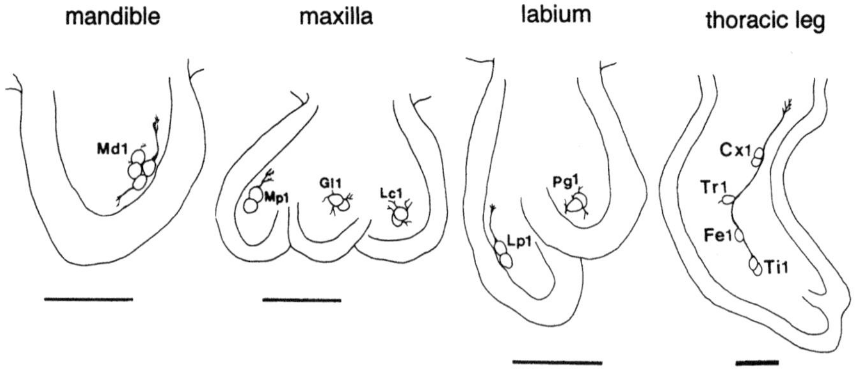

Figure 1. Nerve formation in segmental appendages of the grasshopper embryo (35% developmental stage). The positions of pioneer neurons and guidepost-cells are shown for the three mouthpart appendages, mandible, maxilla, and labium, as well as for thoracic legs. In the thoracic legs, axons of the pioneer neurons (Ti1) follow a well-defined pathway along identifiable guidepost cells (Fe1, Tr1, Cx1) towards the central neuropil. A serially homologous set of pioneer neurons is found in the distal tip of the maxillary and labial appendages (Mp1 and LP1). Axons from these pioneer neurons navigate proximally and orient along identifiable guidepost cells (G11 and Lc1 cells in the maxilla; Pg1 cells in the labium). Nerve formation in the mandibular appendage is also initiated by pioneer neurons (Md1 cells). In contrast to the other appendages, however, axons of the Md1 neurons reach the central neuropile without the help of guidepost cells. Scale bar: 100 µm. Modified after Bentley and Keshishian (1982), Keshishian and Bentley (1983a,b), Meier and Reichert (1991).

Pollack, 1993). In the 10% of development that follows the establishment of the first axonal pathway, additional peripheral neurons and sensory cells arise in the limb bud. Axons from these neurons as well as axons from central efferent neurons selectively fasciculate and form the other major nerve branches of the leg (Keshishian and Bentley, 1983b).

Is this general spatiotemporal sequence of early neurogenesis and nerve pathway formation also found in the highly modified embryonic appendages that differentiate into the mouthparts of the head? Surprisingly, the developmental mechanisms that operate in the segmentally homologous thoracic appendages are preserved to a remarkable degree during the early developmental phase of mouthpart innervation (Meier and Reichert, 1991; Lakes-Harlan and Pollack, 1993). Indeed, the mechanisms of early innervation of the mouthparts are remarkably "leg-like" (Fig. 1). Thus, in all of the developing embryonic mouthparts, the initial nerve pathway is established by distally located neurons that project their afferent axons along a set of more proximally located cell bodies. In the "limb bud" of each mouthpart, specific pioneer neurons initiate axogenesis and project their axons along the inner surface of the "limb bud" epithelium; they also orient their axons along a set of strategically placed guidepost-like cells towards the CNS. Slightly later in embryonic development additional peripherally located cells in these mouthpart "limb buds" grow towards the CNS, fasciculate with efferently growing axons and thus pioneer other nerve pathways. The many striking similarities among the early differentiating neuronal cells in the mouthparts and the legs in features such as structure, position, origin and mode of differentiation, indicate that the pioneer neurons and the guidepost cells in the different thoracic and gnathal segments are serially homologous. A complete homology scheme based on the relative location of these cells in the developing appendages, on the spatial and temporal sequence of differentiation of these cells, and on their role in the formation of peripheral nerve branches is shown in Figure 2.

Considering the overt similarity of developmental mechanisms involved in the early embryonic formation of serially homologous appendage innervation in gnathal and thoracic segments, it is interesting to consider the development of peripheral innervation in the abdominal segments, where embryonic limb bud rudiments are only transiently expressed (Bentley et al., 1979). Are similar, segmentally repeated mechanisms involved in the formation of serially homologous nerves in these abdominal segments? Put differently, are transiently formed embryonic appendage rudiments innervated during early development and, if so, does this innervation persist following the loss of these rudiments? An investigation of the development of the segmental nerves in most of the abdominal segments of the grasshopper shows that in the abdominal limb buds a single pioneer neurons, the limb bud pioneer cell (LbP),

Serial homology among neurons of developing appendages in the grasshopper embryo

Md	Mx	La	T1	T2	T3	proposed function
Md1	Mp1	Lp1	Ti1	Ti1	Ti1	pioneer for posterior nerve root
Md1*	Gl1	Pg1*	Fe1	Fe1	Fe1	guidepost along posterior nerve
Md1*	Lc1*	Pg1*	Tr1	Tr1	Tr1	guidepost along posterior nerve
Md1*	Lc1	Pg1	Cx1	Cx1	Cx1	guidepost along posterior nerve
Md2	Bs1	Pm1	Cx2	Cx2	Cx2	pioneer for anterior nerve root
Md1*	Mp2	Lp2	Fe2	Fe2	Fe2	additional neurons along nerve
Md1*	Mp3	Lp3	Tr2	Tr2	Tr2	additional neurons along nerve

* fused cell group (?)

Figure 2. Serial homology of neurons in the developing mouthpart appendages (mandible: Md, maxilla: Mx, labium: La) and in the developing thoracic legs (T1, T2, T3) of the grasshopper embryo. Criteria for homology are based on the relative position of the cells in the developing appendages, on the spatial and temporal sequence of differentiation of these cells, and on their role in the formation of peripheral nerve branches. For each of the cells or cell clusters, the proposed function during initial nerve formation is indicated. Cell clusters which are likely to be involved in more than one function are marked with an asterisk. This is especially pronounced in the developing mandible. See text for further details. Modified after Caudy and Bentley (1986) and Meier and Reichert (1991).

differentiates from the ectoderm and undergoes axogenesis (Fig. 3). Its axon projects to the next adjacent neuromere without the help of further guidepost cells. Subsequently, other outgrowing axons fasciculate with the LbP axon, and the LbP cell body becomes displaced into the lumen of the body wall. Efferents follow the LbP axon peripherally and extend into more distal parts of the body wall, thus, contributing to the anterior fascicle of the segmental nerve (Meier et al., 1991). In later embryonic stages, the abdominal limb bud disappears after the LbP-cell and associated nerves have become displaced into the body wall. Thus, although most of the abdominal limb buds are only transient embryonic structures, they play an important role in the development of segmental innervation in the abdominal segments. This role clearly reflects the segmentally homologous operation of similar cellular mechanisms for peripheral innervation of all appendages in this insect. Transiently expressed limb-buds still are functionally involved in the generation of a single pioneer cell even when this limb bud has been almost entirely reduced in the course of evolution (Raff and Kaufmann, 1983).

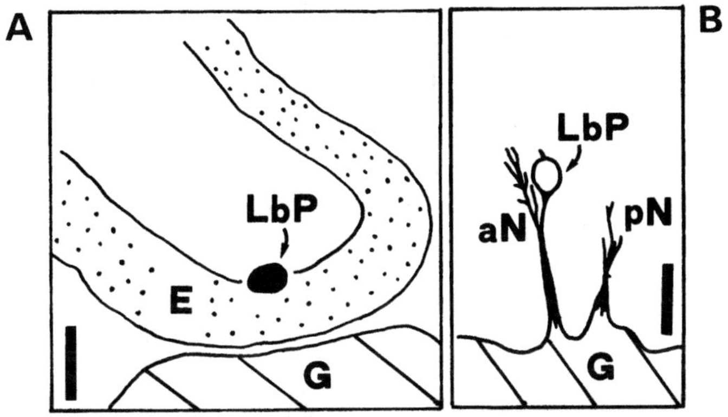

Figure 3. Formation of the anterior fascicle of the segmental nerve in an abdominal segment of the grasshopper embryo. (**A**) Differentiation of the limb bud pioneer neuron (LbP) from the ectoderm (E) of a transiently expressed abdominal limb bud in close proximity to the segmental ganglion neuromere (G). (**B**) A slightly later embryonic stage. The initial connection between the LbP pioneer neuron and the ganglion (G) has been established by the LbP-axon. No guidepost cells are involved in this pathfinding process. This developing anterior nerve fascicle (aN) is subsequently used by efferent motoraxons to project into the body wall. The posterior fascicle of the segmental nerve (pN) is being formed by efferent pioneers alone, without the participation of afferent pioneer neurons. Scale bars: 50 μm (**A**), 25 μm (**B**). Redrawn and modified after Meier et al. (1992).

Serial homology in the development of peripheral sensory structures

The body segments of the grasshopper, as in most other insects, are innervated by a serially repeated set of nerves, the segmental nerves and the intersegmental nerve (see Whitington; this volume). The developmental origin of the segmental nerves is intimately associated with the innervation of appendages as described above. In contrast, the intersegmental nerves are not associated with appendages, rather, they innervate the lateral body wall. How do these nerves develop? The developmental origin of the intersegmental nerves involves a concerted interaction of a serially repeated group of peripheral afferent and central efferent pioneer neurons (Fig. 4). In most segments the distal part of the intersegmental nerves is pioneered by the axons of a group of peripheral body wall (dBw) cells (Meier et al., 1991). The proximal part of the intersegmental nerve is established by axons from efferent pioneer neurons running in the U-fascicle (U_f) (Goodman et al., 1984; Bastiani et al., 1986; du Lac et al., 1986; Whitington, 1989). Peripheral and central axons meet approximately half way. They then fasciculate and overgrow each other in opposing directions. In the grasshopper embryo, this stereotypic mode of intersegmental nerve formation can be seen in the pterothoracic and pregenital abdominal segments. Moreover, this mechanism for the formation of the intersegmental nerve is highly conserved between different insect species. In the *Drosophila* embryo,

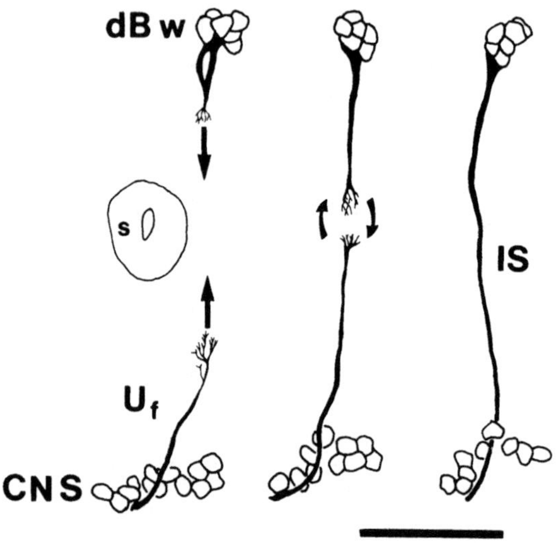

Figure 4. Formation of the intersegmental nerve (IS) in the grasshopper embryo. From left to right, three consecutive stages as seen in three adjacent abdominal segments are shown. Each step represents approximately 1% of developmental time. The axons of the dorsal body wall neurons (dBw) and the axons of the central neurons forming the U-fascicle (Uf) extend in opposite directions. The intersegmental nerve in thoracic and abdominal segments is formed by fasciculation of both axon bundles near the body wall spiracle (s). CNS: lateral border of the central nervous system. Scale bar: 100 µm.

and identical mechanism involving central as well as peripheral axons gives rise to the intersegmental nerve (Ghysen et al., 1986; Ghysen, 1992).

This serially homologous set of segmental and intersegmental nerves form a simple stereotyped scaffold of embryonic nerves. Following the formation of this embryonic nerve scaffolding in the body segments, peripheral sensory neurons differentiate in a well-defined spatiotemporal pattern along the dorsoventral axis of the body. In most segments, this early differentiation of the sensory neurons in the peripheral nervous system occurs in three distinct clusters, a dorsal, a lateral, and a ventral cluster. An overview of these developing sensory cells in different segments of the grasshopper is shown in Figure 5.

The sensory structures in each cluster share similar mechanisms of neurogenesis, morphogenesis, and axogenesis (Meier and Reichert, 1990; Meier et al., 1991). In several cases, this is even true for individual cells within a given cluster. For example, in most of the thoracic and abdominal segments one of the cells in the dorsal cluster, which derives from the dBw cell group, differentiates into an identified single cell sense organ. In the second and third thoracic segments, this cell separates from the dBw cluster soon after formation of the intersegmental nerve (Meier et al., 1991). Subsequently, this cell becomes enlarged, sends out filopodia and begins moving posteriorly. During this movement its axon

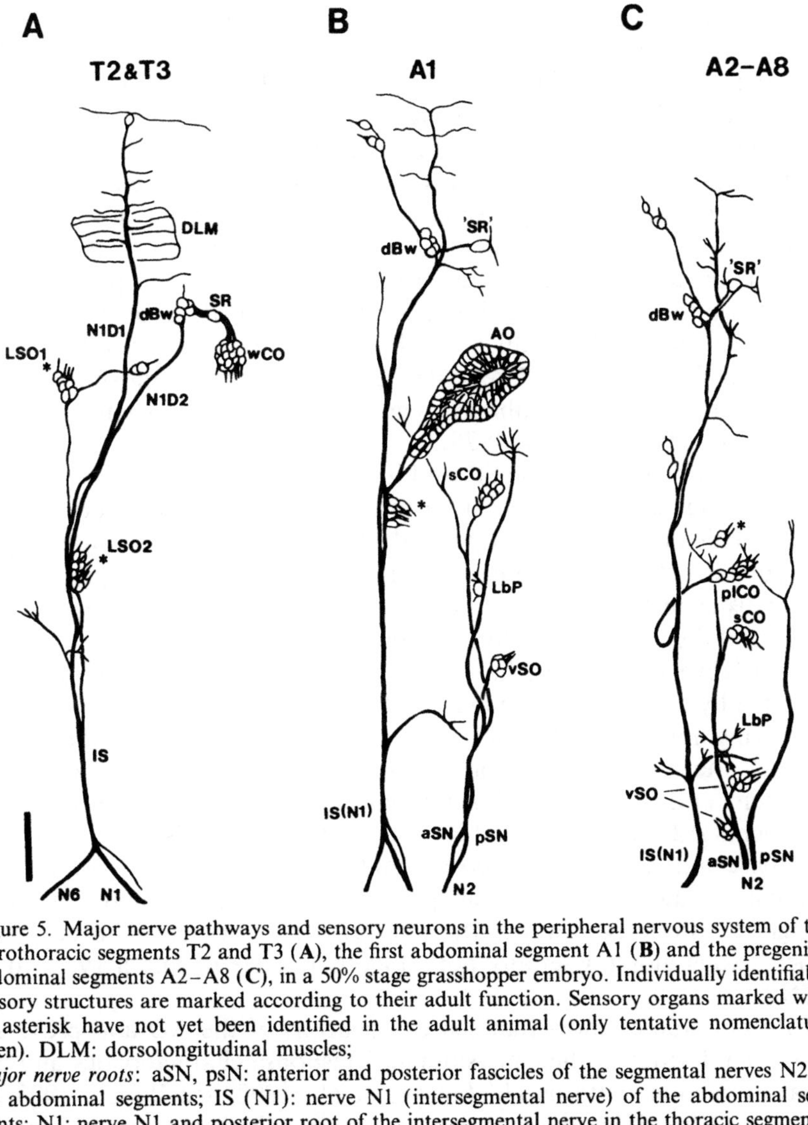

Figure 5. Major nerve pathways and sensory neurons in the peripheral nervous system of the pterothoracic segments T2 and T3 (**A**), the first abdominal segment A1 (**B**) and the pregenital abdominal segments A2–A8 (**C**), in a 50% stage grasshopper embryo. Individually identifiable sensory structures are marked according to their adult function. Sensory organs marked with an asterisk have not yet been identified in the adult animal (only tentative nomenclature given). DLM: dorsolongitudinal muscles;
Major nerve roots: aSN, psN: anterior and posterior fascicles of the segmental nerves N2 in the abdominal segments; IS (N1): nerve N1 (intersegmental nerve) of the abdominal segments; N1: nerve N1 and posterior root of the intersegmental nerve in the thoracic segments; N1D1, N1D2: distal nerve branches of the pterothoracic intersegmental nerve N1; N2: nerve N2 (segmental nerve) in the abdominal segments; N6: nerve N6 and anterior root of the intersegmental nerve in the thoracic segments.
Pioneer cells: dBw: dorsal body wall cells; LbP: limb bud pioneer cell.
Sensory neurons: AO: auditory organ; plCO: pleural chordotonal organ; sCO: sternal chordotonal organ; SR: wing hinge stretch receptor cell; 'SR': serial homolog of the thoracic wing-hinge stretch receptor (SR) in the abdominal segments; vCO: ventral sensory organ; wCO: wing hinge chordotonal organ. Scale bar: 100 μm. Modified after Meier et al. (1992).

remains connected to the intersegmental nerve. Subsequently, the cell reaches the developing wing hinge chordotonal organ (wCO), stops moving at the posterior segment border, and differentiates further into the thoracic wing hinge stretch receptor (SR; Gettrup, 1962; Pearson et al., 1989; Heathcote, 1981). Interestingly, this receptor cell is evolutionary highly conserved in orthopteran insects. The Mexican grasshopper *Barytettix*, which has lost its metathoracic hindwings during evolution, still has a SR-cell with only minor changes in its central branching pattern (Arbas, 1983). In each of the abdominal segments A1–A8, a cell differentiates at approximately the same time, in a similar way and at a segmentally equivalent position. It derives from the dBw cluster, enlarges, moves posteriorly, and stops near the posterior segment boundary while remaining connected with its axon to the intersegmental nerve. Subsequently, this cell, the serial homolog of the thoracic SR ('SR'), differentiates distal T-shaped dendrites that, in the hatched animal, probably register ventilatory movements of the abdomen (Hustert, 1974; 1975).

As demonstrated by the SR and its abdominal homologs, segmentally homologous sensory cells can be involved in different functional contexts. This appears to be a general feature of the serially homologous sensory structures in the insect. Thus, within a given developing sensory cell cluster in the peripheral nervous system, serially homologous sense organs differentiate to perform surprisingly different functions in behaviors as diverse as flight, hearing and respiration. This is amply documented by the fact that the major lateral sensory cell clusters give rise to the wing hinge chordotonal organs (wCO) in the pterothoracic segments, the auditory chordotonal organs (AO) in the first abdominal segment and the pleural chordotonal organs (plCO) in the abdominal segments A2–A8 (Meier and Reichert, 1990). The cells of all three types of sense organs are generated at similar developmental stages by ectodermal invagination near the posterior segment boundary; all three migrate anteriorly towards the intersegmental nerve; all three project their axons along a serially homologous fascicle in this nerve into the CNS.

Recent studies show that sensory neurons of the AO and plCO also have similarities in their physiological response characteristics and central projection patterns (Bickmeyer et al., 1992; see Boyan, 1993, for review). The different number of receptor cells found in auditory and pleural chordotonal organs can also be explained in evolutionary terms. The increased cell number in the AO (about 80 receptor cells in locusts) may be a secondary evolutionary event, since in more primitive grasshoppers, which have not yet evolved auditory organs, a chordotonal organ was found in their first abdominal segment which resembles the plCO of modern grasshoppers in shape and cell number (Meier and Reichert, 1990). The evolutionary relation between wing-hinge chordotonal organs and true auditory organs, on the other hand, also has been postulated for holometabolous moths (Yack and Fullard, 1990). Chordotonal organs, like wCO, AO and plCO, which share developmental similarities have also been discussed in the context of evolutionary biology of insect hearing (Fullard and Yack, 1993; Boyan, 1993).

The overall extent of serial homology that can be uncovered in the sensory cells of the peripheral nervous system in the embryonic grasshopper is astounding. On the basis of classical morphological criteria for homology such as position, innervation, and structural organization, as well as on the basis of developmental criteria such as mode of neurogenesis, relative timing of differentiation and trajectory of cell migration and axonal navigation, it has been possible to identify numerous serially homologous sensory structures in most of the body segments of the grasshopper. Figure 6 summarizes the serially homologous sensory structures in thoracic segments T2–T3 and in abdominal segments A1–A9.

The remarkable degree with which serial homology is preserved in the segments of the grasshopper embryo suggests that the formation of the peripheral nervous system relies on evolutionarily conserved developmental mechanisms (Thomas et al., 1984; Akam et al., 1988; Tear et al., 1988; Meier et al., 1991). Such mechanisms are likely to reflect the origin of modern insects from ancestral myriapod-like animals in which the general body plan consisted of very similar metameric repeats (Raff and Kaufmann, 1983). On the basis of these considerations, one might predict that similar patterns of serial homology occur in the develop-

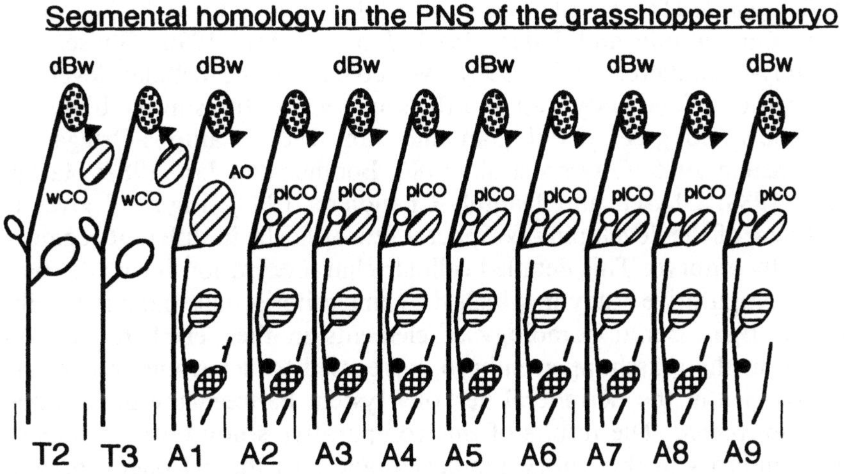

Figure 6. Serial homology of all major sensory cell groups found in the body wall of the grasshopper embryo at 50% of development. Ovals represent clusters of sensory neurons. Homology between sensory clusters is indicated by similar symbols; some of the sensory cell clusters are individually labeled. AO: auditory organ; dBw: dorsal body wall pioneer cells; plCO: pleural chordotonal organ; wCO: wing hinge chordotonal organ. Filled triangle next to the dBw-group: thoracic wing hinge stretch receptor (SR) and its abdominal homologs ('SR'). Small filled circle in the ventral part of segments A1–A9: limb bud pioneer cell (LbP). Modified after Meier and Reichert (1990) and Meier et al. (1991).

ment of the peripheral nervous system in other insects, even in those that are separated from the grasshopper by hundreds of millions of years of divergent evolution. As is discussed below, this is indeed the case.

Drosophila and grasshopper: A comparative analysis reveals evolutionarily conserved developmental mechanisms

Embryogenesis in the grasshopper, as in other hemimetabolous insects, results in a larva which is a miniature replica of the adult. This type of larva has specialized thoracic segments which bear legs and wing buds, and several specialized abdominal segments. Most of the embryonic components of the peripheral nervous system are retained in the adult (Bate, 1978). In contrast, embryogenesis in holometabolous flies like *Drosophila* results in a highly specialized larva with similar thoracic and abdominal segments and no appendages. The larval peripheral nervous system is transient. Most of the adult peripheral nervous system is derived from imaginal discs. Given the radically different modes of postembryonic development of the peripheral nervous systems in grasshopper and fly, an important question is whether the mechanisms that operate during embryonic development in these very different insects are comparable.

A large body of work has focused on the molecular interactions underlying the embryonic development of the peripheral nervous system in *Drosophila* (for recent reviews see: Ghysen and Dambly-Chaudière, 1989; Giangrande and Palka, 1990; Jan and Jan, 1990; Ghysen and Dambly-Chaudière, 1992; 1993). Moreover, at the cellular level, the innervation of the body wall and the sensory structures in the body wall are known in great detail (Jan and Jan, 1982; Campos-Ortega and Hartenstein 1985; Ghysen et al., 1986; Bodmer and Jan, 1987; Hartenstein, 1988; Ghysen and Dambly-Chaudière, 1988). Figure 7 gives an overview of the developing peripheral sensory cells in different segments of the fly embryo. This detailed cellular characterization forms the basis for a comparative study in which the same criteria that make it possible to investigate serial homology of elements in the peripheral nervous system in the grasshopper embryo, were used to determine interspecies homologies in the peripheral nervous system between the grasshopper and *Drosophila*. The results of this comparative study reveal a remarkable similarity in the embryonic development of the peripheral nervous system in both species. Indeed, the existence of a number of homologous structures in the peripheral nervous system of *Drosophila* and grasshopper can be documented.

The basic pattern of peripheral nerves in the body wall of *Drosophila* and grasshopper embryos is extremely similar. Peripheral nerve pathways corresponding to an intersegmental and a segmental nerve branch

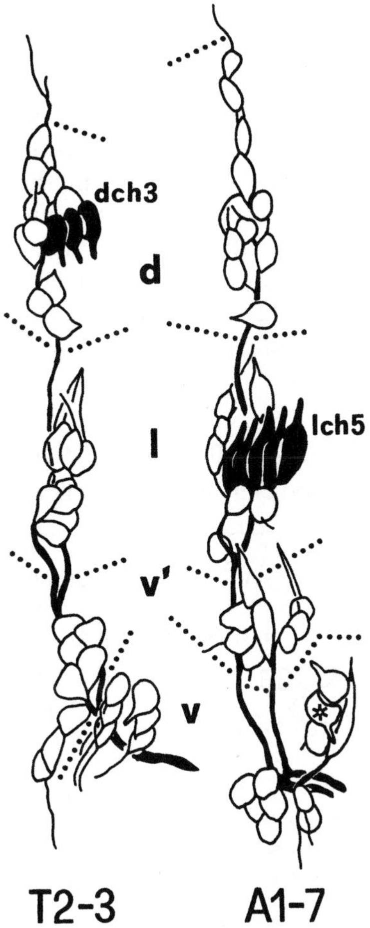

Figure 7. Pattern of peripheral sensory neurons in throacic segments T2 and T3 and abdominal segments A1–A7 of the *Drosophila* embryo. In each of the segments T2–A7, sensory cells are arranged in four distinct clusters as indicated by the dotted lines. Ventral is downwards and anterior is to the left. d: dorsal cluster; l: lateral cluster; v' and v: two distinct parts of the ventral cluster. The dorsal chordotonal organ (dch3) in segments T2–T3 comprises three sensory neurons and the lateral chordotonal (lch5) of segments A1–A7 is composed of five sensory neurons. The vchA neuron of segments A1–A7 is indicated by an asterisk. Modified after Ghysen et al. (1986) and Ghysen and O'Kane (1989).

are found in most of the segments of both insects; in many cases homologies among the individual cells that pioneer these pathways can be established. Obvious similarities also exist in the development of the peripheral sense organs in the homologous segments of both species. Similar dorsal, lateral and ventral clusters as well as similar axonal projection patterns are found in both grasshopper and *Drosophila* (Meier et al., 1991). Moreover, both insects have complex abdominal chordotonal organs that derive from the lateral cluster, are displaced into more anterior positions and finally project axons onto the intersegmental nerve branch. These are the lch5 chordotonal organs in *Drosophila* (Fig. 7) and the pleural chordotonal organs and auditory organ in the grasshopper (Fig. 5). The thoracic dorsal chordotonal organ, dch3, in *Drosophila* shares cellular and developmental characteristics with the grasshopper wCO. In addition, in both animals, a segmentally reiterated single sensory cell in the dorsal cluster of the thoracic ganglia has similar mechanisms of neurogenesis, morphogenesis, and axongenesis. This is the dh1 cell in *Drosophila* and the wing hinge stretch receptor, SR, in the grasshopper; in both animals, these interspecies homologs also have a similar set of serial homologs in the abdominal segments, as summarized in Figure 8.

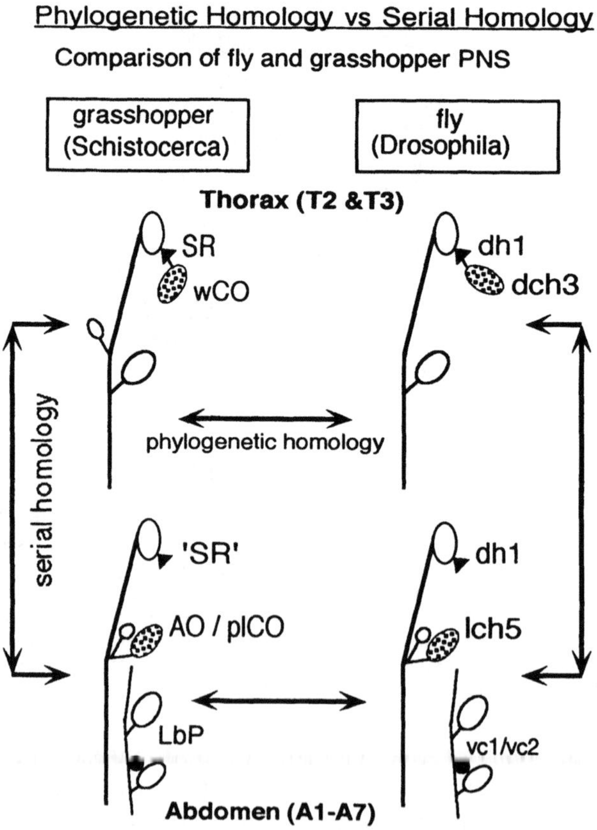

Figure 8. Comparison of individually identified sensory neurons and sense organs in embryonic fly *Drosophila* and grasshopper *Schistocerca*. The peripheral sensory system of the thoracic segments T2 and T3 as well as the abdominal segments A1–A7 are represented separately. Serial homologies in each species (compare vertically) as well as phylogenetic homologies (compare horizontally) are indicated. Grasshopper PNS: symbols and nomenclature follow Figures 5 and 6. Fly PNS: dh1: neuron innervating the dorsal hair; dch3: dorsal chordotonal organ with three receptor neurons; lch5: lateral chordotonal organ with five receptor neurons; vc1/vc2: identified ventral neuron. Modified after Meier et al. (1991).

The extensive interspecies homologies that can be found in the development of the peripheral nervous system of fly and grasshopper indicate that the construction of the embryonic peripheral nervous system, seen in highly specialized holometabolous insects like *Drosophila*, relies on evolutionarily conserved developmental mechanisms. Viewed from this evolutionary perspective, it becomes understandable why the development of the peripheral nervous system in the highly derived fly appears to be based on mechanisms that much more primitive hemimetabolous insects use to generate their adult peripheral nervous system. The underlying mechanisms are evolutionarily conserved. Indeed, given this degree of evolutionary conservation between dipteran and orthopteran species, we postulate that all modern insects might use the same developmental mechanisms for building their peripheral nervous systems embryonically.

Neurogenetic insights into the development of serial homology in the peripheral nervous system

As stated above, an important consequence of the extensive interspecies homologies in the embryonic development of the peripheral nervous system in *Drosophila* and grasshopper, is the implication that the molecular mechanisms which operate in the *Drosophila* embryo also direct nervous system development in other insects. This is important since the molecular genetic control mechanisms of nervous system development can be investigated in great detail in *Drosophila* due to the advanced classical and molecular genetic manipulations that are possible in this insect. Given the genetic tools for manipulation available in *Drosophila*, the question arises whether it is possible to uncover the molecular mechanisms that are responsible for the construction of serially homologous structures in the peripheral nervous system. This question can be addressed initially through a mutational analysis.

One example of how a mutant analysis in *Drosophila* could be used to investigate the development of segmentally homologous neuronal elements has to do with the homology of sensory cells of the lateral cluster of the peripheral nervous system (see Fig. 8). In the grasshopper, a number of cellular and molecular criteria indicates that the thoracic homolog of the auditory organ and the pleural chordotonal organs is the wing hinge chordotonal organ (Meier and Reichert, 1990). Is this serial homology also phylogenetically conserved in *Drosophila*? If so, the thoracic homolog of the abdominal lch5 chordotonal organs in *Drosophila* should be the dch3 chordotonal organ (Fig. 7). However, two gross morphological features are at odds with this notion. First, the dch3 chordotonal organ appears to be associated with the dorsal and not with the lateral sensory cluster. Second, the dch3 chordotonal organ has three sensory neurons instead of five like the lch5 chordotonal organ. A possible explanation for these differences might be based on an evolutionarily ancient situation with two sets of chordotonal organs in each thoracic and abdominal segment combined with a selective loss of a lateral chordotonal organ in segments T2 and T3 and the loss of a dorsal chordotonal organ in the abdominal segments. Interestingly, the first thoracic segment Tl has two chordotonal organs associated with the lateral and dorsal neuronal clusters. Nevertheless, a number of developmental criteria for homology such as precursor identity, lineage, timing of differentiation, and axon pathway trajectory support the hypothesis that at least a part of the abdominal lch5 neuron cluster is serially homologous to the thoracic dch3 neuron cluster (Campos-Ortega and Hartensetein, 1985; Hartenstein, 1988; Ghysen and O'Kane, 1989; Meier et al., 1991; Ghysen, 1992). Thus, morphological data and developmental data are in conflict.

The conflicting data could be resolved by neurogenetics. The homology of the dch3 and lch5 chordotonal organs is supported by a large body of genetic evidence (Fig. 9). For example, a mutation in the gene *rhomboid* transforms the five-cell chordotonal organ lch5 in the abdominal segments into a three-cell chordotonal organ (Bier et al., 1990). This indicates that cell number alone does not necessarily argue in favor of or against homology of sense organs. Along the same lines, overexpression of the gene *atonal* increases the cell number in the lch5 sense organ even beyond the amount seen in wild type embryos (Jarman et al., 1993). Both mutations (*rho* and *atonal*) do not affect the sense organ's positions.

Additionally, a number of mutations that results in parasegmental transformations is in accordance with the proposed homology between dch3 and lch5 organs. I mutants lacking the *abdA* gene, which is part of the segment-identity controlling bithorax complex, the five-cell chordotonal organ in the lateral cluster of abdominal segments A1–A7 is replaced by a three-cell chordotonal organ in the dorsal cluster of the same segments (Pfeiffer et al., 1987; Duncan, 1987; Karch et al., 1990). A mutation in the *engrailed* locus deletes the dch3 and the lch5 chordotonal organs without affecting other sensory cells in the cluster (Hartenstein, 1987). These last findings indicate that both sense organs indeed derive from the same part of the segment (posterior segmental half) irrespective of their final position after cell displacement. Therefore, in this particular case, for the dch3- and lch5- sensory organs, the evidence for serial homology in the peripheral nervous system of *Drosophila*, which was obtained on the basis of comparative structural and developmental data in the grasshopper, is corroborated by developmental genetic data obtained using *Drosophila* wildtype and mutants.

In a more general sense, it is important to realize that by using the neuronal phenotypes of different mutants to demonstrate serial homology in *Drosophila*, one is at the same time uncovering some of the

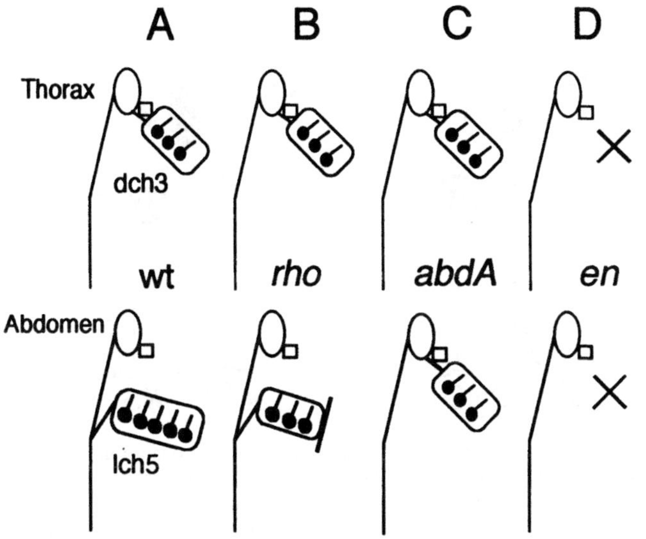

Figure 9. Genetic mutations affecting position and/or cell number of identified chordotonal organ receptors in the embryonic *Drosophila* PNS. In each of the figures (A–D), the position of the thoracic dch3 and abdominal lch5 chordotonal organs is shown in reltion to the intersegmental nerve and dorsal cell cluster (compare to Figs 7 and 8); each receptor cell is represented by ●—. (**A**) Distribution of receptors in wild-type (wt) *Drosophila* dch3 and lch5 chordotonal organ. (**B**) In some rhomboid mutations (*rho*), the number of receptor cells in lch5 is reduced to three (this represents the number of receptors in the unaffected thoracic dch3 organ). The relative position of the mutated lch5 organ is not affected. Data from Bier et al. (1990). (**C**) In certain *abdominal A* mutants (*abdA*) the abdominal lch5 receptor cells take on a thoracic number and arrangement. (**D**) In some *engrailed* mutations (*en*) the dorsal dch3 receptor cells in thoracic segments as well as the lateral lch5 receptor cells are missing entirely, indicating that both receptor clusters are derived from the same (posterior) segmental compartment irrespective of their final segmental position. Data from Hartenstein (1987).

candidate genes for evolutionary specialization of the peripheral nervous system. From this perspective, a developmental analysis using neurogenetic methods might lead to unique insights into the molecular events that may have occurred during evolution.

The usefulness of developmental neurogenetics for an analysis of the molecular mechanisms that result in serially homologous neuronal structures, goes beyond investigations of genes that specify cell position or cell number in the embryonic nervous system. Precise pathfinding processes in the peripheral nervous system that occur in a serially homologous manner can be subjected to a neurogenetic dissection. For example, in *Drosophila* genetic techniques have led to the identification of several neuronal guidance molecules (Grenningloh et al., 1990), like *Fasciclin I*, *Fasciclin II*, and *Fasciclin III*. Mutations in these genes can affect the development of precisely organized axonal tracts in the peripheral nervous system as has been shown recently for the wing sensory system (Whitlock, 1993).

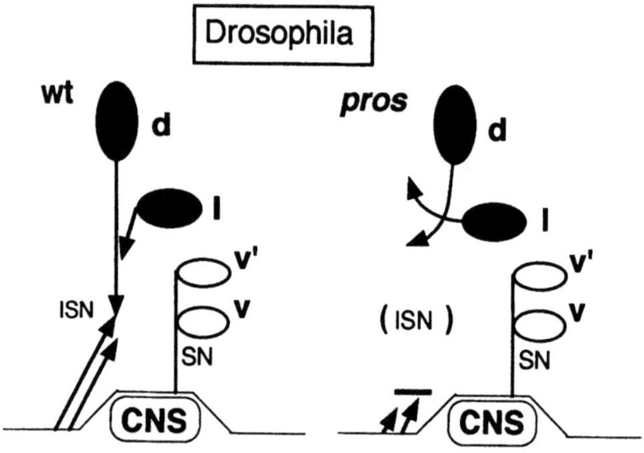

Figure 10. Effects of the *prospero* mutation (*pros*) on the formation of the intersegmental nerve in *Drosophila* embryos. The position of dorsal (d), lateral (l), and ventral neuronal clusters (v, v') along intersegmental (ISN) and segmental (SN) nerves is shown for wild-type (wt) and *pros* flies. The *prospero* mutation specifically affects the formation of the intersegmental nerve due to misguiding of axon fascicles from the dorsal and lateral neuronal clusters and due to alterations in the efferent outgrowth of central axons. The formation of the segmental nerve is not affected in *pros* mutants. Data from Doe et al. (1991), Vaessin et al. (1991) and Chu-LaGraff et al. (1991).

Segmentally repeated defects in axonal pathfinding of structures in the peripheral nervous system are also caused by defects in nuclear regulatory genes. The Drosophila *prospero* gene encodes for a nuclear protein with a divergent homeodomain, and in the CNS the *prospero* gene is thought to be involved in the control of neuronal identity and/or neural differentiation (Doe et al., 1991; Vaessin et al., 1991). In the peripheral nervous system, loss of function in the *prospero* gene leads to a reversal in axon outgrowth polarity for the developing sensory cells in the dorsal and lateral clusters (Fig. 10). In consequence, the affected sensory axons extend dorsally instead of ventrally. This defect, in combination with aberrant pathfinding of motoneuron efferents, prevents the formation of the intersegmental nerve. The result is a serially repeated loss of the intersegmental nerve in all segments. The cell fate of the affected sensory organs is not changed by mutation of the *prospero* gene, only the direction of their outgrowing axons is affected. This is in contrast to the effects of other mutations in other regulatory genes such as *cut*, *numb* or *pox-neural* which affect the entire cell fate of serially homologous sensory neurons in the developing peripheral nervous system (Blochlinger et al., 1988; 1993; Uemura et al., 1989; Dambly-Chaudière et al., 1992; see Ghysen and Dambly-Chaudière (1993) for review).

The *pox-neural* (*poxn*) gene can serve to illustrate the way in which single genes can change the entire cell fate of peripheral neurons including aspects of neuronal pathfinding and axonal termination processes (Dambly-Chaudière et al., 1992; Nottebohm et al., 1992; 1994). The *poxn* gene specifies the formation of poly-innervated chemosensory organs in the peripheral nervous system.

Overexpression of the *poxn* gene using a transgene under the control of a heat-shock promotor leads to the transformation of mono-innervated mechanosensory organs into ectopic poly-innervated chemosensory organs. Additionally, along with changes in the peripheral structure, the central projection patterns of the transformed neurons are also changed (Fig. 11). Transformed neurons no longer follow the projection pattern of mechanosensory neurons in the CNS. Rather, they follow pathways and establish functional connections that are appropriate for chemosensory neurons. The effects of alterations in expression of the *poxn* gene are specific to the mechanosensory neurons that innervate

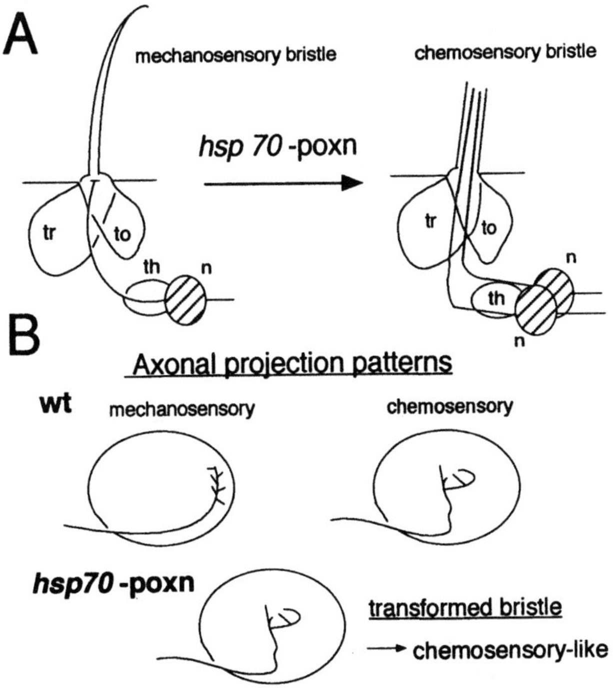

Figure 11. Effect of the *pox neural*-gene (*poxn*) on the morphology and axonal projection patterns of external leg sense-organs in *Drosophila*. (**A**) Ectopic expression of *poxn* induces the morphological transformation of mechanosensory into chemosensory bristles. Mechanosensory bristles are innervated by one neuron which responds to the deflection of the bristle shaft. Chemosensory bristles are innervated by at least two neurons which respond to chemical stimulation. (**B**) Sensory neurons innervating the morphologically transformed bristles show axonal projection patterns in the central neuropile that are appropriate for chemosensory bristles. hsp 70-poxn: the poxn-gene is ectopically expressed under control of a heat-shock promotor (hsp-70); n: sensory neuron; th: thecogen cell; to: tormogen cell; tr: trichogen cell; wt: wild-type situation. Data taken from Dambly-Chaudière et al. (1992) and Nottebohm et al. (1992).

external bristles. These results show that single, "higher order control genes" like the *poxn* gene can directly influence axonal pathways and the establishment of connections of certain classes of PNS neurons, while having no effect on other classes.

Conclusion

Taken together, all of the examples discussed here show that when focused on few neuronal elements, a combined comparative, developmental, and genetic analysis of the insect peripheral nervous system can lead to profound insights into how serially homologous nervous structures are generated. Moreover, by using this type of approach, the underlying developmental mechanisms can be resolved at the cellular and molecular level. However, in addition, this type of analysis can also help formulate hypotheses about how structures like the peripheral nervous system in insects might have changed during evolution. Indeed, some of the candidate genes for the developmental switches that make adaptive evolutionary changes of serially homologous nervous structures possible can be identified by this approach. In this sense, an in-depth analysis of the developmental mechanisms that generate the insect peripheral nervous system not only reveals important structural constraints (Ghysen, 1992), but also has significant importance for understanding the general phylogenetic constraints underlying the evolution of these nervous systems.

Acknowledgements
We thank George Boyan, Alain Ghysen, Hugh Rowell and Paul Whitington for many stimulating discussions during our work on the development and evolution of the insect PNS, and Corey Goodman for providing many of the antibodies used in our studies. This work was supported by the Swiss National Science Foundation.

References

Akam, M., Dawson, I. and Tear, G. (1988) Homeotic genes and the control of segment diversity. *Development* 104 (Suppl.): 123–133.

Arbas, E.A. (1983) Neural correlates of flight loss in a mexican grasshopper, *Barytettix psolus*. I. Motor and sensory cells. *J. Comp. Neurol.* 216: 369–380.

Arbas, E.A., Meinertzhagen, I.A. and Shaw, S.R. (1991) Evolution in nervous systems. *Annu. Rev. Neurosci.* 14: 9–38.

Bastiani, M.J., DuLac, S.L. and Goodman, C.S. (1986) Guidance of neuronal growth cones in the grasshopper embryo. I. Recognition of specific axonal pathway by the pCC neuron. *J. Neurosci.* 6: 3518–3531.

Bate, C.M. (1978) Development of sensory systems in Arthropods. *In:* M. Jacobson (ed.): *Handbook of Sensory Physiology*. Springer Verlag, Berlin, Heidelberg, New York, pp. 1–53.

Bentley, D., Keshishian, H., Shankland, M. and Toroian-Raymond A. (1979) Quantitative staging of embryonic development of the grasshopper, *Schistocerca nitens*. *J. Embryol. Exp. Morph.* 54: 47–74.

Bentley, D. and Keshishian, H. (1982) Pathfinding by peripheral pioneer neurons in grasshoppers. *Science* 218: 1082–1088.

Bickmeyer, U., Kalmring, K., Halex, H. and Mücke, A. (1992) The bimodal auditory-vibratory system of thoracic ventral nerve cord in *Locusta migratoria*. *J. Exp. Zool.* 264: 381–394.

Bier, E., L.Y. and Jan, Y.N. (1990) *rhomboid*, a gene required for dorsoventral axis establishment and peripheral nervous system development in *Drosophila melanogaster*. *Genes Dev.* 4: 190–203.

Blochlinger, K., Bodmer, R., Jack, J., Jan, L.Y. and Jan, Y.N. (1988) Primary structure and expression of a product from *cut*, a locus involved in specifying sensory organ identity in *Drosophila*. *Nature* 333: 629–635.

Blochlinger, K., Jan, L.Y. and Jan, Y.N. (1993) Postembryonic patterns of expression of *cut*, a locus regulating sensory organ identity in *Drosophila*. *Development* 117: 441–450.

Bodmer, R. and Jan, Y.N. (1987) Morphological differentiation of the embryonic peripheral neurons in *Drosophila*. *Roux's Arch. Dev. Biol.* 196: 69–77.

Boyan, G. (1992) Common synaptic drive to segmentally homologous interneurons in the locust. *J. Comp. Neurol.* 321: 544–554.

Boyan, G. (1993) Another look at insect audition: the tympanic receptors as an evolutionary specialization of the chordotonal organ. *J. Insect Physiol.* 39: 187–200.

Boyan, G. and Ball, E.E. (1993) The grasshopper, the fruitfly, and neuronal homology: answers to some fundamental questions in neurobiology. *Progress in Neurobiol*; *in press.*

Campos-Ortega, J.A. and Hartenstein, B. (1985) *The embryonic development of* Drosophila melanogaster. Springer Verlag, Berlin.

Caudy, M., and Bentley, D. (1986) Pioneer growth cone morphologies reveal proximal increase in substrate affinity within leg segments of grasshopper embryos. *J. Neurosci.* 6: 364–379.

Chu-LaGraff, Q., Wright, D.M., McNeil, L.K. and Doe, C.Q. (1991) The *prospero* gene encodes a divergent homeodomain protein that controls neuronal identity in *Drosophila*. *Development* Suppl. 2: 79–85.

Dambly-Chaudière, C., Jamet, E., Burri, M., Bopp, D., Basler, K., Hafen, E., Dumont, N., Spielmann, P., Ghysen, A. and Noll, M. (1992) The paired box gene *pox neuro*: a determinant of poly-innervated sense organs in *Drosophila*. *Cell* 69: 159–172.

Dobzhansky, T., Ayala, F.J., Stebbins, G.L. and Valentine, J.W. (1977) *Evolution*. Freeman, San Francisco.

Doe, C.Q., Chu-LaGraff, Q., Wright, D.M. and Scott, M.P. (1991) The *prospero* gene specifies cell fates in the *Drosophila* central nervous system. *Cell* 65: 451–464.

DuLac, L., Bastiani, M.J. and Goodman, C.S. (1986) Guidance of neuronal growth cones in the grasshopper embryo: II. Recognition of a specific axonal pathway by the aCC neuron. *J. Neurosci.* 6: 3532–3541.

Dumont, J.P.C. and Robertson, R.M. (1986) Neuronal circuits: an evolutionary perspective. *Science* 233: 849–853.

Duncan, I. (1987) The bithorax complex. *Ann. Rev. Genet.* 21: 285–319.

Farrel, S. and Kuhlenbeck, H. (1964) Preliminary computation of the number of cellular elements in some insect brains. *Anat. Rec.* 148: 369–370.

Fullard, J.H. and Yack, J.E. (1993) The evolutionary biology of insect hearing. *Trends Ecol. Evol.* 8: 248–252.

Gettrup, E. (1962) Thoracic proprioceptors in the flight system of locusts. *Nature* 193: 498–499.

Ghysen, A. (1992) The developmental biology in neuronal connectivity. *Int. J. Dev. Biol.* 36: 47–58.

Ghysen, A., Dambly-Chaudière, C., Aceves, E., Jan, L.Y. and Jan, Y.M. (1986) Sensory neurons and peripheral pathways in *Drosophila* embryos. *Roux's Arch. Dev. Biol.* 195: 281–289.

Ghysen, A. and Dambly-Chaudière, C. (1989) Genesis of the *Drosophila* peripheral nervous system. *Trends Genetics* 5: 251–255.

Ghysen, A. and O'Kane, C. (1989) Neural enhancer-like elements as specific cell markers in *Drosophila*. *Development* 105: 35–52.

Ghysen, A. and Dambly-Chaudière, C. (1992) Development of the peripheral nervous system in *Drosophila*. *In*: M. Shankland and E.R. Macagno (eds): *Determinants of neuronal identity*. Academic Press, San Diego, New York, Boston, London, Sydney, Tokyo, Toronto, pp. 226–292.

Ghysen, A. and Dambly-Chaudière, C. (1993) The specification of sensory neuron identity in *Drosophila. BioEssays* 15: 193–298.
Giangrande, A. and Palka, J. (1990) Genes involved in the development of the peripheral nervous system of Drosophila. *Seminars in Cell Biol.* 1: 197–209.
Goodman, C.S., Bastiani, M.J., Doe, C.Q., DuLac, S., Helfand, S.L, Kuwada, J.Y and Thomas, J.B. (1984) Cell recognition during neuronal development. *Science* 225: 1271–1279.
Grenningloh, G., Bieber, A., Rehm, J., Snow, P.M., Traquina, Z., Hortsch, M., Patel, N.H. and Goodman, C.S. (1990) Molecular genetics of neuronal recognition in *Drosophila*: evolution and function of immunoglobulin superfamily cell adhesion molecules. Cold Spring Harbor *Symp. Quant. Biol.* 55: 327–340.
Hartenstein, B. (1987) The influence of segmental compartmentalisation on the development of the larval peripheral nervous system in *Drosophila melanogaster. Roux's Arch. Dev. Biol.* 196: 101–112.
Hartenstein, V. (1988) Development of *Drosophila* larval sensory organs; spatiotemporal pattern of sensory neurons, peripheral axonal pathways and sensilla differentiation. *Development* 102: 869–886.
Heathcote, D.R. (1981) Differentiation of an identified sensory neuron (SR) and associated structures (CTO) in grasshopper embryos. *J. Comp. Neurol.* 202: 1–18.
Ho, R.K and Goodman, C.S. (1982) Peripheral pathways are pioneered by an array of central and peripheral neurons in the grasshopper embryo. *Nature* 297: 404–406.
Hustert, R. (1974) Morphologie und Atmungsbewegungen des 5. Abdominalsegments von *Locusta migratoria. Zool. Jb. Physiol.* 78: 157–174.
Hustert, R. (1975) Neuromuscular coordination and proprioceptive control of rhythmical abdominal ventilation in intact *Locusta migratoria. J. Comp. Physiol.* 97: 159–179.
Jan, L.Y. and Jan, Y.N. (1982) Antibodies to horseradish peroxidase as specific neuronal markers in *Drosophila* and grasshopper embryos. *Proc. Natl. Acad. Sci. USA* 79: 2700–2704.
Jan, Y.N. and Jan, L.Y. (1990) Genes required for specifying cell fates in *Drosophila* embryonic sensory nervous system. *Trends Neurosci.* 13: 493–498.
Jarman, A.P., Grau, Y., Jan, L.Y., and Jan, Y.N. (1993) *atonal* is a proneural gene that directs chordotonal organ formation in the *Drosophila* peripheral nervous system. *Cell* 73: 1307–1321.
Karch, F., Bender, W. and Weiffenbach, B. (1990) *abdA* expression in Drosophila embryos. *Genes Dev.* 4: 1573–1587.
Keshishian, H. and Bentley, D. (1983a) Embryogenesis of peripheral nerve pathways in grasshopper legs. I. The initial nerve pathway to the CNS. *Dev. Biol.* 96: 89–102.
Keshishian, H. and Bentley, D. (1983b) Embryogenesis of peripheral nerve pathways in grasshopper legs. II. The major nerve roots. *Dev. Biol.* 96: 103–115.
Kristensen, N.P. (1981) Phylogeny of insect orders. *Ann. Rev. Entomol.* 26: 135–157.
Lakes-Harlan, R. and Pollack, G.S. (1993) Pathfinding of peripheral neurons in the central nervous system of an embryonic grasshopper (*Chorthippus biguttulus*). *Cell Tissue Res.* 237: 97–106.
Matsuda, R. (1976) *Morphology and evolution of the insect abdomen.* Pergamon Press, New York.
Meier, T. and Reichert, H. (1990) Embryonic development and evolutionary origin of the Orthopteran auditory organs. *J. Neurobiol.* 21: 592–610.
Meier T. and Reichert, H. (1991) Serially homologous development of the peripheral nervous system in the mouthparts of the grasshopper. *J. Comp. Neurol.* 305: 201–214.
Meier, T., Chabaud, F. and Reichert, H. (1991) Homologous patterns in the embryonic development of the peripheral nervous system in the grasshopper *Schistocerca gregaria* and the fly *Drosophila melanogaster. Development* 112: 241–253.
Nottebohm, E., Dambly-Chaudière, C. and Ghysen, A. (1992) Connectivity of chemosensory neurons is controlled by the gene *poxn* in *Drosophila. Nature* 359: 829–832.
Nottebohm, E., Usui, A., Therianos, S., Kimura, K., Dambly-Chaudière, C. and Ghysen, A. (1994) The gene *poxn* controls different steps of the formation of chemosensory organs in *Drosophila. Neuron* 12: 25–34.
Pearson, K.G., Hedwig, B. and Wolf, H. (1989) Are the hindwing chordotonal organs elements of the locust flight pattern generator? *J. Exp. Biol.* 114: 235–255.

Pearson, K.G., Boyan, G.S., Bastiani, M. and Goodman, C.S. (1985) Heterogeneous properties of segmentally homologous interneurons in the ventral nerve cord of locusts. *J. Comp. Neurol.* 233: 133–145.
Pfeifer, M., Karch, F. and Bender, W. (1987) The bithorax complex: control of segmental identity. *Genes Dev.* 1: 891–898.
Raff, R.A. and Kaufmann, T.C. (1983) Embryos, genes and evolution. Macmillan, New York.
Reichert, H. (1994) Molecular correlates of neuronal specificity in the developing insect nervous system. *Mol. Neurobiol.* 7: 349–362.
Robertson, R.M., Pearson, K.G. and Reichert, H. (1982) Flight interneurons in the locust and the origin of insect wings. *Science* 217: 177–179.
Roth, V.L. (1984) On homology. *Biol. J. Linn. Soc.* 22: 13–29.
Sander, K. (1988) Studies in insect segmentation: from teratology to phenogenetics. *Development* (Suppl.) 104: 112–121.
Tear, G., Bate, C.M. and Martinez-Arias, A. (1988) A phylogenetic interpretation of the patterns of gene expression in *Drosophila* embryos. *Development* (Suppl.) 104: 135–145.
Thomas, J.B., Bastiani, M.J., Bate M. and Goodman, C.S. (1984) From grasshopper to *Drosophila*: a common plan for neuronal development. *Nature* 310: 203–207.
Uemura, T., Shepherd, S., Ackerman, L., Jan, L.Y. and Jan, Y.N. (1989) *numb*, a gene required in determination of cell fate during sensory organ formation in *Drosophila* embryos. *Cell* 58: 349–360.
Vaessin, H., Grell, E., Wolff, E., Bier, E., Jan, L.Y. and Jan Y.M. (1991) *prospero* is expressed in neuronal precursors and encodes a nuclear protein that is involved in the control of axonal outgrowth in *Drosophila*. *Cell* 67: 941–953.
Wilson, J.A., Phillips, C.E., Adams, M.W. and Huber, F. (1982) Structural comparison of a homologous neuron in Gryllid and Acridid insects. *J. Neurobiol.* 13: 459–468.
Wagner, G.P. (1989) The biological homology concept. *Annu. Rev. Eco. Syst.* 20: 51–69.
Whitington, P.M. (1989) The early development of motor axon pathways in the locust: the establishment of the segmental nerves in the thoracic ganglia. *Development* 105: 715–721.
Whitlock, K.E. (1993) Development of *Drosophila* wing sensory neurons in mutants with missing or modified cell surface molecules. *Development* 117: 1251–1260.
Yack, J.E. and Fullard, J. H. (1990) The mechanoreceptive origin of insect tympanal organs: A comparative study of similar nerves in tympanate and atympanate moths. *J. Comp. Neurol.* 300: 523–534.

The Nervous Systems of Invertebrates: An Evolutionary and Comparative Approach
ed. by O. Breidbach & W. Kutsch
© 1995 Birkhäuser Verlag Basel/Switzerland

Lineage analysis as an analytical tool in the insect central nervous system: Bringing order to interneurons

G.S. Boyan[1] and J.L.D. Williams[2]

[1]*Zoology Institute, University of Basel, Rheinsprung 9, CH-4051 Basel, Switzerland*
[2]*Arbeitsgruppe Kaissling, Max-Planck Institut für Verhaltensphysiologie, D-82319 Seewiesen, Germany*

Summary. The interneuron is a major neuroarchitectural element within the arthropod central nervous system, both in terms of numbers and diverstiy of form. The interneuron can be defined structurally as having its soma and most or all processes within the central nervous system, and functionally, as receiving information from afferents or other interneurons and then distributing this information to other interneurons or motoneurons. The interneuron is the key integrating unit of the nervous system. The fact that in arthropods many interneurons can be identified as individuals, recognizable from preparation to preparation, has allowed great progress to be made in understanding the contribution of interneurons to behavior. Further, structural and functional comparisons of identified neurons can be made both along the segmented central nervous system of an individual and between species. In fact, a number of interneurons have now been found to bear remarkable structural similarities in different arthropods, and even appear to function in similar sensory pathways. But such structural similarities alone cannot provide proof of an evolutionary relationship, that is, of homology between nervous systems. A far stronger argument for homology comes from lineage analysis, which establishes the ontogeny of any interneuron. It is the comparison of lineages that can reveal homologies between interneurons. In this chapter we examine the properties of a number of lineage-related neurons in different insect species. We describe the developmental events that lead to segmental specializations in the structure and function of homologous interneurons and then show how a comparison of lineages in different insect species suggests that some systems of neurons may have even been conserved across large distances of evolutionary time. We believe that the study of lineage-related and homologous interneurons can provide a fresh insight into the phylogenetic relationships among insect nervous systems.

Introduction

A nervous system is the product of two historical, biological, processes. The first is the evolutionary history of the species, evident in the pattern of gene expression that directs neurogenesis, growth, and connectivity. The second is the developmental history of the individual animal, involving, in addition to differential gene expression, environmental interactions and processes such as activity dependent competition between neurons for synaptic sites which produce inter-individual variations in the structure and function of adult neurons.

The arthropods, and the insects in particular, have proven highly attractive to neurobiologists interested in neuroethology, development, and in the evolution of nervous systems. One reason for this is that most insects possess large, individually identifiable neurons. Some of the estimated 800 000 or so interneurons in the grasshopper CNS, for example, have been structurally and functionally characterized as individuals, and there is now a formal classification scheme for such interneurons (Robertson and Pearson, 1983; Rowell, 1989). Working with such identified interneurons means that structural and functional analyses can be made of neurons located at different levels along the segmented central nervous system of an individual insect, or in different species. This approach has allowed great progress to be made in understanding the neural basis of a range of insect behaviors.

Intra- and interspecific comparisons of neuronal structure and function have also been made with a view to understanding the evolutionary relationships between insect nervous systems. While such comparisons have revealed that a number of interneurons bear remarkable structural similarities, and even appear to function in similar sensory pathways, in different species, these similarities alone are not evidence of a phylogenetic relationship between nervous systems. Central to any understanding of the evolutionary relationships between insect nervous systems is the concept of homology. In general terms, homology in the nervous system implies descent from a common ancestral or precursor cell – the embryonic neuroblast. Within the developmental history of the individual animal, homologous cells are those that derive from an equivalent neuroblast in the embryo – either in each hemisegment (bilaterally homologous) or in each neuromere (serially homologous). Thus, while many insect interneurons can now be grouped into "families" or clusters, with the members of a given cluster sharing descent from the same neuroblast, such clusters of cells in a given neuromere are lineage-related, but not strictly homologous.

In this chapter, we describe firstly how interneurons with diverse structural and functional properties can in fact share descent from a common neuroblast, and that these properties can be directly related to the age of the neuron within the lineage. Secondly, since the arthropod's segmented body plan is mirrored in its segmented nervous system, neuronal lineages repeat serially along the ventral nerve cord. We describe some of the developmental events associated with segmental specializations in the structure and function of homologous interneurons. Finally, we examine the properties of a number of lineage-related neurons in different insect species and show how some systems of neurons may have been conserved across large distances of evolutionary time.

We believe that the study of lineage-related and homologous interneurons allows the basic organization of any arthropod nervous system to be understood ontogentically, and can also provide a fresh insight into the phylogenetic relationships among insect nervous systems.

The groundplan of the arthopod CNS

If it is accepted that modern insects arose from a common ancestor, then in such a hypothetical ancestral insect one might expect each segment to have been more or less similar to the next one in line. During the course of evolution, however, adaptive radiation has resulted in a diversity of insect body forms and segmentation. The central nervous system is among the organ systems showing the greatest variability in degree of fusion, with modern examples ranging from the near-primitive condition found in Diplura to the highly fused condition typical of flies such as *Drosophila* (Fig. 1). Importantly, however, CNS segmentation need not conform to the body segmentation seen externally. This means, for example, that although control of the legs and wings normally resides in the thoracic ganglia, the number of unfused ganglia contributing to control varies between insects. Fusion affects the distribution of neurons within ganglia and can be accompanied by considerable loss of neuron numbers with a consequent reduction in neuromere size.

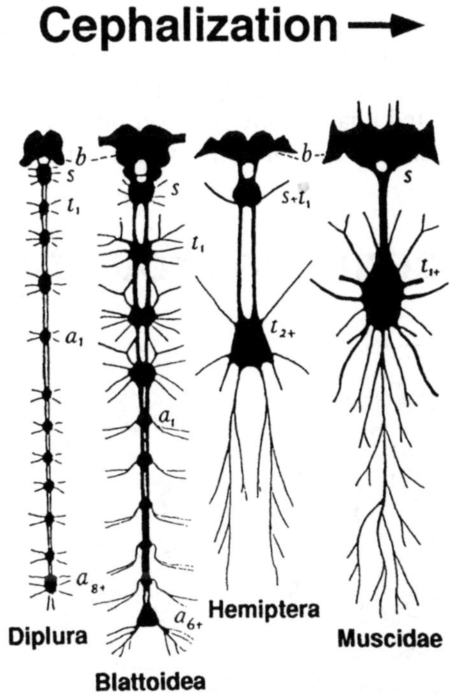

Figure 1. The central nervous systems of modern insects vary considerably in the extent of ganglionic fusion. In the primitive Diplurans there is similar sized ganglion in each segment along the ventral nerve cord (VNC), while advanced insects such as the Muscidae have highly fused systems (cephalization). Abbreviations: b, brain; s, subesophageal ganglion; t, thoracic ganglion; a, abdominal ganglion. Numbers associated with the letters indicate body segments innervated from the ganglion. Modified form Boyan (1993).

Our knowledge of the structure and function of the arthropod nervous system stems largely from the fact that early histological studies showed the nervous system to be resolvable into discrete components – such as the neurons and glia (Ramon y Cajal and Sanchez y Sanchez 1915, 1921; Zawarzin, 1925; Hanström, 1928). While the glia cells are crucial to neuronal development, structure and function, the emphasis of this review is on interneuronal homology, and the glia must therefore be considered beyond the scope of our present study. Comparative studies then revealed the underlying neuroanatomical organization in a wide range of insects and showed that neuroarchitecture could be approached from a phylogenetic perspective (Zawarzin, 1925; Hanström, 1928; Pyle, 1941; Gregory, 1974; Plotnikova, 1979; Tyrer and Gregory, 1982; Pflüger et al., 1988; Breidbach and Kutsch, 1990; Merritt and Murphey, 1992). Such species comparisons show that despite the great diversity of body form and segmentation, the ganglia of the ventral nerve cord in insects as phylogenetically distant as the grasshopper and fly have a very similar basic ground plan (Fig. 2). As seen in transverse section, a ganglion contains an outer "rind" or cortex of cell bodies, and an inner neuropil region which is divided into highly and less highly structured regions. The less structured region is made up of neuron arborizations and is the major site of synaptic interactions. A fundamental building block of this inner neuropil region is the interneuron with its associated glia. An interneuron may be quite simply defined as any neural element with a cell body (and most or all processes) in the

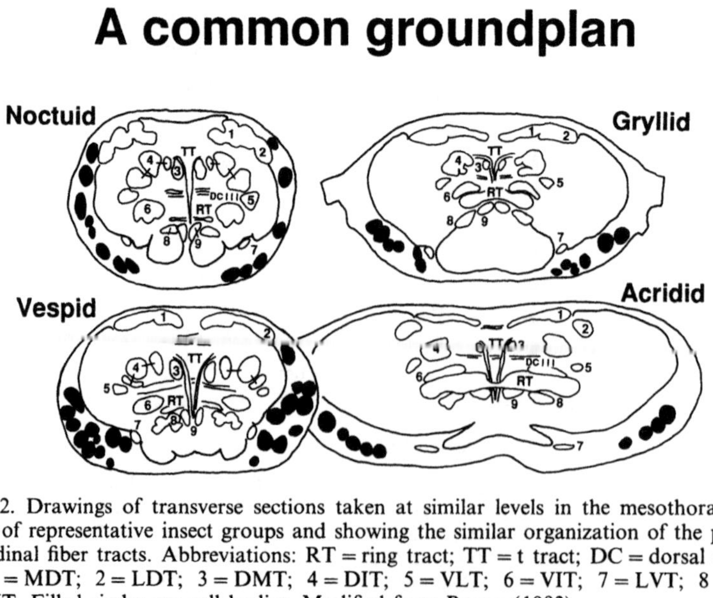

Figure 2. Drawings of transverse sections taken at similar levels in the mesothoracic neuromere of representative insect groups and showing the similar organization of the principal longitudinal fiber tracts. Abbreviations: RT = ring tract; TT = t tract; DC = dorsal commissure; 1 = MDT; 2 = LDT; 3 = DMT; 4 = DIT; 5 = VLT; 6 = VIT; 7 = LVT; 8 = MVT; 9 = VMT. Filled circles are cell bodies. Modified from Boyan (1993).

central nervous system and which synaptically links other neural elements to one another. Thus, an interneuron may link afferents with motoneurons, afferents with other interneurons, motoneurons with other motoneurons, even afferents with other afferents. The diversity of synaptic interactions an interneuron may mediate is immense and renders it a key integrating unit of the nervous system.

The revolution in our understanding of the functioning of interneurons has come first with the advent of single cell recordings (reviewed by Rowell, 1971), and then intracellular recordings and stainings (Crossman et al., 1971, 1972; Burrows, 1973; O'Shea et al., 1974; O'Shea and Williams, 1974; Pearson and Fourtner, 1975; Bacon and Tyrer, 1978) This powerful latter technique demonstrated that some interneurons possess the same morphological and electrophysiological properties in each individual of a species, and so established the "identified neuron" approach in neurobiology. To date, a large number of neurons have been identified in a range of insect preparations (Fig. 3). These interneurons may be sensory, premotor or neurosecretory in function may exhibit spiking or non-spiking electrical properties may occur singly, or as clusters of similar cells, and may extend their processes across many neuromeres (plurisegmental) or be confined to a single neuromere (local). As we shall see, some interneurons bear remarkable structural similarities in different insects, and even appear to function in similar sensory pathways (Bacon, 1980; Wilson et al., 1982; Robertson and Olberg, 1988).

Lineage-related interneurons: Birth, differentiation and determination of properties

While the groundplan of fiber tracts has been very successful in enabling us to organize interneurons into compartments within the CNS, identifying and classifying ever more individual interneurons in different arthropods does not bring a greater understanding of the evolutionary relationships between such nervous systems. The challenge is how to make evolutionary sense of the great array of interneurons identified to date. Luckily for us, paralleling this explosion of information about individual neurons has been an equally impressive advance in our understanding of how insect neurons are related to one another by birth, how they grow, recognize one another, and form the synaptic networks which are the building blocks of behavior (see Goodman and Jessell, 1992). This latter advance has been largely based on determination of a neuron's lineage, that is, its derivation from an identified neuroblast in the early embryo (Wheeler, 1893; Bate, 1976).

The basic outline of the developmental processes leading to the highly ordered adult nervous system is now fairly well understood (see Whiting-

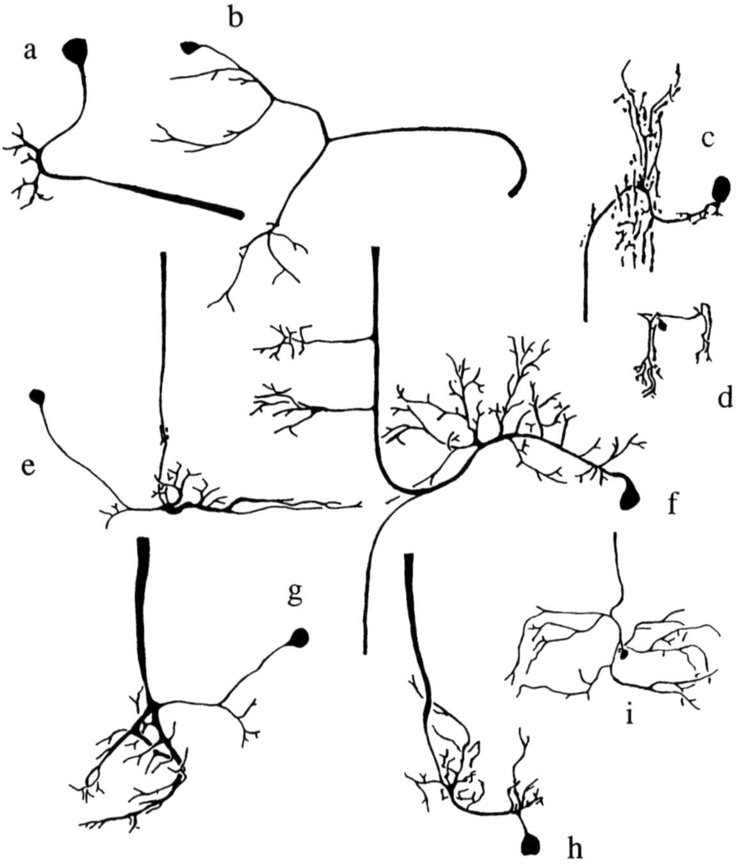

Sensory	Spiking	Singly	Local
Premotor	Non-spiking	Clusters	Plurisegmental
Neurosecretory			

Figure 3. Drawings of representative identified interneurons from the insects. a, DCMD of the grasshopper; b, descending neuron from the brain of *Antherea* spp.; c, lateral white cell in abdominal ganglion of grasshopper (original preparation courtesy of M. Adams and M. O'Shea); d, local neuron in terminal ganglion of grasshopper; e, ascending auditory neuron in a cricket; f, putative G neuron homolog in the mesothoracic ganglion of cricket; g, median giant neuron in terminal ganglion of cricket; h, giant interneuron, GIN 4, in terminal ganglion of grasshopper; i, ascending interneuron with midline cell body from a VNC ganglion in grasshopper. All drawings are originals by the authors. Table summarizes the anatomical configurations and physiological types found among the interneurons illustrated above.

ton, this volume). Although postembryonic development in hemi- and holometabolous insects differs significantly, the birth of neurons and the establishment of fiber pathways during embryonic development is remarkably similar. In the grasshopper, for example, approximately 150

Grasshopper

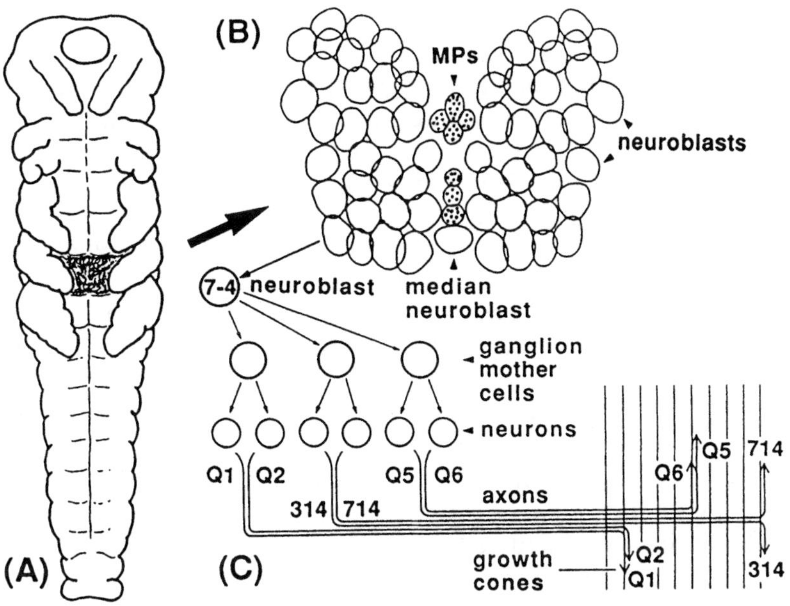

Figure 4. Diagrammatic view of (A) a grasshopper embryo at about 35% of embryonic development. The mesothoracic neuromere is shaded. (B) The pattern of neuroblasts in the mesothorax, and (C) the neuronal progeny of the first three ganglion mother cells produced by neuroblast 7-4 and their projections. Path of outgrowth and selective fasciculation of the progeny of neuroblast 7-4 in the grasshopper are also shown schematically. Abbreviation: MP's = midline precursor cells. Modified from Boyan and Ball (1993) after Goodman and Bastiani (1984).

neuroepithelial cells differentiate from the ectoderm of the ventral midline of each neuromere, from which just over a third differentiate further into stem cells or neuroblasts (Fig. 4; Wheeler, 1893; Bate, 1976; Doe and Goodman, 1985a). These neuroblasts (NBs) are arranged in a regular, symmetrical pattern, and the arrangment is sufficiently ordered for the NBs to be identified according to the row and column in which they are located. Neuroblast segregation is similar in insects as widely separated in evolutionary time as the grasshopper and fly (see Fig. 21; Doe, 1992). Ablation experiments in the grasshopper embryo have revealed that the identity of a given NB results from its position in the array of 30 cells of each hemisegment (Doe and Goodman, 1985b). The developmental history, or lineage, of any interneuron in the CNS can in principle be traced back to one of these NBs. Thus, NB 7-4 (row 7, column 4) produces a different set of adult neruons than NB 1-1 (Bastiani et al., 1984, 1986). Many interneurons can now be grouped into "families" or clusters, with the members of each cluster sharing descent from the same or an equivalent neuroblast – the concepts of lineage-relationship and homology, respectively (Goodman and Spitzer, 1979; Bate et al., 1981). The study of lineage-related neurons therefore allows the basic organization of any insect nervous system to be understood ontogenetically, and for the last decade or so lineage studies and the identified neuron approach have together provided a very powerful tool with which to study the structure, function and evolutionary relationships of neurons in the arthropod nervous system (Goodman and Spitzer, 1979; Bate et al., 1981; Bastiani et al., 1984; Pearson et al., 1985; Thompson and Siegler, 1991a,b).

In insects such as the grasshopper and cricket the cell bodies derived from a given neuroblast tend to remain clumped together in clusters or packets, each cluster encased by its own glial sheath. The number of cell bodies in a given cluster is a reflection of both the number of divisions the NB undertook during embryogenesis and any subsequent cell death. The cell bodies within a cluster may even occupy stereotyped positions with respect to one another, partly reflecting their order of birth. All the members of a given cluster are lineage-related. Intracellular staining of individual cells from a cluster is therefore one method of identifying lineage-related cells in the adult, and this method has been used to identify the progeny of specific neuroblasts and study their properties.

During early embryonic development the progeny of a given NB are dye coupled to one another, and to their contralateral homologs (Goodman and Spitzer, 1979; Shankland and Goodman, 1982). This dye coupling is transient and probably reflects a degree of coordination of metabolic processes between the NB and its progeny. Intracellular injection of current reveals no regenerative processes at this stage, but shows that the neurons are electrically coupled, presumably via gap junctions. If the DUM (dorsal unpaired median) progeny of the median neuroblast (MNB) can be taken as typical, then by about day 12 (50–60%) the siblings and bilateral homologs become electrically uncoupled from one another, and it is around this time that the membranes of insect neurons first appear to be capable of supporting regenerative electrical activity (Goodman et al., 1979; Goodman and Spitzer, 1981a).

A neuron's electrical properties are clearly important for its interactions with other neurons and hence its function in behavior. However, it does not appear possible to make generalizations about the membrane properties of lineally related neurons at present. Abrams and Pearson (1982) compared the membrane properties of the sibling 714 and 314 cells (see Fig. 7) and found that the membranes of these neurons responded very similarly to heating, and differently from other non-related neurons tested. Among sibling progeny of other neuroblasts in the grasshopper, by contrast, the H cell produced by the MP3 precursor in the metathoracic neuromere of the grasshopper (see Fig. 16) has the ability to generate Na^+-dependent action potentials in its axons, and (Na^+-Ca^{2+})-dependent action potentials in its soma. Its sibling, however, can only generate axonal action potentials (Bate et al., 1981; Goodman and Bate, 1981; Goodman and Spitzer, 1981a,b).

There is also a spectrum of electrical properties according to neuronal birthdate in the lineage produced by the median neuroblast (MNB) which generates the dorsal unpaired median (DUM) neurons in the grasshopper (Fig. 5). DUM progeny born early generate axon and soma spikes, those born later support neurite and axon spikes, then come neurons with only axon spikes, and finally non-spiking neurons (Goodman et al., 1980). These differences in electrical properties may relate to cellular morphologies or lie in the kinetics and distribution of Na^+ channels present in each membrane (Goodman and Spitzer, 1979).

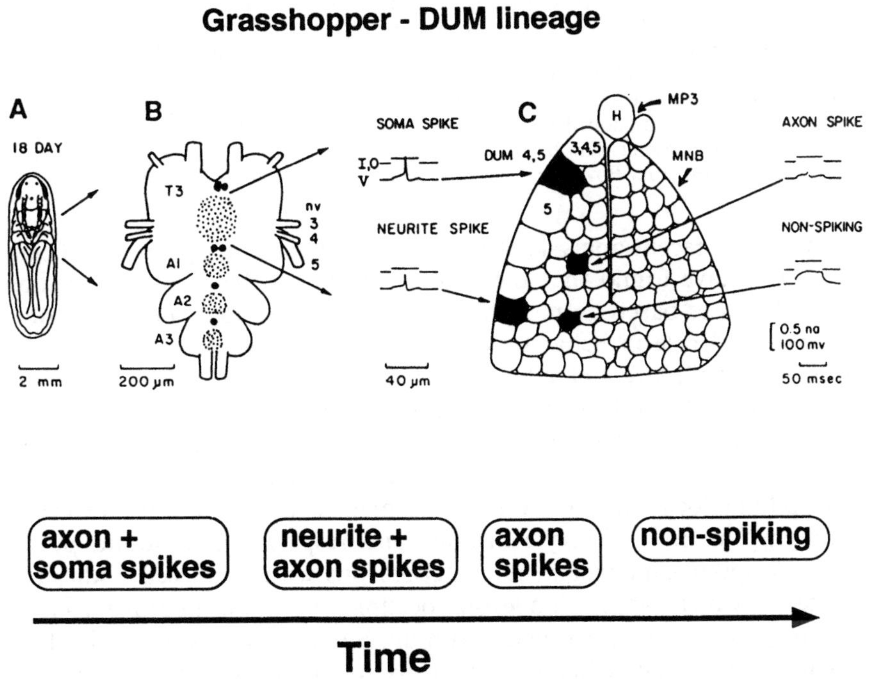

Figure 5. The progeny of the median neuroblast and their electrical properties at 90% of embryonic development. (A) The grasshopper embryo nearly fills the egg at this stage. (B) Fusion of the first three abdominal ganglia into the third thoracic ganglion is nearing completion. Areas occupied by progeny of the MNB are shown stippled, while the surviving progeny of MP3 in each neuromere are shown in black. The progeny of the median neuroblast are all enclosed within a glial sheath, while the progeny of MP3 (H and its sibling) lie just outside this sheath. The temporal distribution of electrical properties found in the various progeny of the median neuroblast are summarized in the schematic. Modified from Boyan and Ball (1993) after Goodman et al. (1980).

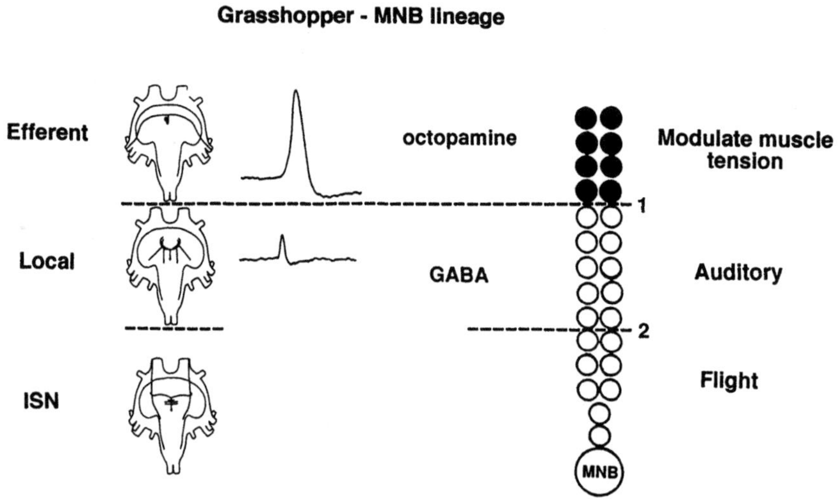

Figure 6. The progeny of the median neuroblast (MNB) have different electrical properties and utilize different transmitters depending on their time of birth. The first-born progeny (top, in black) form large cell bodies, have efferent axons, produce overshooting soma spikes, and are octopaminergic. After a certain time, indicated by the dotted line marked "1", neurons are formed which have small cell bodies and axons which do not leave the ganglion. These neurons produce passive soma spikes, and use GABA as a transmitter. Following another switch, indicated by the dotted line marked "2", still another type of neuron is produced, differing from the second class by having axons extending into the connectives beyond the ganglion. Modified from Boyan and Ball (1993) after Thompson and Siegler (1991b).

In a more recent analysis of this same lineage, Thompson and Siegler (1991a,b) have found that the adult progeny can be divided into three distinct classes based on morphology, electrophysiology, neurotransmitter type and role in behavior (Fig. 6). These classes, in their probable order of birth, are (1) efferent neurons with large cell bodies, and which innervate skeletal muscles, modulate muscle tension by releasing octopamine, and respond to sensory inputs by producing broad overshooting soma spikes characteristic of neurosecretory neurons; (2) spiking local interneurons with small cell bodies, with processes ramifying in the auditory neuropile and responding to sound pulses; and (3) spiking intersegmental interneurons with small cell bodies and processes which ramify in neuropil occupied by the wing sensory inputs and respond to flight-associated stimuli. Since in insects the majority of synaptic interactions between neurons involves chemical information transfer, an obvious and fundamental question is whether all the progeny of a neuroblast express the same transmitter. Siegler et al. (1991) have modified a long-established histological procedure for detecting the presence of endogenous heavy metals in nervous tissue. Their experiments suggest that endogenous copper (which is a co-factor to tyramine beta-hydroxylase, the enzyme synthesizing octopamine) is concentrated in only the large progeny of the neuroblast, such as

DUMETi, which are born first and send axons out the peripheral nerve roots to the musculature (Fig. 6). These results are consistent with immunohistochemical stainings in which the large, but not the small, progeny of the median neuroblast are octopamine immunoreactive (Stevenson et al., 1992). In fact, the smaller progeny, which are born later and probably do not project into the periphery, appear to be GABA immunoreactive (Fig. 6; Thompson and Siegler, 1989). We therefore see two neurotransmitter phenotypes originating from the same lineage. This suggestion that the birthdate of a neuron could play a role in determining the transmitter it is to synthesize is consistent with cell culture studies on clonally derived clusters of neurons in *Drosophila* which show that serotonergic cells are the progeny of early neuroblast divisions, while dopaminergic neurons arise throught the lineage (Huff et al., 1989).

The adult morphology of a given neuron is quite consistent from individual to individual and is a reflection of the navigational decisions made by the growth cone during embryogenesis. The data of Thompson and Siegler (1991a,b) shown in Figure 6, suggest that different progeny of the same neuroblast may make different pathway choices during development. This has been examined in great detail in the case of the progeny of NB 7-4 in the grasshopper. While all the progeny of NB 7-4 identified to date have many anatomical features in common, such as: cell body location (dorso-lateral cell cluster), a crossing segment in the supramedian commissure (SMC), major dendritic arborizations in the ventral intermediate tract (VIT) and anterior ring tract, there are major differences among the progeny involving the direction taken by the axon in the lateral dorsal tract (LDT) to other neuromeres via the connectives of the CNS (Fig. 7). The interneuron, IN 714 (also known as "G" or "TN3"), for example, has two axons, one projects anteriorly to the

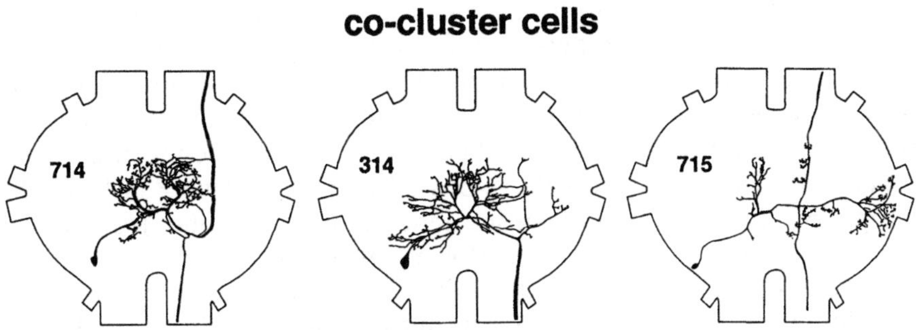

Figure 7. Camera lucida drawings of neurons which belong to the cluster of progeny of neuroblast 7-4 in the mesothoracic ganglion. Interneurons 714 and 314 are sibling cells in that they derive from the same division of the same ganglion mother cell. Neuron 715 derives from a different ganglion mother cell.

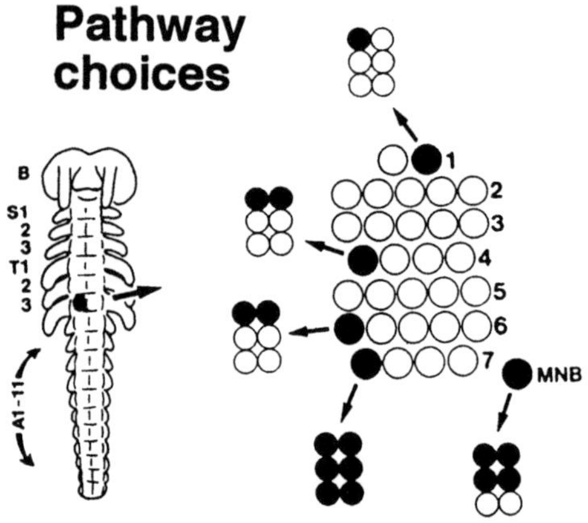

Figure 8. Summary figure showing the pathway choices made by the initial segment of the first six neurons belonging to various lineages in the metathoracic segment of the grasshopper. The neuronal progeny of a given neuroblast which make the same pathway choice are indicated in black.

brain and the other descends to the metathoracic ganglion; its sibling neuron (IN314) is radically different in that it only possesses a single axon which descends to the metathoracic ganglion (Fig. 7; Pearson and Robertson, 1981). Doe and Goodman (1985a) show that even the pathway choice made by the initial segment varies considerably between lineages and can vary within a lineage (Fig. 8). The temporal and spatial diversity of choices made by the growth cones associated with the various branches of IN 714, for example, may reflect the selective expression of cell surface antigens on various parts of the neuron during certain specific periods of embryonic development (Bastiani et al., 1987). The important point here is that a common lineage does not imply a uniformity of structure. Identified cells having the same lineage clearly make divergent choices in the course of development, and the choices they make remain fixed as the adult morphologies we see.

In hemimetabolous insects such as the grasshopper, the dendritic morphology of interneurons in the central nervous system midway through embryogenesis differs little from that seen in the adult (Fig. 9). This means that further development during the various postembryonic instars involves growth of neurons (for example, growth of the axon into the brain), and the establishment of additional synaptic connections: with other central neurons, with the axons of afferents growing in from the periphery, and with motoneurons innervating wing and leg muscles (Shankland and Goodman, 1982; Boyan, 1983; Raper et al., 1983a,b). In holometabolous insects, on the other hand, we see a considerable rebuilding of the nervous system during metamorphosis, reflecting changes in body form and function. The morphological and functional changes occurring at the level of individual cells have been best described for motoneurons and are considered elsewhere in this volume.

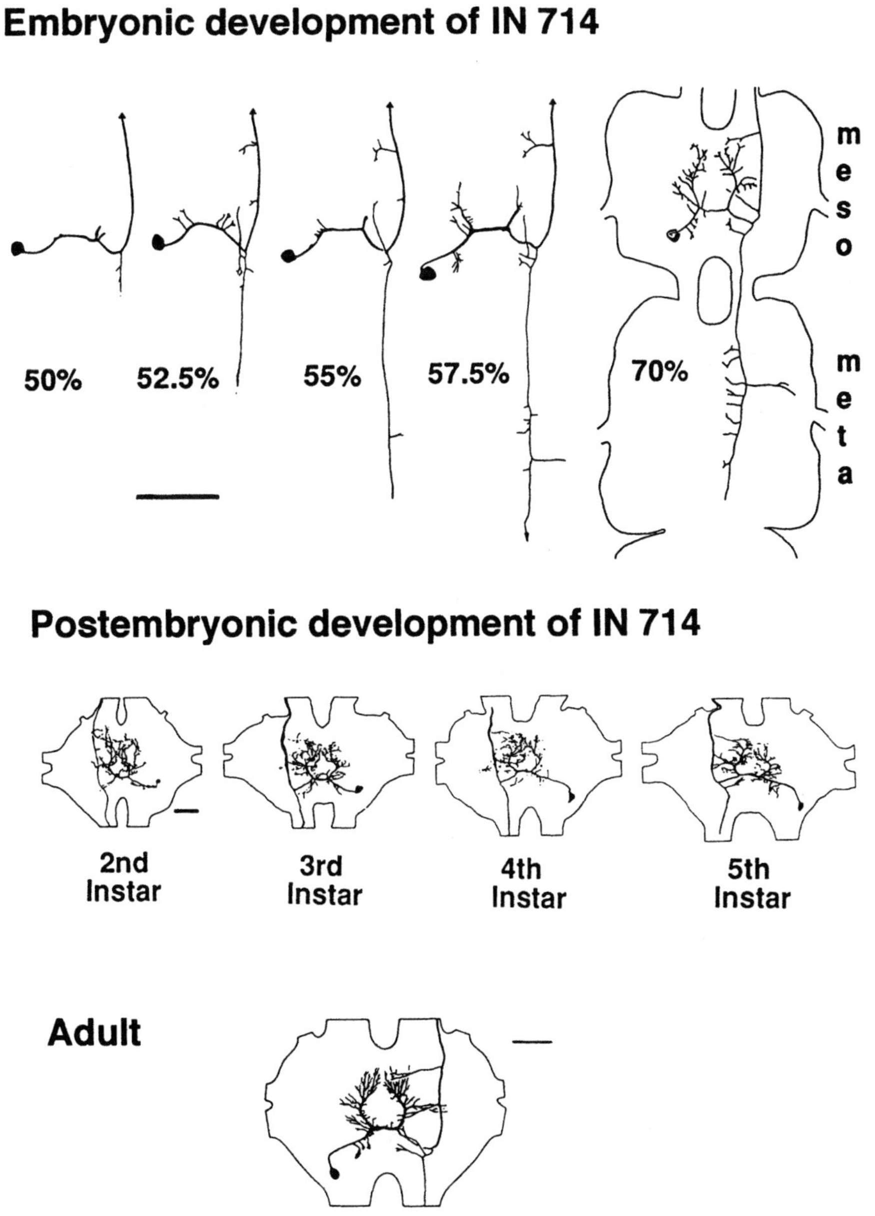

Figure 9. Embryonic development of the branching pattern of interneuron 714 (G) in the mesothoracic and metathoracic neuromeres of the grasshopper (top panel), postembryonic development of the branching pattern (middle panel), and final adult form (bottom panel). Note that the neuron has already reached a form close to that of the adult by 70% of embryonic development. Modified from Raper et al. (1983a) and Boyan (1983). Scale bar: 50 μm (top), 100 μm (middle), 200 μm (bottom).

Serial homology

Since the arthropod's segmented body plan is mirrored in its segmented nervous system, neuronal lineages might be expected to repeat serially along the ventral nerve cord. Indeed, the basic NB pattern is repeated in most neuromeres (Fig. 10). However, the body plan of the adult arthropod is not simply a repetition of identical segments. Segmentally homologous structures (also named homonomous; e.g. legs and mouthparts in insects) may differ strikingly in final form and function and we might expect this specialization to be reflected in the organization and functioning of interneurons in the CNS.

The segmental specificity of the CNS begins with the neuroblasts themselves. Neuroblasts are more mortal, and each dies at a prepro-

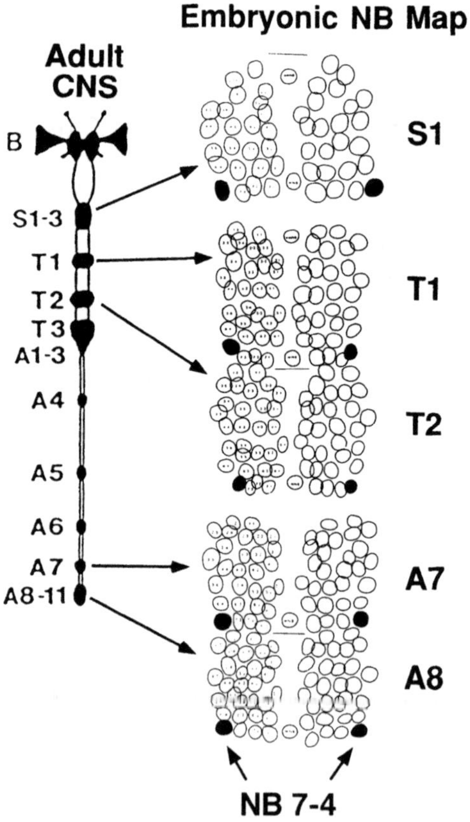

Figure 10. Neuroblast pattern in different segments of the embryonic grasshopper. The neuroblasts are symmetrically distributed on either side of the ganglionic midline, with the median neuroblast (MNB) located in the midline. The basic pattern repeats serially along the ventral nerve cord (neuromeres S1, T1, T2, A7, A8 are shown). Neuroblast 7-4 (row 7 and column 4) which produces the lineage including the "G" neuron (IN 714) is indicated in black. Modified from Boyan (1993) after Doe and Goodman (1985).

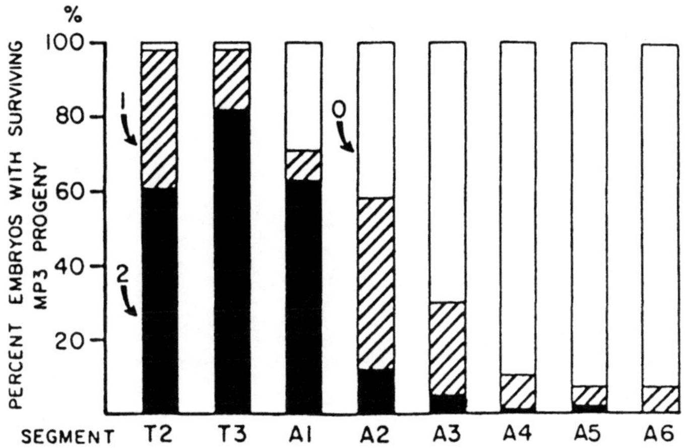

Figure 11. Different numbers of MP3 progeny survive in different segments of the grasshopper embryo. The percentage of embryos with two (solid bar), one (hatched bar), or no (open bar) MP3 progeny surviving per segment is shown for neuromeres T2 to A6. Modified from Bate et al. (1981).

grammed time, depending on its position within a segment and along the body axis. Thus, the abdominal homologs of NB 7-4 commence neurogenesis later than the thoracic ones, and cease dividing earlier; thoracic NBs divide at almost twice the rate of their abdominal counterparts (Shepherd and Bate, 1990). The mesothoracic NB 7-4 dies at approximately 90% of embryonic development, having produced over 100 progeny (Shepherd and Bate, 1990). The survival of progeny is also segment-specific, as demonstrated in the lineage of the MNB Goodman and Bate (1981), and the midline precursor (MP3) (Bate et al., 1981) in the grasshopper by (Fig. 11). Here, the MNB in each of the meso- and metathoracic neuromeres (T2, T3 respectively) produces 100 progeny, all of which survive, whereas the MNB in abdominal neuromeres A1–A3 produce 90 progeny of which only about 50 survive. Although the MP3 precursor is present in all neuromeres of the grasshopper from T2 to A6, segment-specific cell death means that progeny survive only in segments T2 to A3.

The progeny which derive from the same division of the same ganglion mother cell in different segments are considered to be serial homologs. Can we derive any general rules with respect to the structural and functional properties of serially homologous interneurons? At present the answer seems to be "no", since in the grasshopper at least the segment specific properties of surviving progeny may differ greatly, or hardly at all, depending on the lineage we choose. For example, among the many interneurons which are rhythmically active during flight in the grasshopper, several can be shown to have putative homologs in the fused neuromeres comprising the metathoracic ganglion

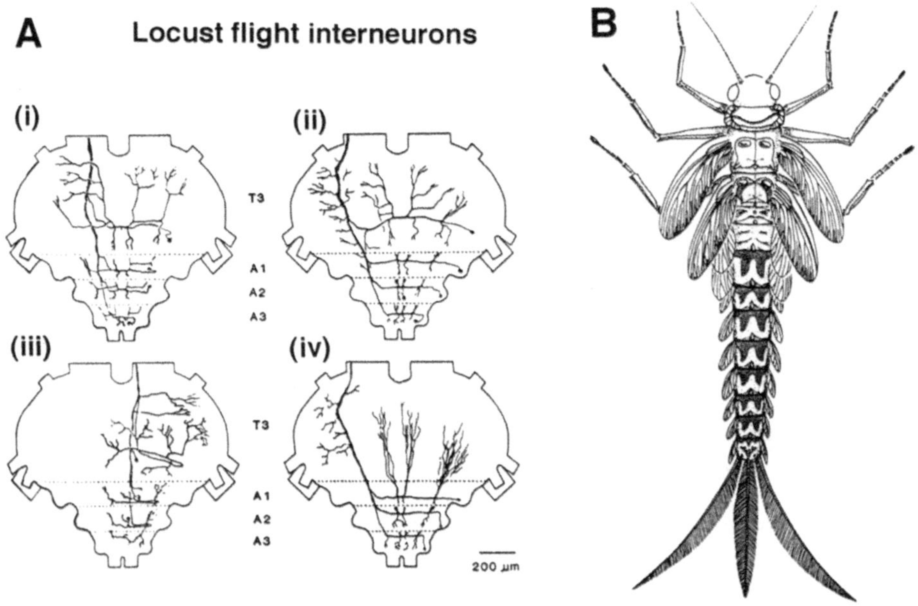

Figure 12. (A) Drawings showing structures of homologous sets of interneurons stained intracellularly with lucifer yellow in the fused neuromeres of the metathoracic ganglion of the grasshopper. All interneurons are phasically active during flight: (i,iv) depresssor interneurons, (ii, iii) elevator interneurons; (B) Drawing showing a reconstruction of a fossil late-instar mayfly nymph (*Kukalova americana*) bearing serially homologous cuticular structures on each thoracic and abdominal body segment and resembling modern wings. Modified from Robertson et al. (1982) and Dumont and Robertson (1986).

including, therefore, segments which do not bear wings in the adult (Fig. 12A; Robertson et al., 1982). The morphologies of all homologs are practically identical, and even their synaptic properties (such as connections to motoneurons and ability to reset the flight rhythm) appear to have been preserved regardless of their location within the serial array. Robertson et al. (1982) interpret this remarkable example of functional conservatism as evidence for a prior evolutionary stage in the organization of the flight motor system, and more specifically as tentative evidence for the pleural appendage theory for the origin of insect wings (Fig. 12B). It is true that in the early embryo each body segment bears a limb bud, some of which are subsequently modified into appendages such as mouthparts or legs, and others are reabsorbed into the body wall. Whatever the final explanation, the data of Robertson et al. (1982) suggest a compelling evolutionary conservatism within part of the CNS in the face of radical adaptive changes in the periphery.

The opposite situation to that described by Robertson et al. (1982) is also found and there are compelling examples at the level of identified neurons where segmental homology does not necessarily imply func-

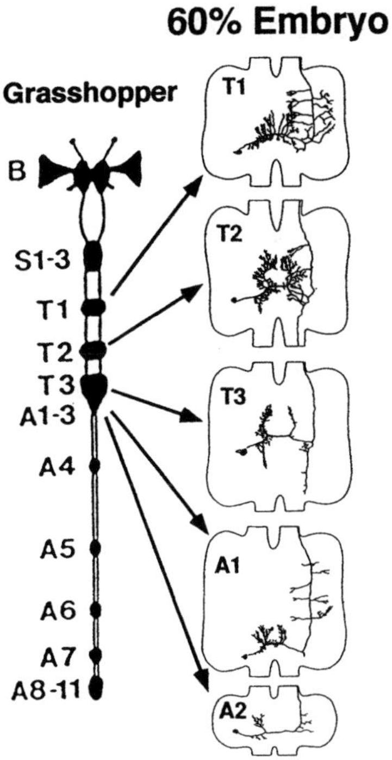

Figure 13. Drawings showing the embryonic morphologies of interneuron 714 (G neuron) and its serial homologs as revealed by intracellular injection of lucifer yellow into neurons in neuromeres T1-A2 in the 60% stage grasshopper embryo. Modified from Bastiani et al. (1984).

tional or morphological identity (grasshopper: Wilson and Hoyle, 1978; Bate et al., 1981; Pearson et al., 1985; crayfish: Kondoh and Hisada, 1983; Miller et al., 1985). Intracellular staining has shown the serial progeny of neuroblast 7-4 (Fig. 13) and of a midline precursor (MP3) (Fig. 16) in the grasshopper to possess clear segment-specific morphological differences – such as the distribution of medially directed dendrites and the direction of growth of the primary axon. Accompanying these structural differences are clear synaptic differences. In the case of the 7-4 progeny there are differences in the inputs these homologs receive from auditory afferents or an identified visual interneuron (Fig. 14), and segment-specific differences in the known outputs to motor pathways along the CNS (Fig. 15). Even within this same lineage, some inputs may be conserved from segment to segment, while others are not (Boyan, 1992). Studies on the progeny of MP3 show that segment-specific influences also play significant role in determining its electrical properties (Bate et al., 1981). While thoracic homologs generate Na^+-dependent and (Na^+-Ca^{2+})-dependent action potentials in their axons and somata respectively, posterior homologs in A1 to A3 produce only Na^+-dependent axon spikes, and do not appear to have an excitable soma membrane (Fig. 16). It appears that the more posterior surviving cells

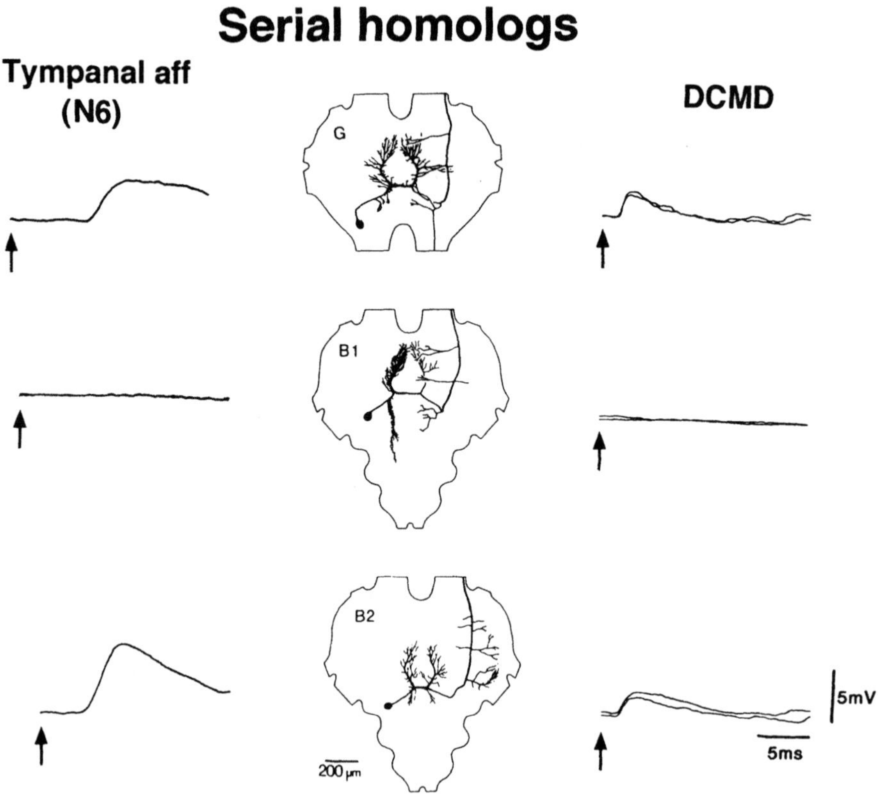

Figure 14. Responses of adult serial homologs belonging to the 7-4 lineage (G = 714, B1 = 531, B2 = 529) to afferent input from the auditory nerve (N6) on the soma side, and visual input from the DCMD neuron on the axon side of the homolog. The various oscilloscope traces (several traces superimposed) were triggered by extracellularly recorded axon spikes in afferent or DCMD respectively at the time indicated by the arrows.

maintain only a fraction of the morphological features of their anterior relatives, and paralleling this, only some of the electrical properties.

The pattern of transmitter expression may also vary among cells of the same lineage along the anterior-posterior axis of the segmentally-organized nervous system. The progeny of NB 7-3 have been investigated for factors determining expression of the neurotransmitter serotonin in the grasshopper. Taghert and Goodman (1984) have shown that within any one neuromere only some progeny of the 7-3 neuroblast express serotonin immunoreactivity (Fig. 17), and these tend to be the cells born earlier in the lineage such as the sibling S1 and S2 cells which derive from the first GMC (ganglion mother cell) (Fig. 17A). Taghert and Goodman (1984) show that while S1 expresses serotonin in all thoracic and abdominal neuromeres, S2 does not do so in the metathoracic and first abdominal neuromeres. The S3 cell derives from the second GMC, and expresses serotonin in only one neuromere – the

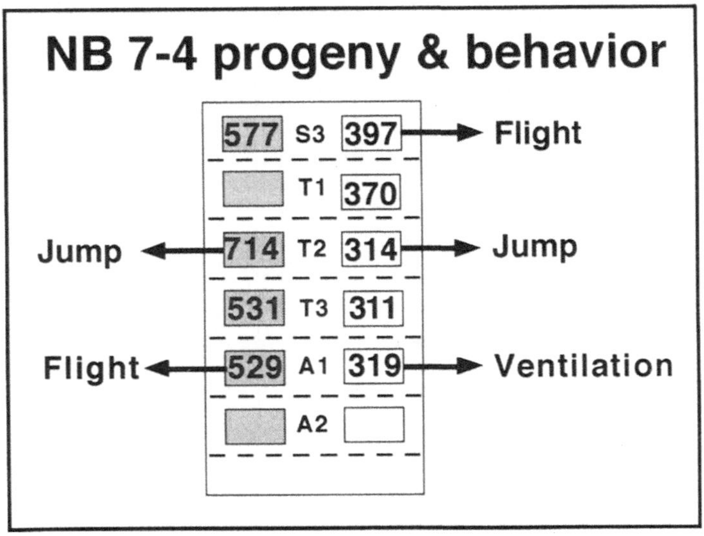

Figure 15. Schematic summarizing segmental specializations in outputs of serially homologous progeny of neuroblast 7-4 in the grasshopper. All the homologs receive input from presynaptic auditory elements. Where outputs have been described, serial homologs can be shown to contribute to different motor pathways. Modified from Boyan (1993).

prothoracic (Fig. 17B). Sibling cells therefore have greater similarities to one another than to co-cluster cells, but show some clear segment-specific differences. These differences are not due to differential cell death. Thus, it is clear that segment-specific factors can modify a basic pattern to produce a spectrum of biochemical phenotypes.

Interspecific homologies

Interspecific comparisons involve the assumption that where neuroblasts can be shown to occupy equivalent positions in their respective arrays and produce progeny with the same patterns of growth, then the two lineages are homologous (see Fig. 21). Such putatively homologous interneurons can now be identified at various levels of the CNS in grasshopper, cricket, cockroach, moth, mantid and fly, and we can begin to examine some of the properties of these interneurons against the background of their species' evolutionary history.

The brain

The arthropod head contains possibly the greatest concentration of sense organs to be found anywhere on the body. Not surprisingly, a

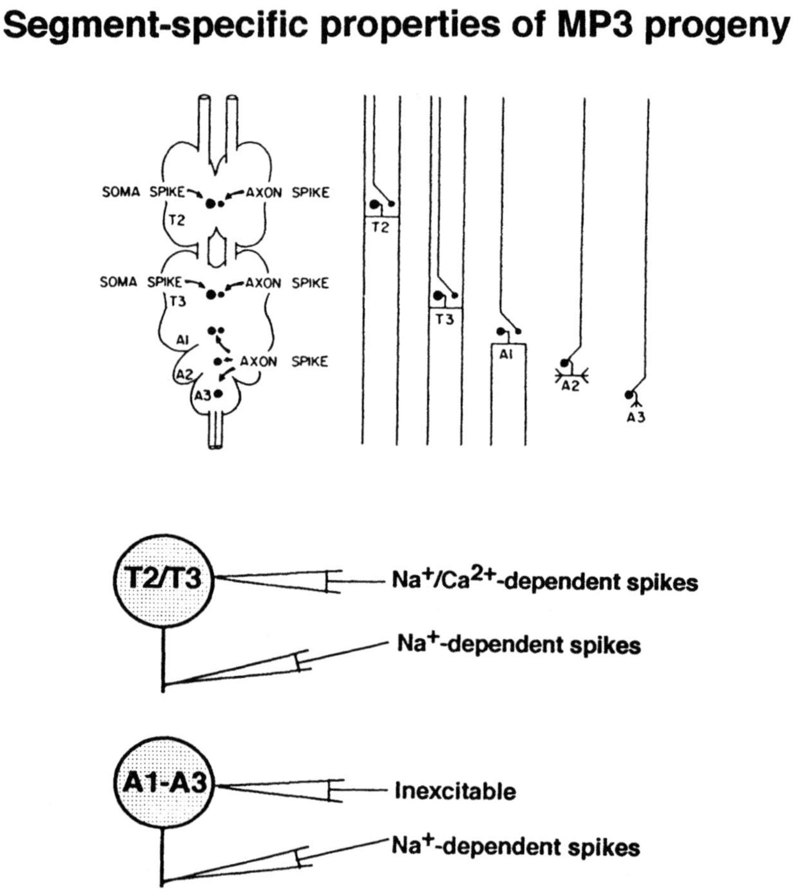

Figure 16. Serial progeny of MP3 progeny exhibit segment-specific morphologies (summarized schematically) and electrical properties. Anterior homologs support Na^+/Ca^{2+}-dependent soma spikes, their axons support Na^+-dependent spikes. Posterior homologs have inexcitable somata and support Na^+-dependent axon spikes. Modified from Bate et al. (1981).

massive sensory input from the compound eyes, ocelli, antennae, head hairs and mouthpart receptors (not to mention inputs from the abdominal cerci and thoracic ears) converges onto the interneurons of the supraoesophageal ganglion and optic lobes in insects – which in the grasshopper, for example, together account for probably 97% of the 800 000 or so interneurons in the central nervous system. This means that there is potentially a wide selection of interneuron classes from which to select examples for interspecific comparisons. There is, however, a significant drawback: to date there is not a single lineage for any brain interneuron in any arthropod. Probably the major reason for this gap in our knowledge is that neuroblast maps for supraoesophoageal ganglia in arthropods are few and far between, compared with say the

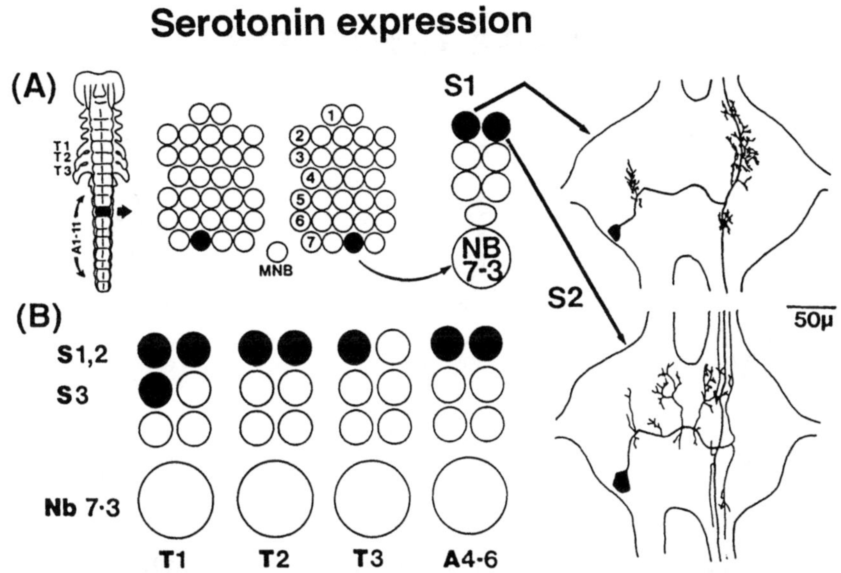

Figure 17. Different numbers of the progeny of neuroblast 7–3 express serotonin in different body segments. (A) Neuroblast map of the fourth abdominal segment showing the location of neuroblast 7-3, which gives rise to serotonin-producing neurons (black) by the lineage shown at the right. At the far right are shown the morphologies of S1 and S2 at 70% of embryonic development. (B) The first-born progeny of neuroblast 7-3 express serotonin (black), but the numbers differ in different body segments. Modified from Boyan and Ball (1993) after Taghert and Goodman (1984), and Taghert et al. (1984).

thoracic neuromeres. Indeed, the first complete embryonic neuroblast map has just been produced for the grasshopper brain (Zacharias et al., 1993). This lack of lineage data means that interspecific comparisons must be made between interneurons which are the most obvious candidates for homology.

Probably the most striking example of putatively homologous interneurons in the brain of a locust, a cricket, and a mantid has been provided by Bacon (1980). The well known TCG (tritocerebral commissure giant) neuron is a prominent interneuron of the posterior brain which descends via the tritocerebral commissure to the thoracic ganglia, and in the locust is part of the flight initiation (Bicker and Pearson, 1983) and flight steering (Bacon and Tyrer, 1978; Möhl and Bacon, 1983) circuitry. The morphology of this interneuron is remarkably consistent across orthopteroid species such as the locust, cricket and mantid (Fig. 18). Complementing this morphological conservatism is the fact that the TCG in each of these three species receives a different weighting of sensory inputs from head hairs, antennae and compound eyes according to the sensory apparatus of the particular animal (Fig. 18). Thus, for example, the cricket's head is much less "hairy" than the locust's, while its antennae are much longer. So in the TCG we

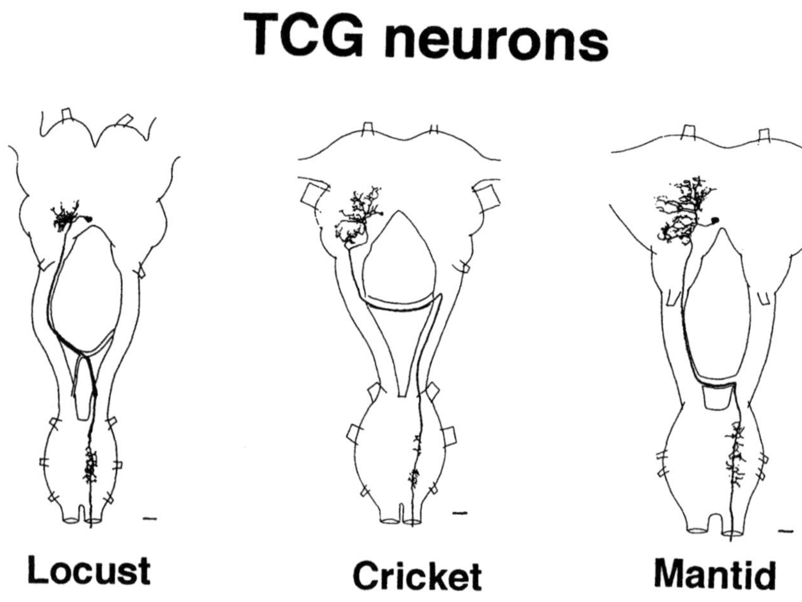

	Head hairs	Antennae	Visual
Locust	+++	++	+
Cricket	-	+++	+
Mantid	-	+	+

Figure 18. Drawings of putatively homologous TCG (tritocerebral commissure giant) neurons in the nervous system of grasshopper (*Schistocerca gregaria*), cricket (*Gryllus campestris*) and mantid (*Sphodromantis lineola*) illustrating the same basic morphology. The cell bodies of each neuron probably belong to the equivalent cell cluster, the crossing segments are located in the equivalent commissures (TCC), and the axons in the equivalent longitudinal fiber bundles (DIT). The relative weighting of synaptic inputs to the TCG neuron of each species from head hairs, antennae and compound eye is indicated in the table, where (+ + +) represents the strongest and (–) no, input. Scale bar: 100 μm. Original data were generously provided by Dr. J.P. Bacon and are unpublished.

may again be seeing an example of central morphological conservatism coupled to an adaptive plasticity in the synaptic inputs from the periphery.

The thorax

If the head represents the great convergence of sensory pathways in the arthropod, the thorax represents the motor powerhouse. Considerable numbers of interneurons in the thoracic region of most arthropods are involved in the generation of motor activity such as running, flight, jumping, swimming, stridulation and ventilation. These interneurons integrate sensory input derived from wing receptors, leg receptors, the ears, and body hairs. Regrettably, very few lineages of identified thoracic interneurons have been elucidated to date in the arthropods. The most obvious examples are NB 1-1 in grasshopper (Bastiani et al., 1986) and fly (Udolph et al, 1993), NB 7-4 in grasshopper (Raper et al., 1983a,b) and fly (Thomas et al., 1984), the MNB in grasshopper (Goodman and Spitzer, 1979; Thompson and Seigler, 1991a), and MP3 in grasshopper (Bate et al., 1981). For the purposes of this review, we will concentrate on the equivalent lineage produced by neuroblast 7-4 in the embryonic thoracic segments of grasshopper, *Drosophila*, and perhaps even *Manduca* (Fig. 19; Thomas et al., 1984). Descriptions of axon outgrowth reveal that initially the same series of events occurs in *Drosophila* as in the grasshopper. Interneurons which are the progeny of the equivalent neuroblast and have the same growth pattern are present at equivalent locations in the embryonic neuropils of all three species and are therefore putative homologs. In the adult, neurons which have

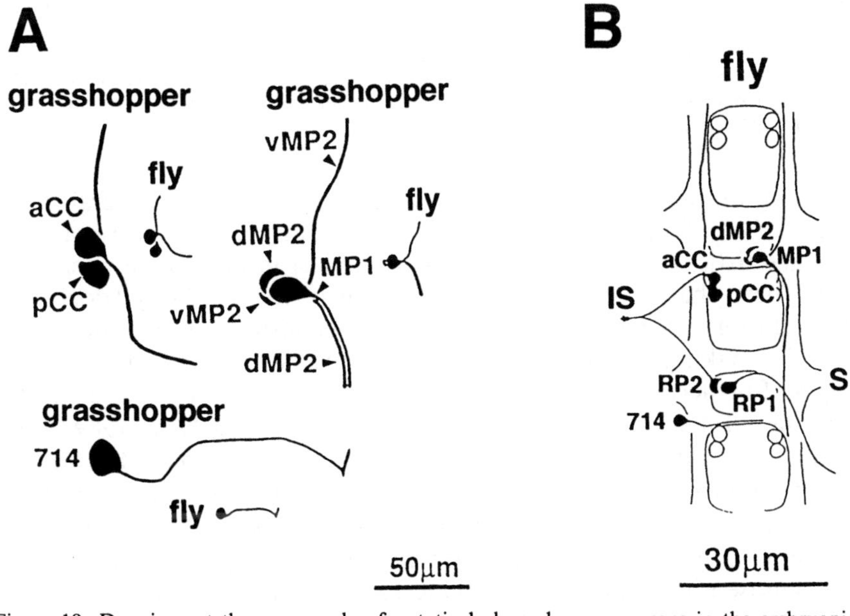

Figure 19. Drawings at the same scale of putatively homologous neurons in the embryonic nervous systems of grasshopper (*Schistocera americana*) and fruitfly (*Drosophila melanogaster*); illustrating the same basic morphology but a vast difference in size. (A) The cell bodies and crossing (commissural) segments of INs 714 in each species are shown in the lower part of this figure. (B) Outgrowth of axonal processes of identified neurons in thoracic neuromeres of embryonic *Drosophila*. Neurons shown are the putative homologs of those in the grasshopper. Abbrevations: S = segmental nerve, IS = intersegmental nerve. Modified from Boyan and Ball (1993) after Thomas et al. (1984).

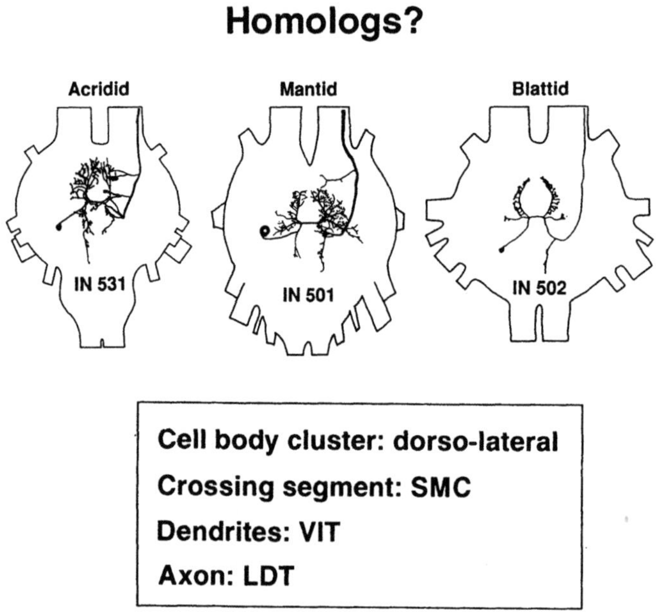

Figure 20. Drawings of neurons from putatively equivalent lineages in the adult nervous systems of an acridid (*Locusta migratoria*), a mantid (*Mantis religiosa*) and a blattid (*Periplaneta americana*). Interneuron 531 in the grasshopper belongs to the lineage of NB 7–4. The neurons shown have their cell bodies in the equivalent cluster, their crossing segments in the SMC. dendrites in VIT on each side of the ganglionic midline, and axons in the LDT. Modified from Boyan (1993) after Boyan (1992), Yager and Hoy (1989), Ritzmann et al. (1991) respectively.

a strong morphological similarity have been described in several insect species (Fig. 20), and these neurons share many of the morphological characteristics of progeny of the 7-4 neuroblast in the grasshopper (Fig. 13). These characteristics include: location of the cell body (dorso-lateral cluster), crossing segment (SMC), dendrites (VIT), and axon (LDT), making it likely that these neurons belong to equivalent lineages. Further, in each case these neurons respond to sound and/or vibrational stimuli (Pearson et al., 1985; Yager and Hoy, 1989; Ritzmann et al., 1991). Among tympanate insects, the axons of tympanal receptors project to equivalent tracts and association areas, even when the ears are located on different parts of the body (Boyan, 1993). In acridids, segments without tympanal receptor projections possess mechanosensitive afferents belonging to the pleural chordotonal system and these then occupy the same areas of neuropil as the tympanal receptors (Hustert, 1978). Appropriately, connections between pleural chordotonal afferents and progeny of NB 7-4 can be demonstrated in the grasshopper (Prier and Boyan, unpublished observations). What is interesting here is that the pleural chordotonal and tympanal receptors have been shown to be serially homologous (Meier and Reichert, 1990) so that from an ontogenetic viewpoint, the afferent input is serially continuous along the body. In atympanate insects such as *Drosophila* mechanosensitive receptors of the pleural chordotonal system are homologous to those in the grasshopper, and therefore to the tympanal receptors as well, and project to the equivalent areas, via the same routes, as auditory receptors in tympanate insects (Meier et al., 1991). We might, therefore, expect the *Drosophila* homologs of the grasshopper interneurons described above to also receive input from the homologous chordotonal receptors.

What we are seeing is a general chordotonal system encompassing a range of receptor types arrayed along the segmented body of the insect. In evolutionary terms it appears that the pathway is primary and the modality of information being conducted is secondary. The progeny of NB 7-4 and their synaptic connections with the chordotonal system offer the real hope that a detailed knowledge of sensory processing in one system may help us understand the circuit properties of evolutionarily equivalent neuropilar regions in other insect nervous systems, even when the modality of the information being processed is different.

The future

The molecular "identity" of interneurons

Over the last decade, we have seen the criteria by which we "identify" an interneuron change. Previously, an identity was based on structure and function; more recently, the definition included lineage, and now the products of gene expression can be added to the list. While molecular characterization of all but the *Drosophila* genome is in its infancy, several genes, notably *engrailed* (Patel et al., 1989), *abdominal A* (Tear et al., 1990), and *antennapedia* (Ball et al., 1991) have similar patterns of expression in both grasshopper and *Drosophila*, first in segmentation and later in the nervous system. The equivalent neuroblasts in the array can now be shown to express the same gene product in both grasshopper and *Drosophila* (Fig. 21): for example, the *engrailed* gene is expressed in NB 7-4, while the *even-skipped* gene is expressed NBs 1-1, 4-2 and 6-2 (Doe, 1992). Identified neurons such as the anterior corner cell, aCC, and posterior corner cell, pCC, and RP2 neurons which are the progeny of such equivalent neuroblasts in grasshopper and *Drosophila* have recently been shown to express the *even-skipped* gene product (Patel et al., 1992). This strongly suggests that homology can now be demonstrated between embryonic neurons on the basis of a molecular characterization. Since it appears that cell fate is genetically regulated, the "identity" of interneurons must now include reference to such patterns of gene expression. Where structure, function, and lineage have

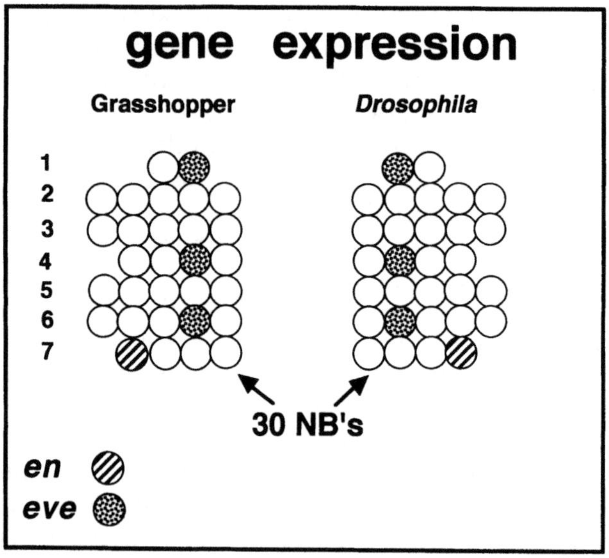

Figure 21. Interspecific comparisons of neuroblasts in the grasshopper and *Drosophila* suggest homologous neuroblasts are present for the following reasons: (1) the same number of neuroblasts (30) are present in thoracic neuromeres of each species, and in the same order; (2) neuroblasts in the same position express the same gene products (*engrailed* or *even-skipped*); (3) equivalent neuroblasts give rise to equivalent lineages (see Fig. 19).

probably reached their limits as criteria in demonstrating homology, it is at this molecular level of cell identity that the greatest advances are going to come in our understanding of the evolutionary relationships between arthropod nervous systems.

Acknowledgments
This work was supported by the Swiss Nationalfonds.

References

Abrams, T.W. and Pearson, K.G. (1992) Effects of temperature on identified central neurons that control jumping in the grasshopper. *J. Neurosci.* 2: 1538–1553.

Bacon, J. (1980) A homologous interneurone in a locust, a cricket and a mantid. *Verh. dt. Zool. Ges.* 1980. 163.

Bacon, J. and Tyrer, N.M. (1978) The tritocerebral commissure giant (TCG): a bimodal interneurone in the locust *Schistocera gregaria. J. Comp. Physiol.* A 126: 317–325.

Ball, E.E., Rehn, E.J., Patel, N.H. and Goodman, C.S. (1991) Evolution of insect segmentation and homeotic genes and their functions during neurogenesis. *Soc. Neurosci. Abstr.* 17: 11.

Bastiani, M., Pearson, K.G. and Goodman, C.S. (1984) From embryonic fascicles to adult tracts: organization of neuropile from a developmental perspective. *J. Exp. Biol.* 112: 45–64.

Bastiani, M.J., du Lac, S. and Goodman, C.S. (1986) Guidance of neuronal growth cones in the grasshopper embryo. III. Recognition of a specific axonal pathway by the pCC neuron. *J. Neurosci.* 6: 3518–3531.

Bastiani, M.J., Harrelson, A.L., Snow, P.M. and Goodman, C.S (1987) Expression of fasciclin I and II glycoproteins on subsets of axon pathways during neuronal development in the grasshopper. *Cell* 48: 745–755.

Bate, C.M. (1976) Embryogenesis of an insect nervous system. I. A map of the thoracic and abdominal neuroblasts in *Locusta migratoria*. *J. Embryol. Exp. Morph.* 35: 107–123.

Bate, M., Goodman, C.S. and Spitzer, N.C. (1981) Embryonic development of identified neurons: segment-specific differences in the H cell homologues. *J. Neurosci.* 1: 103–106.

Bicker, G. and Pearson, K.G. (1983) Initiation of flight by an identified wind sensitive neurone (TCG) in the locust. *J. Exp. Biol.* 104: 289–293.

Boyan, G.S. (1992) Common synaptic drive to segmentally homologous interneurons in the locust. *J. Comp. Neurol.* 321: 544–554.

Boyan, G.S. (1993) Another look at insect audition: the tympanic receptors as an evolutionary specialization of the chordotonal system. *J. Insect Physiol.* 39: 187–200.

Boyan, G.S. (1983) Postembryonic development in the auditory system of the locust. *J. Comp. Physiol. A* 151: 499–513.

Breidbach, O. and Kutsch, W. (1990) Structural homology of identified motoneurones in larval and adult stages of hemi- and holometabolous insects. *J. Comp. Neurol.* 297: 392–409.

Burrows, M. (1973) Physiological and morphological properties of the metathoracic common inhibitory neuron of the locust. *J. Comp. Physiol.* 82: 59–78.

Crossman, A.R., Kerkut, G.A. and Walker, R.J. (1972) Electrophysiological studies on the axon pathways of specified nerve cells in the central ganglia of two insect species, *Periplaneta americana* and *Schistocerca gregaria*. *Comp. Biochem. Physiol.* 43: 393–415.

Crossman, A.R., Kerkut, G.A., Pitman, R.M. and Walker, R.J. (1971) Electrically excitable nerve cell bodies in the central ganglia of two insect species *Periplaneta americana* and *Schistocerca gregaria*. Investigation of cell geometry and morphology by intracellular dye injection. *Comp. Biochem. Physiol.* 40: 579–594.

Doe, C.Q. (1992) Molecular markers for identified neuroblasts and ganglion mother cells in the *Drosophila* central nervous system. *Development* 116: 855–863.

Doe, C.Q. and Goodman, C.S. (1985a) Early events in insect neurogenesis. I. Development and segmental differences in the pattern of neuronal precursor cells. *Dev. Biol.* 111: 193–205.

Doe, C.Q. and Goodman, C.S. (1985b) Early events in insect neurogenesis. II. The role of cell interactions and cell lineage in the determination of neuronal precursor cells. *Dev. Biol.* 111: 206–219.

Dumont, J.P.C. and Robertson, R.M. (1986) Neuronal circuits: an evolutionary perspective. *Science* 233: 849–853.

Goodman, C.S. and Bastiani, M.J. (1984) How embryonic nerve cells recognize one another. *Scien. Am.* 251: 58–66.

Goodman, C.S. and Bate, C.M. (1981) Neuronal development in the grasshopper. *Trends Neurosci.* 4: 163–169.

Goodman, C.S. and Jessell, T.M. (1992) Development: editorial overview. *Current Opinion in Neurobiology* 2: 1–141.

Goodman, C.S. and Spitzer, N.C. (1979) Embryonic development of identified neurones: differentiation from neuroblast to neurone. *Nature* 280: 208–214.

Goodman, C.S. and Spitzer, N.C. (1981a) The development of electrical properties of identified neurones in grasshopper embryos. *J. Physiol.* 313: 385–403.

Goodman, C.S. and Spitzer, N.C. (1981b) The mature electrical properties of identified neurones in grasshopper embryos. *J. Physiol.* 313: 385–384.

Goodman, C.S., Pearson, K.G. and Spitzer, N.C. (1980) Electrical excitability: a spectrum of properties in the progeny of a single embryonic neuroblast. *Proc. Natl. Acad. Sci. USA* 77: 1676–1680.

Gregory, G.E. (1974) Neuroanatomy of the mesothoracic ganglion of the cockroach *Periplaneta americana* (L.). I. The roots of the peripheral nerves. *Phil. Trans. R. Soc. Lond. B* 267: 421–465.

Hanström, B. (1928) *Vergleichende Anatomie des Nervensystems der wirbellosen Tiere*. Springer Verlag, Berlin.

Huff, R., Furst, A. and Mahowald, A.P. (1989) *Drosophila* embryonic neuroblasts in culture: autonomous differentiation of specific neurotransmitters. *Dev. Biol.* 134: 146–157.

Hustert, R. (1978) Segmental and interganglionic projections from primary fibres of insect mechanoreceptors. *Cell Tiss. Res.* 194: 337–351.
Kondoh, Y. and Hisada, M. (1983) Intersegmental to intrasegmental conversion by ganglionic fusion in lateral giant interneurones of cryafish. *J. Exp. Biol.* 107: 515–519.
Meier, T. and Reichert, H. (1990) Embryonic development and evolutionary origin of the orthopteran auditory system. *J. Neurobiol.* 21: 592–610.
Meier T. and Chabaud, F. and Reichert, H. (1991) Homologous pattern in the embryonic development of the peripheral nervous system in the grasshopper Schistocerca gregaria and the fly *Drosophila melanogaster. Development* 112: 241–253.
Merritt, D.J. and Murphey, R.K. (1992) Projections of leg propioceptors within the CNS of the fly, *Phormia* in relation to the generalized insect ganglion. *J. Comp. Neurol.* 322: 16–34.
Miller, L.A., Hagiwara, G. and Wine, J.J. (1985) Segmental differences in pathways between crayfish giant axons and fast flexor motoneurons. *J. Neurophysiol.* 53: 252–265.
Möhl, B. and Bacon, J. (1983) The tritocerebral commissure giant (TCG) wind-sensitive interneurone in the locust. II. Directional sensitivity and role in flight stabilisation. *J. Comp. Physiol.* 150: 452–465.
O'Shea, M. and Williams, J.L.D. (1974) Anatomy and output connections of the lobular giant movement detector neuron (LGMD) of the locust. *J. Comp. Physiol.* 91: 257–266.
O'Shea, M., Rowell, C.H.F. and Williams, J.L.D. (1974) The anatomy of a locust visual interneurone, the descending contralateral movement detector, *J. Exp. Biol.* 60: 1–12.
Patel, N.H., Ball, E.E. and Goodman, C.S. (1992) Changing role of *even-skipped* during the evolution of insect pattern formation *Nature* 357: 339–342.
Patel, N.H., Martin-Blanco, E., Coleman, K.G., Poole, S.J., Ellis, M.C. Kornberg, T.B. and Goodman, C.S. (1989) Expression of engrailed proteins in arthropods, annelids, and chordates. *Cell* 58: 955–968.
Pearson, K.G. and Fourtner, C.R. (1975) Nonspiking interneurons in the walking system of the cockroach. *J. Neurophysiol.* 38: 33–52.
Pearson, K.G. and Robertson, R.M. (1981) Interneurons coactivating hindleg flexor and extensor motoneurons in the locust. *J. Comp. Physiol.* 144: 391–400.
Pearson, K.G. and Boyan, G.S., Bastiani, M., and Goodman, C.S. (1985) Heterogeneous properties of segmentally homologous interneurons in the ventral nerve cord of locusts. *J. Comp. Neurol.* 233: 133–145.
Pflüger, H.J., Bräunig, P. and Hustert, R. (1988) The organization of mechanosensory neuropiles in locust thoracic ganglia. *Phil. Trans. R. Soc. Lond. B* 312: 1–26.
Plotnikova, S.I. (1979) *Structural organization of the insect nervous system.* Nauka, Leningrad, pp. 119.
Pyle, R.W. (1941) The fibre tracts of the fused thoracic ganglia of the adult *Ephestia kuehnilla* Zeller (Lepidoptera: Pyralidae). *Pysche* 48: 123–128.
Ramon y Cajal, S. and Sanchez y Sanchez, D. (1915) Contribución al conocimiento de los centros nervosos de los insectos. Parte I. Rétina y centros ópticos. *Trab. Lab. Invest. Biol. Univ. Madr.* 13: 1–168.
Ramon y Cajal, S. and Sanchez y Sanchez, D. (1921) Sobre la esturctura de los centros nervosos de los insectos. *Rev. Chilena Hist. Nat.* 25: 1–18.
Raper, J.A., Bastini, M. and Goodman, C.S. (1983a) Pathfinding by neuronal growth cones in grasshopper embryos. I. Divergent choices made by the growth cones of sibling neurons. *J. Neurosci.* 3: 20–30.
Raper, J.A., Bastiani, M. and Goodman, C.S. (1983b) Pathfinding by neuronal growth cones in grasshopper embryos. II. Selective fasciculation onto specific axonal pathways. *J. Neurosci.* 3: 31–51.
Ritzmann, R.E., Pollack, A.J., Hudson, S.E. and Hyvonen, A. (1991) Convergence of multimodal sensory signals at throacic interneurons of the escape system of the cockroach, *Periplaneta americana. Brain Res.* 563: 175–183.
Robertson, R.M. and Olberg, R.M. (1988) A comparison of the acitivity of flight interneurones in locusts, crickets, dragonflies and mayflies. *Experientia* 44: 735–738.
Robertson, R.M. and Pearson, K.G. (1983) Interneurons in the flight system of the locust: distribution, connections, and resetting properties. *J. Comp. Neurol.* 215: 33–50.
Robertson, R.M. and Pearson, K.G. and Reichert, H. (1982) Flight interneurons in the locust and the origin of insect wings. *Science* 217: 177–179.
Rowell, C.H.F. (1971) The orthopteran descending movement detector (DMD) neurones; a characterisation and review. *Z. Vergl. Physiol.* 73: 167–194.

Rowell, C.H.F. (1989) The taxonomy of invertebrate neurons: a plea for a new field. *Trends Neurosci.* 12: 169–174.

Shankland, M. and Goodman, C.S. (1982) Development of the dendritic branching pattern of the medial giant interneuron in the grasshopper embryo. *Dev. Biol.* 92: 489–506.

Shepherd, D. and Bate, C.M. (1990) Spatial and temporal patterns of neurogenesis in the embryo of the locust (*Schistocerca gregaria*). *Development* 108: 83–96.

Siegler, M.V.S., Manley, Jr., P.E. and Thompson, K.J. (1991) Sulphide silver staining for endogenous heavy metals reveals subsets of dorsal unpaired median (DUM) neurones in insects. *J. Exp. Biol.* 157: 565–571.

Stevenson, P.A., Pflüger, H.J., Eckert, M. and Rapus, J. (1992) Octopamine immunoreactive cell populations in the locust thoracic-abdominal nervous system. *J. Comp. Neurol.* 315: 382–397.

Taghert P.H. and Goodman C.S. (1984) Cell determination and differentiation of identified serotonin-immunoreactive neurons in the grasshopper embryo. *J. Neurosci.* 4: 989–1000.

Taghert, P.H., Doe, C.Q. and Goodman, C.S. (1984) Cell determination and regulation during development of neuroblasts and neurones in grasshopper embryo. *Nature* 307: 163–165.

Tear, G., Akam, M. and Martinez-Arias, A. (1990) Isolation of an abdominal-A gene from the locust *Schistocerca gregaria* and its expression during early embryogenesis. *Development* 110: 915–925.

Thomas, J.B., Bastiani, M.J., Bate, M. and Goodman, C.S. (1984) From grasshopper to *Drosophilia*: a common plan for neuronal development. *Nature* 310: 203–207.

Thompson, K.J. and Siegler, M.V.S. (1989) Properties of the small dorsal unpaired median (DUM) neurons of the grasshopper. *Soc. Neurosci. Abstr.* 15: 1296.

Thompson, K.J. and Siegler, M.V.S. (1991a) Three classes of neuron in the median neuroblast lineage of the grasshopper. *Soc. Neurosci. Abstr.* 17: 1478.

Thompson, K.J. and Siegler, M.V.S. (1991b) Anatomy and physiology of spiking local and intersegmental interneurons in the median neuroblast lineage of the grasshopper. *J. Comp. Neurol.* 305: 659–675.

Tyrer, N.M. and Gregory, G.E. (1982) A guide to the neuroanatomy of locust suboseophageal and thoracic ganglia. *Phil. Trans. R. Soc. Lond. B.* 297: 91–123.

Udolph, G., Prokop, A., Bossing, T. and Technau, G.M. (1993) A common precursor for glia and neurons in the embryonic CNS of *Drosophila* gives rise to segment-specific lineage variants. *Development* 118: 765–775.

Wheeler, W.M. (1893) A contribution to insect embryology. *J. Morphol.* 8: 1–160.

Wilson, J.A. and Hoyle, G. (1978) Serially homologous neurones as concomitants of functional specialisation. *Nature* 274: 377–379.

Wilson, J.A., Phillips, C.E., Adams, M.E. and Huber, F. (1982) Structural comparison of a homologous neuron in gryllid and acridid insects. *J. Neurobiol.* 13: 459–467.

Yager, D.D. and Hoy, R.R. (1989) Audition in the praying mantis, *Mantis religiosa* L.: identification of an interneuron mediating ultrasonic hearing. *J. Comp. Physiol. A* 165: 471–493.

Zacharias, D., Williams, J.L.D., Meier, T. and Reichert, H. (1993) Neurogenesis in the insect brain: Cellular identification and molecular characterization of brain neuroblasts in the grasshopper embryo. *Development* 118: 941–955.

Zawarzin, A.A. (1925) Der Parallelismus der Strukturen als ein Grundprinzip der Morphologie. *Z. Wiss. Zool.* 124: 118–218.

The Nervous Systems of Invertebrates: An Evolutionary and Comparative Approach
ed. by O. Breidbach & W. Kutsch
© 1995 Birkhäuser Verlag Basel/Switzerland

Comparative aspects of peptidergic signaling pathways in the nervous systems of arthropods

H.-J. Agricola[1] and P. Bräunig[2]

[1]*Institut für Allgemeine Zoologie u. Tierphysiologie, Universität Jena, Erbertstr. 1, D-07743 Jena, Germany*
[2]*Institut für Zoologie d. Technische Universität München, Lichtenbergstr. 4, D-85747 Garching, Germany*

Summary. Comparative aspects of arthropod peptidergic systems – in principle – can be studied on the level of precursor sequences (genes, preprohormones), peptide sequences (peptide families), and peptide expression patterns within the nervous system. The number of known arthropod neuropeptide precursor sequences is as yet far too small to provide a reasonably large basis for extended comparative studies. Comparative studies of peptide sequences have shown that many peptides belong to families with homologous members in both invertebrates and vertebrates. Comparative research on peptide expression has to find out whether phylogenetic necessities lead to "hard wired" neurochemical identities, i.e., a more or less fixed "Bauplan" that not only determines the lineage and morphology of a neuron but also its transmitter(s), or whether these necessities demand greater flexibility (plasticity), and hence cause great variability that would complicate comparative studies.

As will be shown here, both possibilities appear to exist. On the one hand, peptidergic neurons may exist in comparable form in different groups of arthropods. On the other hand, the neurochemical identity of cells may vary in segmented organisms when comparing serially homologous sets of nerve cells in different segments. As a further complication, identical or similar peptides may serve different functions, even in closely related species. In view of these functional aspects in particular, it appears that peptidergic signalling pathways represent rapidly evolving systems. This conclusion, although very interesting in itself, reduces the use of such systems for general comparisons. However, arthropod nervous systems represent excellent model systems for the study of homology. At least for morphological and ontogenetic aspects arthropods provide numerous opportunities to study homology on the level of the individually identified peptidergic nerve cell.

Introduction

Communication between individual nerve cells is usually mediated by chemical messengers. Only a few years ago, the general assumption prevailed that only a few chemical compounds serve as messengers in these pathways. This view was radically changed by the discovery of neuropeptides. These molecules, being composed of single building blocks (amino acids), in principle provide for an unlimited variety of different mediators by simple combinatorics. When considering that there are 20 amino acids in normal tissues, a single peptide chain of seven amino acids, for example could actually exist in 1 280 000 000 different forms (20^7)! In fact, the number of identified "biologically-active" neuropeptides in both vetebrates and invertebrates is large, and growing rapidly, but it appears that only a small fraction of all theoretically possible combinations of amino acids is actually encountered in natural systems. In addition, there appears to be a preferred size range: although peptides may range in length from 2 to 90 amino acids, the great majority of naturally occurring neuropeptides is from 5 to 20 amino acids long. In invertebrates, about 200 peptide sequences have been determined in molluscs and arthropods (Muneoka and Kobayashi, 1992; Keller, 1992; Nässel, 1993). Comparatively less is known about peptide structures in other invertebrate taxa such as coelenterates (Spencer, 1989; Grimmelikhuijzen, 1992), nemathelminths (Brownlee et al., 1993), platyhelminths (Fairweather and Halton, 1991; Shaw and Johnston, 1991), and annelids (Al-Yousuf, 1992). While for all platyhelminths only one peptide sequence is known today (Maule et al., 1993), 38 neuropeptides are known alone for the insect *Locusta migratoria* (Schoofs et al., 1993).

In principle, similarities of peptidergic signaling systems can be studied on various levels. For instance comparing precursor (preprohormone) sequences is a particularly powerful approach, since it is a means for direct and quantifiable determiation of similarities or differences, respectively. This level of comparison, however, is hampered in that only 23 peptide precursor sequences have been determined for all invertebrate taxa (Tab. 1). The structural organization of many precursors resemble that of the precursor for the neuropeptide FMRFamide as characterized in *Aplysia* (Schaefer et al., 1985), *Drosophila* (Nambu et al., 1988; Schneider and Taghert, 1988) and *Caenorhabditis elegans* (Rosoff et al., 1992). A common feature of these precursors is that they code for multiple copies of the same or related (or both) neuropeptides. Similarly, the precursor for the mollusc neuropeptide buccalin contains 31 copies of 19 distinct buccalin-related peptides, one of which, buccalin A, is present in six contiguous copies (Miller et al., 1993;). The precursor for allatostatins from the cockroach *Diploptera punctata* contains messages for 13 structurally related peptides (Donly, 1993). In contrast, the precursors for red pigment concentrating hormone (RPCH/Linck et al., 1993) and molt-inhibiting hormone (MIH/ Klein et al., 1993) of the shore crab, *Carcinus maenas*, each contain only one copy for a specific neuropeptide. In the case of the adipokinetic hormones (AKH) two peptides (AKH I and AKH II) are synthesized by the same neurosecretory cells (NSC) of the corpora cardiaca. Both peptides however, though structurally similar, are formed from two different prohormones (O'Shea and Rayne, 1992).

Homologies have already been studied on the level of peptide structures. The main result of this approach is that most neuropeptides occur in families. Peptides in animals of very different taxa may show sequence homology, or even have identical sequences. Particularly well documentated are the members of the family of the FMRFamide-like

Table 1. Fully or partially characterized cDNAs or genomic DNAs for invertebrate neuropeptides

Peptides	Species	Reference
Cnidaria		
RF-amide	*Calliactis parasitica*	Darmer et al., 1991
Arthropods: Crustaceans		
Molt-inhibiting hormone	*Carcinus maenas*	Klein et al., 1993
Red pigment concentrating hormone	*Carcinus maenas*	Linck et al., 1993
Arthropods: Insects		
Adipokinetic hormones	*Locusta migratoria*	Schulz-Azellen et al., 1989
	Manduca sexta	Bradfield and Keeley, 1989
Bombyxin (PTTH)	*Bombyx mori*	Iwami et al., 1989
	Bombyx mori	Kamito et al. 1992
Bombyxin-related peptide	*Samia cynthia ricini*	Kimura-Kawakami et al., 1992
Eclosion hormones	*Manduca sexta*	Horodyski et al., 1989
	Bombyx mori	Kamito et al., 1992
Prothoracicotropic hormone	*Bombyx mori*	Kawakami et al., 1990
Drosulfakinis	*Drosophila melanogaster*	Nichols et al., 1988
Diapause hormone	*Bombyx mori*	Sato et al., 1992
Pheromone biosynthesis-	*Helicoverpa zea*	Davis et al., 1992
activating neuopeptides (PBAN)	*Bombyx mori*	Kawano et al., 1992
Insulin-like peptide	*Locusta migratoria*	Lagueux et al., 1990
Diuretic hormone	*Manduca sexta*	Digan et al., 1992
Neuropersin A	*Locusta migratoria*	Lagueux et al., 1992
FMRF-amides	*Drosophila melanogaster*	Schneider and Taghert, 1988
		Taghert and Schneider, 1990
Allatostatins	*Diploptera punctata*	Donly et al., 1993
Molluscs		
Egg-laying hormone (ELH)	*Aplysia californica*	Scheller et al., 1982
Vasopressin/Oxytocin	*Lymnaea stagnalis*	Van Kesteren et al., 1992
Molluscan insulin-related peptide	*Lymnaea stagnalis*	Geraerts, et al., 1992
Buccalin-related neuropeptides	*Aplysia californica*	Miller et al., 1993a
Myomodulin-related neuropeptides	*Aplysia californica*	Miller et al., 1993b
FMRF	*Aplysia californica*	Schaefer et al., 1985
SCP_A, SCP_B	*Aplysia californica*	Mahon et al., 1985

peptides (Price and Greenberg, 1989; Platt and Reynolds, 1990; Maule et al., 1993). Another example for a large family is provided by the peptides of the AKH/RPCH- family (AKH = adipokinetic hormone, RPCH = red pigment concentrating hormone; Goldsworthy and Mordue, 1989; Gäde, 1990; Holman et al., 1990). In fact, accumulation of comparative information about invertebrate neuropeptide structures can help to form concepts about how these neuropeptides have evolved. Thus, a tentative scheme of the phylogenetic relationships of known members of the insulin-like superfamily of peptides in vertebrates and invertebrates is given by Ebberink et al. (1989) and Geraerts et al. (1992). In addition to these interspecific familes, intraspecific "families" also exist. For example, at least two isoforms of CHH (crustacean hyperglycemic hormone) occur in the lobster *Homarus americanus* (van Herp, 1992; Keller, 1992). Leucokinins in the cockroach *Leucophaea maderae* exist in at least eight different variants (Holman et al., 1990), and the allatostatins in the cockroach *Diploptera punctata* may exist in as many as 13 isoforms (Donly et al., 1993). In the mollusc *Helix aspersa* 12 different sequences have been determined for members of the MIP family (Mytilus inhibitory peptides), in *Achatina fulica* at least 10 members of this family exist (Muneoka and Kobayashi, 1992). In contrast, no family has yet been found for the pentapeptide proctolin, the first insect neuropeptide to be sequenced (Brown and Starratt, 1975) and one of the most intensively studied (for review see O'Shea, 1985; Orchard et al. 1989). Moreover, this peptide is one of the few examples for a peptide occurring in different taxa with absolute sequence identity. Another example is provided by CCAP (crustacean cardioactive peptide), another peptide for which no obvious family exists and that has been isolated in identical form in both crustacea and insects (Stangier et al., 1987, 1989; Cheung et al., 1992; Lehman et al, 1993; Furuya et al., 1993).

A third level of investigating homologies is to study neuropeptide expression patterns within the nervous system of the various invertebrate taxa. These patterns can be studied using many different neuroanatomical, biochemical, and molecular biological methods. One that is relatively easy to use, but nevertheless very powerful, is immunocytochemistry. Antibodies, particularly monoclonal ones, can be tailored to a very high specificty, and once the investigator knows the pitfalls of the immunohistochemical method, it can be used for "chemical neuroanatomy" at a very high level of resolution. The power of this approach is beautifully illustrated by the recent advances in the comparative neuroanatomy of the insect brain (Nässel, 1993; Homberg, 1994).

In the following, we describe the results of a comparative survey of neuropeptide expression in different invertebrate species. The tools for this survey were antibodies designed to recognize specific peptides belonging to well characterized invertebrate peptide families. Although we shall mention others in passing, we will concentrate on two families: the sulfakinin and allatostatin families. We will look at serial homologies of neuropeptide expression in segmented animals, interspecific similarities (and, of course, differences) in the expression of various neuropeptides, and finally compare the distribution pattern of neuropeptides across species in one particular organ system.

Serial homology in peptidegic neuronal systems

A prominent characteristic common to higher eumetazoan taxa is segmentation. Before addressing the question of homologies of peptidergic neurons and peptidergic signaling systems in different taxa, it is worth-while to focus on intersegmental homologies in individual species. Knowledge about the range of morphological variability of segmentally homologous neuronal systems in different segments of the same species can support the interpretation of comparative data on neuronal phenotypes in different species. A comparison of metamerically repeated neuronal systems is also of interest in view of the tremendous progress recently made in determining the genetic factors involved in segmentation and other aspects of establishing a metameric body plan (for review see Jäckle, 1992; Lawerence, 1992; Gruss and McGinnis, 1993; Movilio, 1993; Wilkinson, 1993). Instead of merely describing differences or similarities between segmentally homologous neurons in adjacent ganglia, the results of comparative studies may ultimately also help to reveal the rules and factors involved in determining segment identity. If, for instance, differences are found between segmentally homologous neurons in adjacent neuromeres, one could ultimately ask which genetic and ontogenetic influences caused these differences. Such questions do not only refer to structural differences, but of course also apply to differences in the chemical identity of the neurons.

An excellent system to study segmentally homologous neuronal systems is provided by the abdominal ganglia in Dictyopteran and Orthopteran insects. Since of this chain of ganglia only a few are fused, the segmental arrangement of neurons is comparatively easy to study. The gross morphology of the unfused ganglia in these insects is identical, so one might perhaps expect extensive similarities in the arrangement of the neurons comprising these ganglia. Antisera specifically directed against members of insect peptide families are an execllent tool to test whether this expectation is sound.

An example for a regular arrangement of segmentally homologous peptidergic neurons is provided by neurons immunoreactive to Dip-allatostatin A1 in *Periplaneta americana.* (Fig. 1). This antiserum intensely labels a pair of ascending interneurons with identical morphological features in each of the abdominal neuromeres 1–9. In addition, a neuron with only slightly different morphology is stained in neuromere 10 (Agricola, unpublished results). An antiserum directed against Lem-leucokinin 1 (a kind gift from D. R. Nässel) in *Periplaneta* also reveals a very regular pattern of neurons. In this case, bilateral groups of three morphologically identical efferent neurons are stained in abdominal neuromeres 1–7 (Agricola, unpublished results). The absence of these efferents in neuromeres 8–10 may be due to segmentally homologous peripheral targets missing in these very derived posterior abdominal segments. However, the antiserum against Dip-allatostatin A1 also reveals differences in the cellular composition in the very same ganglia. Thus, in abdominal ganglia 1–4 of *Periplaneta*, this antiserum stains two cell types each located near the dorsal midline. In all these cells allatostatin-immunoreactivity is colocalized with GABA-immunoreactivity (Fig. 1).

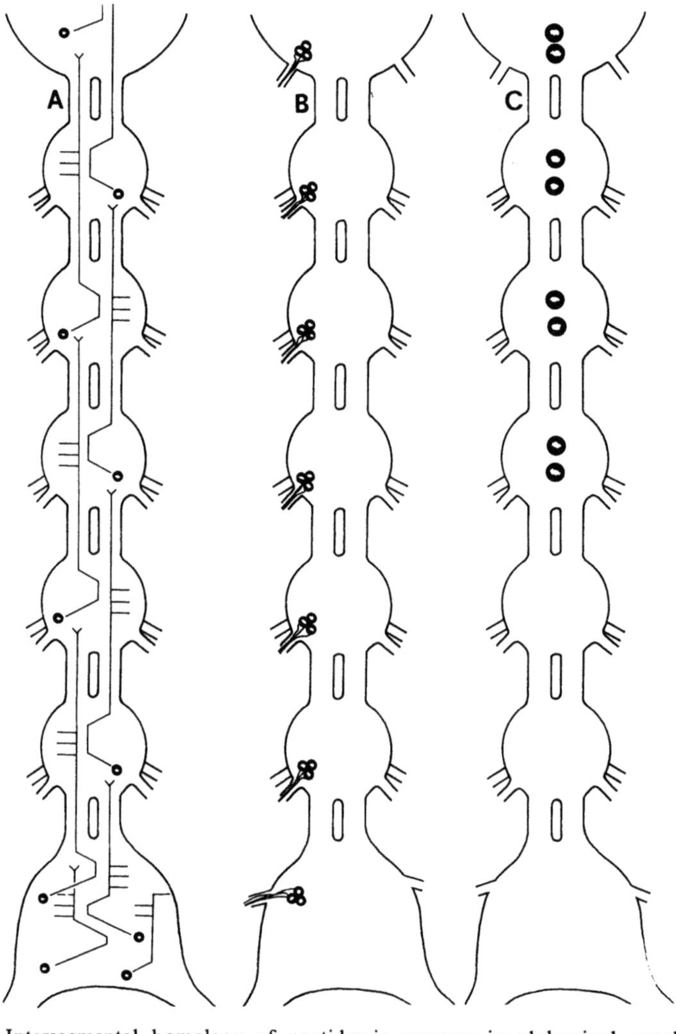

Figure 1. Intersegmental homology of peptidergic neurons in abdominal ganglia of the cockroach *Periplaneta americana.* (A) Schematized representation of serially homologous allatostatin-immunoreactive ascendinag interneurons in abdominal neuromeres 1–9. For reasons of clarity only single members of bilaterally symmetrical cell pairs are shown on alternating sides. (B) Leucokinin-immunoreactive efferents in neuromeres 1–7. (C) Dorsally located cell bodies in neuromeres 1–4 immunoreactive to both allatostatin and GABA.

Interspecific comparisons of peptidergic systems

The examples mentioned above illustrate that segmentally homologous peptidergic neurons exist in segmental ganglia. The last example, however, shows that even in a single species exceptions exist from a regular, metamerically repeated cell pattern, in each ganglion. From this plasticity and variability already observed in the same species one could deduce that the study of homologies becomes more difficult with phylogenetic distance when comparing different species. As we will show in the following examples, this is true in principle, but there are surprising exceptions to this rule. The antiserum against Lem-leucokinin 1, when applied to other insect species, reveals exactly the same pattern of neurons in the abdominal ganglia as in the cockroach. In both Dipteran species and locusts the efferent cells in abdominal neuromeres 1–7 are revealed (Cantera and Nässel, 1992; Dircksen and Nässel, 1993). Neurons stained with an antiserum against CCAP also show similar sets of neurons in different insects species (Dircksen et al., 1991; Breidbach and Dircksen, 1991; Davis et al., 1993). Moreover, this antibody stains segmentally arranged neurons innervating the heart not only in insects, but also in crustacea (Dircksen et al., 1988). However, in crustacea the cell bodies of these neurons are located ipsilateral to their axons, while in insects the cell bodies are located contralateral. This example illustrates the difficulties encountered when we try to homologize neuronal elements in different taxa: In both crustacea and insects, the CCAP immunoreactive neurons are segmentally arranged in thoracic and abdominal ganglia. This means that the first criterion for homology (Remane, 1956; Kutsch and Breidbach, 1994), the criterion of position, is fulfilled. CCAP acts as a cardioactive substance in both groups which means that the criterion of specific quality is fulfilled in terms of a corresponding specific function in both taxa. The soma localization and the arborization patterns of the neurons differs, however, and only additional future comparative studies as well as an investigation of the ontogeny of these neurons can tell whether or not the intermediate forms, as postulated by the criterion of continuity, exist.

Perisulfakinin-immunoreactive neuronal systems in different taxa

CCAP immunoreactivity has so far been studied only in two arthropod groups (three insect and one crustacean species). For the comparative study of homologies of peptidergic systems in invertebrates in general, a much broader basis, including representatives of numerous different taxa, would be desirable. Unfortunately, the scientific literature does not yet provide such a basis. As a first attempt to elucidate similarities and differences in at least one peptidergic system in different species, we used an antiserum against Pea-sulfakinin (Perisulfakinin) in several insect orders, and in spiders, crustacea, centipedes, and annelids. Perisulfakinin, a sulfated myotropic neuropeptide, was isolated from the corpora cardiaca of the American cockroach (Veenstra, 1989a). The peptide induced hindgut contractions in the same species at concentrations as low as 250 pM. It belongs to a family of insects peptides with identical

Table 2. Sequence homologies between identified insect sulfakinins and vertebrate gastro-intestinal-related peptides

Pea. SK	Glu-Gln-Phe-Asp-Asp-Tyr(SO_3)-Gly-His-Met-Arg-Phe	—NH_2
Lom. SK	pGlu-Ser-Asp-Asp-Tyr (SO_3)-Gly-His-Met-Arg-Phe	—NH_2
Lem. SK-I	pGlu-Gln-Phe-Glu-Asp-Tyr (SO_3)-Gly-His-Met-Arg-Phe	—NH_2
Lem. SK-II	pGlu-Ser-Asp-Asp-Tyr(SO_3)-Gly-His-Met-Arg-Phe	—NH_2
Drm. SK-I	Phe-Asp-Asp-Tyr (SO_3)-Gly-His-Met-Arg-Phe	—Gly
Drm. SK-II	Gly-Gly-Asp-Asp-Gn-Phe-Asp-Asp-Tyr(SO_3)-Gly-His-Met-Arg-Phe	—Gly
Neb. SK-I	Phe-Asp-Asp-Tyr($SO_3^?$)-Gly-his-Met-Arg-Phe	—NH_2
Neb. SK-II	X-X-Glu-Glu-Gln-Phe-Asp-Asp-Tyr($SO_3^?$)-Gly-His-Met-Arg-Phe	—NH_2
Cholecystokinin 8	Asp-Tyr (SO_3)-Met-Gly-Trp-Met-Asp-Phe	—NH_2
Human gastrin II[10]	-Glu-Glu-Glu-Ala-Tyr(SO_3)-Gly-Trp-Met-Asp-Phe	—NH_2

C-terminal sequences (sulfakinins) that show structural similarities with the vertebrate cholecystokinin/gastrin family (see Tab. 2), and also to the FMRF-amide-like peptides. At the dilutions used for immunocytochemistry, our antiserum does not detect any FMRFamide related epitopes as was shown by competitive and non-competitive ELISA. In the following, we will describe the results obtained with the antiserum, starting with the species from the peptide originally isolated, *Periplaneta americana*.

In the cockroach brain, there are two main groups of perisulfakinin-immunoreactive neurons (Figs 2a; 3a,b,e; 4). The first group is associated with the neurosecretory system of the *pars intercerebralis* (PI). It consists of 11 to 12 cells on each side. The second group comprises four pairs of interneurons in different parts of the brain – cells A1, A2, B, and C – and is of particular interest. Cells A1 and A2 are located in the posterior part of the PI, and their axons descend into the ventral protocerebrum where they cross to the contralateral side and run all the way into the terminal ganglion of the ventral nerve cord (protocerebral descending sulfakinin neurons, PDS). Cell B lies in the posterior region of the protocerebrum, cell C is located in the posterior deutocerebrum. Both are "centrifugal projection interneurons" with axonal arborizations in the contralateral half of the brain (Figs 2a, insects; 3a,b; 4). In the brain of *Locusta migratoria* there are three cell pairs similar to the cockroach types A and B (Figs 2b; 3c, d). There is only one contralaterally descending interneuron (A), and two protocerebral centrifugal projection interneurons (B1, B2). Unlike the situation in the cockroach, there are no perisulfakinin-immunoreactive neurosecretory cells in the PI, and the prominent single dorsal, bilaterally projecting cell of the cockroach subesophageal ganglion is also missing in the locust. Other striking differences to the cockroach are (1) the numerous neurons with extensive arborizations in the optic lobes of the locust, and (2) up to 26 labeled somata in the frontal ganglion which are entirely absent in the cockroach. On the level of the ventral nerve cord both species show much more similarity: The long descending fibers of the PDS

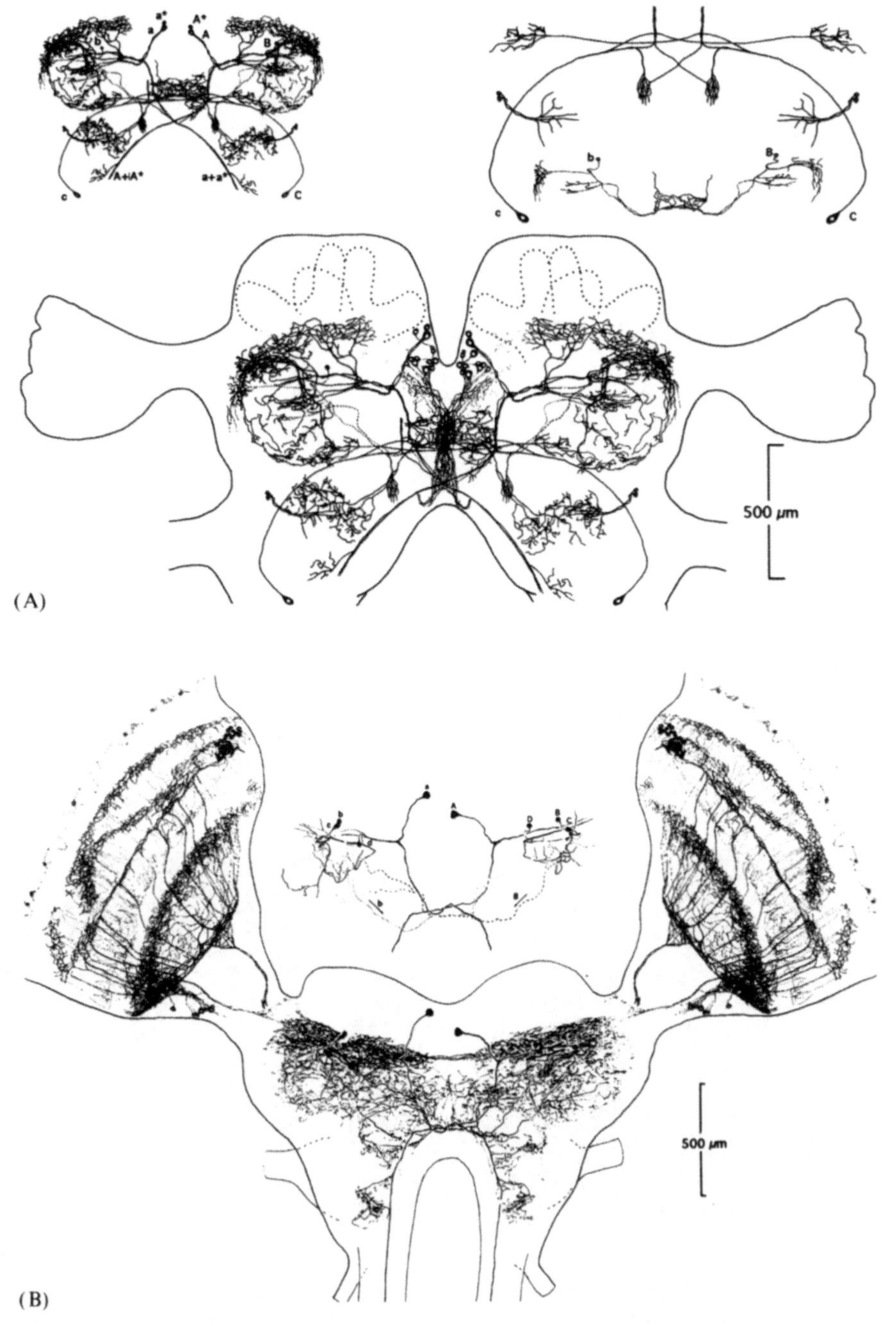

Figure 2. Perisulfakinin-immunoreactivity in the brain of the cockroach *Periplaneta americana* (A) and the locust *Locusta migratoria* (B) (reconstructions of serial vibratome sections). The insets show the course of axons of cells in the optic lobes of *Periplaneta*. Also, there are two descending interneurons in the cockroach (A, A*), while only one is present in the locust (A).

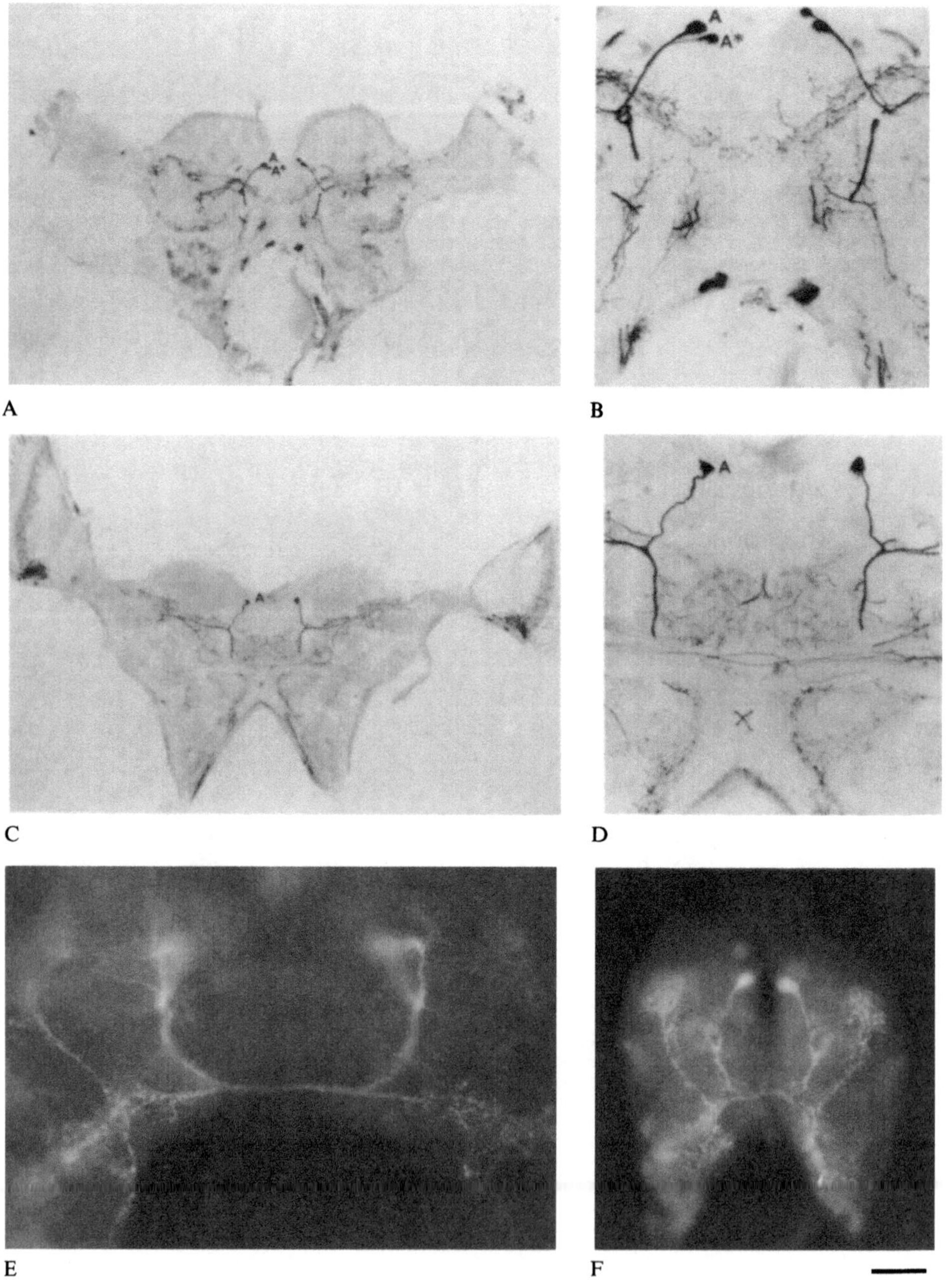

Figure 3. Perisulfakinin-immunoreactivity in the brain of insects (photomicrographs of frontal vibratome sections in A–D, PAP technique; anterior view of wholemount immunofluorescence in E and F). The protocerebral descending sulfakinin (PDS) neurons that send axons as far as the terminal ganglion (cf., Fig. 4) are shown for the cockroach (two pairs of cells, A, A*, in A and B) and the locust (one pair of cells, A in C and D). In F the same cells are shown in the larval brain of the flour beetle, *Tenebrio*. E shows the neurosecretory cells of the PI in the cockroach brain. Scales: 320 μm (A, C), 40 μm (B, D, E, F,).

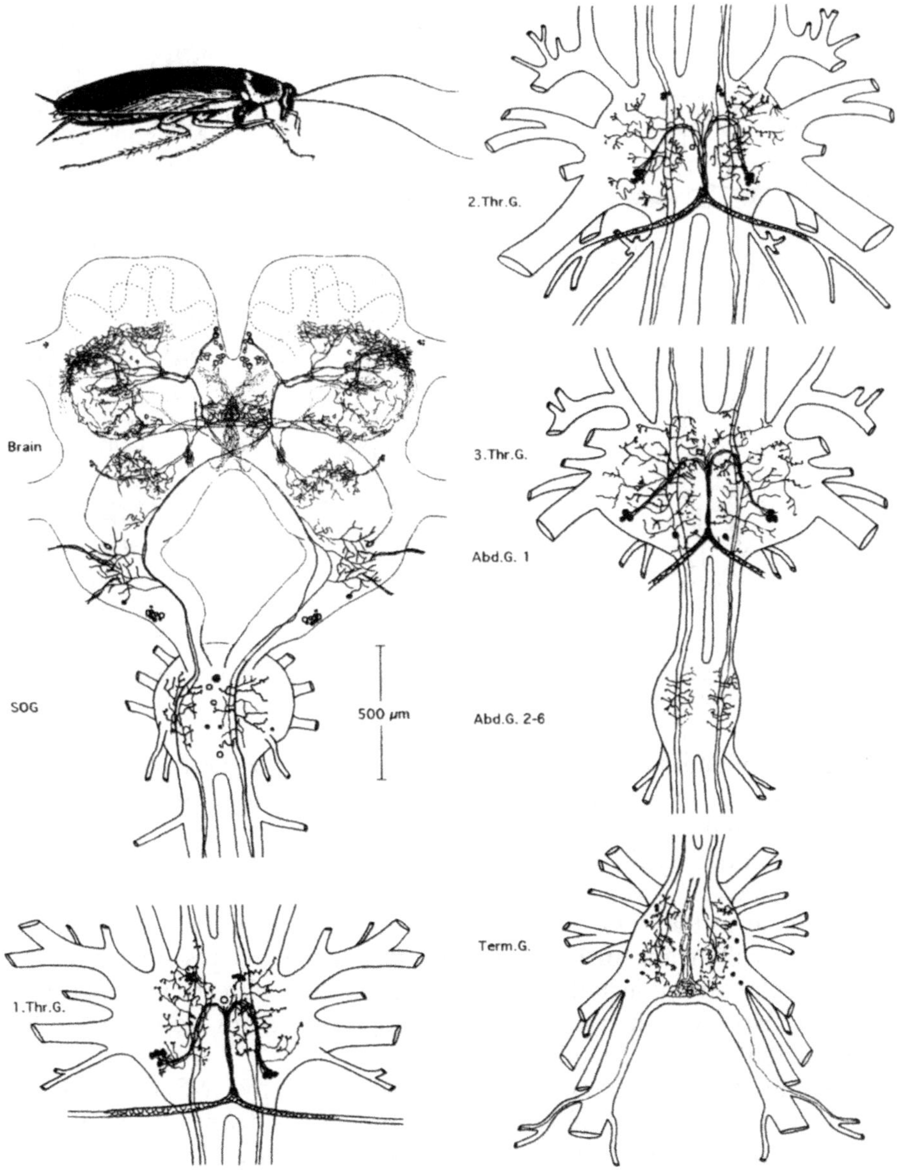

Figure 4. Dorsal view of perisulfakinin-immunoreactive neurons in the CNS of the cockroach *Periplaneta americana* (SOG subesophageal ganglion; 1.–3. Thr. G. pro-, meso-, and metathoracic ganglia; Abd. G. 1–6 abdominal ganglia 1–6; Term. G. terminal ganglion).

interneurons arborize in the dorsal neuropil of each ganglion. Moreover, the thoracic ganglia contain two lateral groups of neurons which send their axons into the neurohaemal regions of the transverse nerves. There is no evidence for perisulfakinin-immunoreactive somata in the

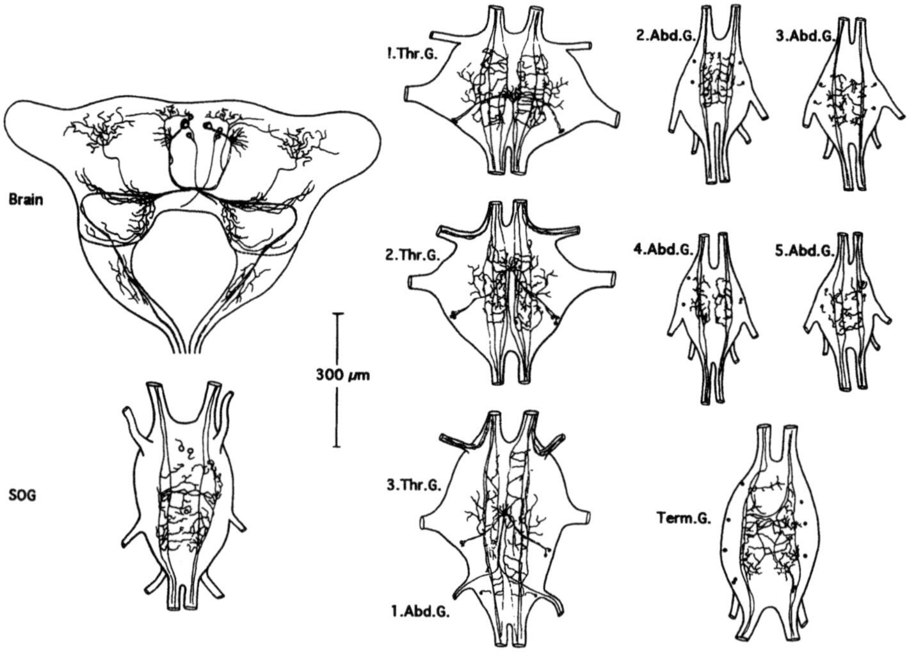

Figure 5. Perisulfakinin-immunoreactive neurons in the CNS of the dark beetle *Blaps gibba* (see text for details; abbreviations as in Fig. 4).

unfused abdominal ganglia of the locust, but in the cockroach we find such neurons in the first abdominal ganglion (fused with the third thoracic ganglion) and in the terminal ganglion.

The situation in the beetles *Blaps gibba* (Fig. 5) and *Tenebrio molitor* (Fig. 3f; see also Urbach et al., 1993) resembles the cockroach arrangement as far as the PDS neurons are concerned. Also in the ventral nerve cord, the cell arrangement in the three thoracic ganglia is similar to the cockroach pattern, including the close association with the neurohaemal system. However, in all abdominal ganglia of *Blaps* additional cells appear that are not found in the cockroach.

Despite the considerable fusion of ganglia in the fly CNS, it is possible to discern most of the cell types seen in the insects discussed so far. The brain of *Musca domestica* contains immunoreactive neurons in the PI and in the posterior protocerbrum. Each of the thoracic neuromeres contains a bilateral pair of labeled cells that is clearly associated with the thoracic neurohaemal system. As shown in *Calliphora vomitoria.* (Duve et al., 1994) two pairs of PDS neurons are present and their descending axons can be followed all the way through the ventral nerve cord.

The brain of the honeybee, *Apis mellifica*, contains three pairs of perisulfakinin-immunoreactive neurons in its posterior part and a few cells in the PI. Due to the small fiber diameters, we were unable to trace them in wholemount preparations. Unlike the situation found in the insects discussed so far, in the ventral nerve cord of the honeybee the neurohemal structures are serially organized: with the exception of the terminal ganglion, all abdominal ganglia show the association between perisulfakinin-immunoreactive neurons and the neurohaemal system.

Further down the evolutionary ladder of insects, in the zygentome *Lepisma*, we again find two pairs of PDS neurons and bilateral clusters of labeled cells in the *pars lateralis* instead of the *pars intercerebralis* (Fig. 6). Remarkably, as in the honeybee, in the ventral nerve cord we find immunoreactive processes in the transverse nerve neurohaemal system of each thoracic and abdominal ganglion. These ganglia also contain somata in the posterior lateral region that might be the source of these processes. The CNS of *Lepisma* also contains weakly labeled cell bodies in the midline of all ganglia except for abdominal ganglia 2 and 3, and the terminal ganglion (this provides an additional example for cell patterns differing in different segments of the same species; see above).

When looking at a representative for the second group of the Tracheata, the centipede *Lithobius forficatus*, the arrangement of perisulfakinin immuno-reactive neurons is quite similar to that observed in insects (Fig. 7). As in insects, there are several pairs of cell bodies in the posterior part of the protocerebrum and axons descending through the entire nerve cord. In our preparations we could not determine from which cell bodies these axons emerge. For this reason, and because of the finding that the axons do not branch at all in the ventral nerve cord (we only observed varicose swellings of the axons in each ganglion), it is uncertain whether any of these protocerebral neurons are the chilopod equivalent of the insect PDS neurons. Similar *Lepisma*, however, cell groups in the pars lateralis exist that innervate the cerebral neurosecretory organs. The third through seventh ventral ganglions (thoracic region?) each contains a conspicuous, intensively labeled pair of efferent cells. These cells are peculiar in that they do not exhibit any arborizations within the CNS.

Representatives of the second major arthropod group, the arachnids, also possess perisulfakinin-immunoreactive neurons in their CNS. In spiders – such as the tropical hunting spider *Cupiennius salei* – the CNS is composed of the brain and a fused subesophageal ganglion complex that includes the neuromeres of all appendages and the abdomen. The brain contains three groups of perisulfakinin-immunoreactive neurons, each consisting of only three cells. Fibers originating from the cells in the ventral group (lying below the "central body") extend ipsilaterally towards the gut region and descend into the subesophageal neuromeres (Agricola et al., in preparation). The neuromeres of the subesophageal ganglion complex contain many more immunoreactive neurons than the equivalent CNS regions of insects and centipedes.

In a crustacean species (the isopod *Porcellio scaber* has been studied as a representative) also two pairs of descending brain neurons exist. As

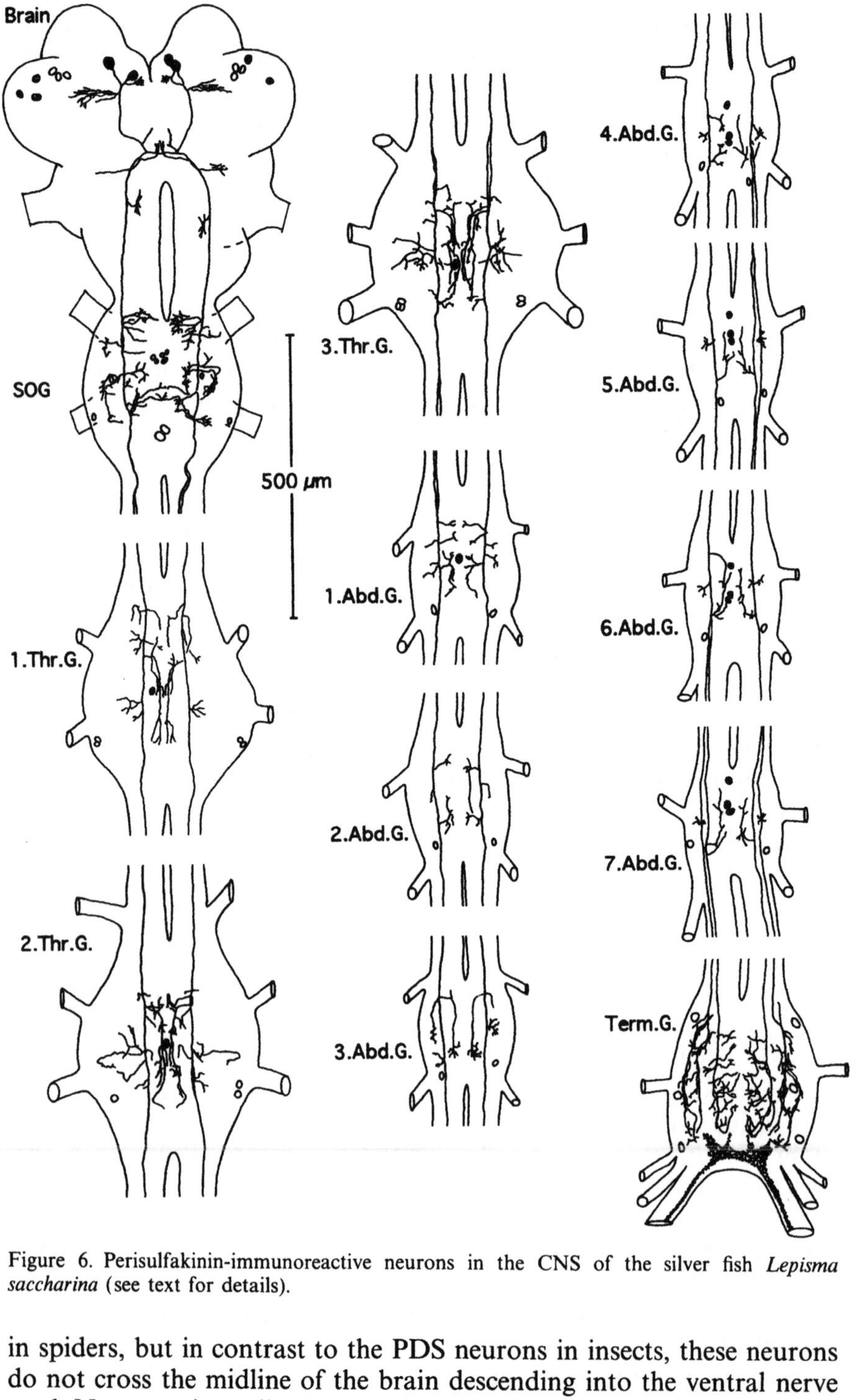

Figure 6. Perisulfakinin-immunoreactive neurons in the CNS of the silver fish *Lepisma saccharina* (see text for details).

in spiders, but in contrast to the PDS neurons in insects, these neurons do not cross the midline of the brain descending into the ventral nerve cord. Non-crossing cells also exist in the brain of the lobster. These send

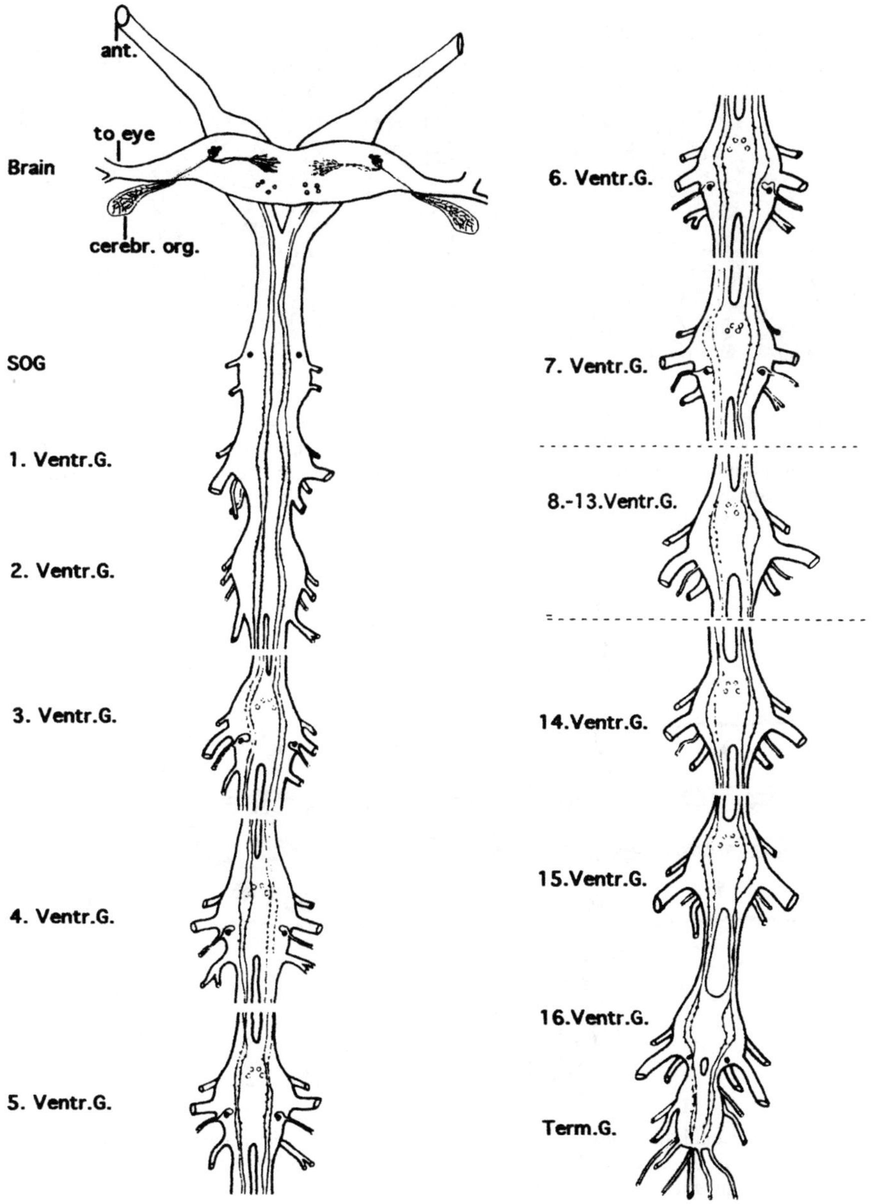

Figure 7. Perisulfakinin-immunoreactive neurons in the CNS of the centripede *Lithobius forficatus* (see text for details; abbreviations as in Fig. 4).

axons at least as far as the subesophageal ganglion (H. Schneider, personal communication).

Moving to a more homogeneous segmented representative of the Articulata, we have also studied the annelid *Aporrectodea caliginosa*. The segmental ganglia of this earthworm contain the highest number of

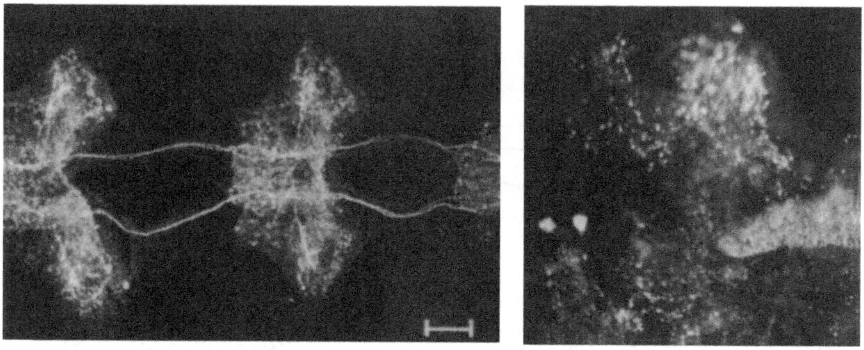

A B

Figure 8. Photomicrographs (dorsal view) of perisulfakinin-immunoreactive cells of the isopod *Porcellio scaber* (A,B). Prominent descending axons are intensely labeled in the CNS. In wholemount preparations they are most clearly visible in the connectives between the thoracic ganglia shown in A. They originate from cells in the brain. The somata of these cells are shown in a brain vibratome section in B. Scale: 40 µm.

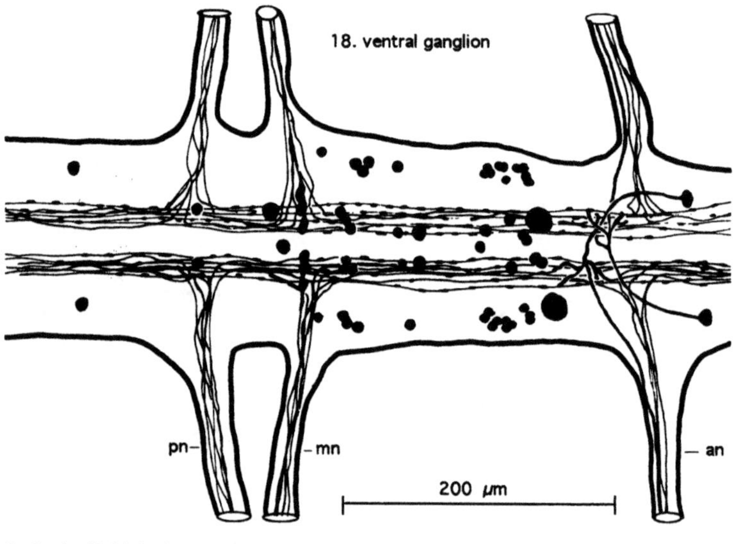

Figure 9. Perisulfakinin-immunoreactive cell bodies and fibers in a segmental ganglion of the earthworm *Aporrectodea caliginosa* (All other ganglia of the posterior ventral nerve cord look like the 18th ganglion shown here).

perisulfakinin immunoreactive cells observed in any of the animals studied here (up to 60 cells per ganglion). The ganglia are connected by numerous immunoreactive fibers running in a discrete bundle. Unfortunately, in the brain only very few and weakly labelled cell bodies were detected. It is not possible at present to use these findings for a comparison with the situations found in the arthropod species described above.

In summary, the comparative study using the perisulfakinin antibody yields the expected result already mentioned above that interspecies comparisons become increasingly difficult with phylogenetic distance. For example within the insects, the pattern of immunoreactive cells projecting into the transverse nerve neurohaemal organs is reasonably consistent. Because of differences, sometimes profound, in the fine morphology of these cells it is very hard to prove that these cells are truly homologous. However, there is the surprising result that in the CNS of all arthropods studied here a group of brain interneurons exists that sends descending axons through the entire nerve cord.

Peptidergic innervation of complex organ systems

So far, we have described and compared invertebrate peptidergic system from the perspective of a single peptide or peptide family. For interspecific comparisons another informative approach is to look at a particular organ system that is innervated by numerous peptidergic neuronal systems, each containing one or more different neuromediators. For this purpose, we chose the gut. In vertebrates, the range of peptides and other potential neurotransmitters in the enteric nervous system is comparable to that in the CNS (Furness et al., 1992). Comparative information about the gut in invertebrates, however, is very scarce. The comparative study of peptidergic innervation of the gut in invertebrates, however, is a fruitful enterprise, since within the Eumetazoa the gut probably represents a phylogenetically very ancient structure and the study of the distribution of "brain-peptides" and "peptide families" in the gut might turn out to be very revealing in terms of homologies and phylogenetic relationships. Within the invertebrates the gut has been probably most intensively studied in insects, at least pharmacologically (Nishiitsutsuji-Uwo, 1981; Osborne et al., 1990; Holman et al. 1991; Zitnan et al., 1993). From these studies it is known that a wide range of neuropeptides influences the gut. As will be shown below, the investigation of the innervation of the insect gut on the cellular level using immunocytochemical methods yields some surprises.

Proctolin, the first insect peptide sequenced, received its name from its association with the cockroach hindgut, the proctodeum. The cellular source of proctolin in the cockroach hindgut has been determined as two groups of neurons located in the terminal ganglion, one of them located in the 8th neuromere (Eckert et al., 1981; Agricola et al., 1993). Quite surprisingly, a recent study of hindgut innervation in *Locusta migratoria* showed that the terminal ganglion contains only one group of proctolin-immunoreactive neurons, also situated in the eighth neuromere. The axons of these neurons bypass the hindgut and establish a dense network of terminal ramifications on the midgut instead (Staufer et al., 1993). Another difference between both insects is that this group of six cells in the cockroach also stains for Dip-allatostatin A1 (Agricola and Weiß,

1993). The cells in the locust do not exhibit this colocalization of the two peptides. In the cockroach, allatostatin-immunoreactivity is colocalized with proctolin-immunoreactivity in the same network on the hindgut. This network is established by the neurons in the terminal ganglion. An allatostatin network (not colocalized with proctolin and of unknown cellular origin) also occurs on the cockroach midgut. In contrast, in locusts an allatostatin-immunoreactive network of terminals is associated with the hindgut exclusively, and the cells establishing this network are located as pairs in abdominal ganglia 4–6. These differences are intriguing and cannot be interpreted properly before getting more information.

In addition to these findings the midgut of both insects is supplied with a large number of endocrine cells of various neurochemical identity: Dip-allatostatin A1, Lom-tachykinin II, Pea-sulfakinin and CCAP (Figs 10; 11). Perisulfakinin-immunoreactive peripheral endocrine cells also occur on the cockroach foregut, but in much smaller numbers than on the midgut. Such cells are not present in the foregut of locusts. In contrast, as mentioned above, locusts possess perisulfakinin-immunoreactive cells in the frontal ganglion which innervate the foregut. These cells do not exist in cockroaches. Thus, perisulfakinin in the foregut appears to derive from different cellular sources in both insects. However, since in insects neurons of both foregut and stomatogastric nervous system derive from neurogenic placodes in the embryonic stomodaeum (Copenhaver, 1990, 1991), it would be of interest to compare the ontogenesis of the perisulfakinin-immunoreactive cells in the two species and possibly find out whether the perisulfakinin-immunoreactive cells have a common origin in both species.

Conclusions

The comparative study of peptidergic signaling pathways in the nervous system of arthropods represents a task of gargantuan dimensions. In view of the existing number of diverse invertebrate taxa, the diversity of known peptide families (and the ones still awaiting discovery), and the diversity of the members within these families, the examples presented above can only provide a first survey. One should also bear in mind that there are methodological problems. When immunocytochemistry fails to reveal a particular type of cell in a particular segment or a particular species, this does not neccessarily mean that cells corresponding to homologues in neighboring segments or related species do not exist. Homologous cells might have shown or might show up in other ontogenetic phases. They might also have (or have switched to) a different neurochemical phenotype (e.g., express other neuropeptide genes). Such a switch in neuropeptide gene expression is illustrated by abdominal

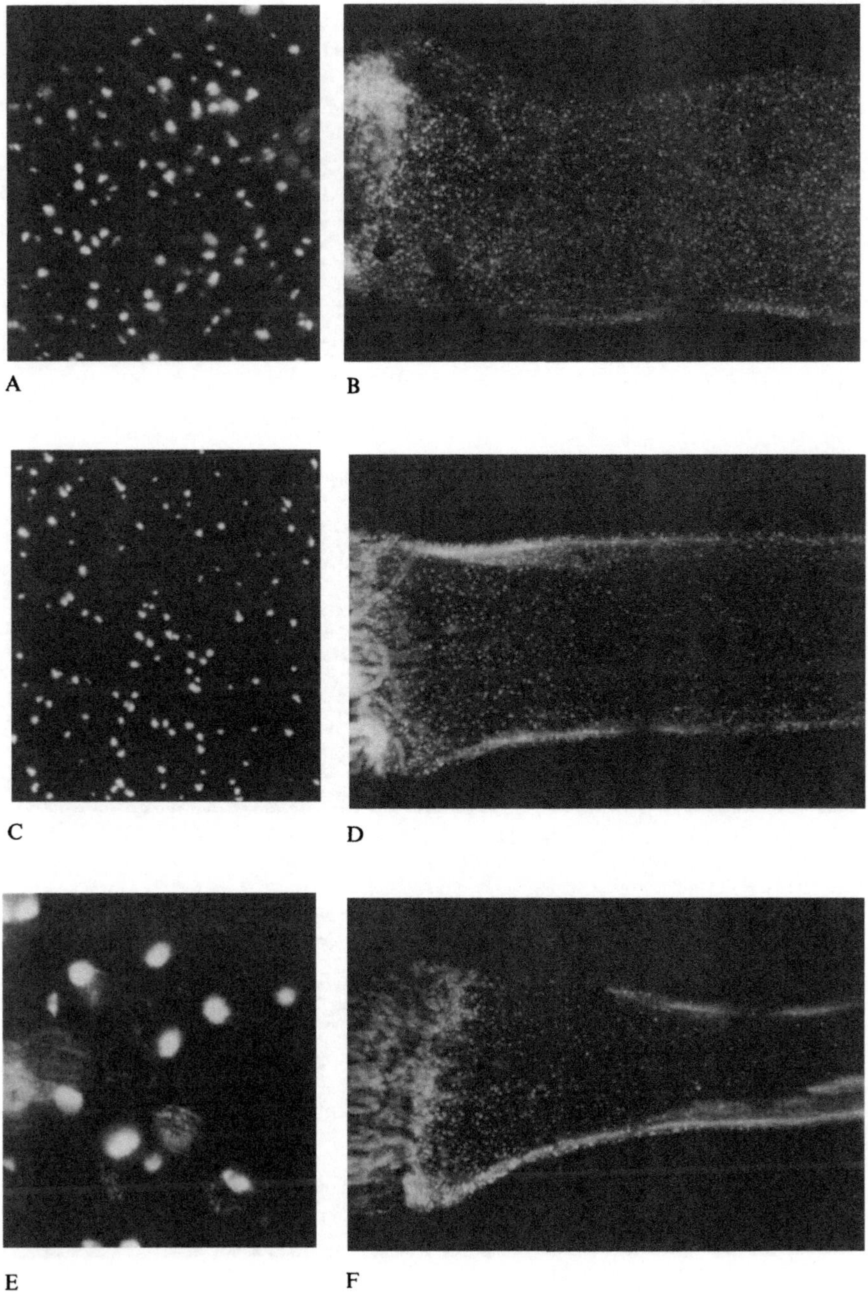

Figure 10. Immunoreactive cells in the locust midgut. Peripheral endocrine cells are revealed by antisera against Lom-tachykinin II (top pair of photomicrographs), Pea-sulfakinin (middle pair), and Dip-allatostatin A1 (bottom pair), please note uneven distribution of immunoreactive cells. The posterior region of the midgut is shown on the right (Malpighian tubules are visible near the left margin). All cell types are shown at higher magnification on the left. Scale: 320 μm (B, D, F), 40 μm (A, C), 20 μm (E).

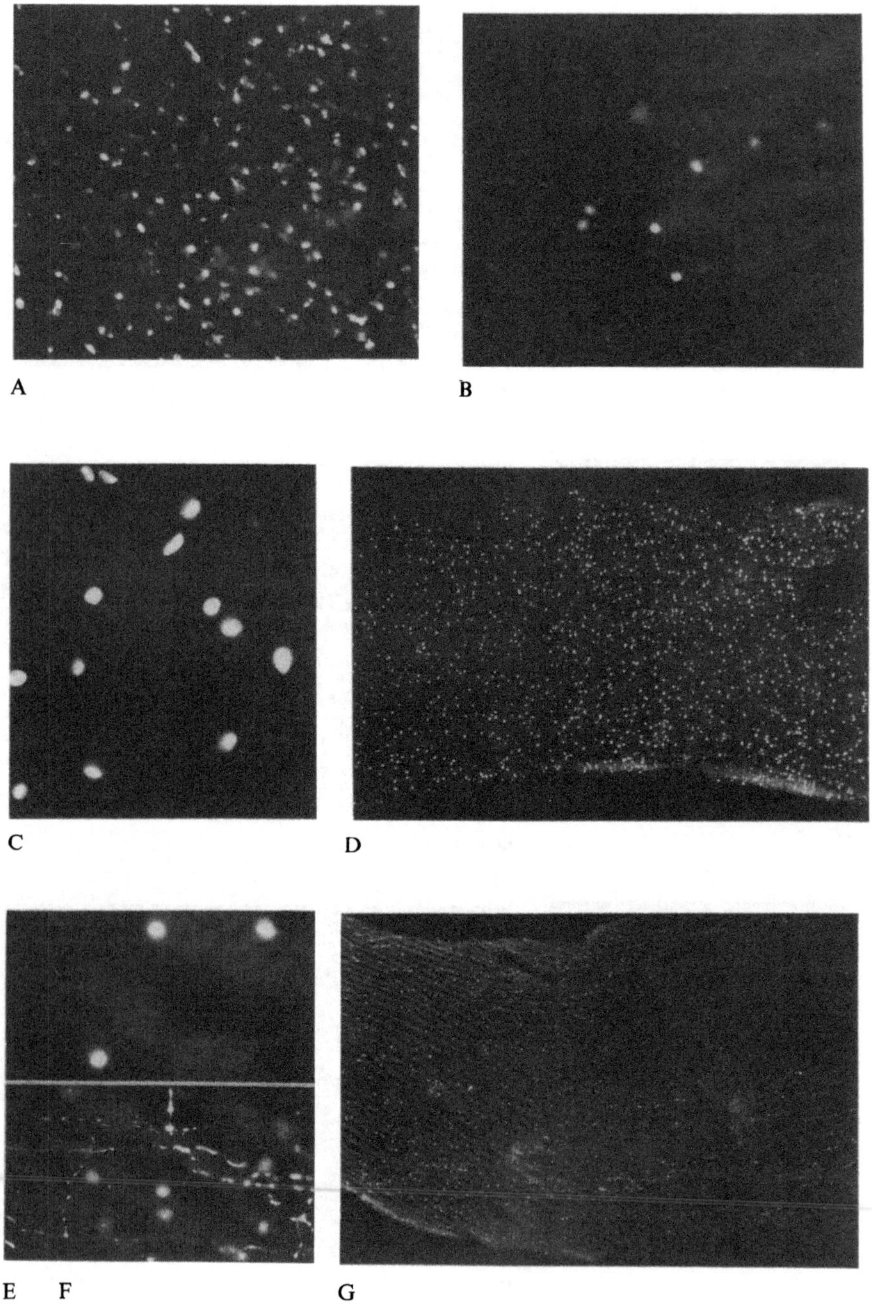

Figure 11. Immunoreactive cells in the cockroach midgut. Peripheral endocrine cells are revealed by antisera against Pea-sulfakinin (A), Proctolin (B), Loc-Tachykinin (middle pair, C and D), and Dip-allatostatin A1 (bottom photomicrographs E, F and G). Note uneven distribution of immunoreactive cells in G and the net work of immunoreactive fibers in F and G. Malpighian tubules (not shown here) are located near the left margin (D,G). Scale: 320 μm (D, G), 40 μm (A, B, F), 20 μm (C, E).

neurosecretory cells of the tobacco hawk moth *Manduca sexta*. These cells switch from the predominant expression of a cardioactive peptide in the caterpillar to the predominant expression of the tanning factor bursicon in pupae and adults (Tublitz, 1993). A second example is provided by the cerebral neuroendocrine light green cells (LGCs) of the freshwater snail, *Lymnaea stagnalis*. These cells express the molluscan insulin-related peptides (MIPs), deriving from six different genes in a stimulus-dependent manner (Geraerts et al., 1992).

As these examples show, part of the diversity of neuropeptide signaling systems is already generated by the flexibility of the organization and expression of the genes coding for the signal molecules. The reader should bear in mind that a similar, perhaps even greater amount of variability and/or plasticity is to be expected for the genes coding for the receptor proteins expressed by cells situated at the receiving end of peptidergic signaling pathways. Such diversity at the receiving end is the most likely explanation for the differences in function observed for similar or even identical peptides. Thus, the callatostatins, Dipteran homologues of the cockroach allatostatins, have no potency to inhibit juvenile hormone synthesis in flies. An extreme example for functional diversity even in closely related species is provided by corazonin, a peptide that was originally isolated in the cockroach *Periplaneta* and named for its cardioactive effects (Veenstra, 1989b). When tested in eight other cockroach species (Predel et al., 1994) this peptide showed absolutely no cardioexcitatory effects.

It is surprising that, despite all these difficulties and the ones mentioned earlier, cell populations that are good candidates for truly homologous systems can be found when comparing different species. One example is provided by the system of Pea-sulfakinin-immunoreactive neurons located in the brain of various arthropod taxa and descending through the entire ventral nerve cord (PDS neurons). In the case of CCAP, similarities are not only obvious when structural features are compared between taxa, but even functional similarities exist. Future studies have to show whether these structures really match all criteria for homology; immunocytochemistry at least provides a useful tool to answer the question of where to look. It also provides hints for the existence of peptides structurally related to known peptide families in species from which no such peptides have yet been isolated. Further characterization of more neuropeptide and neuropeptide precursor sequences in representative invertebrate groups (perhaps also with the aid of antibodies) should greatly enhance the study of homologies. With our increasing knowledge about cell fate allocation during ontogeny, there is hope that we will understand the factors that cause individual neurons to acquire their specific phenotype. With this knowledge, we might finally be able to solve the interesting question of whether the neurochemical identity of neurons is at least partially influenced by phylogenetically conserved mechanisms.

Acknowledgments
The authors are indebted for excellent technical assistance to Mrs. Angelika Schmidt, Mrs. Rosemarie Meißner and Mr. Bernd Mäusezahl. Supported by research grants from Bundesministerium für Forschung und Technologie (0316919 A).

References

Agricola, H. and Weiß, T (1993) Immunocytochemical characterisation of hindgut-innervating neurons of the cockroach *Periplaneta americana* (L.). *In:* N. Elsner and M. Heisenberg (eds): *Gen-Brain-Behaviour*. Thieme Verlag, Stuttgart-New York, p. 604.

Breidbach, O. and Dircksen, H. (1991) Crustacean cardioactive peptide-immunoreactive neurons in the ventral nerve cord and the brain of the meal beetle *Tenebrio molitor* during postembryonic development. *Cell Tissue Res.* 263: 129–144.

Brown, B.E. and Starratt, A.N. (1975) Isolation of proctolin, a myotropic peptide, from *Periplaneta americana* (Blatt, Blattidae). *J. Insect Physiol.* 21: 1879–1881.

Brownlee, D.J.A., Fairweather, I., Johnston, C., Smart, D., Shaw, C. and Halton, D.W. (1993) Immunocytochemical demonstration of neuropeptides in the central nervous system of the roundworm, *Ascaris suum* (Nematoda, Ascaroidea). *Parasitology* 106: 306–316.

Cantera, R. and Nässel, D.R. (1992) Segmental peptidergic innervation of abdominal targets in larval and adult dipteran insects revealed with an antiserum against leucokinin I. *Cell Tissue Res.* 269: 459–471.

Cheung, C.C., Loi, P.K, Sylwester, A.W., Lee, T.D. and Tublitz, N.J. (1992) Primary structure of a cardioactive neuropeptide from the tobacco hawkmoth, *Manduca sexta. FEBS Lett.* 313: 165–168.

Copenhaver, P.F and Taghert, P.H (1990) Neurogenesis in the insect enteric nervous system: Generation of premigratory neurons from an epithelial placode. *Development* 109: 17–28.

Copenhaver, P.F and Taghert, P.H (1991) Origins of the insect enteric nervous system: Differentiation of the enteric ganglia from a neurogenic epithelium. *Development* 113: 1115–1132.

Damer, D., Schmutzler, C., Diekhoff, D. and Grimmelikhuizen, C.J.P. (1991) Primary structure of the precusor for the sea anemone neuropeptide Antho-RFamide (<Glu-Glu-Arg-Phe-NH2). *Proc. Natl. Acad. Sci. USA* 88: 2555–2559.

Davis, M.-T.B., Vakharia, V.N., Henry, J., Kempe, T.G. and Raina, A.K. (1992) Molecular cloning of the pheromone biosynthesis-activating neuropeptide in *Helicoverpa zea. Proc. Natl. Acad. Sci. USA* 89: 142–146.

Davis, N.T., Homberg, U., Dircksen, H., R.B. and Hilderbrand, J.G. (1993) Crustacean cardioactive peptide-immunoreactive neurons in the hawkmoth *Manduca sexta* and changes in their immunoreactivity during postembyronic development. *J. Comp. Neurol.* 338: 612–627.

Digan, M.E., Roberts, D.N., Enderlin, F.E., Woodworth, A.R. and Kramer, S.J. (1992) Characterization of the precursor for *Manduca sexta* diuretic hormone Mas-DH. *Proc. Natl. Acad. Sci. USA* 89: 11704–11078.

Dircksen, H. and Keller, R. (1988) Immunocytochemical localization of CCAP a novel crustacean cardioactive peptide, in the nervous system of the shore crab, *Carcinus maenas* L. *Cell Tissue Res.* 254: 347–360.

Dircksen, H., Müller, A. and Keller, R. (1991) Crustacean cardioactive peptide in the nervous system of the locust, *Locusta migratoria*: an immunocytochemical study on the ventral cord and peripheral innervation. *Cell Tissue Res.* 263: 439–457.

Dircksen, H. and Nässel, R. (1993) Immunoreactive leucokinin-like peptides and crustacean cardioactive peptide show similar, but distinct neuronal distributions in neurohaemal systems of the locust abdomen. *In:* N. Elsner and M. Heisenberg (eds): *Gen-Brain-Behavior*. Thieme Verlag, Stuttgart-New York, p. 19.

Donly, B.C., Ding, Q, Tobe, S.S. and Bendena, W.G. (1993) Molecular cloning of the gene for the allatostatin family of neuropeptides from the cockroach *Diploptera punctata*. Proc. *Natl. Acad. USA* 90: 8807–8811.

Duve. H., Rehfeld, J.F., East, P. and Thorpe, A. (1994) Localization of sulfakinin neuronal pathways in the blowfly *Calliphora vomitoria. Cell Tissue Res.* 275, 1: 177–186.

Ebberink, R.H.M., Smit, A.B. and Van Minnen, J. (1989) The insulin family: Evolution of structures and function in vertebrates and invertebrates. *Reference Biol. Bull.* 177: 176–182.

Eckert, M., Agricola, H. and Penzlin, H. (1981) Immunocytochemical identification of proctolin-like immunoreactivity in the terminal ganglion and hingut of the cockroach *Periplaneta americana. Cell Tissue Res.* 217: 633–645.

Fujita, T. and Kobayashi, S. (1979) Current views on the paraneurone concept. *Trends Neurosci.* 2, 2: 27–30.

Furness, J.B., Bornstein, J.C., Murphy, R., Pompolo, S. (1992) Roles of peptides in transmission in the enteric nervous system. *Trends Neurosci.* 15: 66–71.

Furuya, K. Liao, S. Reynolds, S.E., Ota, R.B., Hackett, M. and Schooley, D.A. (1993) Isolation and identification of a cardioactive peptide from *Tenebrio molitor* and *Spodoptera eridania. Biol. Chemi. Hoppe-Seyler* 374: 12: 1065–1074.

Gäde, G. (1990) The adipokinetic hormone/red pigment-concentrating hormone peptide family: Structures, interrelationships and functions. *J. Insect. Physiol.* 36: 1–12.

Geraerts, W.P.M., Smit, A.B., Li, K.W. and Hordijk, P.L. (1992) The light green cells of Lymnaea: A neuroendocrine model system for stimulus-induced expression of multiple peptide genes in a single cell type. *Experientia* 48: 464–473.

Goldsworthy, G. and Mordue, W. (1989) Adipokinetic hormones: functions and structures. *Reference Biol. Bull.* 177: 218–224.

Grimmelikhuijzen, C.J.P., Carstensen, K., Darmer D., Moosler, A., Nothacker, H.-P., Reinscheid, R.K., Schmutzler, C., Vollert, H., McFarlane, I. and Rinehart, K.L. (1992) Coelenterate neuropeptides: Structure, action and biosynthesis. *Am. Zool.* 32: 1–12.

Holman, G.M., Nachman, R.J. and Wright, M.S. (1990) Insect neuropeptides. Review. *Ann. Rev. Entomol.* 35: 201–217.

Holman, G.M., Nachman, R.J., Schoofs, L., Hayes, T.K., Wright, M.S. and DeLoof, A. (1991) The *Leucophaea maderae* hindgut preparation: A rapid and sensitive bioassay tool for the isolation of insect myotropins of other insect species. *Insect Biochem.* 21: 107–112.

Horodyski, F.M., Riddiford, L.M. and Truman, J.W. (1989) Isolation and expression of a eclosion hormone gene from the tobacco hornworm moth *Manduca sexta. Proc. Natl. Acad. Sci. USA* 86: 8123–8127.

Iwami, M, Kawakami, A., Ishizaki, H., Takahashi, S.Y., Adachi, T., Suzuki, Y., Nagasawa, H. and Suzuki, A. (1989) Cloning of a gene encoding bombyxin, an insulin-like brain secretory peptide of the silkmoth *Bombyx mori* with prothoracicotropic activity. *Dev. Growth Differ.* 31: 31–37.

Kamito, T., Tanaka, H., Sato, B., Nagasawa, H. and Suzuki, A. (1992) Nucleotide sequence of cDNA for the eclosion hormone of the silkworm, *Bombyx mori*, and the expression in a brain. *Biochem. Biophys. Res. Commun.* 182: 514–519.

Kawakami, A., Kataoka. H., Oka, T., Mizoguchi, A., Kimura-Kawakami, M., Adachi, T., Iwami, M., Nagasawa, H., Suzuki, A. and Ishizaki, H. (1990) Molecular cloning of the *bombyx mori* prothoracicotropic hormone. *Science* 247: 1333–1335.

Keller, R. (1992) Crustacean neuropeptides: Structures, functions and comparative aspects. *Experientia* 48: 439–448.

Kimura-Kawakami, M., Iwami, M., Kawakami, A., Nagasawa, H., Suzuki, A. and Ishizaki, H. (1992) Structure and expression of bombyzin-related peptide genes of the moth *Samia cynthia ricini. Gen. Comp. Endocrinol.* 86: 257–268.

Klein, J.M., Mangerich, S., Dekleijn, D.P.V., Keller, R. and Weidemann, W.M. (1993) Molecular cloning of crustacean putative molt-inhibiting hormone (MIH) precursor. *FEBS Letters* 334: 1: 139–142.

Kutsch, W. and Breidbach, O. (1994) Homologous structures in the nervous systems of arthropoda. *Adv. Insect Physiol.* 24: 1–113.

Lagueux, M., Lwoff, Meister, M., Goltzene, F. and Hoffman, J.A. (1990) cDNAs from neurosecretory cells of brains of *Locusta migratoria* (Insecta, Orthoptera) encoding a novel member of the superfamily of insulins. *Eur. J. Biochem.* 187: 249–254.

Lagueux, M., Kromer, E. and Girardie, J. (1992) Cloning of a *Locusta* cDNA encoding neuroparsin A. *Insect Biochem.* 22: 511–516.

Lehman, H.K., Murgiuc, C.M., Miller, T.A., Lee, T.D. and Hildebrand, J.G. (1993) Crustacean cardioactive peptide in the sphinx moth, *Manduca sexta. Peptides* 14: 4: 735–74.

Linck, B., Klein, J.M., Mangerich, S., Keller, R. and Weidemann, W.M. (1993) Molecular cloning of crustacean red pigment concentrating hormone precursor. *Bioch. Biophys. Res. Comm.* 195, 2: 807–813.
Mahon, A.C., Lloyd, P.E., Weiss, K.R., Kupfermann, I. and Scheller, R.H. (1985) The small cardioactive peptides A and B of *Aplysia* are derived from a common precursor molecule. *Proc. Natl. Acad. Sci. USA* 82: 3925–3929.
Maule, A., Shaw, Ch., Halton, D. and Thim, L. (1993) GNFFRFamide: A novel FMRFamide-immunoreactive peptide isolated from the sheep tapeworm, *Moniexia expansa. Biochem. Biophys. Res. Comm.* 193:3: 1054–1060.
Miller, M.W., Beushausen, S., Cropper, E.C., Eisinger, K., Stamm, S., Vilim, F.S., Vitek, A., Zajc, A., Kupfermann, I., Brosius, J. and Weiss, K.R. (1993a) The buccalin-related neuropeptides: Isolation and characterization of an *Aplysia cDNA* clone encoding a family of peptide cotransmitters. *J. Neurosci.* 13: 8: 3346–3357.
Miller, M.W., Beushausen, S., Vitek, A., Stamm, S.A., Kupfermann, I., Brosius, J. and Weiss K.R. (1993b) The myomudulin-related neuropeptides: Characterization of gene encoding a family of peptide cotransmitters in *Aplysia. J. Neurosci.* 13: 8: 3358–3367.
Muneoka, Y., Kobayashi, M. (1992) Comparative aspects of structure and action of molluscan neuropeptides. *Experientia* 48: 448–456.
Nässel, D.R. (1993) Neuropeptides in the insect brain: a review. *Cell Tissue Res.* 273: 1–29.
Nichols, R., Schneuwly, S.A. and Dixon, J.E. (1988) Identification and charaterization of a *Drosophila* homologue to the vertebrate neuropeptide cholecystokinin. *J. Biol. Chem.* 263: 12167–12170.
Nishiitsutsuji-Uwo, J. and Endo, Y. (1981) Gut endocrine cells in insects – The ultrastructure of the endocrine cells in the cockroach midgut. *Biomed. Res.* 2: 30–44.
O'Shea, M., Adams, M.E., Bishop, C., Witten, J. and Worden, M.A. (1985) Model peptidergic systems at the insect neuromuscular junction. *Peptides* 6: 417–424.
Pearse, A.G.E. (1968) Common cytochemical and ultrastrucutal characteristics of cell producing polypeptide hormones (the APUD series) and their relevance to thyroid and ultimobranchial C cells and calcitonin. *Proc. Roy. Soc. Lond. B. Biol. Sci.* 170: 71–80.
Platt, N. and Reynolds, St. E. (1990) Invertebrate neuropeptides. *In:* Lunt, G.G. and Olsen, R.W. (eds): *Comparative Invertebrate Neurochemistry.* Croom Helm, London, Sydney, 175–226.
Price, D.A. and Greenberg, M.J. (1989) The hunting of the FaRPs: The distribution of FMRFamide-related peptides. *Reference Biol. Bull.* 177: 198–205.
Rao, R.K. and Riehm, J.P. (1989) The pigment-dispersing hormone family: chemistry, structure-activity relations, and distribution. *Biol. Bull.* 177: 225–229.
Remane, A. (1956) *Die Grundlagen des natürlichen Systems der vergleichenden Anatomie und der Phylogenetik.* Akademische Verlagsgesellschaft, Geest and Porting, Leipzig.
Sato, Y., Nakazawa, Y., Menjo, N., Imai, K., Komiya, T., Saito, H., Shin, M., Ikeda, M., Sakakibara, K., Isobe, M. and Yamashita, O. (1992) A new diapause hormone molecule of the silkworm, *Bombyx mori. Proc. Japan Acad.* 68, Ser. B: 75–79.
Schaefer, M., Picciott, M.R., Kreiner, T., Kaldany, R.P., Taussig, R. and Scheller, R.H. (1985) Aplysia neurons express a gene encoding multiple FMRF-amide neuropeptides. *Cell* 41: 457–467.
Scheller, R.H., Jackson, J.F., McAllis, L.B., Schwartz, J.H., Kandel, E.R. and Axel, R. (1982) A family of genes that codes for ELH, a neuropeptide eliciting a stereotyped pattern of behavior in *Aplysia. Cell* 28; 707–719.
Schneider, L.E. and Taghert, P.H. (1988) Isolation and characterization of a *Drosophila* gene that encodes multiple neuropeptides related to Phe-Met-Arg-Phe-NH_2 (FMRFamide). *Proc. Natl. Acad. Sci. USA* 85: 1993–1997.
Schoofs, L., Broeck, J.V. and DeLoof, A. (1993) The myotropic peptides of *Locusta migratoria*: structures, distribution, functions and receptors. *Insect. Biochem. Molec. Biol.* 23: 859–881.
Schulz-Aellen, M.F., Roulet, E., Fischer-Lougheed, J. and O'Shea, M. (1989) Synthesis of a homodimer neurohormone precursor of locust adipokinetic hormone studied by *in vitro* translation and cDNA cloning. *Neuron* 2: 1369–1373.
Spencer, A.N. (1989) Neuropeptides in the Cnidaria. *Amer. Zool.* 29: 1213–1225.
Stangier, J. Hilbich, C. Beyreuther, K. and Keller, R. (1987) Unusual cardioactive peptide (CCAP) from pericardial organs of the shore crab, *Carcinus maenas. Proc. Natl. Acad. Sci. USA* 84: 575–580.

Stangier, J., Hilbich, C. and Keller, R. (1989) Occurrence of crustacean cardioactive peptide (CCAP) in the nervous system of an insect, *Locusta migratoria*. *J. Comp. Physiol. B.* 159: 5–11.

Staufer, B., Bräunig, P. and Agricola, H. (1993) Innervation of the hindgut in the locust *Locusta migratoria*. *In:* Elsner, N. and Heisenberg, M. (eds): *Gen-Brain-Behaviour*. Thieme Verlag, Stuttgart-New York, p. 609.

Taghert, P.H. and Schneider, L.E. (1990) Interspecific comparison of a *Drosophila* gene that encodes FMRFamide-related neuropeptides. *J. Neurosci.* 10: 1929–1942.

Tublitz, N.J. (1993) Steroid-induced transmitter plasticity in insect peptidergic neurons. *Comp. Biochem. and Physiol. C* 105: 147–154.

Urbach, R., Agricola, H. and Breidbach, O. (1993) Perisulfakinin-immunoreactive neurons in the embryo and larva of a holometabolous insect. *In:* Elsner, N. and Heisenberg, M. (eds): *Gen-Brain-Behaviour*. Thieme Verlag, Stuttgart-New York, p. 737.

Van Herp, F. (1992) Inhibiting and stimulating neuropeptides controlling reproduction in Crustacea. *Invertebrate Reproduction and Develop.* 22, 1–3: 21–30.

Van Kesteren, R.E., Smit, A.B., Dirks, R.W., De With, N.D., Geraerts, W.P.M. and Joosse, J. (1992) Evolution of the vasopressin/oxytocin superfamily: Characterization of a cDNA encoding a vasopressin-related precursor, preproconopressin from the mollusc *Lymnaea stagnalis*. *Proc. Natl. Acad. Sci. USA* 89: 4593–4597.

Veenstra, J.A. (1989a) Isolation and structure of two gastrin/CCK-like neuropeptides from the American cockroach homologous to the leukosulfakinins. *Neuropeptides* 14: 145–149.

Veenstra, J.A. (1989b) Isolation and structure of corazonin, a cardioactive peptide from the American cockroach. *FEBS Lett.* 250: 231–234.

Zitnan, D., Sauman, I. and Sehnal, F. (1993) Peptidergic innervation and endocrine cells of insect midgut. *Archiv. of Insect Biochem. Physiol.* 22: 113–132.

The Nervous Systems of Invertebrates: An Evolutionary and Comparative Approach
ed. by O. Breidbach & W. Kutsch
© 1995 Birkhäuser Verlag Basel/Switzerland

Ground plans, evolutionary changes and homologies in decapod crustacean brains

D. Sandeman[1] and G. Scholtz[2]

[1]*School of Biological Science, University of N.S. Wales, Kensington, NSW 2033, Australia*
[2]*Institut für Zoologie, Freie Universität Berlin, Königin-Luise-Strasse 1-3, D-14195 Berlin Dahlem, Germany*

Summary. Brain ground plans of the Decapoda and Eureptantia are constructed within the framework of the phylogenetic trees of these groups. Changes that have occurred during the evolution of the decapod brains are described and neural homologies both at the neuropil and single neuron level are discussed. The possibilities that the appearance of new structures in the brain are correlated with a change in habit, and that evolutionary success is correlated with a reduction in brain complexity are considered.

Comparative studies and the rules of the game

Comparative studies of animal nervous systems are of interest because they can provide clues to understanding similarities and evolutionary divergences of neuronal systems. They also complement studies on neuronal function, in relation to changes in general morphology, behaviour and lifestyles.

A prerequisite for comparative studies dealing with the evolutionary change of structures is the knowledge of the phylogenetic relationships of the group under investigation. The most valid approach to constructing family trees is the cladistic method of Hennig (1966). The pivotal procedure of this method is the discrimination between convergent and homologous characters, and between primitive (plesiomorphic) and novel or derived (apomorphic) characters. Shared derived characters (synapomorphies) are used to establish monophyletic groups which have an ancestor that is common, and exclusive, to that group. In this way the sequence of branch points in the genealogy of a given group can be reconstructed.

Given such a phylogenetic tree, constructed using non-neural characters to avoid circularity, neuronal structures can be analysed concerning their appearance and distribution among the members of the phylogenetic group. Central to such comparisons is the concept of homology. Establishing the homology of characters is not a trivial exercise and depends for its plausibility on the complexity of characters being compared (e.g., Remane, 1952; Riedl, 1975; Dohle, 1989). The appearance of a similar complex character in several organisms makes it unlikely that this character has evolved independently several times, and supports the argument for such a character being homologous. The distribution of similar characters among members of a given monophyletic group can support the case for homology even when the characters are not particularly complex. The arthropods, for example, most likely represent a monophyletic group (e.g., Ax 1987; Wheeler et al., 1993) and all arthropods from onychophorans to the insects possess a "central body" neuropil in the brain. There is little reason to suppose that the central body has evolved several times in the different groups despite differences in its substructure. It is more parsimonious to suggest that the last common ancestor of the arthropods already had a brain that contained a central body and that this structure is therefore considered to be homologous.

The possibility of identifying neurological homologies has always been difficult because of the apparent flexibility of nervous systems. There are also some astounding examples of complex similarities at the functional level (for example the receptors and central neurons associated with the senses of olfaction, vision and balance in vertebrates and arthropods) that cannot possibly have been derived from a common ancestral character. We nevertheless believe that neuronal structures can be homologised like all other organismal traits that are the product of a certain evolutionary history.

We are less interested here in the collection of homologous characters to support phylogenies, but rather in the use of established family trees to test the possible homologies of similar neural structures found in the crustacean central nervous system. The phylogenies of the Decapoda, and in particular the Eureptantia within the decapods, that we use here are based on the work of Abele (1991) and Scholtz and Richter (in press).

Who are the decapods?

Although this paper is confined to the decapods, it is useful to include here an overview of all the Eumalacostraca (as envisaged by Hessler, 1983) because it provides us with a valuable perspective in which the separation between the Hoplocarida, Decapoda, Euphausiacea and Syncarida can be appreciated (Fig. 1A). The eumalacostracans probably constitute a monophyletic taxon, with the Leptostraca representing their sister group (Siewing, 1956; Dahl, 1987), although aspects of this phylogeny are still being worked out (Richter, 1993).

The Decapoda as a group (Fig. 1B) are regarded to be monophyletic and contain the Dendrobranchiata and the Pleocyemata, mainly characterized and separated from the Dendrobranchiata by the novel attachment of their eggs to the pleopods and the loss of a free-living nauplius

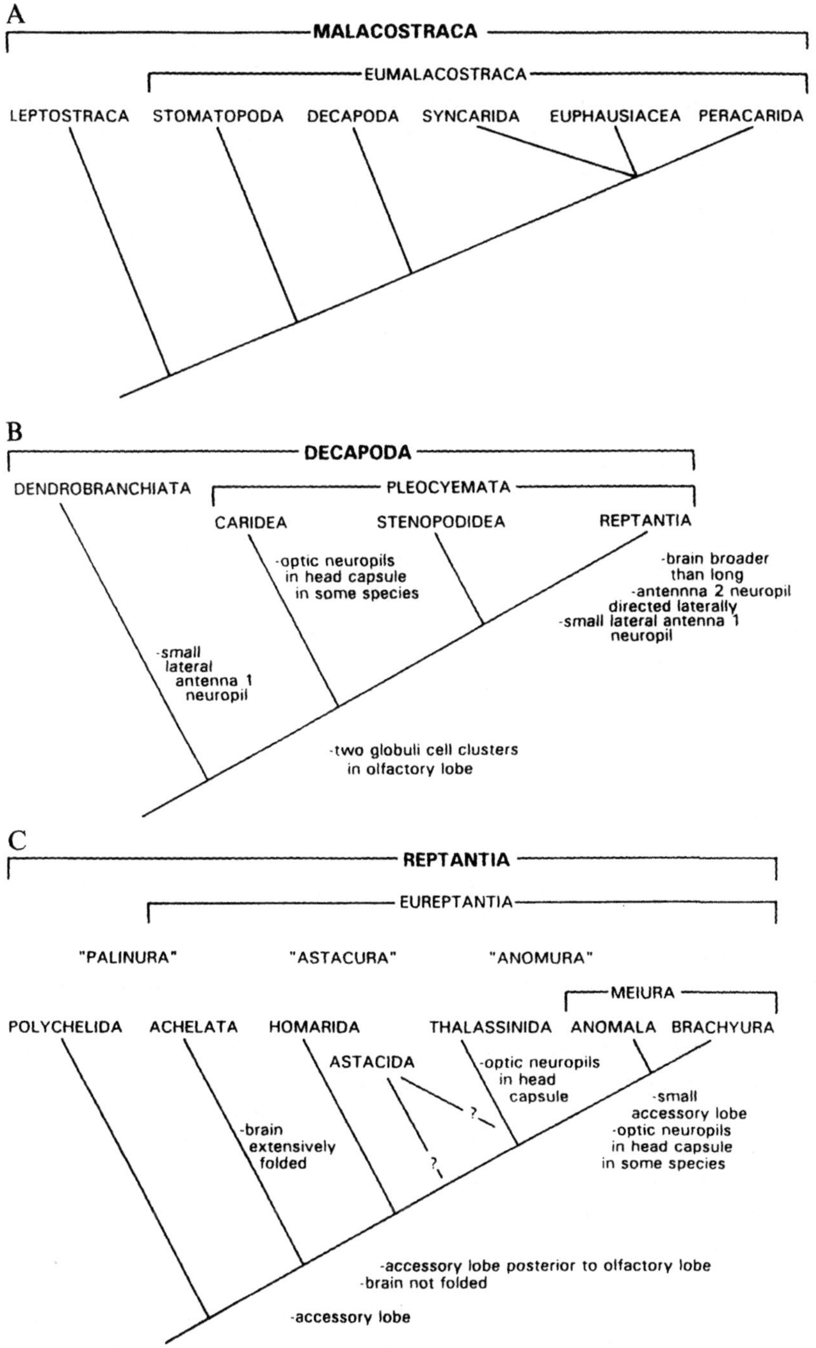

Figure 1. (A) Phylogenetic trees of the Malacostraca (combined from Siewing, 1956; Dahl, 1987; Richter 1993), Decapoda (Abele, 1991) and Reptantia (Scholtz and Richter, in press). Brackets indicate the different monophyletic groups. Neural apomorphies characteristics for the groups are placed at the appropriate branch points of the trees of the decapods and reptants (B and C after Sandeman et al., 1993)

larva (Burkenroad, 1963; Christoffersen, 1988). The Pleocyemata are therefore a monophyletic group within the Decapoda in that they all share a common ancestor but one that is not shared with the Dendrobranchiata. Figure 1C shows the detailed partition of the Reptantia (another monophyletic group) into the "Palinura", "Astacura" and "Anomura" traditional, but not monophyletic groups as can be seen immediately in the "Palinura" where the Achelata and the Polychelida do not share an ancestor with an accessory lobe. In this sense the Achelata are monophyletic with the Homarida, Astacida, Thalassinida, Anomala and Brachyura and are grouped together as the Eureptantia (Scholtz and Richter, in press.)

The decapod brain ground plan

By brain ground plan (*Grundplan*), sometimes also referred to as ground pattern (*Grundmuster*, Ax, 1987), we mean a ground plan that combines phylogenetically old, or plesiomorphic characters with the apomorphies of the investigated group and that represents the brain of the stem species of that group. We do not mean a basic plan (or *Bauplan*) exemplified by a "typical" brain that represents a combination of characters which do not appear in the same combination in recent or older forms (Hanström, 1947). Some authors use the terms "Bauplan" and "Grundplan" interchangeably (e.g., Brusca and Brusca, 1990; Sudhaus and Rehfeld, 1992).

The original brain was longer than broad and slightly bent along the longitudinal axis so that the protocerebrum lay dorsal to the deutocerebrum but still anterior to it. The optic neuropils were located in the eyestalks and the lateral protocerebrum was linked to the olfactory lobes in the deutocerebrum by the olfactory globular tract which formed a chiasm in the centre of the brain.

The deutocerebrum was dominated by large olfactory lobes with columnar synaptic glomeruli associated with a single cluster of posteriorly located, small cell somata. The primary neurites of these cell somata projected into the neuropil of the olfactory lobes and their axons made up the olfactory globular tract.

A large cylindrical antenna II neuropil with its long axis and nerve root directed posteriorly was the most prominent component of the tritocerebrum (Fig. 2).

The decapod ground plan that we have reconstructed does not differ from that of most members of the eumalacostracans. A similar distribution and relative size of the neuropil regions is found within the Stomatopoda, the Euphausiacea (Hanström, 1947; Johansson and Hallberg, 1992; Sandeman et al., 1993) and the Mysidacea within the Peracarida (Hanström, 1947; Johansson and Hallberg, 1992). These

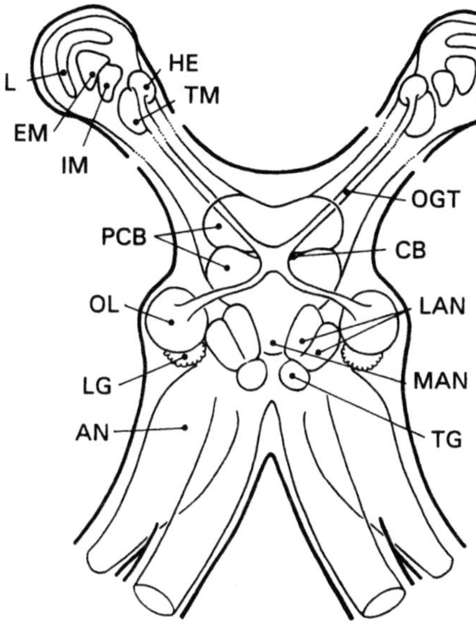

Figure 2. Decapod brain ground plan: The brain is elongated with dominant caudally directed antenna II neuropils, small olfactory lobes and one cluster of globuli cells located caudal to the olfactory lobe. The olfactory globular tract links the olfactory lobe with the hemi-ellipsoid body which is located with the terminal medulla and the three optic ganglia, in the eyestalk. The precise location and size of the neuropils in this ground plan are conjectural. N.B. The neuropils, nerve tracts and cell bodies of interest are labelled in this and in other figures in this account using the following abbreviations: L – lamina: EM – external medulla; IM – internal medulla; TM – terminal medulla; HE – hemi-ellipsoid body; PCB – median proctocerebrum; CB – central body; OL – olfactory lobe; AL – accessory lobe; LAN – lateral antenna I neuropil; MAN – medial antennular neuropil; DCN – deutocerebral commissure neuropil; AN – antenna II neuropil; OGT – olfactory globular tract; DC – deutocerebral commissure. MG – medial globuli cells; LG – lateral globuli cells; TG – tegumentary neuropil. The diagrams (except when otherwise stated) are derived from camera lucida reconstructions of silver impregnated and sectioned material. All views are of the dorsal aspect except that of *Jasus* (Fig. 6) which is more caudal. The left side of some diagrams are depicted as though the tissues were transparent, the right side as they appear to the observer.

similarities also extend, with some exceptions, to the Syncarida (Hanström, 1947; this paper). The leptostracans, on the other hand, exhibit a clearly different brain organization in the position of the olfactory lobes (Hanström, 1928; 1947) and the presence of two instead of three optic neuropils (Elofsson and Dahl, 1970; Scholtz, 1992) characteristics that resemble those occurring in lower Crustacea ("entomostracans") and which might be original (plesiomorphic) for malacostracans. Thus, regardless of interpretations of the relationships within the Malacostraca, the original decapod brain represents, to a considerable degree, that of the ancestor shared by all members of the Eumalacostraca (Fig. 3).

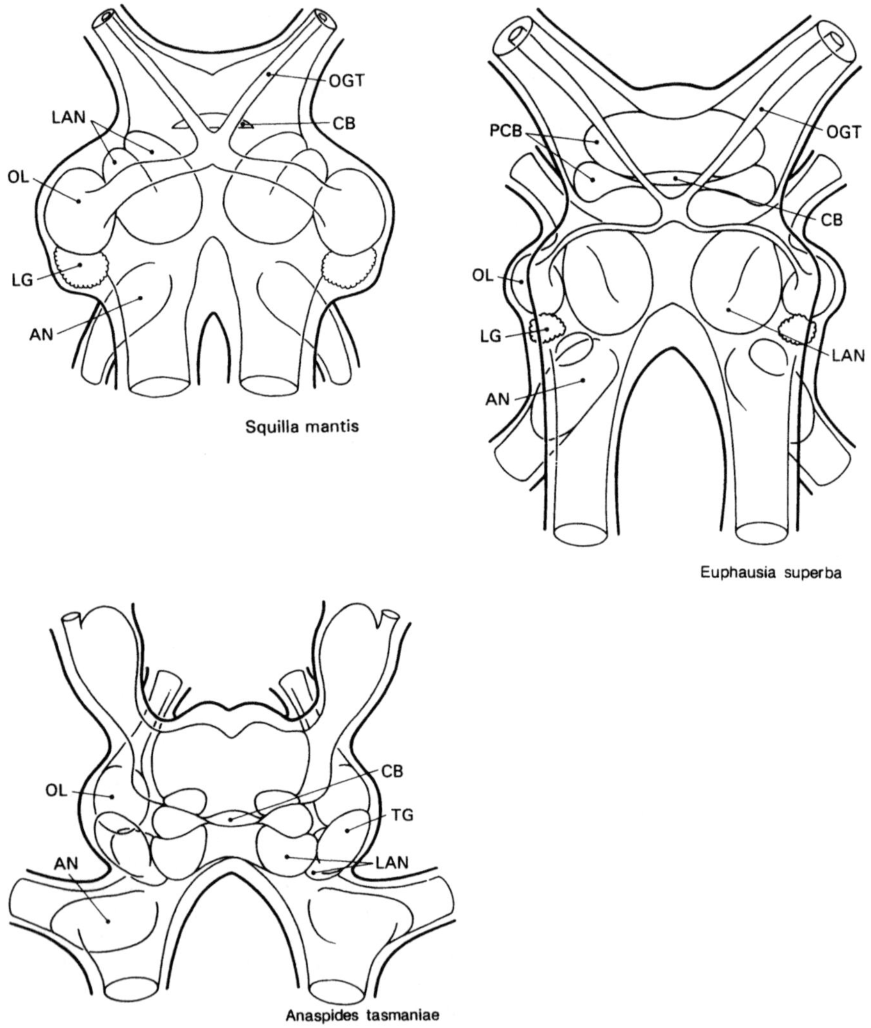

Figure 3. The brains of *Squilla* (Stomatopoda) (redrawn from Hanström, 1947), *Anaspides* (Syncarida) and *Euphausia* (Euphausiacea) in which similarities to the decapod brain ground plan can be detected in the presence of the olfactory lobes (OL) with a single cluster of caudally located globuli cells (LG) and an olfactory globular tract (OGT). Olfactory lobes can be determined in Anaspides but not the olfactory globular tracts. The medial protocerebral neuropils (PCB), divided lateral antenna I neuropils (LAN) and antenna II neuropils (AN) are represented in all three examples. The antenna II of Anaspides is directed laterally, a character that is convergent with a similar change that occurs in the reptants (see below). (Please see previous figure legends for abbreviations not defined here.)

From decapod to eureptant

A distinguishing apomorphy in the brain that can be detected in the line between the Dendrobranchiata and the Pleocyemata is the appearance of two clusters of small cell bodies associated with the olfactory lobes, a feature found in all forms of the Caridea, Stenopodida and Reptantia (Sandeman et al., 1993) (Fig. 4). Given the large number of cell somata that surround the brains of crustaceans, this may be considered to be a rather dubious character to select. The small cells in question though, are unique in that they are always associated with the olfactory lobes, and are not only small but contain nuclei that almost entirely fill the cell body, leaving a very thin layer of cytoplasm between the nucleus and the outer cell membrane. Nerve cell bodies with identical characteristics are found in association with the mushroom bodies of insects and *Limulus* (Fahrenbach, 1979) and have been collectively referred to as "globuli cells" (Bullock and Horridge, 1965).

In some of the Caridea, the three optic neuropils that in the original form were contained in the stalked eye, just behind the retina, have migrated medially and are found fused into the brain with the lateral and medial protocerebral neuropils (Hanström, 1947). This feature appears again in the Reptantia where it is also associated with an enlargement of the hemiellipsoid bodies (see below) (Fig. 8).

A marked alteration in the overall form of the brain took place in the ancestor shared by the reptants: The elongated shape of the eumalacostracan ground plan was replaced by a brain that was broader than long, a laterally directed antenna II neuropil, and a small lateral antenna I neuropil. A ground plan for the reptant brain therefore incorporates this characteristic although there are no other differences to the decapod

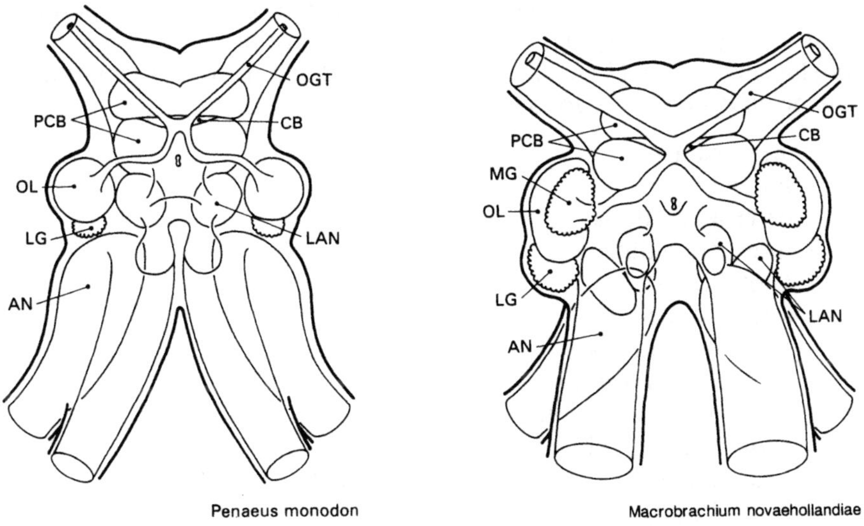

Figure 4. The brains of *Penaeus* (Dendrobranchiata) and *Macrobrachium* (Caridea). *Penaeus* has maintained the original character of a single cluster of globuli cells (LG) and a long brain with the antenna II neuropil (AN) directed caudally. In *Macrobrachium* the brain has broadened and a second cluster of globuli cells (MG) is found associated with the olfactory lobe (OL). (Please see previous figure legends for abbreviations not defined here.)

ground plan in the numbers of neuropils or in their overall arrangement.

The most striking and remarkable event to occur in the evolution of the reptantian brain was the appearance of a completely new neuropil, the accessory lobes. Characterised by round glomeruli and close association with the olfactory lobes, the accessory lobes can be positively identified, and have been found in no other group of arthropods. Neurites from the small cell bodies, associated in earlier forms with the olfactory lobes, branch in both the olfactory and accessory lobes and the olfactory globular tract is constituted from axons of neurons originating in both lobes (Mellon and Alones, 1993). The similar location of the accessory lobes, their connection across the brain through the deutocerebral commissure, their glomerular structure and their exclusive presence in all forms of the Eureptantia, allow the accessory lobes to be recognised as truly homologous neuronal structures (Helm, 1928; Sandeman and Luff, 1973; Blaustein et al., 1988).

The eureptant ground plan

The Eureptantia are the largest and best known of the decapods and have been favourite animals for neurobiologists. The brain ground plan of the eureptants has also undergone significant changes from that of the original decapods and for these reasons it is described here in some detail (Fig. 5). Like all ground plans, that of the eureptants consisted of a combination of apomorphic and plesiomorphic characters. Characters that are synapomorphies in the eureptant groups are shown in the phylogenetic trees (Fig. 1). An inventory of the neuropils contained in the ground plan is as follows:

Protocerebrum
- lamina
- external medulla
- internal medulla
  (optic tract)
- terminal medulla
- hemiellipsoid body
  (lateral protocerebral tract)
- anterior medial protocerebral neuropil
- protocerebral bridge
- posterior medial protocerebral neuropil
- central body
  (olfactory globular tract)

Deutocerebrum
- olfactory lobe
- accessory lobe (apomorphic)

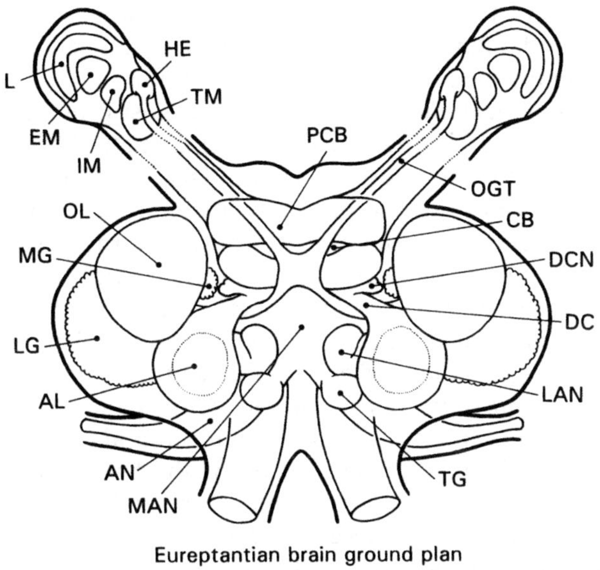

Figure 5. The eureptant brain ground plan: The brain is broader than long with relatively small antenna II neuropils (AN), directed laterally. The olfactory lobes (OL) are linked to the hemi-ellipsoid bodies (HE) that are located with the terminal medulla (TM) and optic ganglia in the eyestalk. Two clusters of globuli cells are associated with the olfactory lobes (MG, LG), the originally caudally lying group now containing many more cells and located laterally (LG). The divided lateral antenna I neuropils (LAN) are represented. The novel feature is the presence of the accessory lobes (AL) lying posterior to the olfactory lobes (OL). These were probably about the same size as the olfactory lobes and organized into outer and inner layers of glomeruli. As in the ground plan of the decapods, the precise size and location of the neuropils is conjectural. (Please see previous figure legends for abbreviations not defined here.)

- lateral antennular neuropil
- medial antennular neuropil
- deutocerebral commissure neuropil (apomorphic)

Tritocerebrum

- antenna neuropil
- tegumentary neuropil

The first three *optic neuropils* have a geometrical organization imposed on them by the photoreceptor array which is preserved down to the third neuropil.

The two neuropils that belong to the lateral protocerebrum, the *terminal medulla* and the *hemiellipsoid body*, do not show the regular patterning of the first three neuropils and are located at the base of the neuropil mass contained in the eyestalk. The hemi-ellipsoid body is the target of the axons in the olfactory globular tracts.

The *proctocerebral bridge* is roughly boomerang-shaped with the apex pointing caudally.

The cigar-shaped *central body* of the reptantians lies across the protocerebrum dividing the anterior medial protocerebral neuropils from the posterior medial protocerebral neuropils.

The nearly spherical *olfactory lobes* are the most prominent neuropils and are characterised by their density, small diameter fibres and columnar glomeruli. Also characteristic are the large, fine fibred tracts that project from the centre of the olfactory lobes to the lateral protocerebrum. These olfactory globular tracts always form a chiasm in the centre of the brain, immediately dorsal to the central body.

The large *accessory lobes* contain many small round glomeruli, are situated medial and posterior to the olfactory lobes and are linked across the brain by a large bundle of axons, the deutocerebral commissure.

The *deutocerebral commissure neuropils* are associated with the presence of the accessory lobes and a large deutocerebral commissure. These two small neuropils on the ventral side of the brain receive a strong projection from the deutocerebral commissure and also from axons in the protocerebrum. Like the accessory lobes, they appear to have been newly evolved, perhaps as part of the accessory lobe complex.

The *lateral* and *medial antenna I neuropils* do not exhibit a geometric patterning of fibres. The lateral antennular neuropils are divided along their length resulting in a horseshoe-shaped appearance when they are sectioned in some planes.

The *antenna II neuropils* are elongated towards the nerve root and have a barred appearance with tracts of fibres running at right angles to the long axis of the neuropil. This is also true for the ground plan of the decapods.

The *tegumentary neuropils* lie dorsal to the medial ends of the antenna II neuropils and receive afferent input from the carapace.

Evolution in eureptant brains

The changes that have occurred during the evolution of the eureptants can be followed with the help of the phylogeny depicted in Figure 1C.

The broadened brain of the last common ancestor of the Reptantia was further modified in the Achelata, a division including the spiny lobsters (Fig. 6). The slight folding of the original decapod brain in which the medial protocerebral neuropils lay dorsal but still anterior to the deutocerebrum has been greatly exaggerated in the Achelata (Sandeman et al., 1992; 1993). The neuropils of the medial protocerebrum have come to lie dorsal to the deutocerebrum and the tritocerebrum. All other members of the Eureptantia have unfolded the original decapod brain to result in a situation in which the medial protocerebal neuropils lie directly anterior to the deutocerebrum (Helm, 1928; Sandeman et al., 1992), a synapomorphy of these groups (Sandeman et al., 1993). A possibly apomorphic

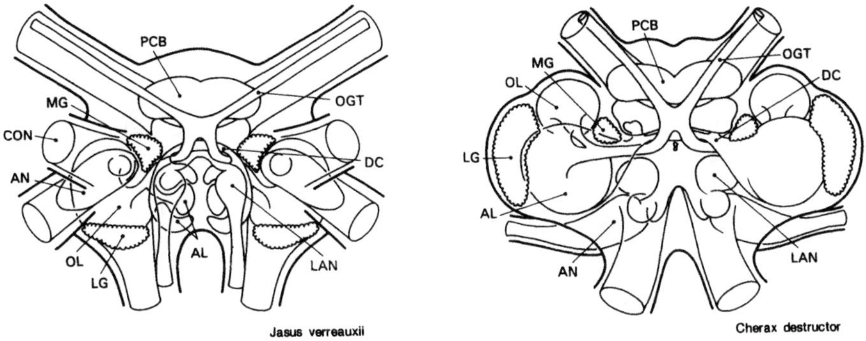

Figure 6. The brains of *Jasus* (Achelata) and *Cherax* (Astacida) showing the extreme folding of the brains in the spiny lobsters where the median protocerebrum (PCB) has come to lie dorsal to the deutocerebral neuropils. The view of the *Jasus* brain is from behind so that the oesophageal connectives (CON) are directed towards the viewer. The complexly layered accessory lobes (AL) lie medial to the olfactory lobes (OL) and are about the same size. The brain of *Cherax* has become less folded than the original condition and the neuropils of the median protocerebrum, deutocerebrum and tritocerebrum all lie one behind the other. The accessory lobes (AL) in *Cherax*, though, are significantly larger than the olfactory lobes (OL). (Please see previous figure legends for abbreviations not defined here.)

character in the Achelata is the presence of two pairs of small neuropils with a glomerular structure, associated with the lateral antenna I neuropils and not found in the brains of any other eureptants (Blaustein et al., 1988).

The accessory lobes were at first large. This can be deduced from the presence of large accessory lobes in the Achelata, Homarida, Astacida and Thalassinida which together constitute a paraphyletic assemblage. It is not clear whether the complexly layered organization of the neuropil seen in the Achelata represents the original condition or is an apomorphic character of the Achelata. The less obvious layering of the accessory lobe neuropil in *Homarus* might indicate that layers were originally present but not as pronounced as now seen in the Achelata. The accessory lobes became considerably reduced in the ancestral lineage leading to the Anomala and Brachyura, a very small accessory lobe representing a synapomorphy of these two groups (Fig. 7).

The deutocerebral commissure neuropils are most clearly defined in the Astacida, Achelata and Homarida. They have yet to be positively identified in the Anomala where the lack of large accessory lobes makes tracing the few fibers in the deutocerebral commissure uncertain.

A characteristic structural change that has occurred in the brains of the Thalassinida (Scheuring, 1923; Hanström, 1924; Sandeman et al., 1993) and convergently in some anomalan (Hanström, 1924; 1947; Helm, 1928; Sandeman et al., 1993) and brachyuran (Helm, 1928) species, is the relocation of the optic neuropils and lateral protocerebral neuropils from the original position in the stalked eye to lie fused medially with the brain. The lateral protocerebral neuropils are also

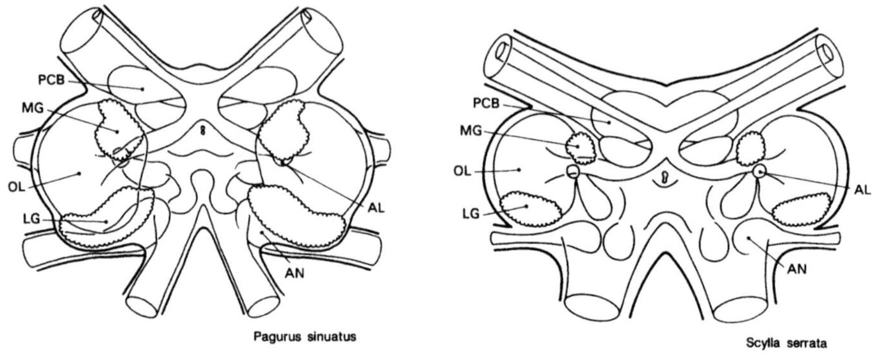

Figure 7. The brains of *Pagurus* (Anomala) and *Scylla* (Brachyura) are very similar. They are dominated by large olfactory lobes (OL). The accessory lobes, while still maintaining their glomerular structure have been greatly reduced. Other neuropils are represented but are characterised by the indistinct boundaries between them. (Please see previous figure legends for abbreviations not defined here.)

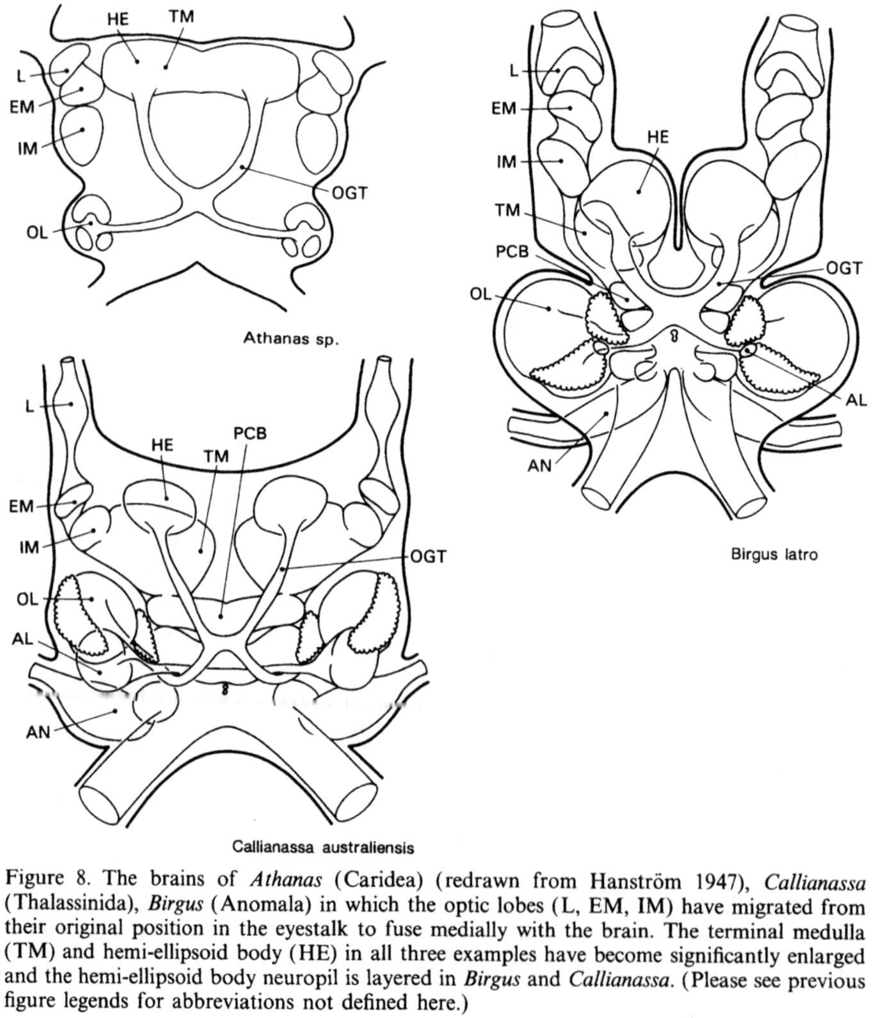

Figure 8. The brains of *Athanas* (Caridea) (redrawn from Hanström 1947), *Callianassa* (Thalassinida), *Birgus* (Anomala) in which the optic lobes (L, EM, IM) have migrated from their original position in the eyestalk to fuse medially with the brain. The terminal medulla (TM) and hemi-ellipsoid body (HE) in all three examples have become significantly enlarged and the hemi-ellipsoid body neuropil is layered in *Birgus* and *Callianassa*. (Please see previous figure legends for abbreviations not defined here.)

enlarged in these animals and the hemi-ellipsoid body, not normally structured, takes on a layered appearance. Migration of optic neuropils from the eyestalks to the brain is also found in some Caridea (Helm, 1928; Hanström, 1947) (Fig. 8).

Neural homologues in eureptantian brains

The various neuropil areas in the decapods can be identified by the segment in which they are found and in most cases also by the afferent inputs they receive and there is little doubt about their being homologous throughout the Decapoda or even Eumalacostraca. Within the Eureptantia the accessory lobe fulfils all the criteria of a true apomorphy.

At the level of single neurons, however, we are less certain using common architecture as a reliable character to establish homology because we are unsure of the nature of the developmental or functional constraints placed on individual interneurons within the brain. Some integrative properties of neurons do depend on their shape and size, but the function of the neuron is determined by its connectivity in a neuropil and this is governed more by propinquity than by shape. The application of immunocytochemical methods to the brains of eureptantians has nevertheless provided some new perspectives: Individual cells with the same connectivity, architecture and immunoreactivity to specific antibodies, have been found to occupy similar locations in the brains of different eureptants.

The paired giant serotonin containing cells in the crayfish and lobster accessory and olfactory lobes are good examples. These large interneurons have large dorsally situated cell bodies which send thick primary neurites ventrally to branch repeatedly in the accessory lobes. Each branch ends in a large number of spherical glomerular endings. A single axon extends to the olfactory lobes and branches extensively to the column-shaped glomeruli there. While there are a number of serotonin-IR neurons in the brains of both crayfish and lobsters, only one pair have this particular form and distribution (Sandeman and Sandeman, 1987; Sandeman et al., 1988; Beltz et al., 1990).

The giant 5HT neurons would seem to be an important component of the accessory lobes of crayfish and lobsters and if true homologues may be expected to be detectable in all the reptant brains. This does not appear to be the case in the brachyurans where labelling with antibodies to 5HT has so far not revealed a neuron of this kind. On the other hand, the accessory lobes of the brachyurans are extremely small and may have also lost their original function in which the giant serotonin neurons perhaps played a critical role. It would be of considerable interest to look at the brains of one of the thalassinids (e.g., *Callianassa*) in this respect as their accessory lobes are well developed.

A second example of probably homologous neurons is provided by a pair of large cells that label with an antibody against Substance P. Present in the brains of crayfish and crab, their immunoreactivity is by no means the only similarity. The cell bodies of these neurons are located lateral to the protocerebrum at the level of the central body in both animals from where they appear to obtain their input (Fig. 9). The cells project both ipsilaterally and contralaterally to end among the small cell bodies associated with the olfactory lobes (Sandeman et al., 1990). This is their only projection and is unusual in that there are few examples known of arthropod neurons projecting to cell somata instead of neuropil. That features of such complexity are shared by the cells of the two animals makes a strong argument for being homologous and again it would be of significant interest to search for these neurons not only in the Thalassinida and Anomala, where they should be, but in the Homarida and the Achelata where they may be, depending at which stage in the evolution of the Reptantia they appeared.

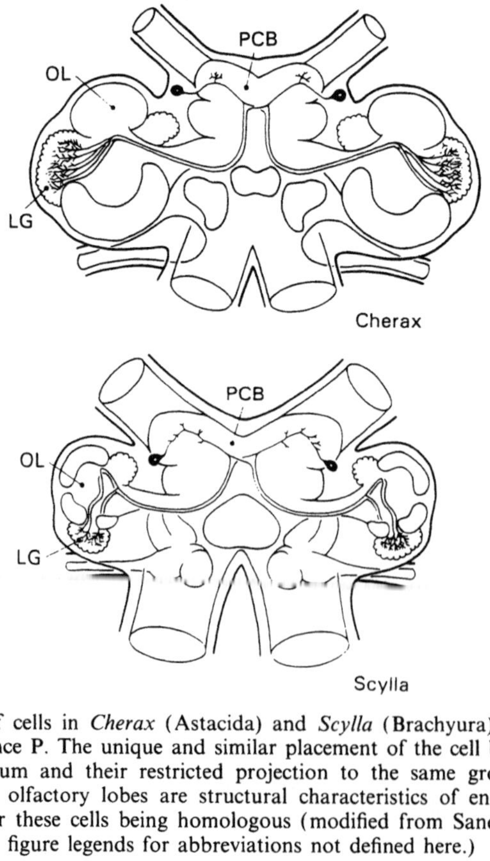

Figure 9. A pair of cells in *Cherax* (Astacida) and *Scylla* (Brachyura) that label with an antibody to Substance P. The unique and similar placement of the cell bodies lateral to the median protocerebrum and their restricted projection to the same group of globuli cells associated with the olfactory lobes are structural characteristics of enough complexity to support the case for these cells being homologous (modified from Sandeman et al., 1990). (Please see previous figure legends for abbreviations not defined here.)

Behavioural correlates with structural evolution?

It is tempting to correlate the appearance of large and thus apparently important regions of the brain with some obvious change in either the behaviour or environment in which the animals now live in relation to that occupied by their ancestors. Such correlations may not provide us with a causal explanation but they can lead to the erection of testable hypotheses (Miklos, 1993).

One of the most interesting apomorphies to be found in the brains of the Eureptantia is the accessory lobe and a common habit that links these animals is their primarily benthic lifestyle. Early larval forms of many of these are pelagic and in lobsters the development of the olfactory lobes occurs early whereas that of the accessory lobes is completed only when they descend from the water column and adopt a benthic habit. The development of the two lobes in freshwater crayfish, on the other hand, that have no pelagic stage in their development, occurs almost simultaneously (Helluy et al., 1993).

One can speculate on the increased complexity of the spatial and olfactory world of the benthic dwellers and that the accessory lobes may be concerned with the neural integration of these modalities. Recent work on the crayfish has shown that some interneurons in the deutocerebral commissure that end in the accessory lobes are driven by changes in light intensity (Sandeman, Beltz and Sandeman, in press). Also, a close functional association between the olfactory and accessory lobes is demonstrated by the large 5HT neurons (see above) with their branches confined almost exclusively to the olfactory and accessory lobes (Sandeman and Sandeman, 1994). The accessory lobes of freshwater crayfish contain about ten times as many glomeruli as the marine lobster, perhaps reflecting the broader spectrum of dissolved organic substances confronting the freshwater animal.

The possession of an accessory lobe though, is not a prerequisite for successful benthic living. Accessory lobes are not present in the brains of stomatopods, nor in the isopods or amphipods, which are all successful benthic animals. The evolution of the accessory lobe within the Eureptantia in relation to a benthic habit has to be viewed in relation to the many other malacostracans without accessory lobes that have adopted a benthic existence. The evolution of the second set of globuli cells in the Pleocyemata may have been the prerequisite for the appearance of the eureptantian accessory lobe, particularly as it is the primary neurites of many of these cells that project to the accessory lobes. The line leading to the benthic isopods, for instance, did not include this character.

The secondary reduction of the accessory lobe in the Anomala Brachyura and a simplification of their brain, in that clear boundaries between the different neuropil areas are no longer to be seen, is difficult to explain or relate to changes in lifestyles of the crabs. Members of these groups are found in the deepest parts of oceans, on the shores and in estuaries, in freshwater and on land, where they extend into the Australian deserts and are found in sago palms on tropical islands. A unique morphological character is shared by the meiurans: The first antennae has become compact and carries an increased number of olfactory sensilla on the outer branch. It is also highly mobile. These changes are coupled with an enlargement of the olfactory lobes in relation to other brain areas.

Brain reduction and the success of the simple-minded

The evolutionary developments within the decapod crustaceans present us with a certain paradox: Parts of the brains of the anomalans and brachyurans have undoubtedly become reduced or simplified and yet we are dealing with an extraordinarily successful group of animals. More than half of the decapod species belong to the Meiura and of these about 80% are brachyurans. Viewed from a morphological point of view, the anomalans and the brachyurans exhibit the widest range of forms found among any group of decapods. They have also occupied habitats ranging from the deepest parts of the oceans through the littoral and fresh water to land. They exhibit a wide range of complex behaviours, including such well documented cases like signalling to one another both visually and through ground borne vibrations, cleverly selecting empty mollusc shells in which to live, decorating themselves with sponges or anemones, and stripping the husks off coconuts before opening them. In spite of this diversity, all the members of the meiurans share a brain that is virtually identical and, the phylogenetic analysis suggests, was simplified before the massive radiation of the group took place.

Interesting parallels can be drawn between the situation in the decapods and the amphipods and, independently, the isopods. Both groups have a brain that is reduced in many aspects when compared with the original ground plan of the Peracarida (Hanström, 1947) and yet consititute the most successful of all peracaridans in terms of the number of species and the diversity of the habitats they have colonised (Bousfield, 1983; Wägele, 1989).

Another example of the apparent correlation between evolutionary success and a reduced brain is provided by the plethodontid salamanders which, according to Roth and his colleagues (Roth et al., 1993) are very likely to have undergone secondary simplification with respect to their ancestors and have reduced central nervous systems. As a group though, they are highly successful in evolutionary terms, consisting as they do of a large number of species that has radiated into a wide range of habitats and increased the size of their genome.

The question can be posed as to whether the simplification of the nervous system in all these animals is to be seen as an adaptation that enabled their evolutionary success. The alternative is that the evolutionary success of these animals is dependent on some other character and the reduction of the brain, as a non-adaptive event, is simply tolerated. Roth et al., (1993) postulate the latter. They point out that the plethodontids are characterised by an increased genome, leading to larger cell and size, and the simplified brain is the result of paedomorphic development. The secondary reduction of the brain is therefore the consequence of internal constraints and not an adaptation.

One factor can be excluded from situation in the Meiura: Comparative data on the size of the genome in the decapods (Chow et al., 1990) reveal no significant differences between the Meiura and other decapods. That the brain reduction in the meiurans has been accompanied by a large number of morphological changes may indicate that it does have an adaptive value which only expressed itself fully with the advent of a shortened and laterally broadened brachyuran body. Overall, there seems to be no general evolutionary tendency towards more complex brains and evolutionary success is not correlated with an increase in brain complexity.

References

Abele, L.G. (1991) Comparison of morphological and molecular phylogeny of the Decapoda. *Mem. Qld. Mus.* 31: 101–108.

Ax, P. (1987) *The Phylogenetic System. The Systematization of Organisms on the Basis of Their Phylogenesis.* J. Wiley and Sons, New York.

Beltz, B.S., Pontes, M.S., Helluy, S.M. and Kravitz, E.A. (1990) Patterns of appearance of serotonin and proctolin immunoreactivities in the developing nervous system of the American lobster. *J. Neurobiol.* 21: 521–542.

Blaustein, D.N., Derby, C.D., Simmons, R.B. and Beall, A.C. (1988) Structure of the brain and medulla terminalis of the spiny lobster *Panulirus argus* and the crayfish *Procambarus clarkii,* with an emphasis on the olfactory centers. *J. Crust. Biol.* 8: 493–519.

Bousfield, E.L. (1983) An updated phyletic classification and palaeohistory of the Amphipoda. *In:* F.R. Schram (ed.): *Crustacean Phylogeny.* Balkema, Rotterdam, pp. 257–277.

Brusca, R.C. and Brusca, G.J. (1990) *Invertebrates.* Sinauer Associates, Sunderland, Maryland.

Bullock, T.H. and Horridge, G.A. (1965) *Structure and Function in the Nervous Systems of Invertebrates.* Vol. II, Freeman & Co, San Francisco.

Burkenroad, M.D. (1963) The evolution of the Eucarida, (Crustacea, Eumalacostraca), in relation to the fossil record. *Tulane Stud. Geol.* 2: 2–17.

Chow, S., Dougherty, W.J. and Sandifer, P.A. (1990) Meiotic chromosome complements and nuclear DNA contents of four species of shrimps of the genus *Penaeus. J. Crust. Biol.* 10: 29–36.

Christoffersen, M.L. (1988) Phylogenetic systematics of the Eucarida (Crustacea Malacostraca). *Revta bras. Zool.* 5: 325–351.

Dahl, E. (1987) Malacostraca maltreated – the case of the Phyllocarida. *J. Crust. Biol.* 7: 721–726.

Dohle, W. (1989) Zur Frage der Homologie ontogenetischer Muster. *Zool. Beitr.* (*N.F.*) 32: 355–389.

Elofsson, R. and Dahl, E. (1970) The optic neuropiles and chiasmata of Crustacea. *Z. Zellforsch.* 107: 343–360.

Fahrenbach, W.H. (1979) The brain of the horseshoe crab (*Limulus polyphemus*) III. Cellular and synaptic organization of the corpora pedunculata. *Tissue Cell* 11: 163–200.

Hanström, B. (1924) Untersuchungen über das Gehirn, insbesondere die Sehganglien der Crustaceen. *Ark. Zool.* 16: 1–117.

Hanström, B. (1928) *Vergleichende Anatomie des Nervensystems der Wirbellosen Tiere.* Springer Verlag, Berlin.

Hanström, B. (1947) The brain, the sense organs and the incretory organs of the head in the Crustacea Malacostraca. *Kungliga Fysiografiska Sallskapets Handlingar. N.F.* 58/9: 1–44.

Helluy, S. Sandeman, R. Beltz, B. and Sandeman, D.(1993) Comparative brain ontogeny of crayfish and clawed lobster: Implications of direct and larval development. *J. Comp. Neurol.* 335: 343–354.

Helm, F. (1928) Vergleichend-anatomische Untersuchungen über das Gehirn, insbesondere das "Antennalganglion" der Dekapoden. *Z. Morph. Ökol. Tiere* 12: 70–134.

Hennig, W. (1966) *Phylogenetic Systematics.* University of Illinois Press, Urbana.

Hessler, R.R. (1983) A defense of the caridoid facies: wherein the early evolution of the Eumalacostraca is discussed. *In:* F.R. Schram (ed.): *Crustacean Phylogeny.* Balkema, Rotterdam, pp. 145–164.

Johansson, K.U.I. and Hallberg, E. (1992) The organization of the olfactory lobes of Euphausiacea and Mysideacea (Crustacea, Malacostraca). *Zoomorphology* 112: 81–89.

Mellon, DeF. and Alones, V. (1993) Cellular organization and growth-related plasticity of the crayfish olfactory midbrain. *Microsc. Res. Tech.* 24: 231–259.

Miklos, G.L.G. (1993) Molecules and cognition: the latterday lessons of levels, language, and *lac. J. Neurobiol.* 24: 842–890.

Remane, A. (1952) *Die Grundlagen des natürlichen Systems, der vergleichenden Anatomie und der Phylogenetik.* Akademische Verlagsgesellschaft Geest & Portig, Leipzig.

Richter, S. (1993) Die Eucarida sind nicht monophyletisch – Ommatidienstruktur deutet auf nähere Verwandtschaft von Euphausiacea, Syncarida und Peracarida (Crustacea). *Verh. Dtsch. Zool. Ges.* 86: 144.

Riedl, R. (1975) *Die Ordnung des Lebendigen.* Parey Verlag, Hamburg.

Roth, G., Nishikawa, K.C., Naujoks-Manteuffel, C., Schmidt, A. and Wake, D.B. (1993) Paedomorphosis and simplification in the nervous system of salamanders. *Brain Behav. Evol.* 42: 137–170.

Sandeman, D.C. and Luff, S.E. (1973) The structural organization of glomerular neuropile in the olfactory and accessory lobes of an Australian freshwater crayfish, *Cherax destructor. Z. Zellforsch. mikrosk. Anat.* 142: 37–61.

Sandeman, R.E. and Sandeman, D.C. (1987) Serotonin-like immunoreactivity of giant olfactory interneurons in the crayfish brain. *Brain Res.* 403: 371–374.

Sandeman, D.C. and Sandeman, R.E. (1994) Electrical responses and synaptic connections of giant 5HT-IR neurons in crayfish olfactory and accessory lobes. *J. Comp. Neurol.* 341: 130–144.

Sandeman, D.C. Sandeman, R.E. and Aitken, A.R. (1988) Atlas of serotonin-containing neurons in the optic lobes and brain of the crayfish, *Cherax destructor. J. Comp. Neurol.* 269: 465–478.

Sandeman, R.E. Sandeman, D.C. and Watson, A.H.D. (1990) Substance P antibody reveals homologous neurons with axon terminals among somata in the crayfish and crab brain. *J. Comp. Neurol.* 294: 569–582.

Sandeman, D.C., Scholtz, G. and Sandeman, R.E. (1993) Brain evolution in decapod Crustacea. *J. Exp. Zool.* 265: 112–133.

Sandeman, D.C., Beltz, B. and Sandeman, R.E. (1994) Crayfish brain interneurons that converge with serotonin giant cells in accessory lobe glomeruli *J. Comp. Neurol.*; *in press.*

Sandeman, D., Sandeman, R., Derby, C. and Schmidt, M. (1992) Morphology of the brain of crayfish, crabs, and spiny lobsters: a common nomenclature for homologous structures. *Biol. Bull.* 183: 304–326.

Scheuring, L. (1923) Die Augen von *Gebia* und *Callianssa.* Ein Beitrag zur Kenntnis des rückgebildeten Facettenauges. *Zool. Jb. Anat.* 44: 389–438.

Scholtz, G. (1992) Sind die optischen Loben von Insekten und Höheren Krebsen (Malacostraca) homolog? *Verh. Dtsch. Zool. Ges.* 85: 90.

Scholtz, G. and Richter, S. (1994) Phylogenetic systematics of the Decapoda Reptantia. *Zool. J. Linn. Soc*; *in press.*

Siewing, R. (1956) Untersuchungen zur Morphologie der Malacostraca (Crustacea). *Zool. Jb. Anat.* 75: 39–176.

Sudkaus, W. and Rehfeld, K. (1992) *Einführung in die Phylogenetik und Systematik*. Gustav Fischer Verlag, Stuttgart.

Wägele, J.-W. (1989) Evolution und phylogenetisches System der Isopoda. *Zoologica* 140: 1–262.

Wheeler, W.C., Cartwright, P. and Hayaski, C.Y. (1993) Arthropod phylogeny: A combined approach. *Cladistics* 9: 1–39.

The Nervous Systems of Invertebrates: An Evolutionary and Comparative Approach
ed. by O. Breidbach & W. Kutsch
© 1995 Birkhäuser Verlag Basel/Switzerland

The arthropod mushroom body: Its functional roles, evolutionary enigmas and mistaken identities

N.J. Strausfeld, E.K. Buschbeck and R.S. Gomez

Arizona Research Laboratories Division of Neurobiology, University of Arizona, Tucson, AZ 85721, USA

Summary. Insects share the same physical environment as terrestrial vertebrates. However, insects are vastly more successful in terms of their adaptive radiation and colonization. The biomass of insects alone is said to outweigh all other organisms and it can be speculated that their evolutionary success is a consequence, not only of their environmental colonization, but also of their behavioral versatility which sets them apart from other terrestrial arthopods. If this hypothesis is correct, it might be expected that there exist neural organizations that are peculiar to, or greatly elaborated in, the Insecta alone. This chapter describes the occurrence among selected arthropods of a brain region that, in eusocial insects, reaches great complexity and is implicated in learning and memory. Its identification in various taxa is assessed against the background of current theories of arthropod evolution and brain segmentation.

Introduction

Insect brains are minute. Yet, parts of them are structurally similar to regions in vertebrate brains and are used by many scientists for fundamental research in neurobiology. Indeed, the experimental accessibility of the insect brain, and its use as a "model" of vertebrate-like neural systems, much reduces the need to use vertebrates for basic research and teaching. It surprises many to learn that the insect brain contains neuronal densities comparable or even greater than found in ourselves and other primates: counts of the neurons contributing to a cockroach's higher brain center (the paired mushroom bodies) reveal nearly 1 million nerve cells contained in a volume of brain less than one-fifteenth of a cubic millimeter (Neder, 1959). This compares to the density of granule cells supplying our cerebellum (the highest nerve cell density in primates), which amounts to 10 million neurons per cubic millimeter (Blinkov and Glezer, 1968).

The versatility of insect behavior is matched by their sensory capabilities. Organizational and functional similarities between the brains of insects and vertebrates include neuronal networks underlying the control of locomotion and posture (Burrows and Laurent, 1989), networks computing motion and depth perception from achromatic cues (Lehrer et al., 1988; Hausen and Egelhaaf, 1989), and parallel networks that are involved in computing information about form and color (Zhang and

Horridge, 1992; Zhang et al., 1992; O'Carroll, 1993; Hertel, 1980; Menzel et al., 1986; Backhaus et al., 1987; Strausfeld and Lee, 1991). Insects can learn the organization of visual cues around them and relate visual objects to each other (Cartwright and Collett, 1987; Collett et al. 1993). They can rapidly compute interception courses (Wagner, 1986). These capabilities all attest to a highly sophisticated visual brain. Likewise, olfactory centers and behaviors reach great elaboration in insects (Christensen et al., 1993; Homberg et al., 1989a,b; Kanzaki et al., 1991), and in many groups, such as the ants, behavioral repertoires are driven almost exclusively by complex chemical signals (Hölldobler and Wilson, 1990). These abilities are reflected by the insect's olfactory neuropil where cellular organizations and nerve cell shapes are remarkably like those found in the mammalian olfactory bulb. As a consequence, those working on olfactory processing in vertebrates and insects find easy common ground (Shepherd, 1992). The convergent evolution of many insect and vertebrate neural systems suggests that certain computational tasks necessary for analyzing the sensory world are performed best by a specific neural architecture.

Uniquely identifiable neuropils thought to be involved in learning and memory

Evidence suggests that insects, like vertebrates, possess networks localized within special neuropils that associate, store, and retrieve multimodal sensory information, thus providing the organism with cognition, memory, and predictive behavior. In mammals, such functions have been ascribed both to specific brain regions, as well as to distributed circuits involving many different centers. An example of the first is the hippocampus which, as is now known in mammals, is involved in learning and storage of information about the organism's position within its sensory world (O'Keefe and Nadel, 1978; Wilson and McNaughton, 1993).

The notion that special neuropils in insects are involved in learning and memory originated in the 19th century from comparative observations of social insects. The French natural historian Dujardin (1850) published a study on comparative brain anatomy based on tissues cleared in oils, thereby revealing the shapes of neuropils by virtue of their different optical densities. Dujardin named a bilateral pair of centers the *corps fongiformes*, to indicate their characteristic mushroom-like shapes. Posteriorly, on each side of the brain, a paired cap-like region (calyx) surmounts a stalk (peduncle) extending anteriorly and then dividing into two lobes. Because these centers were exaggerated in eusocial insects, Dujardin suggested that these neuropils were the centers of intelligence, thus echoing sentiments of post-revolutionary France that intelligence epitomized a well-ordered society.

The internal structure of the mushroom body was revealed 20 years after the discovery of the method that became synonymous with the discipline of neuroanatomy. Almost 100 years ago, the American anatomist, F.C. Kenyon, used the Golgi method to describe the principal types of neurons characterizing the honey bee's mushroom bodies (Kenyon, 1896a,b). By demonstrating their major relationship with the olfactory antennal lobes, Kenyon suggested that mushroom bodies were higher olfactory centers. Curiously, until a comprehensive study of the fly's brain (Strausfeld, 1976) and the bee's mushroom bodies (Mobbs, 1982), Kenyon's study was one of only a few that focused on the cellular organization of the insect brain as opposed to descriptions of its gross features. Other important works included Zawarzin's (1913) methylene blue analysis of the optic lobes, Golgi studies on the insect visual system (Vigier, 1908; Cajal, 1909; Cajal and Sanchez, 1915), a sketchy account of neurons in the heteropteran brain (Pflugfelder, 1937), and at least 11 important but little known works by Sanchez. These included studies on the mushroom bodies of cockroaches, bee mechanosensory centers, and bee mid-brain neuropils (Sanchez, 1933, 1937, 1940). Most other descriptions of insect brain anatomy relied either on stains that were not designed to stain neurons, or on reduced silver stains that revealed only cytoarchitecture as well as glial and neuronal cell bodies (Viallanes, 1887a,b; Bretschneider, 1913; Bierbrodt, 1942; Jawlowski, 1948, 1958; Power, 1943, 1946). These unspecific methods supported a general view of the mushroom bodies that departed little from Kenyon's original description.

Neuroanatomical observations on a variety of insects demonstrated important interspecific differences of mushroom body structure. These included: (1) species-specific differences related to the size and extent of pairing of the mushroom body calyces (Pandazis, 1930); (2) differences of calycal volume and concomitant diameter of the pedunculus (Witthöft, 1967; Goll, 1967; Neder, 1959); and (3) differences in the number of lobes arising from the pedunculus (Böttger, 1910; Bretschneider, 1924; Pflugfelder, 1937; Schürmann, 1973; Pearson, 1971; Vilinsky et al., 1994). It is now recognized that in pterygote insects, each calyx consists of a cup-like neuropil composed of dendrites of intrinsic neurons (termed Kenyon cells by Strausfeld, 1976) whose parallel axons and their bifurcations form the pedunculus and its lobes. Intrinsic neurons are supplied by interneurons from the antennal lobes, via the antennoglomerular tract. In eusocial and semisocial insects (such as cockroaches), mechanosensory and visual interneurons also supply the calyx and overlap olfactory terminals there (Mobbs, 1982; Gronenberg, 1986; Mizunami et al., 1993). Outputs from the pedunculi and lobes are represented by "extrinsic" neurons. Some of these provide feedback to the calyces (Gronenberg, 1987). Others project from the lobes to various regions of the central brain (Rybak and Menzel, 1993). An impor-

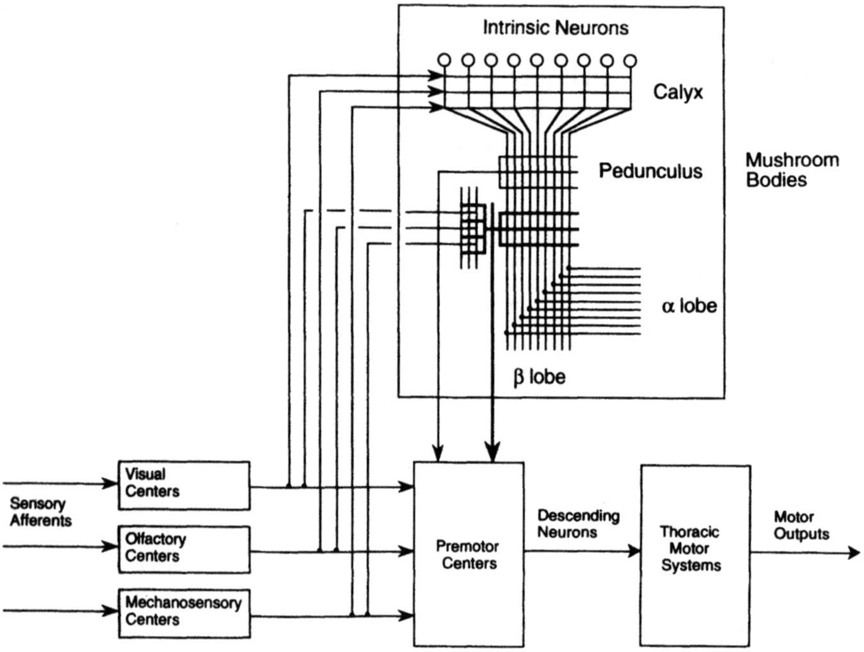

Figure 1. Schematic of the insect mushroom body and its relationship with other supra-esophageal brain areas. Sensory afferents (*left*) supply sensory neuropils (e.g., antennal lobes; optic lobe lobula complex) that connect to premotor neuropils of the ventrolateral and posterolateral deutocerebrum from which arise descending neurons supplying thoracic motor- and local interneuron circuits. Pathways from sensory regions also provide two sets of tributaries. One extends to the calyx of the mushroom body where sensory afferent maps are transformed into distributed and partially overlapping domains that provide convergence onto many thousands of intrinsic neurons. Intrinsic neuron (Kenyon cell) axons supply the pedunculus and lobes. Dendrites of output neurons ("extrinsic neurons") intersect and are visited by Kenyon cell axons. Certain outputs have dendrites so as also to receive, outside the lobes, direct afferent supply from sensory neuropils. Mushroom body output neurons are here figured as projecting to premotor centers. Recurrent axons from the lobes back to the calyces, demonstrated by Gronenberg (1987), are here omitted for sake of clarity as are the many outputs from the α and β lobes.

tant connection is mediated by extrinsic neurons that arise in the mushroom body lobes and project to the ventrolateral and posterolateral deutocerebra, regions containing the initial segments of descending interganglionic interneurons that eventually supply motor centers of the thoracico-abdominal ganglia (Mizunami et al., 1993). This general scheme of organization is diagrammed in Figure 1.

Intrinsic neurons arise from a population of densely packed neuronal cell bodies (perikarya) that partly extend into the calycal cup, as well as over and around its rim. Kenyon cell bodies are chromatin rich, have little cytoplasm, and have the smallest perikarya in the central nervous system. Their traditional name, "globuli cells," has been used by neuroanatomists since early this century and may have been coined by Retzius (1895). Throughout the classical comparative studies on arthropods and annelids, globuli cells have been used as a diagnostic criterion for identifying centers (even in Turbelleria; Hadenfeldt, 1929) that were claimed to be homologous with insect mushroom bodies.

The functional roles of the insect mushroom bodies

Mushroom bodies have long been suspected of controlling complex behavioral repertoires. It has been claimed that electrical stimulation of neuropil in the vicinity of the mushroom bodies in cricket brains elicits extended episodes of courtship behavior (Huber, 1960; Wadepuhl, 1983). In gynandromorph bees, the paired mushroom bodies must have worker-type morphology for the expression of worker-type behavior (Buitkamp-Möbius, 1975). In *Drosophila* gynandromorph mosaics, neuropils associated with the mushroom bodies are required to be male for the individual to demonstrate male behavior (Hall, 1979). In honey bees, mushroom body α-lobes have been specifically implicated in an associative learning task, where the proboscis extension reflex is conditioned to gustatory stimuli using a specific odor or color (Erber et al., 1980). Focused cooling of the α-lobe shortly after the conditioning stimulus blocks short term memory (Erber et al., 1980). Efferent activity in uniquely identified neurons originating from the α-lobes can be modified by olfactory conditioning (Mauelshagen, 1993). This evidence for a specific olfactory role for the α-lobe in insects is corroborated by recordings from the cockroach's α- and β- lobes showing that the α-lobe associates sensory events whereas activity in the β-lobe anticipates motor actions and may be involved in patterning them (Mizunami et al., 1993).

Mutations that give rise to structural defects of the mushroom bodies (Heisenberg et al., 1985) can show defects in olfactory conditioning. The behavioral mutant *dunce* (*dnc*) is deficient in conditioning (Dudai et al., 1976; Tully and Quinn, 1985), habituation (Duerr and Quinn, 1982), and learning-dependent courtship (Hall, 1986). These defects are the result of lesions in a gene encoding cAMP-specific phosphodiesterase (PDE; Davis and Kauvar, 1984). Antibodies raised against *dnc* PDE reveal intense staining in wild-type *Drosophila* mushroom body lobes and calyces (Nighorn et al., 1991). *In situ* hybridization using *dnc* RNA demonstrates intense enrichment in the cell bodies of neurons intrinsic to the mushroom bodies.

Anatomical studies suggest fascinating parallels between the cellular organization of mushroom bodies (Mobbs, 1982) and the neuronal architecture of the mammalian hippocampus (Leonard and McNaughton, 1990), a region that mediates spatial memory (O'Keefe and Nadel, 1978). Moreover, evidence suggests that place memory is not an exclusive attribute of vertebrates. Classical behavioral observations demonstrate that hymenopterous insects use landmarks for locating a site, such as a nest or burrow, which does not itself act as a visual cue (Tinbergen, 1935; Collett and Cartwright, 1983; Cartwright and Collett, 1987). Experiments on cockroaches have demonstrated that mushroom bodies are necessary for place memory in situations where the animal uses far-field visual or olfactory cues to locate a hidden target. Bilateral lesions in one part of the mushroom bodies abolish this ability although the ability to learn the location of a target by direct association between it and a visual cue at the same site is not lost (Mizunami et al., 1993). A similar finding has been found from *Drosophila*, where the chemical ablation of mushroom body neuroblasts, and hence the entire adult structures, perturbs associative olfactory learning but not visual association learning (de Belle and Heisenberg, 1994). In conclusion, attributes ascribed to the insect mushroom bodies must now include the multiplex nature of this brain region in place memory, in the control of complex motor repertoires, and in associative olfactory learning.

The mushroom body as an indicator of arthropod phylogeny: Classical studies

Interest in invertebrate brain evolution was probably inspired by Retzius in the late 19th century, and was continued as a Swedish tradition by Holmgren and then Hanström, both of whom attempted to infer arthropod evolution from brain structures. Holmgren (1916) wrote a concussive treatise comparing the brains of annelids, onychophorans, xiphosurans, chelicerates, crustaceans, and hexapods. Hanström (1928) elaborated on many of Holmgren's ideas, eventually formulating his own theory of arthropod brain evolution. He proposed that the singly segmented preoral brain of polychaetes represents an ancestral form of protoarthropod that was hallmarked by a pair of large mushroom bodies, each associated with glomeruli in the same segment and supplied by chemosensory afferents from the palps. He also proposed that visual centers serving the single lens eyes were connected to the main lobe of the polychaete mushroom body. Thus, according to Hanström, the ancestral mushroom body was bimodal. Hanström suggested that this archaic condition gave rise to the antennated arthropod (hexapod/crustacean) brain, which consisted of two preoral segments where the first segment contained the mushroom bodies. These also received visual and antennal relays. Hanström (1923) stated that in arthropods the antennal supply arose from a separate glomerular neuropil situated in the second preoral segment. Hanström suggested that the chelicerate evolutionary line had secondarily lost chemosensory appendages, as well as the second preoral neuromere. Using araneans to exemplify this, he proposed that the annelid-like mushroom bodies of chelicerates now receive only visual afferents.

Holmgren (1916) described mushroom bodies in the onychophoran *Peripatus* sp. Both his and Hanström's criteria for naming a structure the "mushroom body" relied on features thought to be mutually complementary: a dense cluster of globuli cells from which elongated a homogenous and compact stem of neuropil that subsequently branched into two or more lobes extending towards the midline. Holmgren and Hanström assembled an impressive volume of anatomical evidence demonstrating the existence of this combination of features in annelids, chelicerates, and insects. Holmgren also suggested that mushroom bodies occurred in Myriapoda and apterygote insects. Holmgren and Hanström identified globuli cell clusters in Crustacea, but ascribed unbranched neuropil to them. In some Crustacea (e.g., the amphipod *Callomysis maculata*, and decapod *Pachygrapsus crassipes*), globuli cells and their neuropil were shown to reside in the eyestalk, close to the optic lobe neuropils. In others (e.g., the crab *Calocaris* sp.) globuli cells and their neuropils occur medially in the brain.

Given such obvious differences between crustaceans, chelicerates, and insects, can arthropod mushroom bodies be used as synapomorphic phylogenetic characters? Or is there evidence for homoplasy, with mushroom bodies having independently emerged in different and unrelated invertebrate groups? If they function as association centers, do the mushroom bodies of insects differ characteristically from those of other arthropods? To shed some light on these questions, it is first necessary to review current theories of the phylogeny of the Arthropoda, and to consider basic features of their prognathous blueprint.

Arthropod monophylogeny

Fossil evidence

One of the most active debates about arthopod phylogeny has centered on the interpretation of limb and mouthpart morphologies, with discussions about head segmentation being almost as disputatious (see Rempel, 1975). Most famously, Tiegs and Manton (1958; see also Manton, 1964, 1973), and Anderson (1973) proposed that "arthropods" must be divided into at least two groups, the "biramia" and "uniramia," which cannot possibly have arisen from a common ancestor. The former were declared unified by biramous limbs in which the leg consists of a main shaft giving rise to a second outgrowth from one of its segments. In contrast, the diagnostic feature of uniramia was an unbranched appendage typical of onychophorans, myriapods, and hexapods. Biramia were also distinguished by mouthparts, typical of chelicerates and crustaceans, which bite or masticate with gnathobasic endites of a biramous multisegmented mandible. Uniramians were supposed to have an unbranched mouthpart that bit with the tip. Support for a separate uniramian line (Sawyer, 1984) proposed that uniramian ancestry might have arisen from terrestrial clitellid-like ancestors. Sawyer argued that because certain modern land leeches possess unbranched limb-like protuberances and single piece mouthparts they accord with Manton's criteria for uniramia. Sawyer cites the terrestrial leeches *Bogabdella diversa* and *Colombella ringueleti* (Richardson, 1969; Mané-Garzon, 1973) as having evolved lobopod-like branchia whose internal musculature and hydrostatic eversion suggest a model for an evolutionary precursor of the onychophoran lobopod. Sawyer suggested that further affinities between onychophorans and clitellids were indicated by their unsegmented entognathous mandibles, an arrangement distinct from the Myriapoda but reminiscent of the so-called "lower" hexapods and Collembola. Further similarities claimed between Clitellata and Onychophora, Myriapoda, and Hexapoda are internal fertilization, egg yolk, and an absence of a plankton-like larval stage. Similarities in early embryogenesis of clitellids and onychophorans were also emphasized by Anderson (1973).

Recent studies by Kukalova-Peck (1991) on paleozoic fossil insects assert that the division of "arthropods" into bi- and uniramia is spurious. Observations of the paleozoic diaphonopteroid insects *Uralia* and *Paruralia,* have resolved mandibles homologous to the sliding gnathobase typical of crustaceans, as well as leg-like mandibles and labial palps, polyramous thoracic legs and abdominal leglets. Kukalova-Peck (1991) and Smith (1988) also stress that because most studies of leg segmentation have ignored or missed the primitive condition of 11 "segments," evolutionary speculations based on misinterpreted leg morphology can simply be discounted.

For many students of external morphology, head segmentation is held to be as important in defining relationships as are the appendages. Much debate pivots around the existence of an acron, a presegmental structure anterior to the first metameric segment of annelids, which "...stops food from falling out of the mouth" (Kukalova-Peck quoting Manton, 1977). In insects and in crustaceans, there is no evidence of an acron (Smith, 1988; Kukalova-Peck, 1991). It is the first somite that provides innervation to a structure termed the clypeolabrum, primitively derived from appendages of the first head segment (Smith, 1988; Kukalova-Peck, 1991). Kukalova-Peck has pointed out that atavistic (homeotic) mutants such as *pterophthalmia* and *ophthalmopedia* (Goldschmidt and Ledermann-Klein, 1958) possibly revive this ancestral state in expressing epicoxal structures that interpose between the compound eye and lateral ocellus. However, Sharov (1966) has suggested that the eyes reflect an extra somite at this level (see below).

Arthropod monophylogeny is strongly supported from embryological observations on crustaceans (Snodgrass, 1938) and on chelicerates (Sharov, 1966; Weygoldt, 1979), suggesting that homologous coxopodites arise from the first segmental Anlagen. Kukalova-Peck (1991) and Smith (1988) concluded that, in insects, the second cephalic segment gives rise to antennae (modified legs), which are expressed as true mesothoracic legs in homeotic mutants and regenerates (Stocker and Lawrence, 1981; Edwards et al., 1989). Insect antennae are homologous to that segment of crustaceans bearing the first antennal pair (the antennules). In crustaceans, the second antennal pair (the antennae) arise from the third cephalic segment and are homologized with the chelicerate chelicerae (Smith, 1988). Crustacean antennae and chelicerate chelicerae are suggested to be represented in the hexapodan third (tritocerebral) segment as the superlinguae. These, with other modified third leg segments, comprise the hypopharynx (Smith, 1988). Thus, external morphology indicates three prognathous head segments which could be reflected by the three traditional preoral (prognathous) neuromeres: the pro-, deuto-, and tritocerebrum (Strausfeld and Bacon, 1983). Sharov (1966) has suggested a fourth prognathal somite representing the compound eyes, ranking them as ancestral appendages situated between the antennae and clypeolabrum. This idea is not so bizarre as it first seems, and is supported by atavistic mutants in *Drosophila* in which leg and wing vestiges are expressed instead of compound eye (Goldschmidt and Ledermann-Klein, 1958). The same evolutionary heritage may also be supported in insects and crustaceans by the growth of ectopic antennae or antennules after compound eye removal (Krizeneky, 1913; Sandeman and Luff, 1974). Sharov's idea that the eye is representative of an additional somite is attractive when considering the neuronal relationships of the eye's retinotopic neuropils with the rest of the preoral brain. In adult insects, neurons carrying information from the optic lobe project to neuropils lateral and posterior to the antennal lobe (the neuropil classically termed the deutocerebrum): the lateral, and posterolateral deutocerebrum of Strausfeld (1976), and the dorsal deutocerebral lobes of Sanchez (1937).

Irrespective of arguments concerning the segmental affiliations of the eye, arthropod visual systems have been used in support of monophylogeny. Paulus (1979) makes a strong case in arguing that even though many arthropods show loss or great modification of eye structures, lateral compound eyes are present in fossil arthropods (e.g., trilobites) as well as in the major extant subphyla: hexapods, chelicerates (xiphosurans), and crustaceans. Paulus argues that despite a secondary loss of facets (as in araneans) the fundamental organization of the photoreceptor, and its arrangement into clusters (rhabdoms), is a unifying feature.

Combined character analysis

A relatively new method for investigating conserved characters is molecular analysis of conserved ribonucleic acid sequences. Two recent papers, one analyzing mitochondrial 12S RNA (Ballard et al., 1992), the other nuclear 18S DNA (Wheeler et al., 1993), have provided cladograms that, while not entirely in agreement, support arthropod monophylogeny. In Ballard et al.'s (1992) study, a consensus tree derived from maximum parsimony analysis (Swofford, 1991) supports the view that arthropods, including the Onychophora, are monophyletic. However, annelids are monophyletic, as are the mollusks. This analysis promotes the view that Onychophora are not a primitive link between annelids and arthropods, but are a sister taxa of the Arthropoda along with chelicerates, crustaceans, and insects. The myriapods, rather than the onychophorans, exemplify the earliest evolutionary branch of the arthropods, a view which Ballard et al. (1992) suggest is supported by a mid-Cambrian marine myriapod-like fossil (Robison, 1990).

An even broader taxonomic range and a larger set of diagnostic characters underpin Wheeler et al.'s (1993) analysis. Derived from 100 characters (including embryological features), two maximum parsimony cladograms support a common ancestry for polychaetes, onychophorans, and all other arthropods (Crustacea, Myriapoda, and Hexapoda being sister groups, divergent from the pycnogonids and chelicerates). This evolutionary tree differs little from a cladogram based on molecular analysis alone and on a consensus cladogram based on both molecular and morphological data from which Wheeler et al. (1993) demonstrate six phylogenetic branches originating from an assumed annelid-like ancestor. Monophylogeny thus includes annelid-like protoarthropods, onychophorans, trilobites, pycnogonids, chelicerates, crustaceans, and hexapods. As Wheeler et al. (1993) point out, their cladogram is almost identical to that proposed purely on morphological grounds by Snodgrass in 1938.

Evidence for head segmentation in insect preoral neuropil

In *Drosophila*, four preoral and three postoral segments have been inferred from expression patterns of the segment polarity genes *engrailed* and *wingless* during embryogenesis (Schmidt-Ott and Technau, 1992). Four prognathal centers of *engrailed* expression contribute to the supraesophageal brain, and are suggested to be segmental markers, each contributing to a neuromere of the central nervous system. *Engrailed* expression located in a pair of anterior dorsal lobes has been attributed by Schmidt-Ott and Technau (1992) to the clypeolabrum, the anteriormost "legged" segment of Kukalova-Peck and Smith. However,

Schmidt-Ott and Technau (1992) suggest that clypeolabral cell expression does not contribute to the central nervous system. The *engrailed* head spot is attributed to an ocular segment that later becomes fused to the secondarily derived optic lobes (Campos-Ortega and Hartenstein, 1985). The *engrailed* antennal spot is attributed by Schmidt-Ott and Technau (1992) to the antennal or deutocerebral segment. The *engrailed* intercalary spot lies immediately anterior to the mandibular neuromere and presumably corresponds to the adult tritocerebrum.

A crucial finding by these authors is that five neuroblasts adjacent to the lower anterior edge of the *engrailed* head spot contribute to the mushroom body/antennal lobe complexes of the *Drosophila* brain. Four neuroblasts contribute to the intrinsic cells (Kenyon cells) of the mushroom body (Ito and Hotta, 1992; Prokop et al., 1991), whereas the fifth neuroblast provides the lineage for cell bodies of the antennal lobes (its intrinsic neurons and projection neurons, the latter supplying axons to the mushroom body calyces).

Schmidt-Ott and Technau's (1992) observations support other lines of evidence (Sharov, 1966; Rohrschneider, 1968) suggesting four prognathal segments in both insects and crustaceans where, in the latter, remnants of two pairs of coelomic sacks reside anterior to the first antennal segment. Ablation studies of early insect embryos have contributed to the notion of four prognathal "proliferation centers," each providing segmental components (Wada, 1966a): anteriorly the clypeolabrum, followed by the eyes, optic lobes, mushroom bodies, and the central body. The third and fourth centers relate to the antennal segments (and its associated neural structures, the antennal lobes), and the intercalary segment, or tritocerebrum.

While postulating four prognathal segments, Wada's (1966a,b) and Schmidt-Ott and Technau's (1992) studies ascribe only three of them to the preoral brain, placing the optic lobes at the same somite level as the protocerebrum, a feature that is difficult to reconcile with neuroanatomical observations of adult insects and that may also be contradicted by recent studies of neuroblast organization in embryonic brain. Observations on the organization and proliferation of brain neuroblasts in grasshoppers (Zacharias et al., 1993) suggest that the complex arrangements typical of the adult insect brain arise from a very simple intial arrangement of precursors arranged as five proliferation groups: a pars intercerebralis group, a protocerebral group, and a "lateral protocerebral" group externally flanking a deutocerebral group, itself preceding the tritocerebral group along the neuraxis. The "lateral protocerebral" group may correspond in adults to the areas termed the ventro- and dorsolateral deutocerebrum (Strausfeld and Gronenberg, 1990), which in adult holo- and hemimetabolous insects is supplied by optic lobe outputs.

Evolutionary position of mushroom bodies and antennal lobes

Evidence for arthropod monophylogeny, and current views on the tri- or quadripartite organization of insect brain segmentation, provides the background for phylogenetic studies of specific brain regions.

Here, we summarize our initial comparisons between an annelid and six arthropods representing six major groups: onychophorans, chelicerates, myriapods (diplopods and chilopods), crustaceans, and insects. We compare these with a polychaete annelid, *Nereis bicolor*, to test Hanström's suggestion that polychaetes may represent an ancestral form of protoarthropod.

All observations are from reduced silver preparations, using Bodian's (1937) original method on paraffin-embedded material, fixed either in AAF (alcohol, acetic acid, formalin) or in Carnoy's fixative (alcohol, formalin, chloroform). Onychophorans (see acknowledgements) had been fixed entirely in 70% alcohol, and were postfixed in AAF. Drawings in Figures 3 and 4 are coded as follows: globuli cells are stippled; pedunculi and lobes are carmine, separate antennal lobes are green, and the antennal glomerular tract linking them to the pedunculus or calyx is blue. Chemosensory afferents are indicated in red from the antennal (or palp) nerve(s). Antennal lobe projection neurons are shown as large filled circles. Visual neuropil supplying the calyx is shown brown with its tract to the calyx in gray. Kenyon's (and then Holmgren's and Hanström's) original criteria for identifying mushroom bodies have been expanded to include their morphological association with chemosensory pathways, again reflecting a relationship originally suggested by Hanström. Figure 2 illustrates in an insect (*Drosophila melanogaster*) the diagnostic features for identifying a neuropil region as a "mushroom body:" lobed neuropil comprising parallel intrinsic neurons arising from a cap of globuli cells supplying dendrites to the head or "calyx" of a pedunculus and visited by afferents from second order chemosensory interneurons. The calyx in insects is sometimes entirely divided into two components, and in social insects receives relays from visual and mechanosensory neuropils. Output neurons insert their dendrites at specific locations down the pedunculi and lobes, and send axons to other specific brain areas.

A polychaete annelid: Nereis bicolor

In the polychaete *Nereis bicolor* (Fig. 3A), perikarya supplying the supra- and subesophageal ganglia have a range of diameters. This contrasts with the onychophoran and diplopod brains, both of which contain neurons of roughly similar diameters. *Nereis* also differs from the onychophoran and diplopod representatives in showing clear evi-

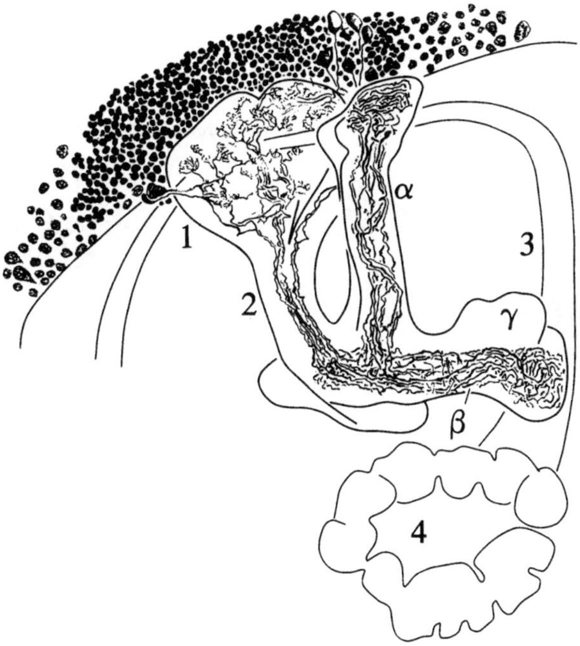

Figure 2. Diagnostic anatomical features of an insect mushroom body, illustrated from *Drosophila melanogaster*. Cell bodies and outline are derived from a Bodian preparation. A double calyx (*1*) merges into a pedunculus (*2*) that then gives rise to several lobes (α, β, γ *lobes*). The main input to the mushroom body is to the calyces, and is supplied by a tract of projection neurons (*3*) arising from the glomerular olfactory neuropil of the antennal (olfactory) lobe (*4*). A dense cluster of basophilic globuli cells above the calyx gives rise to about 2400 intrinsic neurons (Kenyon cells), some of which are shown impregnated by the Golgi method. Their dendrites are restricted to the calyces and their (approximately) parallel axons invade one or more of the branching (α, β) or parallel (γ) lobes.

dence of lateralized neuropils: clearly demarcated regions of homolateral neuropil are connected heterolaterally by tracts. Some other neuropils extend across the mid-line. On each side of the brain, and dorsal in the preoral (supraesophageal) lobes, an extensive cap of globuli cells provides many hundreds of small fibers to an undivided pedunculus which projects ventro-posteriorly and then divides into parallel lobes that extend towards the mid-line, but do not cross it. Contrary to Hanström's figures of nereids, there is no distinct olfactory lobe in the present species. Instead, bundles of afferents from the palp nerve terminate as neuroma-like protoglomeruli. These flank the lateral margins of the pedunculus from which arise branches of intrinsic neurons that extend out to invade or enwrap the olfactory protoglomeruli. Some protoglomeruli also arise from a more medial chemosensory nerve. These abut the inner edge of the pedunculus, just before its division into medial lobes. In *Nereis*, therefore, four criteria for mushroom bodies have been met: (1) an abundance of globuli cells arranged as a dense cluster; (2) many hundreds of small diameter parallel fibers comprising

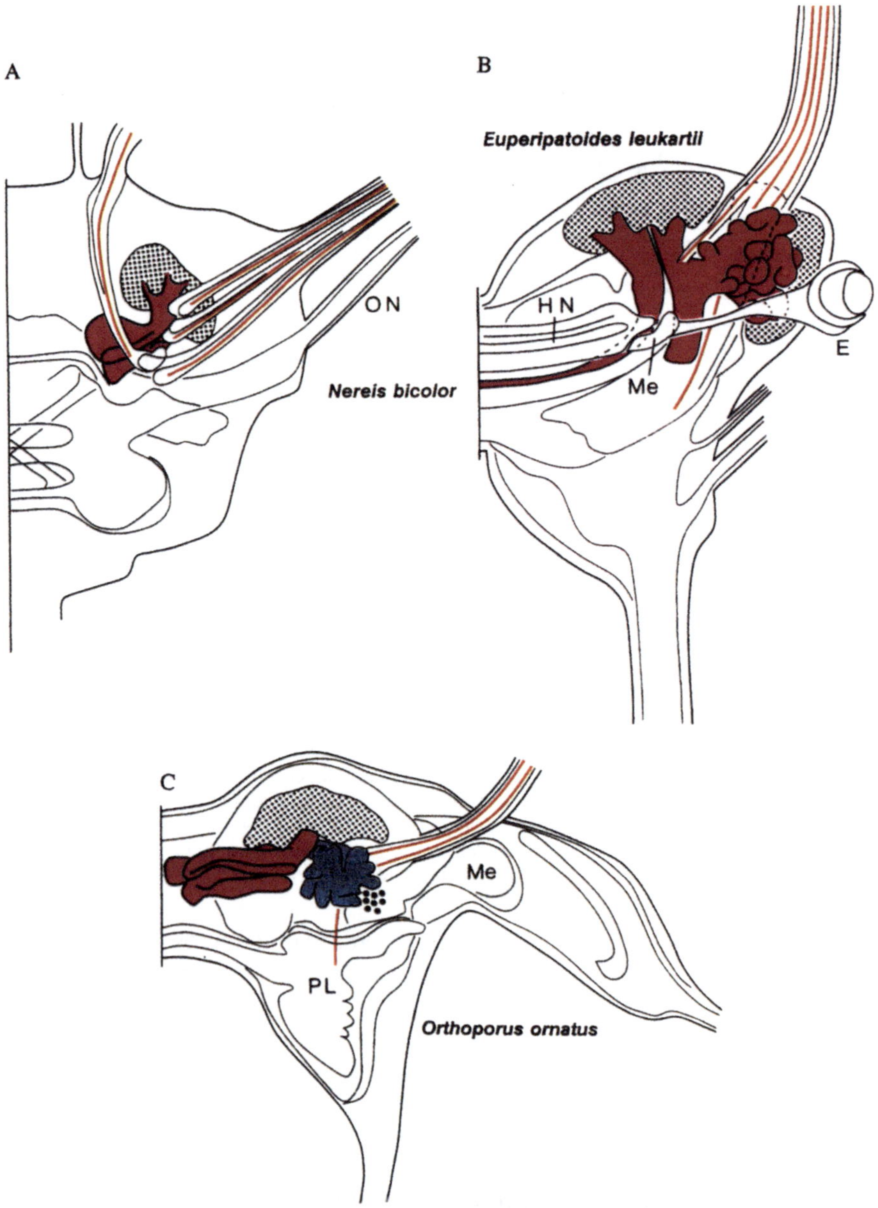

Figure 3. Mushroom bodies in representative arthropod species. The right half of each brain (preoral segments) is shown from its dorsal aspect. Globuli cells *stippled*, pedunculi and lobes *carmine*, antennal lobes *green*, glomerular projection neuron perikarya *filled cicrles*, antenno-glomerular tract *blue*, optic lobe and inputs to mushroom bodies *brown* and *gray*, respectively. Other explanations in text.

(3) a pedunculus dividing into several lobes with (4) the entire structure associated with the chemosensory supply.

An onychophoran: Euperipatoides leukartii

In this species, the brain lies entirely above the esophagus (Fig. 3B). Two large nerve cords extend along either side of the gut, but do not adopt a convincingly ventral position. Commissures extend over and beneath the digestive tract and are organized segmentally. As originally shown by Holmgren (1916), clusters of cell bodies surround each nerve cord, becoming numerous at the origin of peripheral nerve roots. There are no obvious ganglia.

The cell body rind of the brain mainly comprises some thousands of uniformly small perikarya (8–10 μm diameter). We have observed fewer than six pairs of neurons having diameters greater than 15 μm. Most neuropils appear to extend across the mid-line, there being less lateralization in this brain than in the brain of *Nereis*. However, anteriorly in the preoral neuropil, there are two large islets of extremely small and intensely basophilic perikarya, ranging from 5–8 μm in diameter. The medial islet supplies many hundreds of axons to a neuropil that consists of several parallel lobes, each comprising dense neuropil. Unlike *Nereis*, there is no single pedunculus which then divides. Instead, each lobe arising from the globuli cells extends posteriorly and then medially. The lobes are clearly distinguished from the surrounding neuropils and, together with their cell bodies, conform to the Holmgren-Hanström criteria for identifying mushroom bodies. Again, as in the polychaete, there is no distinct olfactory lobe. Instead, a second antero-lateral islet of globuli cells is associated with a glomerular outswelling of the most lateral and largest lobe to arise from the medial globuli cell cluster. The glomerular nature of this mushroom body component is that the lateral group of globuli cells represents a specialized subset of neuronal processes that, instead of loosely enwrapping olfactory terminals, as is observed in *Nereis*, indeed comprise a specialized receptive neuropil into which chemosensory afferents invade. Many thousands of olfactory afferents originate from a massive antennal nerve, which also sends some tributaries directly into one other mushroom body lobe. In addition, some lateral branches of the antennal nerve extend posteriorly into the lateral posterior brain neuropil.

A parallel, and possibly multisensory organization for the onychophoran mushroom body, is further suggested by the smallest and anatomically most clearly defined medial lobe. This appears to be continuous with its contralateral partner. A small kidney-shaped neuropil (Me), receiving visual relays from the single lens eye (E), abuts the medial lobe and also appears to be connected to a second heterolateral neuropil (HN). The layered organizaton of this second heterolateral neuropil is reminiscent of the arachnid "central body" (Strausfeld et al., 1993). Holmgren (1916), describing *Peripatus capensis*, also suggested more than one group of globuli cells, each providing a parallel pedunculus-like lobe. However, in his description of that species, the lobes did not cross the mid-line. Another departure from our observations of *Euperipatoides* is evident in Holmgren's figures of a separate glomerular neuropil linked to the mushroom body by nerve bundles.

A diplopodan: Orthoporus ornatus

The brain of this species (Fig. 3C) is divided into supra- and suboesophageal components, the latter supplying a ventral nerve cord linking segmental ganglia. It is again difficult to identify many discrete homolateral neuropils. Like the onychophoran brain, several "undifferentiated" neuropils appear to be arranged as successive heterolateral partitions from anterior to posterior, preorally, possibly reflecting the segmental framework of this brain. Also, like the onychophoran brain, the cell body rind is characterized by thousands of small perikarya. Very few large cell bodies have been identified.

A prominent islet of minute globuli cells (4–6 μm diameter) supplies axons to an elongated dense neuropil. This consists of a short and thick unbranched pedunculus which then extends laterally, dividing into several parallel lobes and reaching the midline but not crossing it. A region of the pedunculus just beneath the globuli cells is supplied by a prominent tract of axons arising from a separate glomerular neuropil that is massively invaded by afferents from the antennal nerve. This peripheral neuropil is reminiscent of the insect antennal lobe in having associated with its glomeruli several clusters of larger cell bodies (6–10 μm diameter) that supply projection neuron (second order) relays to the mushroom body's pedunculus. As in the onychophoran, a lateral branch of the antennal nerve projects into a posterior lobe (PL), possibly representing the second preoral segment. In addition to separate antennal (olfactory) lobes, *Orthoporus* also has a well-defined visual lamina and medulla (Me) the neuropils of which are arranged as columns.

A chilopodan: Lithobius variegatus

The supra- and subesophageal ganglia differ from those of diplopods in having many discrete homolateral neuropil regions, as is seen in insects (Fig. 4A). The cell body rind consists of distinctly separated clusters of perikarya varying greatly in size, the largest giving rise to prominent interganglionic neurons that are reminiscent of large descending neurons typical of insects.

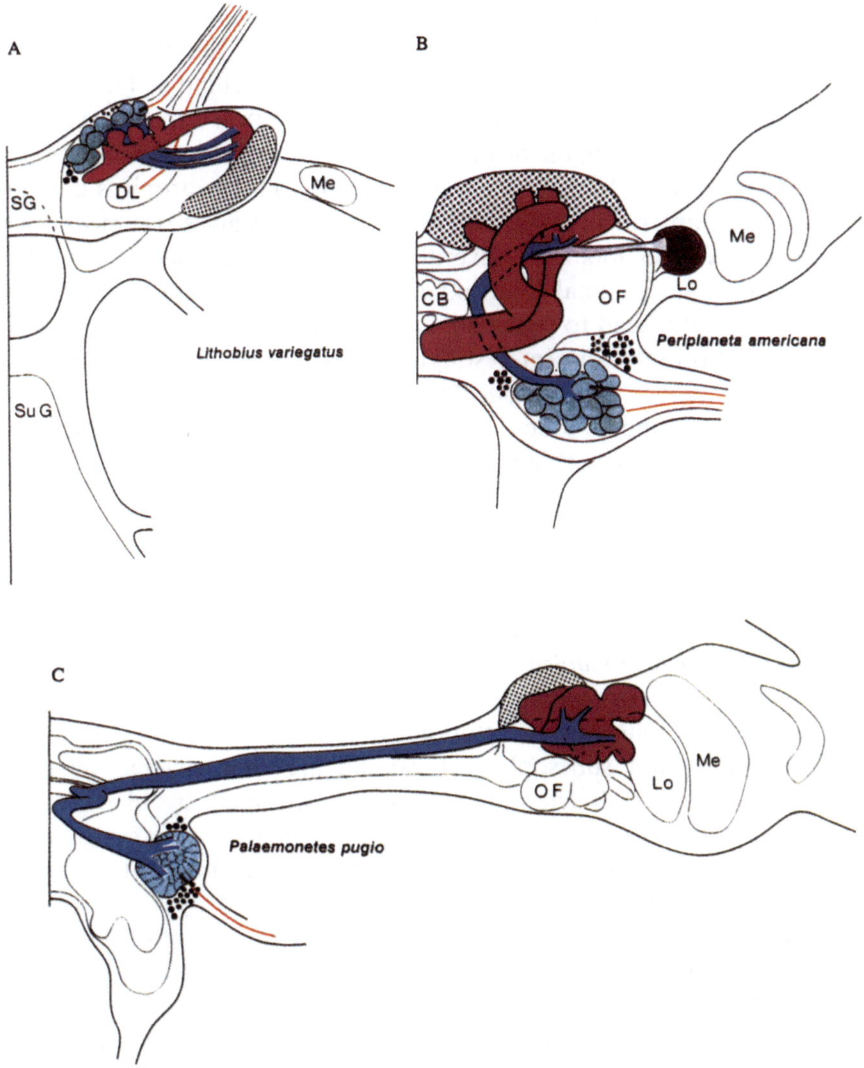

Figure 4. Mushroom bodies in representative arthropod species. Color coding as in Figure 3. Explanations in text.

An antennal nerve innervates individual glomeruli of an anterior antennal lobe. A lateral strand of the antennal nerve bypasses the glomeruli and projects to a region of neuropil behind it (DL: dorsal lobe). Glomeruli give rise to a diffuse arrangement of axons that project dorsolaterally to a region of neuropil lying directly beneath a postero-lateral cluster of extremely small globuli cells. This neuropil is contiguous with an anteriorly projecting pedunculus that gives rise medially to a second parallel lobe and several glomerulus-like outswellings. This ensemble of an antennal lobe, antennal glomerular tracts, globuli cells, and a calyx-like head to the pedunculus suggests close affinities with the insect mushroom body. Interestingly, an early study by Böttger (1910) on the apterygote *Lepisma saccharina* shows a similar organization of the mushroom body lobes, where the mushroom body pedunculus provides several outswellings along its length.

An insect: Periplaneta americana

So far, our own studies have surveyed some pterygote species (*Apis mellifera*, *Vespa germanica*, *Periplaneta americana*, *Sarcophaga bullata*, *Hoplia farinosa*, *Drosophila melanogaster*, *Locusta migratoria*, *Manduca sexta*) and the following description generalizes.

The supra- and subesophageal ganglia are characterized by many discrete homolateral synaptic neuropils, connected by characteristic tracts of axons and associated with groups of perikarya varying greatly in size (Fig. 4B). The central area of the protocerebrum is distinguished by several interlinked neuropils that are dominated by a heterolateral scallop-shaped region comprising a dramatic arrangement of fan-shaped dendrites and terminals (Williams, 1975; Strausfeld, 1976; Hanesch et al., 1989). The entire ensemble, which includes a heterolateral torus-shaped neuropil, accompanied by a pair of compact ball-like centers, is called the central body complex. Its possible relationship to heterolateral neuropils in other arthropods is the subject of a separate study (Strausfeld, in preparation).

On each side of the brain, globuli cells are grouped dorsally, at the rear of the protocerebrum, towards its lateral margins. The perikarya are usually minute (diameters of 4–6 μm in *Drosophila* and 8–10 μm in *Periplaneta*), and their number in each mushroom body ranges from less than 1000 (in *Notonecta*) to nearly half a million in a large adult *Periplaneta*. Globuli cells provide dendrites to the calyx. As reviewed early in this chapter, different types of insects have greater or lesser elaboration of their mushroom bodies. However, one character seems to be ubiquitous and unique. This is the duplication of the mushroom bodies, there being a pair of calyces on each side of the brain. The double nature of the mushroom body is selectively revealed by the ethyl gallate staining method, which, in *Apis*, for example, shows a dense osmiophilic barrier separating the pedunculi and lobes into halves longitudinally. In bees, the medial calyx supplies the medial half of the pedunculus, the upper half of the medial lobe (termed the β lobe by Vowles, 1955) and the medial half of the dorsal lobe (Vowels's α lobe). The lateral calyx provides intrinsic neurons to the lateral pedunculus, lower half of β, and lateral half of α. In certain orders of insects, the medial lobe is clearly divided into several discrete subunits, one of which, in Lepidoptera, has been termed the γ-lobe (Pearson, 1971). it receives contributions from both calyces. Its homologue has been identified in Orthoptera (Schürmann, 1973) and in Diptera (Strausfeld, 1976).

In certain insects, there is irrefutable evidence for other sensory relays to the mushroom bodies: from the optic lobes (e.g., Gronenberg, 1986), from areas receiving optic lobe outputs (Strausfeld, 1976), from the subesophageal (gustatory) neuropils (Strausfeld, 1976), and from mechanosensory neuropils of the deutocerebrum (Mizunami et al., 1993). Thus, in many insects, the mushroom bodies are multimodal: terminals of neurons carrying different modalities occupy specific domains in the calyces (Mobbs, 1982) where they establish connections onto the dendrites of intrinsic cells (Schürmann, 1987). In some areas of the calyx Kenyon cell dendrites are supplied by a single modality, in others their dendrites extend through successive layers of neuropil, each receiving a separate modality. Thus, in the most complex of mushroom bodies, such as in bees and wasps, certain subsets of Kenyon cells may compute relationships within only a single modality, whereas others compute relationships among several modalities.

A crustacean: Palaemonetes pugio

The supra- and subesophageal ganglia comprise many discrete homolateral neuropils with prominent tracts linking them (Fig. 4C). Cell bodies are clustered into discrete groups within the rind, and their dimensions vary greatly, some of the largest belonging to descending neurons. Although small medial heterolateral neuropils have been identified in Crustacea (see Sandeman, this volume), there is no evidence for a central neuropil comparing in complexity to the central body complex of insects.

The wedge-shaped glomeruli of the crustacean olfactory lobe provide multiglomerular relay neurons that form a conspicuous tract (Mellon et al., 1992). This projects first to the mid-line, and then courses anteriorly and outwardly to end in dense neuropil of the hemiellipsoid body. In many decapod crustaceans the hemiellipsoid body resides in the eyestalks. However, as shown by Hanström and others, this is not the case in all decapods. In some crabs, the hemiellipsoid body resides medially and anteriorly in the protocerebrum.

The hemiellipsoid body is associated with many thousands of minute globuli cells, the axons of which contribute to its internal matrix. This, however, does not consist of elongated lobes, and thus does not conform to the original criterion for mushroom body identification. Instead, the region is usually folded into hillocks and sulci. Possibly, this difference may be trivial, or may have some deeper evolutionary significance.

A chelicerate: Cupiennius salei

Thus far, our own studies have sampled araneans and a solipugid. In the wandering spider *Cupiennius salei*, the supraesophageal brain is small, and is continuous with a much larger and metamerically fused prosomal ganglion. This ventral nervous system contains the bulk of integrative neuropils (Babu and Barth, 1984).

Although the supraesophageal neuropil of *Cupiennius* is clearly divided into discrete centers linked by fiber tracts, there is no evidence for a mushroom body associated with chemosensory pathways (Strausfeld and Barth, 1993). Instead, as described below, globuli cells situated anteriorly in the brain provide third order retinotopic neurons that connect the secondary eye medullas to the distal head of a lobed structure (Hanström's "bridge") that is continuous across the mid-line (Fig. 5).

Discussion

Mushroom bodies as a character for evolutionary analysis

We have employed a limited set of characters to identify certain preoral (prognathous) neuropils as mushroom bodies. Extending Dujardin's original description of paired mushroom-shaped centers, we have added to this the feature of compact neuropils arising from dense aggregations of chromatin-rich cell bodies supplying its compact neuropil. To these criteria have also been added the elongated nature of the neuropil, and its subdivision into parallel lobes and outswellings. The paired nature of the mushroom bodies reflects their position as bilaterally symmetric homolateral centers, as opposed to neuropils shared by the left and right brain halves.

The presence of mushroom bodies in *Nereis bicolor* supports the notion that these centers might have been present in polychaete-like

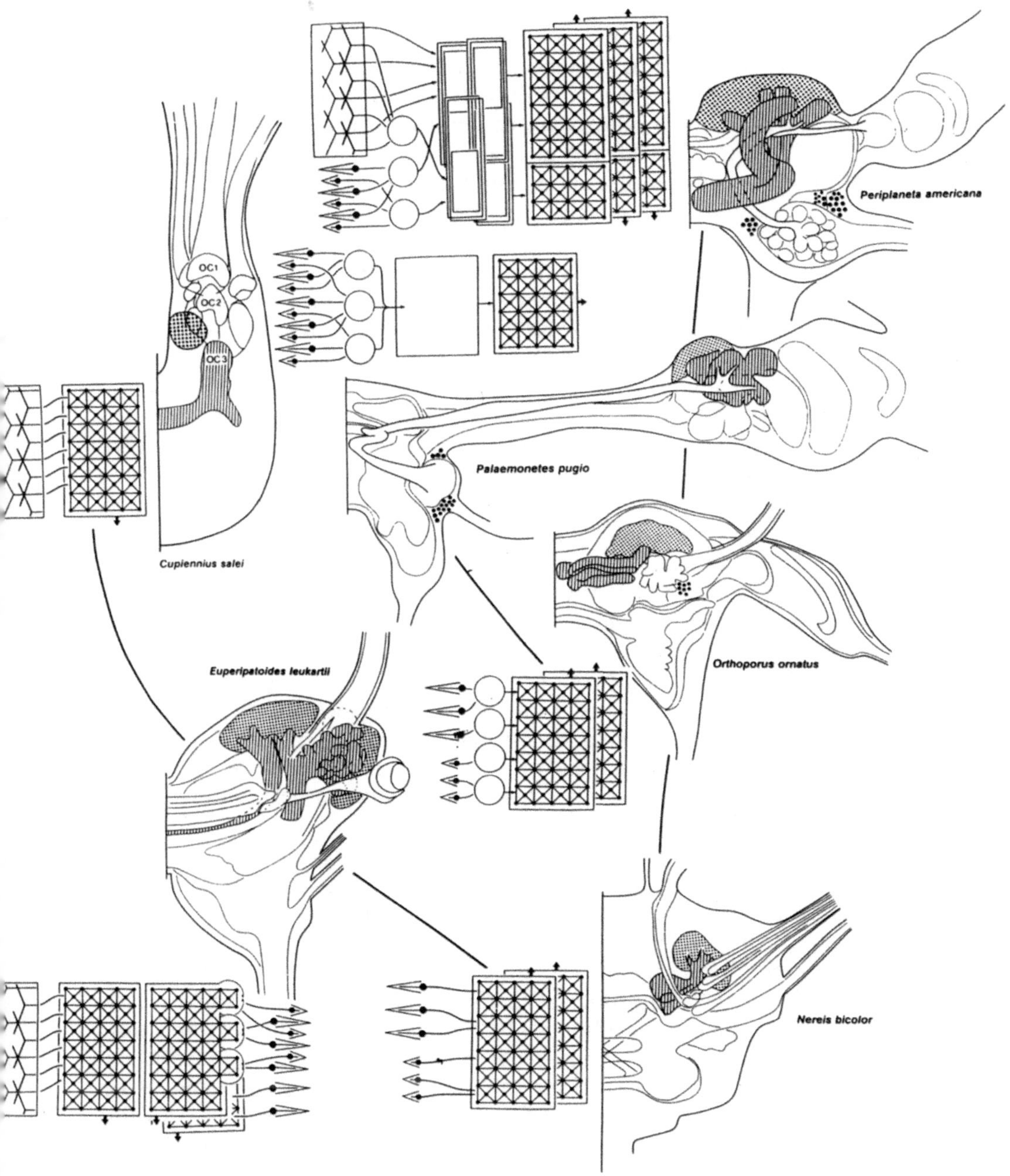

Figure 5. Schematic suggesting progressive elaboration of associative pathways, and their possible modifications in crustaceans. A possible modification can be suggested in araneans, by elimination of the onychophoran-type chemosensory component. The diagram shows modern examples representing a progression of characters, not an evolutionary sequence of species.

protoarthropods, conforming to Wheeler et al.'s (1993) placement of annelids as a sister group to the arthropods. However, in modern arthropods, we see clear evidence of divergent evolution in brain organization, with the mushroom body represented in myriapods and hexapods, but probably modified as the hemiellipsoid body in crustaceans, and greatly altered or absent in chelicerates (see below).

Many more species of each group must be studied and compared to reach a consensus about the characteristic features of each taxon. However, the Onychophora may play a key role in interpreting the position of the chelicerates with respect to arthropod brain evolution. Using 6 taxa, and 20 characters derived from our own observations (Tab. 1), we have performed the first phylogenetic analysis using parsimony (PAUP: Swofford, 1991) on arthropod central nervous tissue. Maximum parsimony analysis reveals 10 shortest and equally parsimonious trees, six of which are shown in Figure 6A–C. The strict consensus of all 10 trees (Fig. 6D) suggests that the representative annelid, onychophoran, and aranean (chelicerate) together comprise three groups, each being a sister to a super-group consisting of hexapods, myriapods, and crustaceans. The several possible cladograms suggested by parsimony analysis (Figs 6A–C) show that if the few extant species used so far are representative of annelid and arthropod orders, then these are probably derived from a common protoarthropod-like ancestor, with the position of onychophorans being ambiguous. In half of the 10 best trees, mushroom body morphology suggests an onychophoran-like ancestor as a point of divergence of the chelicerate from the hexapod, myriapod, crustacean branch. In the other half, the Onychophora and Chelicerata are shown to be more closely related, comprising a sister group to the crustacean-hexapod-myriapod assembly.

The evolutionary position of the Onychophora has long been debated. Recent molecular studies (Ballard et al., 1992) place it as a sister group of the chelicerates, with the myriapods forming a group separate from all other arthropod ensembles (including the onychophorans). The annelids are suggested to be a monophyletic group distinct from the arthropods. Wheeler et al. (1993), and some classic evolutionists such as Snodgrass (1938), suggest that annelids and onychophorans arise successively from branches of a monophyletic progression. Our own observations support Holmgren's (1916) studies on *Peripatus* suggesting that onychophorans share features of annelids and chelicerates. Holmgren describes a multiply-lobed mushroom body, each lobe originating from separate islets of globuli cells. Holmgren also identified a dorsal heterolateral center he termed the "central body." Unlike our observations on *Euperipatoides*, he did not ascribe a visual pathway to this or any part of the onychophoran mushroom body. Although our present preparations of *Euperipatoides* are not ideal (prepared from material stored in alcohol), the close juxtaposi-

Table 1. Matrix table of characters scored against taxa. 1 = character present; 0 = character absent; 2 = not applicable; ? = not known

		1	2	3	4	5	6	7
		Annelids	Araneans	Onychoph.	Hexapods	Chilopods	Diplopods	Crustaceans
1	Globuli cells	1	1	1	1	1	1	1
2	Homolateral lobes	1	0	1	1	1	1	1
3	Chemosensory afferents	1	0	1	1	1	1	1
4	Supply to pedunculus	1	0	1	0	0	0	0
5	Supply to glomerular pedunculus	0	0	1	0	0	0	0
6	Supply to antennal lobes	0	0	0	1	1	1	1
7	Antenno-glomerular tract	0	0	0	1	1	1	1
8	Glomeruli ball-like	0	2	1	1	1	1	0
9	Glomeruli wedge-shaped	0	2	0	0	0	0	1
10	Lobes elongated	1	1	1	1	1	1	0
11	Lobes folded	0	0	0	0	0	0	1
12	Multiple lobes	1	0	1	1	1	1	1
13	Mid-line confluence	0	1	1	0	0	0	0
14	Visual supply only	0	1	0	0	0	0	0
15	Mixed modalities	0	0	1	1	0	0	?
16	Modalities separated in (calyx)	2	2	1	0&1	2	2	?
17	Modalities overlap (calyx) lobe	2	2	0	1	2	2	?
18	Presence of calyx	0	0	0	1	1	0	?
19	Double calyx	0	2	0	1	0	0	?
20	Intrinsic neurons	1	0	1	1	1	1	1

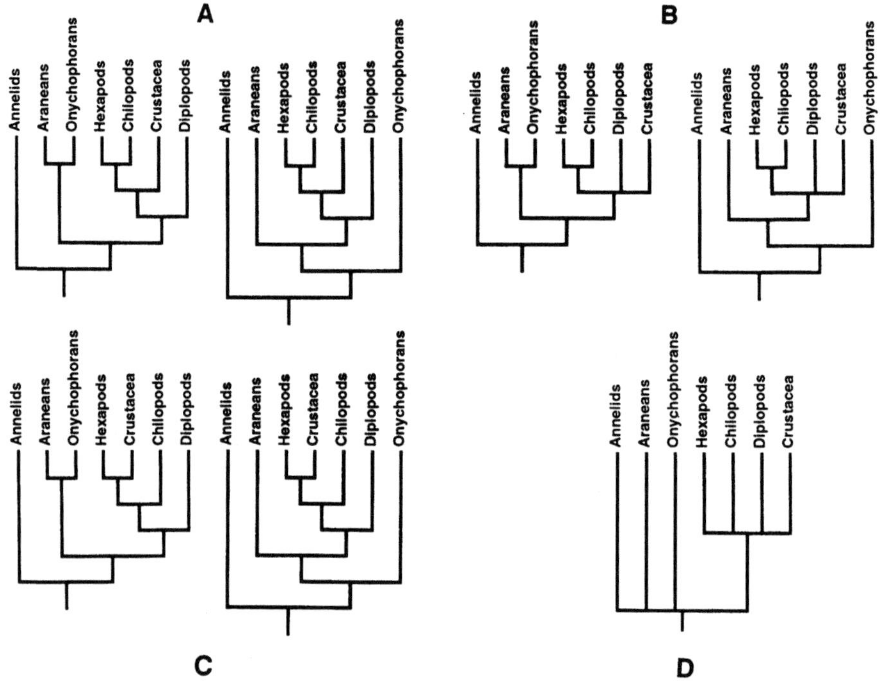

Figure 6A–D. Cladograms, contructed with MacClade (Maddison and Maddison, 1992), based on parismony analysis of characters listed in Table 1. (D) Strict consensus trees based on cladograms in (A–C). Cladograms are arranged in pairs (A, B, C) to show two possible alternative position of the Onychophora. Other differences between cladograms are arrangements amongst the representative Hexapoda, Myriapoda, and Crustacea.

tion of an optic neuropil with the innermost (and heterolateral) mushroom body lobe, as well as with a dorsal unpaired and layered neuropil, is reminiscent of the association between the eyes and so-called "mushroom" and "central bodies" of spiders (Strausfeld et al., 1993). Our observations suggest that by sharing features of annelids and diplopods with features of chelicerates, extant onychophorans may reflect a point of divergence of the chelicerates from the crustacean and hexapod progression.

The reason chelicerates are difficult to interpret is suggested by Hanström's (1926, 1928) own attempts to use the mushroom bodies as a uniting character for all arthropods. By ascribing visual *and* olfactory functions to annelid and insect mushroom bodies, it was simple to modify the aranean mushroom body (which he used as the chelicerate archetype) by removing the olfactory component. In view of the onychophoran mushroom body's possible relationship with the eye, and the apparent segregation between visual and olfactory mushroom body lobes in *Euperipatoides*, Hanström's hypothesis might still seem tenable for the arachnids alone.

The central issue concerning chelicerates is, then, whether similarities seen between aranean and insect mushroom bodies reflect this center's synapomorphic character uniting arthropods, or suggest that the observed similarities are a consequence of homoplasy: evolutionary convergence of structural but not functional similarity.

An argument for independent evolution of mushroom body-like structures in chelicerates rests on the internal organization of the lobes and the function of the globuli cells. In araneans, these provide third order visual interneurons, rather than intrinsic cells. The bridge therefore constitutes a third order visual neuropil, receiving retinotopic relays from the secondary eye medullas, themselves supplied by achromatic retinas, which presumably mediate detection of wide-field motion (Land and Barth, 1992). Retinotopic axons from the globuli cells cross the mid-line and are in register with globuli cell axons from the other side of the brain (Strausfeld and Barth, 1993). The "mushroom body" bridge contains large dendritic field efferent neurons visited by globuli cell axons. Strausfeld and Barth (1993) argue, that by comparison with insects, the spider mushroom body is analogous to achromatic motion-sensitive neuropil (lobula plate) characteristic of Diptera (Hausen and Egelhaaf, 1989; Strausfeld and Lee, 1991).

There are problems in comparing chelicerates. Different classes vary in their cerebral morphology. In *Limulus*, the compound eye is described by Hanström (1926) as supplying, in succession, a lamina, medulla, and a crescent-shaped heterolateral "central body." This structure has been identified in scorpions (Heinrichs and Fleissner, 1987) spiders (Strausfeld et al., 1993), and opilionids (Breidbach and Wegerhoff, 1993) as being associated with the primary eyes. In *Limulus* a massively folded neuropil, occupying much of the supraesophageal neuropil, has been termed the "mushroom body," presumably because of the abundance of globuli cells associated with it (Hanström, 1926). However, it bears no other resemblance to aranean structures of the same name: its structure and cellular composition are more comparable to antennal glomerular neuropil in insects. In certain opilionids (*Phalangium*; Hanström, 1923), globuli cell-like clusters are described as supplying islets of glomerular neuropil, yet lateral eyes are absent in this group. What then is the sensory relationship of their globuli cells? In Amblypygi (e.g., *Heterophrynus*), the supraesophageal brain is crammed with glomerular and barreloid neuropils (Strausfeld, in preparation) supplied by ascending afferents from the modified antennoform first leg pair. Yet, we have failed to identify a mushroom body-like center associated with the glomeruli. In contrast, Babu's (1965) illustration of an amblypigid brain (Bullock and Horridge, 1965) depicts a pair of globuli cell clusters associated with an arabesque of intertwining lobes. All these seem to depart from the protoarthropod archetype.

Chelicerates may still, however, reveal crucial evidence as to whether higher order chemosensory centers might have evolved independently in different groups. In scorpions and solpugids, antenna-like appendages (pectens and malleoli, respectively) arise from body segments. In scorpions they are associated with the ninth postgnathal somite, posterior to the fourth leg pair and presumably homologous to legs, whereas in solpugids they arise from the sixth somite, accompanying the fourth leg pair and reflecting the biramous nature of chelicerate limbs (Brownell, in preparation; Brownell, 1989, 1990; Brownell and Farley, 1974). Pectens and malleoli chemoreceptors supply olfactory afferents to knot- or glomeruli-like regions of the fused prosomal ganglia (Brownell, 1990). Should these glomeruli be found to supply mushroom body type neuropils in the prosomal ganglia, then cladistic analysis based on central nervous specializations must consider independent origins of similar neural arrangements, even within homologous segments.

Elaboration of chemosensory neuropils to multimodal association centers

If modern polychaetes, myriapods, and insects represent ancestral groups of an evolutionary progression, then there appears to have been an elaboration of the mushroom body from a primary olfactory neuropil in annelids to a multimodal association neuropil in higher insects. Two extremes of this progression are schematized in Figure 7, where the simplest network consists of interconnected synapses (see Eichenbaum, 1993) representing an association matrix (parametric relationships) which, in polychaetes, is directly supplied by chemosensory receptors (parametric filters) that are assumed to encode different odorants. A hypothetical progression is schematized in Figure 5, each stage being exemplified in an extant species.

In chelicerates, araneans exemplify a possible evolution of globuli cells to supply an exclusively visual neuropil (the third optic center of the lateral eyes. OC 3), the medial lobes of which are confluent heterolaterally. Interestingly, although the medial lobes of myriapods and hexapods are strictly homolateral, one structural mutant of *Drosophila melanogaster* is characterized by confluence between left and right β-lobes without any apparent defect in olfactory associative memory (Heisenberg et al., 1985).

In myriapods, insects, and crustaceans, chemoreceptors of the antennae no longer directly innervate an association network, but are first grouped into glomerular partitions (parametric convergence) where

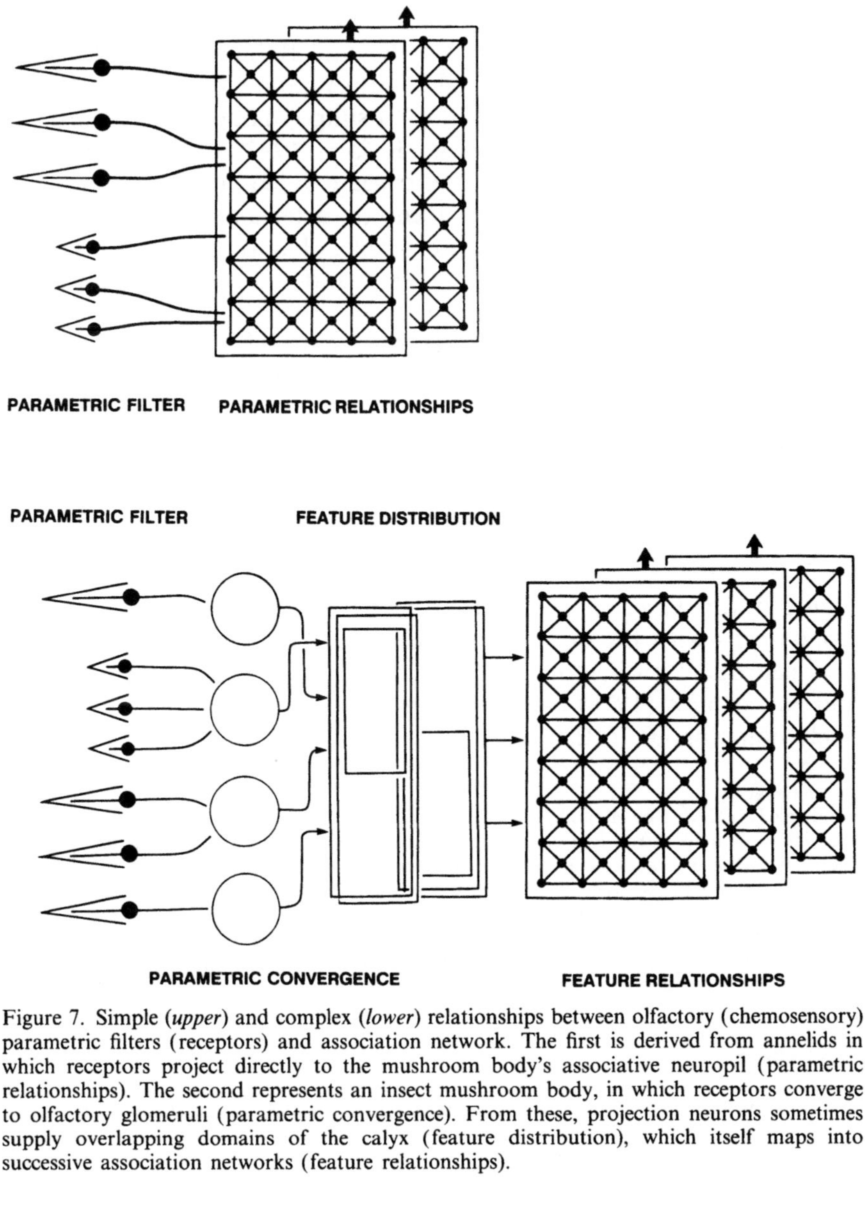

Figure 7. Simple (*upper*) and complex (*lower*) relationships between olfactory (chemosensory) parametric filters (receptors) and association network. The first is derived from annelids in which receptors project directly to the mushroom body's associative neuropil (parametric relationships). The second represents an insect mushroom body, in which receptors converge to olfactory glomeruli (parametric convergence). From these, projection neurons sometimes supply overlapping domains of the calyx (feature distribution), which itself maps into successive association networks (feature relationships).

afferents synapse onto interneurons that project to globuli cell dendrites. In insects, projection neurons from different modalities invade different zones of the calyx (feature distribution), overlapping in some, and remaining segregated in others (Mobbs, 1982). A major evolutionary step from chilopod-like to hexapod mushroom body may have been the doubling up of these centers at each side of the brain, to provide four identical networks. A longitudinal division of the mushroom bodies (multiplexing) in insects is indicated by amino acid immunocytochemistry (Schäfer and Bicker, 1986; Schäfer and Rehder, 1989; Schäfer et al., 1988). This reveals longitudinal bands of immunospecificity that can only be interpreted as corresponding to subpopulations of Kenyon cell axons. Enhancer trap technology (Kaiser, 1993) demonstrates that mushroom bodies of *Drosophila* are indeed divided into discrete longitudinal partitions. It will be important to determine if these mediate different types of computation, for example, olfactory association in one division of the α lobes (Erber et al., 1980), and place memory and motor orchestration in another (Mizunami et al., 1993).

The position of the crustacean hemiellipsoid body is difficult to assess without a thorough neuroanatomical review of members of this extraordinarily diverse group. It could be suggested that crustaceans also departed early from the annelid-myriapod-insect line, and have evolved association neuropils that are characteristic of this group alone. Indeed, they show some peculiarities. In decapods, the olfactory lobes exclusively contain wide-field projection neurons with wide-field endings in the hemiellipsoid body (Mellon et al., 1992), formally represented in Figure 5 for *Palaemonetes*. Nevertheless, the obvious association of globuli cells with the densest neuropil in the brain, and the supply of that neuropil by olfactory interneurons, supports Holmgren's (1916) and Hanström's (1928) contention that the hemiellipsoid body is most likely homologous to the insect mushroom body.

Acknowledgements
We thank David M. Rowell, and Ron R. Hoy for stimulating discussions and encouragement. We thank Camilla Strausfeld and Ron R. Hoy for reading and commenting on the manuscript. Carol Arakaki and Charles Hedgecock, R.P.B., provided expert photographic assistance. David M. Rowell, at the Division of Botany and Zoology, Australian National University, Canberra, generously provided fixed specimens of *Euperipatoides*. The work was supported by a grant from the National Science Foundation (IBN 9316729) and an award from the Office of International Programs at the University of Arizona.

References

Anderson, D.T. (1973) *Embryology and Phylogeny in Annelids and Arthropods*. Pergamon Press, New York.

Babu, K.S. (1965) Anatomy of the central nervous system of arachnids. *Zool. Jb. Abt. Ontog. Tiere. Anat.* 82: 1–154.

Babu, K.S. and Barth, F.G. (1984) Neuroanatomy of the central nervous system of the wandering spider, *Cupiennius salei* (Arachnida, Araneida). *Zoomorphol.* 104: 344–359.

Backhaus, W., Werner, A. and Menzel, R. (1987) Color vision in honeybees: metric, dimensions, constancy, and ecological aspects. *In:* R. Menzel and A. Mercer (eds): *Neurobiology and Behavior of the Honeybee*. Springer Verlag, Heidelberg, Berlin, New York, pp. 83–97.

Ballard, J.W.O., Olsen, G.J., Faith, D.P., Odgers, W.A., Rowell, D.A. and Atkinson, P.W. (1992) Evidence from 12S ribosomal RNA sequences that onychophorans are modified arthropods. *Science* 258: 1345–1348.

Belle, J.S. de. and Heisenberg, M. (1994) Associative learning in *Drosophila* abolished by chemical ablation of mushroom bodies. *Science* 263: 692–695.

Bierbrodt, E. (1942) Der Larvenkopf von *Panorpa communis* L. und seine Verwandlung, mit besonderer Berücksichtigung des Gehirns und der Augen. *Zool. Jb. Anat.* 68: 49–136.

Blinkov, S.M. and Glezer, I.I. (1968) *Human Brain in Figures and Tables. A Quantitative Handbook.* Plenum Press, New York.

Bodian, D. (1937) A new method for staining nerve fibers and nerve endings in mounted paraffin sections. *Anat. Rec.* 69: 153–162.

Böttger, O. (1910) Das Gehirn eines niederen Insektes (*Lepisma saccharina* L.) *Jena. Z. Naturwiss.* 46: 801–844.

Breidbach, O. and Wegerhoff, R. (1993) Neuroanatomy of the central nervous system of the harvestman, *Rilaena triangularis* (Herbst 1799) (Arachnida; Opiliones) – Principal organization, GABA-like and Serotonin-immunochemistry. *Zool. Anz.* 230: 55–81.

Bretschneider, F. (1913) Der Centralkörper und die pilzförmigen Körper im Gehirn der Insekten. *Zool. Anz.* 41: 560–569.

Bretschneider, F. (1924) Über das Gehirn eines Bärenspinners (*Callimorpha dominula*, die Jungfer). *Jena. Z. Naturwiss.* 60: 147–173.

Brownell, P.H. (1989) Neuronal organization and function in the pectinal sensory system in scorpions. *Soc. Neurosci. Abstr.* 15: 1289.

Brownell, P.H. (1990) Structural features of ventral chemosensory organs in scorpions and solpugids suggest common evolutionary origins. *Chem. Sens.* 15: 547.

Brownell, P.H. and Farley, R.D. (1974) The organization of the malleolar sensory system in the solpugid *Chanbria* sp. *Tissue Cell* 6: 471–485.

Buitkamp-Möbius, K. (1975) *Strukturuntersuchungen an den Pilzkörpern im Oberschlundganglion von Apis mellifica-Gynandromorphen unter Berücksichtigung ihres Verhaltens.* Ph.D Dissertation. Universität Bonn.

Bullock, T.H. and Horridge, G.A. (1965) *Structure and Function in the Nervous Systems of Invertebrates.* Freeman, San Francisco, London.

Burrows, M. and Laurent, G. (1989) Reflex circuits and the control of movement. *In:* R. Durbin, C. Miall and G. Mitchison (eds): *The Computing Neuron.* Addison Wesley, Wokingham, pp. 224–261.

Cajal, S.R. (1909) Nota sobre la estructura de la retina de la Mosca. *Trab. Lab. Invest. Biol. Univ. Madrid* 7: 217–257.

Cajal, S.R. and Sanchez y Sanchez, D. (1915) Contribucion al conocimiento de los centros nerviosos de los insectos. Parte I. retina y centros opticos. *Trab. Lab. Invest. Biol. Univ. Madrid* 13: 1–168.

Campos-Ortega, J.A. and Hartenstein, V. (1985) *The Embryonic Development of Drosophila melanogaster.* Springer Verlag, Berlin, Heidelberg.

Cartwright, B.A. and Collett, T.S. (1987) Landmark maps for honeybees. *Biol. Cybern.* 57: 85–93.

Christensen, T.A., Waldrop, B.R., Harrow, I.D. and Hildebrand, J.G. (1993) Local interneurons and information processing in the olfactory glomeruli of the moth *Manduca sexta. J. Comp. Physiol.* (A) 173: 385–399.

Collett, T.S. and Cartwright, B.A. (1983) Eidetic images in insects: their role in navigation. *Trends Neurosci.* 6: 101–105.

Collett, T.S., Fry, S.N. and Wehner, R. (1993) Sequence learning by honeybees. *J. Comp. Physiol.* (A) 172: 693–706.

Davis, R.L. and Kauvar, L. (1984) *Drosophila* cyclic nucleotide phosphodiesterases. *In:* S. Strada and W. Thompson (eds): *Advances in Nucleotide Research.* Raven Press, New York, pp. 393–402.

Dudai, Y., Jan, T.N., Byers, D., Quinn, W.G. and Benzer, S. (1976) *dunce*, a mutant of *Drosophila* deficient in learning. *Proc. Natl. Acad. Sci. USA* 73: 1684.

Duerr, J.S. and Quinn, W.G. (1982) Three *Drosophila* mutations that block associative learning and also affect habituation and sensitization. *Proc. Natl. Acad. Sci. USA* 79: 3646–3650.

Dujardin, F. (1850) Memoire sur le systeme nerveux des insects. *Ann. Sci. Nat. Zool.* 14: 195–206.

Edwards, J.S., Reddy, G.R. and Rani, M.U. (1989) Central projections of a homoeotic regenerate, Antennapedia, in a stick insect, *Carausius morosus* (Phasmida). *J. Neurobiol.* 20: 101–114.

Eichenbaum, H. (1993) Thinking about brain assemblies. *Science* 261: 993–994.

Erber, J., Masuhr, T. and Menzel, R. (1980) Localization of short-term memory in the brain of the bee, *Apis mellifera*. *Physiol. Entomol.* 5: 343–358.

Goldschmidt, E. and Ledermann-Klein, A. (1958) Reoccurrence of a forgotten homeotic mutant in *Drosophila*. *J. Hered.* 49: 262–266.

Goll, W. (1967) Strukturuntersuchungen am Gehirn von *Formica*. *Z. Morphol. Oekol. Tiere.* 59: 143–210.

Gronenberg, W. (1986) Physiological and anatomical properties of optical input-fibers to the mushroom body of the bee brain. *J. Insect Physiol.* 32: 695–704.

Gronenberg, W. (1987) Anatomical and physiological properties of feedback neurons of the mushroom bodies in the bee brain. *Exp. Biol.* 46: 115–125.

Hadenfeldt, D. (1929) Das Nervensystem von *Stylochoplana maculata* und *Notoplana atomata*. *Z. wiss. Zool.* 133: 586–638.

Hall, J.C. (1979) Control of male reproductive behavior by the central nervous system of *Drosophila*; dissection of a courtship pathway by genetic mosaics. *Genetics* 92: 437–457.

Hall, J.C. (1986) Learning and rhythms in courting mutant *Drosophila*. *Trends Neurosci.* 9: 414–418.

Hanesch, U., Fischbach, K.F. and Heisenberg, M. (1989) Neuronal architecture of the central complex in *Drosophila melanogaster*. *Cell Tissue Res.* 257: 343–366.

Hanström, B. (1923) Further notes on the central nervous system of arachnids: scorpions, phalangids, and trap door spiders. *J. Comp. Neurol.* 35: 249–274.

Hanström, B. (1926) Das Nervensystem und die Sinnesorgane von *Limulus polyphemus*. *Lunds Univ. Arsskr.* 22: 2–78.

Hanström, B. (1928) Vergleichende Anatomie des Nervensystems der Wirbellosen Tiere unter Berücksichtigung seiner Funktion. Springer Verlag, Berlin, Heidelberg, New York.

Hausen, K. and Egelhaaf, M. (1989) Neural mechanisms of visual course control in insects. *In:* G.G. Stavenga and R.C. Hardie (eds): *Facets of Vision*. Springer Verlag, Heidelberg, New York, Berlin, pp. 391–424.

Heinrichs, S. and Fleissner, G. (1987) Neuronal components of the circadian clock in the scorpion, *Androctonus australis*: central origin of the efferent neurosecretory elements projecting to the median eyes. *Cell Tissue Res.* 250: 277–285.

Heisenberg, M., Borst, A., Wagner, S. and Byers, D. (1985) *Drosophila* mushroom body mutants are deficient in olfactory learning. *J. Neurogenetics* 2: 1–30.

Hertel, H. (1980) Chromatic properties of identified interneurons in the optic lobes of the bee. *J. Comp. Physiol.* (A) 137: 215–232.

Hölldöbler, B. and Wilson, E.O. (1990) *The Ants*. Belknap, Harvard.

Holmgren, N. (1916) Zur vergleichenden Anatomie des Gehirns von Polychaeten, Onychophoren, Xiphosuren, Arachniden, Crustaceen, Myriapoden, und Insekten. *K. Svenska Vetensk Akad. Handl.* 56: 1–303.

Homberg, U., Christensen, T.A. and Hildebrand, J.G. (1989a) Structure and function of the deutocerebrum in insects. *Annu. Rev. Entomol.* 34: 477–501.

Homberg, U., Montague, R.A. and Hildebrand, J.G. (1989b) Anatomy of antenno-cerebral pathways in the brain of the sphinx moth *Manduca sexta*. *Cell Tissue Res.* 254: 255–281.

Huber, F. (1960) Untersuchungen über die Funktion des Zentralnervensystems und insbesondere des Gehirns bei der Fortbewegung und der Lauterzeugung der Grillen. *Z. Vergl. Physiol.* 44: 60–132.

Ito, K. and Hotta, Y. (1992) Proliferation pattern of postembryonic neuroblasts in the brain of *Drosophila melanogaster*. *Dev. Biol.* 149: 134–148.

Jawlowski, H. (1948) Studies on the insect brain. *Ann. Univ. M. Curie-Sklodowska* (C) 3: 1–37.

Jawlowski, H. (1958) Nerve tracts in bee (*Apis mellifica*) running from the sight and antennal organs to the brain. *Ann. Univ. M. Curie-Sklodowska* (C) 12: 307–323.

Kaiser, K. (1993) Second generation enhancer traps. *Current Biol.* 8: 560–562.

Kanzaki, R., Arbas, E.A. and Hildebrand, J.G. (1991) Physiology and morphology of protocerebral olfactory neurons in the male moth *Manduca sexta*. *J. Comp. Physiol.* (A) 168: 281–298.

Kenyon. F.C. (1896a) The brain of the bee. A preliminary contribution to the morphology of the nervous system of the Arthropoda. *J. Comp. Neurol.* 6: 133–210.

Kenyon, F.C. (1896b) The meaning and structure of the so-called "mushroom bodies" of the hexapod brain. *Amer. Nat.* 30: 643–650.

Krizeneky, J. (1913) Über Restitutionseerscheinungen an Stelle von Augen bei *Tenebrion*-Larven nach Zerstöung der optischen Ganglien. *Arch. Entwickl. Mech. Org.* 37: 629–634.

Kukalova-Peck, J. (1991) The "Uniramia" do not exist: the ground plan of the Pterygota as revealed by Permian Diaphanopterodea from Russia (Insecta: Paleodictyopteroidea). *Can. J. Zool.* 70: 236–255.
Land, M.F. and Barth, F.G. (1992) The quality of vision in the ctenid spider *Cupiennius salei. J. Exp. Biol.* 164: 227–242.
Lehrer, M., Srinivasan, M.V., Zhang, S.W. and Horridge, G.A. (1988) Motion cues provide the bee's visual world with a third dimension. *Nature* 332: 356–357.
Leonard, B. and McNaughton, B.L. (1990) *In:* R.P. Kesner and D.S. Olton (eds): *Neurobiology of Comparative Cognition.* Lawrence Erlbaum Associates, Hillsdale, New Jersey, pp. 348–363.
Maddison, W.P. and Maddison, D.R. (1992) *MacClade, Version 3. Analysis of Phylogeny and Character Evolution.* Sinauer Association Inc., Sunderland.
Mane-Garzon, F. (1973) Un nuevo tipo de Hirudinae *Colombobdella ringueleti* n. gen. n. sp. parasito de una tortuga de Colombia. *Trab. V. Congr. Latinamerica Zoologica.* 1: 129–137.
Manton, S.M. (1964) Mandibular mechanisms and the evolution of arthropods. *Phil. Trans. R. Soc. Lond. B* 247: 1–183.
Manton, S.M. (1973) Arthropod phylogeny – a modern synthesis. *J. Zool.* 171: 111–130.
Manton, S.M. (1977) *The Arthropoda: Habits, Functional Morphology, and Evolution.* Clarendon, Oxford.
Mauelshagen, J. (1993) Neural correlates of olfactory learning paradigms in an identified neuron in the honey bee brain. *J. Neurophysiol* 69: 609–625.
Mellon, De F., Alones, V. and Lawrence, M.D. (1992) Anatomy and fine structure of neurons in the deutocerebral projection pathway of the crayfish olfactory system. *J. Comp. Neurol.* 321: 93–111.
Menzel, R., Ventura, D.F., Hertel, H., de Souza, J. and Greggers, U. (1986) Spectral sensitivity of photoreceptors in insect compound eyes: comparisons of species and methods. *J. Comp. Physiol.* (A) 158: 165–177.
Mizunami, M., Weibrecht, J.M. and Strausfeld, N.J. (1993) A new role for the insect mushroom bodies: place memory and motor control. *In:* R.D. Beer, R. Ritzmann and T. McKenna (eds): *Biological Neural Networks in Invertebrate Neuroethology and Robotics.* Academic Press, Cambridge, pp. 199–225.
Mobbs, P.G. (1982) The brain of the honeybee *Apis mellifera.* 1. the connections and spatial organizations of the mushroom bodies. *Phil. Trans. R. Soc. Lond. B* 298: 309–354.
Neder, R. (1959) Allometrisches Wachstum von Hirnteilen bei drei verschiedenen größen Schabenarten. *Zool. Jb. Anat.* 4: 411–464.
Nighorn, A., Healy, M.J. and, Davis R.L. (1991) The cyclic AMP phosphodiesterase encoded by the *Drosophila* dunce gene is concentrated in the mushroom body neuropil. *Neuron* 6: 455–467.
O'Carroll, D. (1993) Feature-detecting neurons in dragonflies. *Nature* 362: 541–542.
O'Keefe, J. and Nadel, L. (1978) *The Hippocampus as a Cognitive Map.* Clarendon, Oxford.
Pandazis, G. (1930) Über die relative Ausbildung der Gehirnzentren bei biologisch verschiedenen Ameisenarten. *Z. Morph. Ökol. Tiere* 18: 114–169.
Paulus, H.F. (1979) Eye structure and the monophyly of the Arthropoda. *In:* A.P. Gupta (ed.): *Comparative Insect Morphology and Arthopod Phylogeny.* van Nostrand Reinhold, New York, pp. 299–383.
Pearson, L. (1971) The corpora pedunculata of *Sphinx ligustri* L. and other Lepidoptera: an anatomical study. *Phil. Trans. R. Soc. Lond. B* 259: 477–516.
Pflugfelder, O. (1937) Vergleichende-anatomische, experimentelle, und embryologische Untersuchungen über das Nervensystem und die Sinnesorgane der Rhynchoten. *Zoologica. Stuttgart.* 34: 1–102.
Power, M.E. (1943) The brain of *Drosophila melanogaster. J. Morph.* 72: 517–559.
Power, M.E. (1946) The antennal centers and their connections within the brain of *Drosophila melanogaster. J. Comp. Neurol.* 85: 485–517.
Prokop, A., Barleben, F., Fischbach, K.-F. and Technau, G.M. (1991) The *mushroom body defect* gene affects proliferation behavior of postembryonic neuroblasts. *In:* N. Elsner and H. Penzlin (eds): *Synapse, Transmission, Modulation.* Thieme Verlag, Stuttgart, p. 544.
Rempel, J.G. (1975) The evolution of the insect head: the endless dispute. *Quaest. Entomol.* 11: 7–25.
Retzius, G. (1895) Das sensible Nervensystem des Crustaceen. *Biol. Untersuchungen. Neue Folge* 7: 12–18.

Richardson, L.R. (1969) The family Ozobranchidae redefined, and novel ozobranchiform leech from Murray River turtles. *Proc. Linn. Soc. N.S.W.* 94: 61–80.
Robison, R.A. (1990) Earliest-known uniramous arthropod. *Nature* 343: 163.
Rohrschneider, I. (1968) Beitrag zur Entwicklung des Vorderkopfes und der Mundregion von *Periplaneta americana. Zool. Jb. Anat.* 85: 537–578.
Rybak, J. and Menzel, R. (1993) Anatomy of the mushroom bodies in the honey bee brain: the neuronal connections of the alpha-lobe. *J. Comp. Neurol.* 334: 444–465.
Sanchez y Sanchez, D. (1933) Contribution a la connaissance de la structure des corps fongiformes (calices) et leurs pédicules chez la blatte commune (*Stylopyga (Blatta) orientalis*). *Trab. Lab. Invest. Biol. Univ. Madr.* 28: 149–185.
Sanchez y Sanchez, D. (1937) Sur le centre antenno-moteur ou antennaire postérieur de l'abeille. *Trav. Lab. Rech. Univ. Madr.* 31: 245–269.
Sanchez y Sanchez, D. (1940) Contribution à la connaisance des centres nerveux des abeilles (Apis mellifica). *Trab. Lab. Inst. Cajal Invest. Biol.* 32: 123–210.
Sandeman, D.C. and Luff, S.E. (1974) Regeneration of the antennules in the Australian fresh water crayfish, *Cherax destructor. J. Neurobiol.* 5: 475–488.
Sawyer, R.T. (1984) Arthropodization in the Hirudinae: evidence for a phylogenetic link with insects and other Uniramia? *Zoll. J. linn. Soc.* 80: 303–322.
Schäfer, S. and Bicker, G. (1986) Distribution of GABA-like immunoreactivity in the brain of the honeybee. *J. Comp. Neurol.* 246: 287–300.
Schäfer, S. and Rehder, V. (1989) Dopamine-like immunoreactivity in the brain and suboesophageal ganglion of the honeybee. *J. Comp. Neurol.* 280: 43–58.
Schäfer, S., Bicker, G., Ottersen, O.P. and Storm-Mathisen, J. (1988) Taurine-like immunoreactivity in the brain of the honeybee. *J. Comp. Neurol.* 265: 60–70.
Schmidt-Ott, U. and Technau, G.M. (1992) Expression of *en* and *wg* in the embryonic head and brain of *Drosophila* indicates a refolded band of seven segment remnants. *Development* 116: 111–125.
Schürmann, F.W. (1973) Über die Struktur der Pilzkörper des Insektengehirns. III. Die Anatomie der Nervenfasern in der Corpora pedunculata bei *Acheta domesticus* L. (Orthoptera): eine Golgi-Studie. *Z. Zellforsch. Mikr. Anat.* 145: 247–285.
Schürmann, F.W. (1987) The architecture of the mushroom bodies and related neuropils in the insect brain. *In:* A.P. Gupta (ed.): *Arthropod Brain.* Wiley Interscience, New York, pp 231–264.
Sharov, A.G. (1966) *Basic Arthropodan Stock with Special Reference to Insects.* Pergamon Press, New York, Oxford.
Shepherd, G.M. (1992) Modules for molecules. *Nature* 358: 457–456.
Smith, E.L. (1988) Morphology and Evolution of the Hexapod Head. *Proc. 18th Cong. Entomol.* 18: 81.
Snodgrass, R.E. (1938) Evolution of the Annelida, Onychophora, and Arthropoda. *Smiths. Misc. Coll.* 97: 1–159.
Stocker, R.F. and Lawrence, P.A. (1981) Sensory projections from normal and homoeotically transformed antennae in *Drosophila. Dev. Biol.* 82: 224–237.
Strausfeld, N.J. (1976) *Atlas of an Insect Brain.* Springer Verlag, Heidelberg, Berlin, New York.
Strausfeld, N.J. and Bacon, J.P. (1983) Multimodal convergence in the central nervous system of dipterous insects. *Fortschr. Zool.* 28: 47–76.
Strausfeld, N.J. and Barth, F.G. (1993) Two visual systems in one brain: neuropils serving the secondary eyes of the spider. *Cupiennius salei J. Comp. Neurol.* 328: 43–62.
Strausfeld, N.J. and Gronenberg, W. (1990) Descending neurons supplying the neck and flight motor of Diptera: organization and neuroanatomical relationships with visual pathways. *J. Comp. Neurol.* 302: 954–972.
Strausfeld, N.J. and Lee, J.K. (1991) Neuronal basis for parallel visual processing in the fly. *Vis. Neurosci.* 7: 13–33.
Strausfeld, N.J., Welzien, P. and Barth, F.G. (1993) Two visual systems in one brain: neuropils serving the principal eyes of the spider *Cupiennius salei. J. Comp. Neurol.* 328: 63–75.
Swofford, D.L. (1991) *PAUP: Phylogenetic analysis using parsimony. Version 3.0q.* Illinois Natural History Survey, Champagne, Illinois.
Tiegs, O.O. and Manton, S.M. (1958) The evolution of the Arthropoda. *Biol. Rev.* 33: 255–337.

Tinbergen, N. (1935) Über die Orientierung des Bienenwolfes (*Philanthus triangulum* Fabr.). *Z. vergl. Physiol.* 21: 699–716.

Tully, T. and Quinn, W.G. (1985) Classical conditioning and retention in normal and mutant *Drosophila melanogaster. J. Comp. Physiol.* A 157: 263–277.

Viallanes, H. (1887a) Études histologiques et organologiques sur les centres nerveux et les organes des sens animaux articulés. 4. Le cerveau de la guêpe (*Vespa crabro* et *V. vulgaris*). *Ann. Sci. nat. (Zool)* 2: 5–100.

Viallanes, H. (1887b) Études histologiques et organologiques sur les centres nerveux et les organes des sens des animaux articulés. Cinquième mémoire. Le cerveau du criquet (*Oedipoda coerulescence* et *Caloptenys italicus*). *Ann. Sci. nat. (Zool)* 4: 1–98.

Vigier, P. (1908) Sur l'existence réelle et le rôle des appendices piriformes des neurones. La neurone périoptique des Diptères. *C. R. Soc. Biol.* 64: 959–961.

Vilinsky, I., Hansen, L.C., Kloppenburg, P., Yang, M.Y., Armstrong, J.D., Kaiser, K. and Strausfeld, N.J. (1994). The comparative organization of insect brain centers involved in learning, memory, and motor control. *Soc. Neurosci. Abstr.* 20: 816.

Vowles, D.M. (1955) The structure and connections of the corpora pedunculata in bees and ants. *Q. J. Microsc. Sci.* 96: 239–255.

Wada, S. (1966a) Topographie der Anlagenkomplexe der Cephalregion von *Tachycines* (Saltatoria) bein Keimstreif. *Naturwiss.* 53: 414.

Wada, S. (1966b) Analyse der Kopf-Hals-region von *Tachycines* (Saltatoria) im morphogenetische Einheiten. II. Mitteilung: Experimentell-teratologische Befunde am Kopfskelett mit Berücksichtigung des zentralen Nervensystems. *Zool. Jb. Anat.* 83: 235–326.

Wagner, H. (1986) Flight performance and visual control of flight of the free-flying housefly (*Musca domestica*). II. Pursuit of targets. *Phil. Trans. Roy. Soc. Lond. B* 312: 527–551.

Wahdepuhl, M. (1983) Control of grasshopper singing behavior by the brain: responses to electrical stimulation. *Z. Tierpsychol.* 63: 173–200.

Weygoldt, P. (1979) Significance of later embryonic stages and head development in arthropod phylogeny. *In:* A.P. Gupta (ed.): *Arthropod Phylogeny*. Van Nostrand Reinhold. New York, pp. 107–135.

Wheeler, W.C., Cartwright, P. and Hayashi, C.Y. (1993) Arthropod phylogeny: a combined approach. *Cladistics* 9: 1–39.

Williams, J.L.D. (1975) Anatomical studies of the insect central nervous system: A groundplan of the midbrain and an introduction to the central complex of the locust, *Schistocerca gregaria* (Orthoptera). *J. Zool. Lond.* 176: 67–86.

Wilson, M.A. and McNaughton, B.L. (1993) Dynamics of the hippocampal ensemble code for space. *Science* 261: 1055–1058.

Witthöft, W. (1967) Absolute Anzahl und Verteilung der Zellen im Hirn der Honigbiene. *Z. Morph. Tiere.* 61: 160–184.

Zacharias, D., Williams, J.L.D., Meier, T. and Reichert, H. (1993) Neurogenesis in the insect brain: cellular identification and molecular characterization of brain neuroblasts in grasshopper embryo. *Development* 118: 941–955.

Zawarzin, A.A. (1913) Histologische Studien über Insekten. IV. Die optischen Ganglien der *Aeschna*-larven. *Z. Wiss. Zool.* 108: 175–257.

Zhang, S.W. and Horridge, G.A. (1992) Pattern recognition in bees: size of regions in spatial layout. *Phil. Trans. Roy. Soc. Lond. B* 337: 65–71.

Zhang, S.W., Srinivasan, M.V. and Horridge, G.A. (1992) Pattern recognition in honeybees: local and global analysis. *Phil. Trans. Roy. Soc. Lond. B* 248: 55–61.

The Nervous Systems of Invertebrates: An Evolutionary and Comparative Approach
ed. by O. Breidbach & W. Kutsch
© 1995 Birkhäuser Verlag Basel/Switzerland

Is the evolution of arthropod brains convergent?

O. Breidbach

Institut für Angewandte Zoologie der Universität Bonn, An der Immenburg 1, D-53121 Bonn, Germany and Institut für Mathematik der Universität Bochum, Universitätsstr. 150, NA 5-26, D-44801 Bochum, Germany

Summary. Within the nervous systems of different arthropod groups, homologous neurons were identified that show high degrees of structural invariances. Analyses of the development and the projection patterns of such homologous neurons established criteria by which neuropil areas of different species within arthropod groups like insects or arachnids can be homologized.

Such a comparison of principle brain organizations is extended to include species of different arthropod subphyla. It is shown that main central brain neuropil areas such as the central body and the corpora pedunculata, formerly seen as common characteristics of all arthropod groups, are non-homologous in Chelicerata and Mandibulata. Consequently, the central brains of these two arthropod subphyla have to be regarded as the outcome of convergent developments.

Introduction

What is the scope of a comparative analysis of the arthropod brains? Structural and behavioral diversifications in the different arthropod groups are enormous (Boudreaux, 1979). Thus, already one group of insects, the beetles, constitute more than one-fourth of all animal species known (Muller and Cambell, 1953). Because of this great diversity, the arthropods are an ideal group in which to study principal features in the evolution of nervous systems. Furthermore, in arthropods, the evolution of the nervous systems can be described on a very fine scale: on the level of identified homologous neurons (Dumont and Robertson, 1986; Shaw and Meinertzhagen, 1986; Breidbach, 1990a; Breidbach and Kutsch, 1990; Sandeman et al., 1990; Arbas et al., 1991; Kutsch and Heckmann, this volume). Such homologous neurons were described for certain arthropod groups, e.g., various insect and arachnid species (for review see Kutsch and Breidbach, 1994). Thereby, in phylogenetically distant groups with different functional and behavioural adaptations (Fig. 1), motoneurons and interneurons were characterized that proved to be invariant in their structural characteristics. It could be demonstrated that not only the shapes, but also the developmental pathways of such neurons are conserved in different species (Bastiani et al., 1985; Thomas et al., 1984; Levine, 1986; Breidbach, 1987, 1990a; Whitington, this volume). Thus, the analysis of the evolution of the nervous systems of arthropods allows to resolve principal mechanisms of neuropil evolutons on the level of identified neurons.

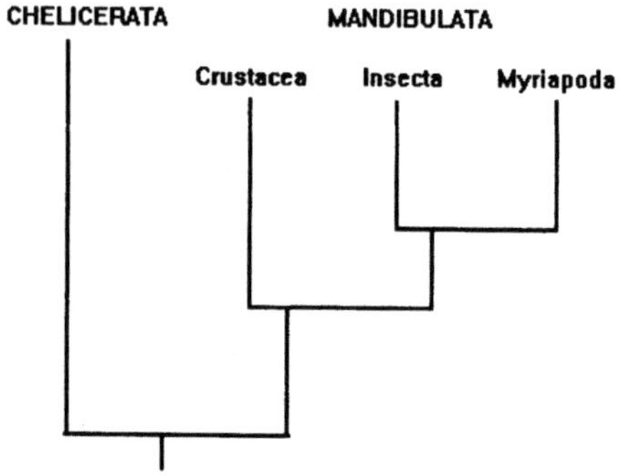

Figure 1. Scheme of the phylogenetic relationships of the main arthropod groups; first terrestrial forms from both Chelicerata (Acarida-like *Protacarus crani*; Bergström 1979), and Mandibulata (Collembola-like *Rhyniella praecursor*; Hennig 1969) were present in upper Devonian times.

Several of these homologous neurons have conserved their essential structural features throughout phylogenesis (Breidbach, 1990b; Kutsch and Breidbach, 1994). The persistent morphologies of these neurons were shown to be regulated by cell–cell interactions throughout ontogenetic development (Raper et al, 1984; Doe et al., 1985; Harrelson and Goodman, 1988; Snow et al, 1988; Breidbach, 1990c; Grenningloh et al, 1990). The invariant structural characteristics of these neurons, thus, are not due to invariant endogeneous programs of single cells. Their structural characteristics are stabilized within a cascade of cell–cell interactions. As the outcomes of such interactions are so similar as to allow identification of homologous branching characteristics in phylogenetically persisting, homologous neurons, it can be concluded that the programs of interactions that determine the neuronal shapes, likewise, are similar in two species with structurally conserved homologous neurons (Whitington, this volume).

Structurally invariant homologous neurons were found in the ventral nerve cord and in the brain of pterygote species (Kutsch and Breidbach, 1994). Up to now, however, in contrast to the wealth of information about the development of segmental ganglia (Boyan and Williams, this volume; Whitington, this volume), very little is known about the embryonic development of the insect brain (Zacharias et al., 1993).

Accordingly, identification of homologous neurons in the insect brain has to rely on comparisons of structural and biochemical characteristics of certain neurons, e.g., their main pathway characteristics, arborization patterns or immunoreactivities. At least some of these homologous brain interneurons are structurally conserved to the level of their arborization patterns (Breidbach, 1990a). Accordingly, it is possible to identify homologous brain neuropil areas in closer related species, where homologous neurons have been identified, using structurally conserved arborizations and axonal projections of these homologous neurons as marker structures. A neuropil area is defined as being homologous in two species, if it is innervated by corresponding arborizations (axons or dendritic branches) of such homologous neurons that are structurally similar. By such precise identifications of common neuropil characteristics in insect brains, a ground plan of the insect brain neuropil can be described, allowing to decode common characteristics of the commissural architectures of different insect species (Kutsch and Breidbach, 1994). Phylogenetically conserved neurons like the deutocerebral ascending serotonin immunoreactive neurons were described. The latter were identified in *Periplaneta*, *Manduca* and *Tenebrio* (Kent et al., 1987; Breidbach, 1990b; Salecker and Distler, 1990). Regarding other arthropod groups, it needs to be determined whether their brain architectures are only variations of the design which is apparent in insect species, or whether their brain neuropil organizations are completely different.

Synapomorphic neuropil structures in arthropods

Due to the classical studies of Holmgreen (1916) and Hanström (1921, 1923, 1924, 1926, 1928a,b,c, 1935), in comparative zoology the nervous system is regarded as the evolutionary most conserved tissue in arthropods. According to Ax (1984), one of the few synapomorphies of the arthropod stock is to be seen in the internal organizations of the brain neuropils in the different groups, e.g., the possession of a central body (Tab. 1).

The present study, however, presents arguments that dispute the hypothesis of a homologous arthropod brain organization. Main neuropil centers like the "corpora pedunculata" or the "central body" are non-homologous in chelicerate and mandibulate species and – consequently – are not common characteristics of the arthropod brains. Based on comparative anatomical and developmental biological studies, this account argues for a parallel evolution of arachnid and insect central brain neuropil centers. The brains of these groups are formed according to two different kinds of "Bauplan".

However, within each of these different groups, the insects and the arachnids, the principal brain organizations are invariant up to the level of identified neurons. Thus, within these two groups, the brain structures of different species do not necessarily reflect functional morphological or behavioral adaptations (Dumont and Robertson, 1986). A study of invariant structural characteristics of the homologous neurons in the ventral nerve cord (VNC) of insects demonstrated that structural invariant homologous neurons share not only morphological characteristics but possess comparable life histories (Thomas et al., 1984). Thus, motoneurons or neuropeptidergic interneurons of the ventral nerve cord

Table 1. Neuropil areas in the central brain of arthropoda, according to Gupta (1987). Relative size of the respective areas, from small (X) to large (XXX) is indicated; the so-called "corpora penduncculata" and "central bodies" of arachnids are regarded as non-homologous to the respective structures (corpora pedunculata, central body) in mandibulate species. The neuropil structure of crustaceans correlated with the corpora pedunculata of insects is the hemiellipsoid body (for discussion, see Strausfeld et al., this volume)

	Corpora pedunculata	Central body
Trilobitomorpha	"(X?)"	"(X?)"
Chelicerata		
Xiphosura	"XXX"	"X"
Scorpionida	"X-XXX"	"X-XXX"
Pedipalpida	"X"	"X"
Araneida	"X-XXX"	"X-XXX"
Opilionida	"X"	"XX"
Pseudoscorpionida	"X"	"XXX"
Solifugida	"X"	"X"
Acarina	"O"	"X"
Crustacea		
Entomostraca	X	XXX
Malacostraca	?	X
Myriapoda		
Symphyla	O	XXX
Diplopoda	O	X
Chilopoda	?	XX
Pauropoda	?	?
Hexapoda		
Endognatha	*O-XXX*	*XXX*
non-pterygot Ectognatha	*O-X*	*XXX*
Paurometabola	*X-XXX*	*O-XXX*
Paraneoptera	*X-XXX*	*X-XXX*
Holometabola	*X-XXX*	*X-XXX*

of locusts and *Tenebrionid* beetles share developmental characteristics (Breidbach and Kutsch, 1990; Breidbach and Dircksen, 1991; Urbach et al., unpublished results).

Morphological similarities of homologous neurons are due to conservations of their modes of development. If the developmental pathways of homologous neurons differ, or if the modes of interactions of cells vary in the development of two species or in homonomous (serial homologous) neuronal clusters of one individual, the respective homologous or homonomous neurons will show remarkable structural differences (Wilson and Hoyle, 1978; Boyan and Williams, this volume). Interestingly, neuropeptidergic or serotonin-immunoreactive brain interneurons mapped in the late embryonic and larval holometabolan *Tenebrionid* beetles, which show remarkable changes in the morphology and behavior in the larval and adult stages (Kutsch et al., 1993; Urbach et al., 1994), persist throughout ontogenesis and maintain their basic structural characteristics (Breidbach, 1990c; Breidbach and Dircksen, 1991; Breidbach and Wegerhoff, 1994). Comparison of the neuroanatomies of *Tenebrionids* with that of *Drosophila* or locusts shows that these ontogenetically persisting neurons also persist phylogenetically. Thereby, these homologous neurons show a remarkable conservation of structural features within the different species (Kutsch and Breidbach, 1994). The shapes of this subset of neurons are maintained irrespective of functional or behavioral specifications in the different life forms of one individual during ontogenesis. These highly invariant structures allow to suggest a ground plan of an insect brain (Edwards, 1977). Studied on a broader phylogenetic scaling, the characteristics of such homologous structures will allow to discern conserved and varying brain neuropil subunits.

Structurally similar neuropil areas like the optic lobes of various invertebrates may have been formed convergently. Such convergently evolved neuropils reflect common adaptational constraints leading to similar neuroarchitectural characteristics (e.g., the optic or olfactory lobes in various arthropods (see Strausfeld et al., this volume). To enable more detailed analyses of such convergencies, at least on the level of arthropod species, reliable criteria are needed that allow to define homologues in the arthropod brains.

The arthropod neuromer

The common annelid-like ancestor of chelicerate and mandibulate species possessed a repetitively organized ventral nerve cord (VNC; Bullock, and Horridge, 1965): Recent annelids possess a chain of ganglia in the VNC with neuron-free connectives between them. However, in some groups: the Oligochaeta, among certain Polychaeta, and in some archiannelids, the whole ventral nerve cord is a medullary cord, and the nerve cell bodies are scattered throughout its length (Bullock and Horridge, 1965). According to Bullock and Horridge (1965), however, the distribution of this medullary formed nervous system among the annelid classes shows that the medullary cord is not to be regarded as a primitive, but rather as a secondarily derived condition. In earthworms, the ganglia are gradually and inconspiciously bunched in the cord (Retzius, 1892). There is a central neuropil which occupies the greater part of the ganglion. Superficial to this lies the cell body rind which is covered by a peritoneum. A glial framework develops (Coggeshall and Fawcett, 1964; Pentreath, 1989; Riehl and Schlue, 1989, 1990). Terminations of sensory and motor fibers are segregated. Analysis of serotonin-immunoreactive neurons showed a repetitive pattern of interneurons in the ganglia of the ventral nervous system (Spörhase-Eichmann et al., 1987a,b). Thus, the principal organizations of the earthworm VNC neuromers resemble structural characteristics of the nervous system of tracheate species. In spite of some studies on leech neuroembryogenesis (Kuwada and Kramer, 1983; Kuwada, 1984; Gao and Macagno, 1987; DeRiemer and Macagno, 1989), it is not clear how far the developments of central axon pathways correspond between annelids and arthropods. The expression patterns of genes like *engrailed* (Patel et al., 1989b) or those of the *Antennapedia*-class homeobox group (Aisemberg et al., this volume) show significant differences between arthropod and annelid species. Consequently, it is not yet possible to correlate the developmental patterns of annelid and arthropod species.

For insects, comparative analyses of the neuroembryogenesis of thoracic neuromers have shown the existence of a common basic Bauplan (Bastiani et al., 1985). Based on comparisons of the expression pattern of the *engrailed* genes, Patel et al. (1989a,b) extended the claim for such a common organization to include all mandibulate groups. Recent comparative investigations of gene expression patterns and morphogenesis in different arthropods have confirmed this claim (Scholtz et al., 1993; Whitington et al., 1991, 1993): It seems possible to describe a common neuronal framework of the organization of the mandibulate ventral nerve cord neuromer (see Whitington, this volume).

Comparative neuroanatomical studies of free abdominal ganglia in Chelicerata, albeit scanty, indicate that even in these species at least the basic neuromeral organization of the ventral nerve cord ganglia is structurally similar to that of the Mandibulata (see Wegerhoff and Breidbach, this volume). There are serial repetitive commissures, by which identified neurons pass contralaterally. Tracts are formed, by which single neuromers are connected. The sensory and the motoric neuropils are separated, forming a dorsal motoric and a ventral sensoric neuropil. Homonomous motoneurons and interneurons were identified (Wegerhoff and Breidbach, this volume). Detailed embryological data on the development of chelicerate nervous systems are lacking up to now, thus, the claim for a similar organization of arachnid and insect free abdominal ganglia is not substantiated by neuroembryological studies (Anderson, 1973). Gross morphologically, the mode of arachnid neuronal development differs fundamentally from that of mandibulate species (Chabaud et al., 1990).

Nevertheless, both chelicerate and mandibulate species derive from a common ancestor that already possessed a chain of homonomous neuronal structures, which later were condensed in the head region to form the arthropod brains (Ax, 1984). Can we describe basic characteristics of such an "ancestral" neuronal segment – a basic neuromer? Tests with molecular markers that might allow to compare basic segmental characteristics across the arthropod phylum are still lacking for chelicerates. Thus, we can only compare morphological features of the brain and ventral nerve cord (VNC) organizations in the different arthropod subphyla.

All arthropod brains were formed by fusion of a series of neuromers. Consequently, the brains of all arthropod species should share those characteristics that are the common and essential features of an arthropod neuromer.

The "classical" view

In a gross morphological analysis of brain neuroanatomy, the principal brain structures of Chelicerata and Mandibulata show such common characteristics as a more or less centrally located neuropil area with a dense network of columnarily organized neurites and two bilaterally symmetrical sets of tightly packed projections, derived from clusters of small globular cells. Until recently, such gross histological similarities in the brain organization seemed sufficient to define homologous brain neuropil areas (Tab. 1; Gupta, 1987). Reconsidering the scrupulous analyses of the principle morphology of arthropod legs (Manton, 1977), it seems surprising that these comparable vague criteria were thought to be sufficient to discern structural similarities in the arthropod brains. This might be due to the fact that structural details within the nervous system are far less obvious than the specifications of the sclerites of an arthropod appendage, but it may also reflect a principal difficulty in an application of the homology concept to the nervous system (Campbell and Hodos, 1970).

Homology

What is a homologous structure? In Owen's extended definition (1848), he assigned two criteria for the identification of homologous structures: relative position and connectivity. More or less precise, this definition gave the guideline for comparative zoology to apply the homology concept over the next century. In 1956, Remane defined homology more strictly, listing three criteria to define possible homologies:

- The criterion of position: a structure is homologous if it is found in an equivalent position within a comparable framework.
- The criterion of specific quality: similar structures may be homologized when they match in several specific features, irrespective of their position in an organism.
- The criterion of continuity: dissimilar and differently placed structures are homologous as long as there are intermediate stages satisfying one or both of the aforementioned criteria (Kutsch and Breidbach, 1994).

All these criteria rely on a definition of a grid of topological relations between structural characteristics of an organism. What, however, are the marker structures by which two maps of structures in the neuropils of two organisms can be correlated? Regarding a leg, already its relative position to the thoracic sclerites gives a primary orientation for such a comparison. Yet, how can the criteria of position be applied to brain neuroanatomy? Descriptions like "mid-dorsal region of the central brain" are too unspecific to give reliable criteria to relate certain areas in different arthropod brains. Already common gross orientations such as "anterior" or "dorsal" cannot so easily be compared in different arthropod brains (Zacharias et al., 1993). Definitive landmarks are only found in the network of fibers that constitute the commissural and tract architectures of an arthropod brain. The position of a neuropil area has to be defined in relation to this commissural architecture (Breidbach and Wegerhoff, 1994). Thus, an attempt to homologize brain areas has to start with a homologization of the commissural and tract organizations of different species (Boyan et al., 1993). Using the projections of homologous neurons, homologous tracts and commissures can be reliably identified in arthropod brains. Thus, we can characterize marker structures which will allow to define homologous positions in the brains of different arthropod species irrespective of morphometric changes in the commissural and tract patterns in those different brains.

What are the arguments against a general homologization of arthropod brain neuropil areas?

According to this body of evidence, neuropil areas such as those commonly described as central body can be defined for both insect and arachnid species. The insect central body neuropil is embedded in a superior and a ventral commissure (Fig. 2A; Hanesch et al., 1989; Wegerhoff and Breidbach, 1992), but the chelicerate "central body" is situated most dorsal and posterior in the central brain, and above a number of inferior commissures (Strausfeld and Barth, 1993). Thus, the commissural surroundings of these two neuropils are totally different in

---

Figure 2A–B. Central brain neuropil of the larval beetle *Tenebrio molitor*. (**A**) Frontal section, Bodian stain, showing localization of the fan-shaped body (FB), which constitutes the larval central body; note the commissures that enclose the fan-shaped body neuropil, its glomerular organization, and situation of the neuropil of the dorsal lateral triangle (DLTR) dorsally to the fan-shaped body neuropil; βL: beta-lobe, D: dorsal, DC: dorsal commissure, DCFB: dorsal commissure of the fan shaped body, PED: pedunculus, V: ventral, VCFB: ventral commisure of the fan shaped body; (**B**) Serotonin-immunoreactive neurons in the central brain, an axon of an identified immunoreactive neuron running through the posterior dorsal commissure (PDC) is shown; note glomerular pattern of the fan-shaped body neuropil (FB); PB: protocerebral bridge (for details see Wegerhoff and Breidbach, 1992); scale bars: 50 μm.

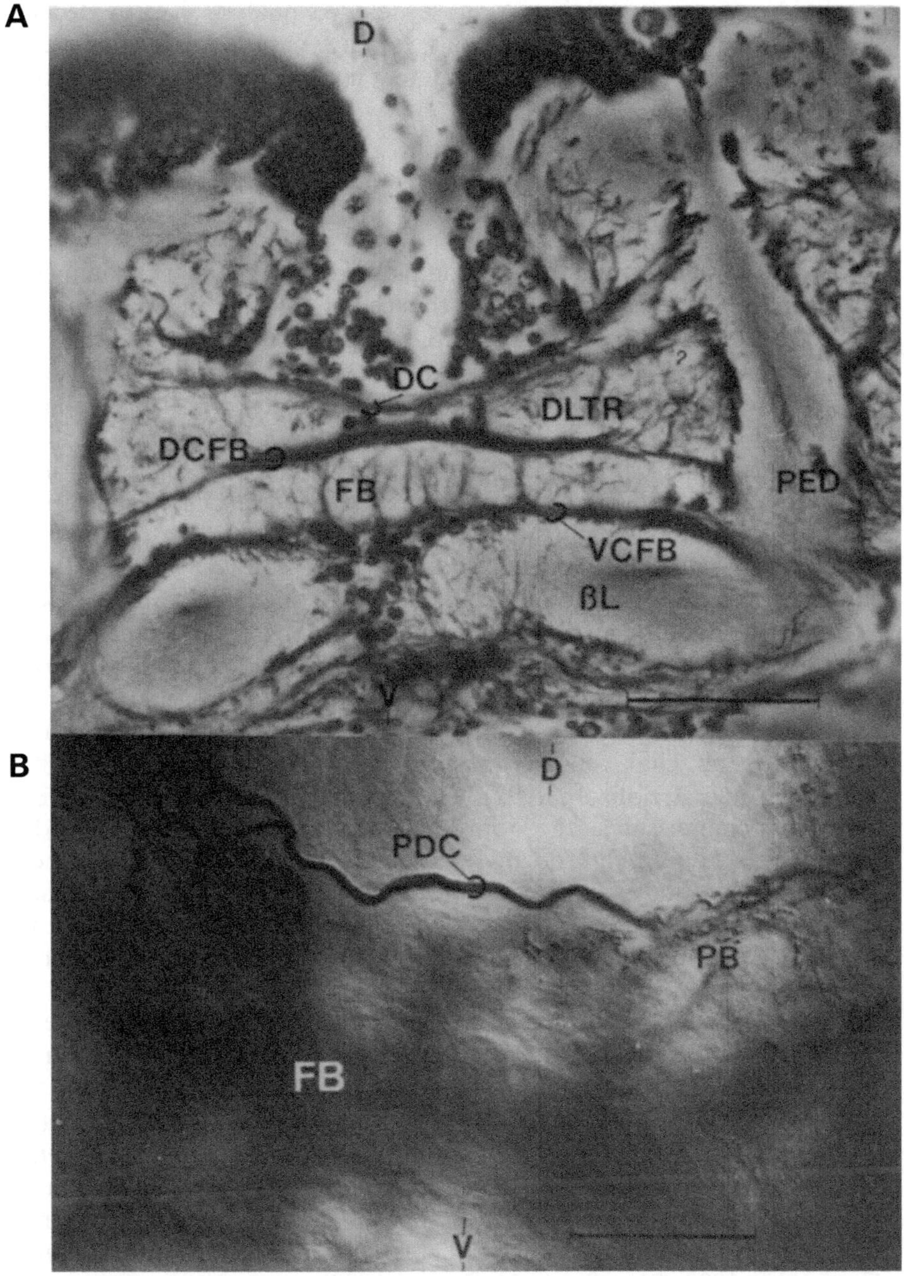
A
D
DC
DLTR
DCFB
FB
PED
VCFB
ßL
V
B
D
PDC
PB
FB
V

arachnid and insect species. A more detailed analysis has to decide whether these different situations only vary a synapomorphic characteristic or whether they reflect totally different modes of brain development.

The insect central complex

Morphology

In the insect midbrain, the central complex is one of the most prominent neuropil areas. It is situated in the center of the protocerebrum, bounded laterally and ventrally by the corpora pedunculata (Fig. 2A; Homberg, 1987). According to Williams (1975), the term central complex is here used as a collective term for the protocerebral bridge and the central body. The protocerebral bridge is the most anterior part of the central complex (Figs 2B, 3, 4). It is a cylindrical neuropil with a transverse longitudinal axis and endings which are curved posterior ventral. The more caudally situated central body is a compact neuropil structure found ventral caudally to the protocerebral bridge. It is composed of several subunits: Besides two small, globular-shaped noduli, it consists of an upper division (including a fan-shaped body and the more dorsally situated superior arch) and a lower division (ellipsoid body) (Fig. 3). The fan-shaped body shows a characteristic repetitive organization with glomerular-like subcompartments that are surrounded by glial cells (Strausfeld, 1976; Hanesch et al., 1989; Wegerhoff and Breidbach, 1992). The fan-shaped body is enclosed by two main commissures, the dorsal and the ventral fan-shaped body commissures (Fig. 2A). Dorsally adjacent to the dorsal commissure of the fan-shaped body, a paired neuropil structure is found, the dorsal lateral triangles (Fig. 2A). Dorsally to these, a "dorsal commissure" is situated that connects the median dorsal neuropils of the protocerebral hemibrains.

Physiological and anatomical investigations of central complex neurons have shown the absence of direct sensory input into the central body (Homberg, 1987). For the adult, connections from this neuropil to interneurons of the optic pathway are present (Homberg, 1985). Genetic dissections of central complex functions in *Drosophila* using single gene mutations which produce structural abnormalities in this neuropil have shown that the central complex is involved in control, maintenance, and initiation of walking, and affects flight behavior (Strauß and Heisenberg, 1991, 1992; Strauß and Hofbauer, 1993).

Development of the insect central body

The central body develops from interhemispheric commissures (Fig. 4). Initially, the central complex merely consists of commissural tracts (Hanesch et al., 1989). In dipteran species, this neuropil is established in

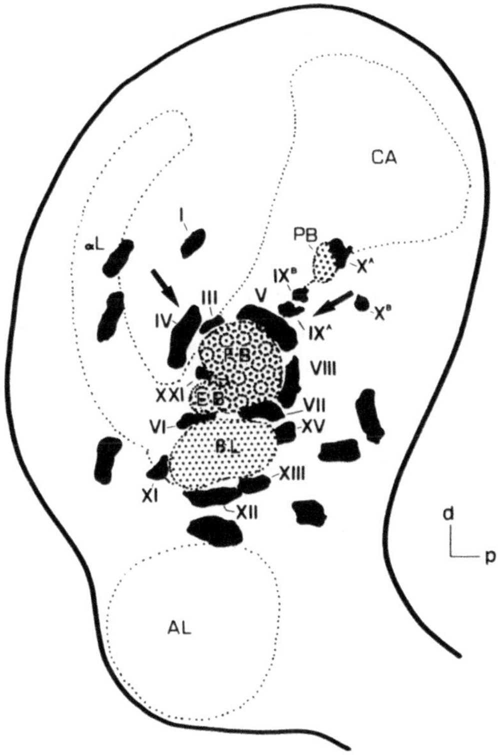

Figure 3. The insect central body (FB, EB) is ensheathed by commissures (black). Sagittal section through the central brain of a 24-h-old pupa of the beetle *Tenebrio molitor*, contours of associated main neuropil areas (AL, αL, CA) are shown for orientation; AL: antennal lobe, αL: alpha lobe, βL: beta lobe, CA: calyx, d: dorsal, EB: ellipsoid body, FB; fan-shaped body, p: posterior, PB protocerebral bridge, I: dorsal median commissure, III: anterior commissure of the fan-shaped body, IV: ventral commissure of the fan-shaped body, V: dorsal commissure of the fan-shaped body, VI: ventral commissure of the central body, VII: median commissure of the central body, VIII: posterior commissure of the central body, IX^A: anterior part of the dorsal commissure IX^B; posterior part of the dorsal commissure, X^A: anterior part of the posterior median commissure, X^B: posterior part of the posterior median commissure, XI: anterior commissure of the ventral bodies, XII: connective of the ventral bodies, XIII: posterior commissure of the ventral bodies, XV: posterior median commissure, XXI: anterior median commissure of the central body (modified from Weiß, 1994).

later larval stages, whereas in Lepidoptera and Coleoptera, already the larvae possess a central complex neuropil (Bretschneider, 1913, 1914; Jawlowski, 1936; Panov, 1959; Granger et al., 1989). In the beetle *Tenebrio molitor*, such a fan-shaped neuropil is already formed in the late embryo and becomes gradually extended throughout larval development (Breidbach et al., 1992; Wegerhoff and Breidbach, 1992). Intially, this fan-shaped body neuropil is formed by side projections of extrinsic neurons of both the dorsal and the ventral body commissures of the fan-shaped body. Intrinsic interneurons were placed into this preformed neuropil architecture. A subdivision into a lower and a superior part (fan-shaped body and superior arch) is not present in the larva, but is developed during metamorphosis.

"Central body" neuropils of other mandibulata

The brain structures of Pauropoda, Symphyla, and Chilopoda outline significant differences in their shapes, the relative dimensions of their neuropils and the organizations of their sensory neuropil regions (Joly and Deschamps, 1987). All these different Myriapoda show their central brain to be crossed by an array of commissures. Various authors identified a myriapod central body within this network of commissures (Gupta, 1987). The criterium for this, up to now, is only its principal situation in the median brain neuropil. As the commissural organization of the Myriapoda is not homologized with that of insects, the claim for a homologization of myriapod and insect central bodies can be regarded only as a working hypothesis.

The discussion about dissimilarities in myriapod and insect embryogenesis, and comparative analyses of rRNA sequences might indicate that Myriapoda and Insecta are not so closely related as generally thought, when classifying these groups as Tracheata (Field et al., 1988), and separating these from the Crustacea. Regarding their neuroembryogenesis, the latter seem to be more closer related to insects than the Myriapoda (Whitington et al., 1993).

In Crustacea, a central complex consisiting of a medially situated central body and a more dorsal anterior placed protocerebral bridge is present in most of the groups studied so far (Sandeman, 1982; Sandeman and Scholtz, this volume). A central body is present in all Crustacea, whereas the protocerebral bridge cannot be recognized in some non-malacostracan crustaceans (Nässel and Elofsson, 1987). The criteria for an identification of a central body neuropil in Crustacea are

---

Figure 4A–B. Postlarval development of the central body of the beetle *Tenebrio molitor*. (**A**) Schematic representation of the neuropil areas associated with the central complex of the last larval stage of the meal beetle, frontal view, 1: dorsal commissure of the fan-shaped body, 2: ventral commissure of the fan shaped body, 3: ventral body commissure, aL: alpha-lobe, bL: beta-lobe, CA: calyx, d: dorsal, DLTR: dorsal lateral triangle, FB: fan-shaped body, P: pedunculus, PB: protocerebral bridge, v: ventral; (**B**) Development of the central body. In the 90% embryo (E) side arborizations are formed by the initial dorsal (DCFB) and ventral (VCFB) commissures of the fan-shaped body (FB). These commissures are extended throughout the first larval instars; in later larval stages new differentiating local interneurons invade the fan-shaped body neuropil and extend the glomerular pattern of the fan-shaped body neuropil. At the end of the larval development the initial ellipsoid body neuropil (EB) is formed which differentiates during the beginning of pupation. L1–L7: larval instars 1–7; P: early pupa (5% pupation), PB: protocerebral bridge, PP: prepupa, VCBC: ventral commissure of the central body.

morphological ones (Schürmann et al., 1991). In this neuropil, no terminations of primary afferents or motor effector neurons are known. Characterization of serotonin-immunoreactive interneurons showed that eight serotonergic neurons form a commissure that gives off small tuft-like branches into crustacean central body neuropils (Sandeman et al., 1988).

Structural details that show similarities in the organizations of insect and crustacean central body neuropils (Kutsch and Breidbach, 1994)

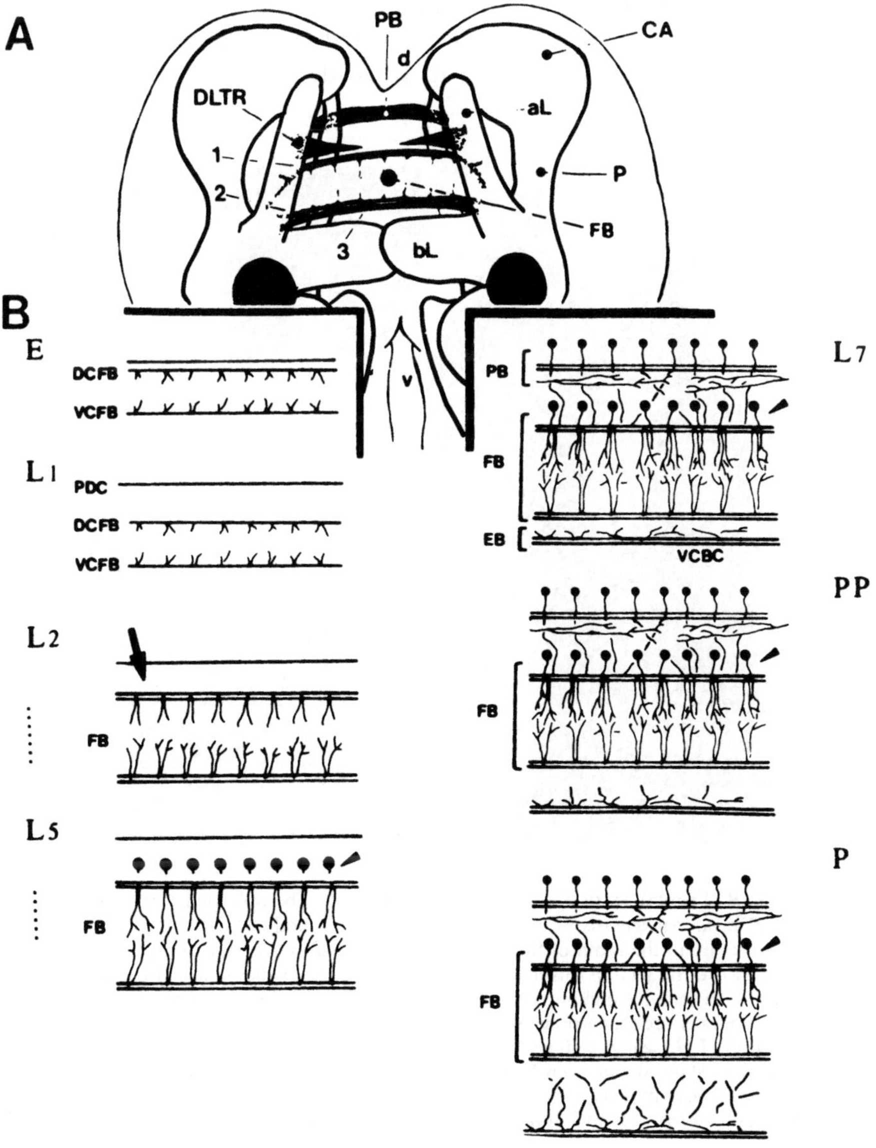

have not yet been traced down to the level of homologous commissural architectures. Identification of such common characteristics might, however, be possible in the near future.

Arachnid "central body" neuropil

In spiders, the term "central body" refers to the most posterior, dorsal neuropil (Figs 5, 6). The aranean "central body" consists of two sausage-like lobes that extend horizontally in the dorsal posterior protocerebrum, which are subdivided into horizontal strata (Babu, 1985; Weltzien, 1988). Apically, they are covered by a dense layer of small neuronal somata which give rise to small descending fibers that pene-

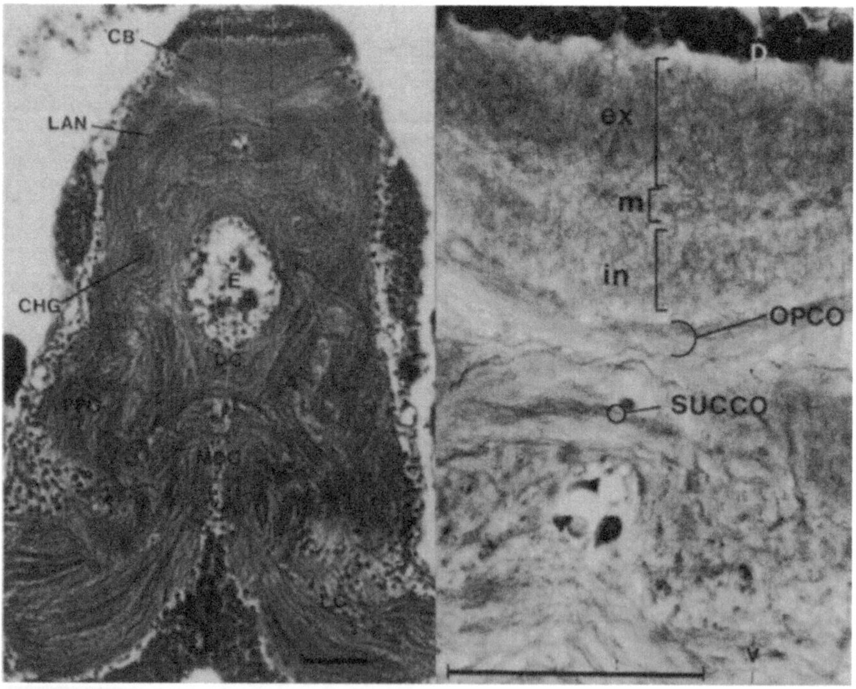

Figure 5A–B. Frontal section through the brain and subesophageal nerve mass of the arachnid *Rilaena triangularis* (Opilionidae). The central body ("CB") is the most dorsal neuropil structure in the brain of this arachnid. Commissures are found only ventrally to this neuropil. (A) Frontal section through the anterior supra- and subesophageal ganglion, Bodian stain; CHG: cheliceral ganglion, DC: dorsal commissure (subesophageal ganglion), MDC: mid-dorsal commissure (subesophageal ganglion), LAN: lateral associative neuropil ("corpora pedunculata"), LG: leg ganglion, PPG: pedipalpal ganglion. (B) "Central body" frontal section, Bodian stain, detail from Figure 5A, the insert shows its position; note stratification of this neuropil (external (ex), median (m) and inferior (in) neuropil); OPCO: optic commissure, SUCCO: superior caudal commissure; scale bars 50 µm.

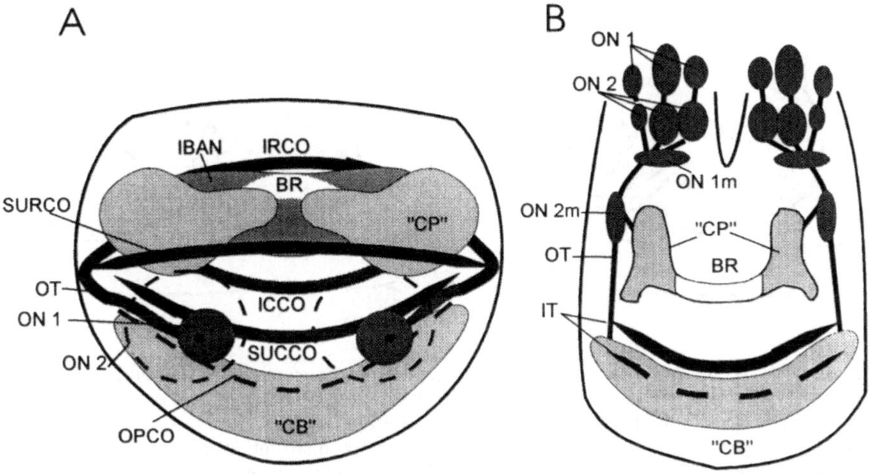

Figure 6A–B. Principal morphology of the arachnid brain. (A) Harvestman *Rilaena triangularis*, dorsal view, modified from Breidbach et al., 1994; (B) Spider *Cupiennius salei*, dorsal view, modified from Babu 1985; terminologies according to Breidbach and Wegerhoff, 1993; Strausfeld and Barth, 1993, and Strausfeld et al., 1993; BR: bridge, "CB": central body, "CP": corpora pedunculata, IBAN: inferior bilateral associative neuropil ICCO: inferior caudal commissure, IRCO: inferior rostral commissure, IT: inferior tract, ON1,2: optic neuropils 1, 2, OPCO: optic commissure, OT: optic tract, SUCCO: superior caudal commissure, SURCO: superior rostral commissure, ON 1: secondary eye lamina, ON 2: secondary eye medulla, ON 1m: primary (median) eye lamina, ON 2m: primary (median) eye medulla.

trate the "central body" neuropil, and pass rostrally, constituting a columnar organization of the "central body" neuropil (Strausfeld et al., 1993). This pattern is similarly found in Opilionidae, whose "central body" is also formed by two stratified horizontal lobes (Breidbach and Wegerhoff, 1993). Ventral anterior, the spider's "central body" is underlaid by a set of commissures, whereas posterior dorsal or anterior dorsal no commissures are present (Figs 5B, 6). Dorsally, it is superimposed by a dense layer of small neuronal somata whose axons descend anterior ventrally into the central body neuropil (see Wegerhoff and Breidbach, this volume).

Bullock and Horridge (1965) regarded the "central bodies" of arachnids as visual centers. This view is supported by the analysis of Strausfeld et al. (1993) that shows the spider's "central body" to be a third-order optic neuropil, comparable to the complex retinotopic neuropils of Insecta and Crustacea (see Wegerhoff and Breidbach, this volume).

Using crustacean cardioactive peptide (CCAP) immunohistochemistry, a set of homologous interneurons was identified that innervates the central body neuropil in opilionids and various spiders (Breidbach et al., 1994). Fourteen clusters of CCAP-immunoreactive neurons were labeled in the protocerebrum of the spider *Tegenaria atrica*, *Nephila clavipes*, and the opilionid *Rilaena triangularis*. In all species, these

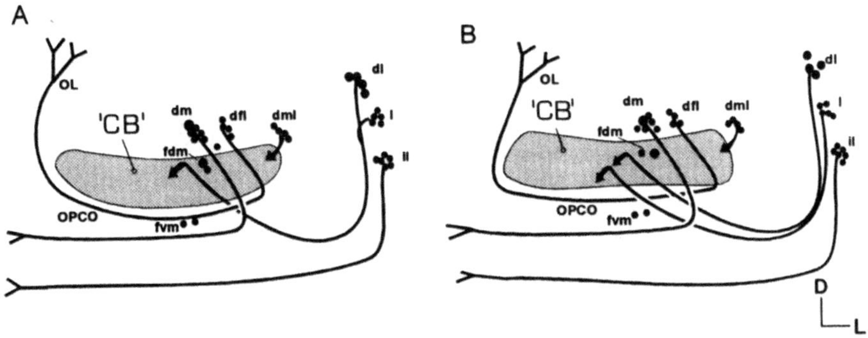

Figure 7A–B. Schemes of the distribution and projections of CCAP-immunoreactive (ir) neurons in the brains of arachnids, frontal views; only the CCAP-ir cells of one hemibrain are shown. The "central body" ("CB") is indicated in gray; (A) Harvestman *Rilaena triangularis*; (B) Spider *Nephila clavipes*; note similarities in number, distribution, and main pathways of CCAP-ir neurons in both species; dfl: dorso-frontal-lateral cluster, dl: dorso-lateral cluster, dm: dorso-median cluster, dml: dorso-medio-lateral cluster, fdm: frontal dorso-median cluster, fvm: frontal ventro-median cluster, il: inferior lateral cluster, 1: lateral cluster, OL: optic lobes, OPCO optic commissure; modified from Breidbach et al., 1994.

clusters possess the same number of neurons and share structural and topological characteristics (Fig. 7). Two sets of bilaterally symmetrical neurons associated with the optic lobes and the arachnid "central body" were analyzed in detail comparing one opilionid and five aranean species. Thereby, 16 neurons were identified that show structural similar axonal pathways and arborization patterns. Thus, these neurons were characterized as homologous neurons that have preserved all their major structural characteristics as discernable at the level of light microscopy (Fig. 8). These neurons demonstrate a high degree of structural invariance in the neuropils of different arachnid species, which is irrespective of their different behavioral and functional adaptations (Breidbach and Wegerhoff, 1993).

Comparison between arachnid and insect central bodies

Comparison of the arachnid and the insect central body shows principal differences in the commissural framework in which these brain neuropils are embedded. For insects, it has been shown that the central body neuropil derives from arborizations of neurons that pass the dorsally and ventrally adjacent interhemispheric commissures. That definition is certainly not applicable to the spider's "central bodies". The spider's "central body" ("CB") is not "embedded," but rather is superimposed on the most dorsal brain commissures. Embryologically, it is formed by a mode totally different from that of insects (see Wegerhoff and Breidbach, this volume). Furthermore, the coupling characteristics of

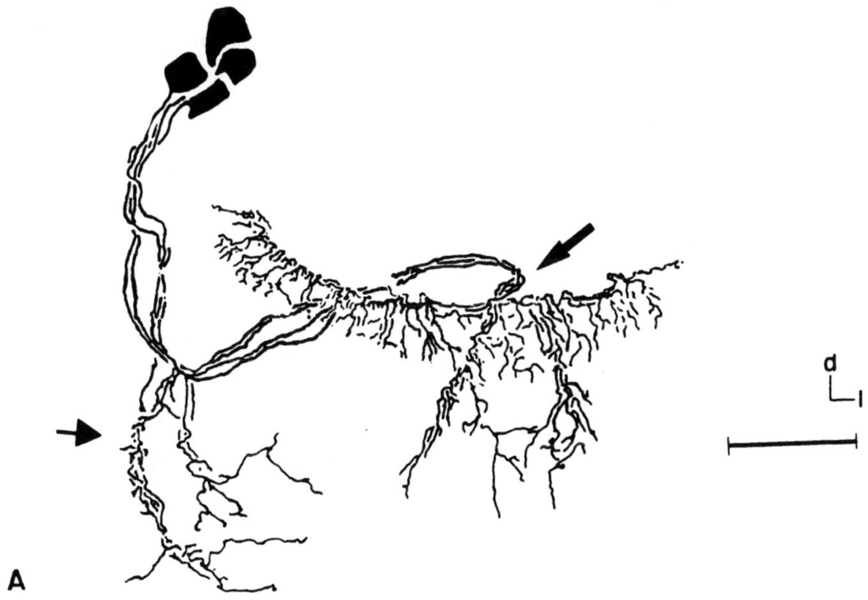

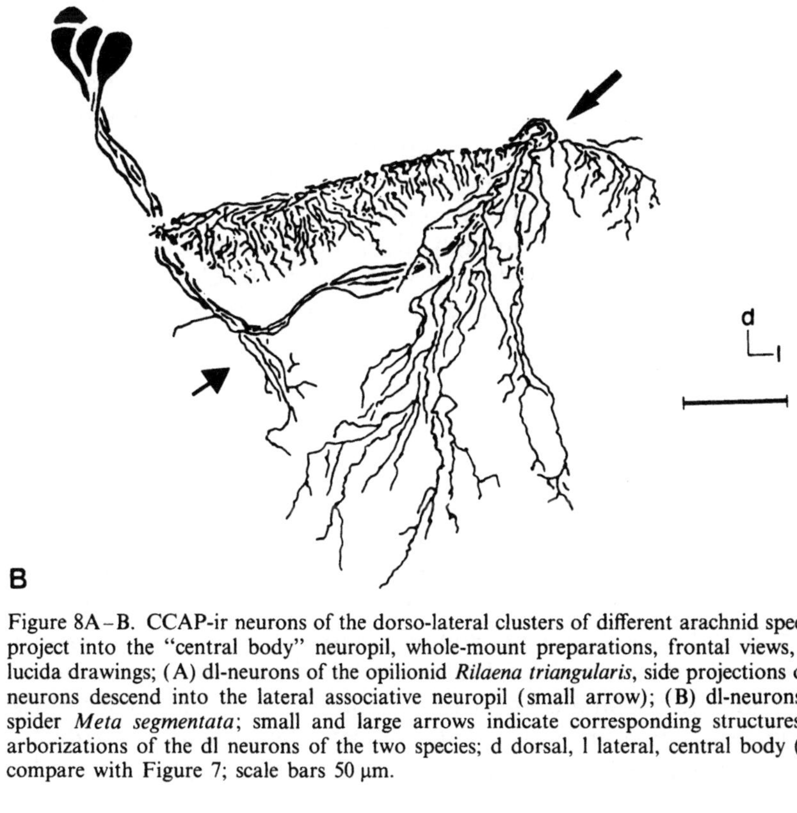

Figure 8A–B. CCAP-ir neurons of the dorso-lateral clusters of different arachnid species that project into the "central body" neuropil, whole-mount preparations, frontal views, camera lucida drawings; (A) dl-neurons of the opilionid *Rilaena triangularis*, side projections of the dl neurons descend into the lateral associative neuropil (small arrow); (B) dl-neurons of the spider *Meta segmentata*; small and large arrows indicate corresponding structures in the arborizations of the dl neurons of the two species; d dorsal, l lateral, central body ("CB"); compare with Figure 7; scale bars 50 μm.

the insect CB and the arachnid "CB" are different (Strausfeld et al., 1993). Accordingly, none of the criteria to define homologies (Remane, 1956, Kutsch and Breidbach, 1994) is applicable in a comparison of the insect CB and the arachnid "CB". Consequently, they are to be regarded as non-homologous neuropil structures.

The same is true for the corpora pedunculata of insects and the "corpora pedunculata" of arachnids (Breidbach and Wegerhoff, 1993; Strausfeld and Barth, 1993): Coupling characteristics, topological organizations and situations in the commissural architectures of the various arthropods are different. Consequently, these neuropil structures are regarded as non-homologous in the two main arthropod groups (Strausfeld and Barth, 1993; Strausfeld et al., this volume).

What are the consequences of such a non-homology of mandibulate and chelicerate brain neuropils? The central-body-like and the corpora pedunculata-like structures are the most prominent neuropil areas in the central brains of both arachnids and insects. In chelicerates, both the "central body" ("CB") and the "corpora pedunculata" ("CP") are found in all species analyzed so far (see Wegerhoff and Breidbach, this volume). The insect brain is characterized by a central body (CB) and the corpora pedunculata (CP) (Strausfeld, 1976; Mobbs, 1985). Whereas the CP are not found in all Mandibulata, the CB might be a common characteristic of the mandibulate brains (Tab. 1; Gupta 1987). Accordingly, the CB is a characteristic of the insect brain and – according to the hypothesis of the homology of the myriapod and the insect CB – seems to be a synapomorphic feature of the mandibulate brain. The "CB" of arachnids is different form the CB of mandibulata. This "CB" and the "CP" that might be secondarily lost in some acarine groups (Binnington, 1987) are synapomorphic characters of the brain neuropil of chelicerates. Accordingly, the principal characteristics of the chelicerate and the mandibulate central brains differ. Consequently, the central brains of both groups have to be regarded as results of convergent developments.

Brain segmentation

How different are the organizations of the chelicerate and the mandibulate brains? Both share characteristics of an arthropod neuromer, but it is not known how many and which of these neuromers have been fused in the chelicerate and the mandibulate brains. As has been reviewed by Wegerhoff and Breidbach (this volume), hypotheses about the segmental characters of the arachnid brains are based on controversial histological findings: Nothing is known about the segmental character of the arachnid protocerebrum. Some data indicate the presence of a deutocerebral segment, but these are not commonly accepted. Even the correlation of the chelicerate ganglion with a mandibulate tritocerebrum is not definitely established.

Based on the expression pattern of segmental polarity genes (*engrailed* and *wingless*) during embryonic development, Schmidt-Ott and Technau (1992) gave evidence that the head of *Drosophila* consists of four supraesophageal neuromers. They found four patches of *engrailed* expression in the brain. They attributed these to an intercalary, an antennal, an ocular, and a labral segment. New findings with gap gene

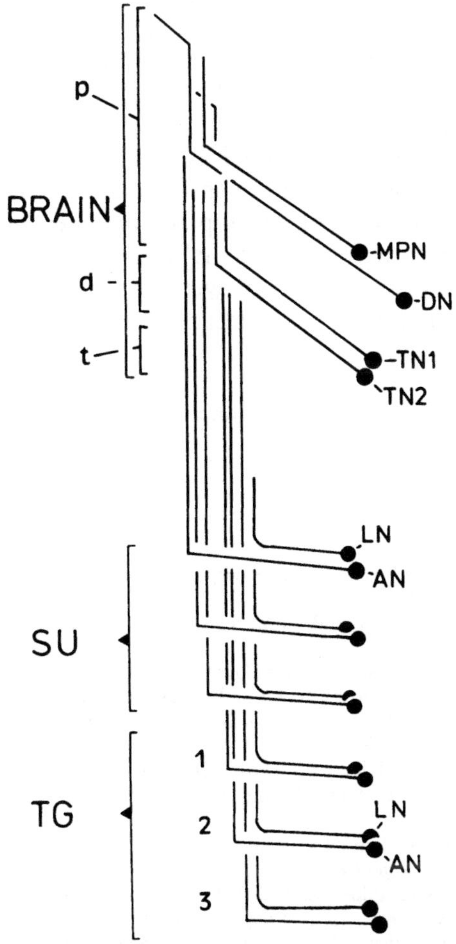

Figure 9. Homonomy (serial homology) of VNC, trito- and deutocerebral serotonin immunoreactive neurons. Scheme of projection patterns of one-half of the total set of immunoreactive neurons in the thoracic (TG) and subesophageal (SU) neuromeres, and of putative serial homologues in the supraesophageal ganglion (BRAIN) of the beetle *Tenebrio molitor*. AN: ascending neurons, d: deutocerebrum, DN: deutocerebral neuron, LN: local interneuron, MPN median posterior neuron, p: protocerebrum, t: tritocerebrum, $TN_{1,2}$: tritocerebral neurons. 1: prothoracic neuromer, 2: mesothoracic neuromer, 3: methathoracic neuromer (modified from Breidbach, 1990b).

mutations support this evidence (Schmidt-Ott et al., 1993). After reviewing the literature about mandibulate brain neurogenesis, Schmidt-Ott and Technau (1992) concluded that even the basic plan of a mandibulate brain consists of these four neuromers.

The identification of insect brain neuroblasts that express homonomous molecular markers enables to characterize homonomous thoracic and brain cell lineages (Zacharias et al., 1993). Thus, in the near future, it will be possible to identify homonomous brain neurons and, thereby, to describe the internal segmental organization of the insect brain neuropil. First attempts to identify homonomous interneurons in the insect brain excluded the pre-antennal segments (Fig. 9, Breidbach 1990b). Identification of internal organization of this brain region is crucial, however, for the central body and the corpora pedunculata are located in this area.

A correlation of certain neuromeral characteristics common to ventral nerve cord neuromers of both insect and arachnid species will be possible only if the analysis of developmental genes and their expression patterns will also be applied for the chelicerate nervous systems. Eventually, these kinds of studies might even allow to identify homologous neuromeral structures in chelicerate and mandibulate nervous systems.

Acknowledgement
This study was supported by the Deutsche Forschungsgemeinschaft.

References

Anderson, D.T. (1973) *Embryology and Phylogeny in Annelids and Arthropods.* Pergamon Press, Oxford, New York.

Arbas, E.A., Meinertzhagen, I.A. and Shaw, S.R. (1991) Evolution in nervous systems. *Annu. Rev. Neurosci.* 14: 9–38.

Ax. P. (1984) *Das Phylogenetische System.* Gustav Fischer Verlag, Stuttgart.

Babu, K.S. (1985) Patterns of arrangement and connectivity in the central nervous system of arachnids. *In:* F.G. Barth (ed.): *Neurobiology of Arachnids.* Springer Verlag, Berlin, Heidelberg, New York, pp. 3–19.

Bastiani, M.J., Doe, C.Q., Helfand, S.L. and Goodman, C.S. (1985) Neuronal specificity and growth cone guidance in grasshopper and *Drosophila* embryos. *Trends Neurosci.* 4: 257–266.

Bergström, J. (1979) Morphology of fossil arthropods as a guide to phylogenetic relationship. *In:* A. Gupta (ed.): *Arthropod Phylogeny.* Van Nostrand Reinhold, New York, pp. 3–56.

Binnington, K.C. (1987) Histology and ultrastructure of the acarine synganglion. *In:* A.P. Gupta (ed.): *Arthropod Brain.* John Wiley & Sons, New York, pp. 95–111.

Boudreaux, H.B. (1979) *Arthropod Phylogeny with Special Reference to Insects.* Wiley, New York.

Boyan, G., Williams, L. and Meier, T. (1993) Organization of the commissural fibres in the adult brain of the locust. *J. Comp. Neurol.* 332: 358–377.

Breidbach, O. (1987) The fate of persisting thoracic neurons during metamorphosis of the meal beetle *Tenebrio molitor* (Insecta, Coleoptera). *Roux's Arch. Dev. Biol.* 196: 450–459.

Breidbach, O. (1990a) Constant topological organization of the Coleopteran metamorphosing nervous sytem, analysis of persistent elements in the nervous system of *Tenebrio molitor. J. Neurobiol.* 21: 990–1001.

Breidbach, O. (1990b) Serotonin-immunoreactive brain interneurons persist during metamorphosis of an insect. A developmental study of the brain of *Tenebrio molitor* L. (Coleoptera) *Cell Tissue Res.* 259: 345–360.

Breidbach, O. (1990c) Reorganization of persistent motoneurons in a metamorphosing insect (*Tenebrio militor* L., Coleoptera). *J. Comp. Neurol.* 302: 173–196.

Breidbach, O. and Dircksen, H. (1991) Crustacean cardioactive peptide-immunoreactive neurons in the ventral nerve cord and the brain of the meal beetle *Tenebrio molitor* during postembryonic development. *Cell Tissue Res.* 256: 129–144.

Breidbach, O. and Kutsch, W. (1990) Structural homology of identified motoneurones in larval and adult stages of hemi- and holometabolous insects. *J. Comp. Neurol.* 297: 392–409.

Breidbach, O. and Wegerhoff, R. (1993) Neuroanatomy of the central nervous system of the harvestman, *Rilaena triangularis* (HERBST 1799) (Arachnida; Opiliones) – Principal organization, GABA-like and Serotonin-immunohistochemistry. *Zool. Anz.* 230: 55–81.

Breidbach, O. and Wegerhoff, R. (1994) FMRFamide-like immunoreactive neurons in the central brain of the beetle *Tenebrio molitor*: Constancies and variations in development from embryo to adult. *Int. J. Insect Morphol. Embryol.* 23: 383–404.

Breidbach, O., Dennis, R., Marx, J., Görlach, C. Wiegandt, H. and Wegerhoff, R. (1992) Insect glial cells show differential expression of a glycolipid-drived, glucuronic acid-containing epitope throughput neurogenesis: detection during postembryogenesis and regeneration in the central nervous system of *Tenebrio molitor* L. *Neurosci. Lett.* 144: 5–8.

Breidbach, O., Dircksen, H. and Wegerhoff, R. (1994) Common bauplan of peptidergic neurons in the arachnid brain – Curstacean cardioactive peptide-immunoreactive neurons in the protocerebrum of seven arachnid species. *Cell Tissue Res; in press.*

Bretschneider, F. (1913) Der Centralkörper und die pilzförmigen Körper im Hirn der Insekten. *Zool. Anz.* 41: 560–569.

Bretschneider, F. (1914) Über die Gehirne der Küchenschabe und des Mehlkäfers. *Jenaische Z.* 52: 269–362.

Bullock, T.H. and Horridge G.A. (1965) *Structure and Function in the Nervous Systems of Invertebrates.* Freeman, San Francisco.

Campbell, C.B.G. and Hodos, W. (1970) The concept of homology and the evolution of the nervous system. *Brain Behav. Evol.* 3: 353–367.

Chabaud, F., Seyfarth, E.-A. and Reichert, H. (1990) Neuronal development in the spider nervous system. *In:* N. Elsner and G. Roth (eds): *Brain-Perception-Cognition. Proceedings of the 18th Göttingen Neurobiology Conference.* Thieme Verlag, Stuttgart, p. 386.

Coggeshall R.E. and Fawcett, D.W. (1964) The fine structure of the central nervous system of the leech, *Hirudo medicinalis. J. Neurophys.* 27: 229.

DeRiemer, S.A. and Macagno, E.R. (1989) Quantitative studies of the growth of neuronal arbors. *In:* T.J. Carew and D.B. Kelley (eds): *Perspectives in Neural Systems and Behavior*, Plenum Press, New York, p. 11–31.

Doe, C.Q., Kuwada, J.W. and Goodman, C.S. (1985) From epithelium to neuroblasts to neurons: the role of cell interactions and cell lineage during insect neurogenesis. *Phil. Trans. R. Soc. Lond. B* 312: 67–81.

Dumont, J.P.C. and Robertson, R.M. (1986) Neuronal circuits: an evolutionary perspective. *Science* 233: 849–853.

Edwards, J.S. (1977) One organism, several brains: Evolution and development of the insect central nervous system. *Ann. New York Acad. Sci.* 299: 59–71.

Field, K.G., Olsen, G.J., Lane, D.J., Giovanni, S.J., Ghiselin, M.T., Raff, E.C., Pace, N.R. and Raff, R.A. (1988) Molecular phylogeny of the animal kingdom. *Science* 239: 748–753.

Gao, W.-Q. and Macagno, E.R. (1987) Extraction and retraction of axonal projections by some developing neurons in the leech depends upon the existence of neighbouring homologues. I. The HA cells. *J. Neurobiol.* 18: 43–59.

Granger, N.A., Homberg, U., Henderson, P., Towle, A. and Lauder, J.M. (1989) Serotonin-immunoreactive neurons in the brain of *Manduca sexta* during larval development and larval-pupal metamorphosis. *Int. J. Dev. Neurosci.* 7: 55–72.

Grenningloh, G., Bieber, A., Rehm, E.J., Snow, P., Traquina, Z., Hortsch, M., Patel, N. and Goodman, C.S. (1990) Molecular genetics of neuronal recognition in *Drosophila*: Evolution and function of immunoglobulin superfamily cell adhesion molecules. *Cold Spring Harbor Symp. Quant. Biol.* 55: 327–340.

Gupta, A.P. (1987) Evolution trends in the central and mushroom bodies of the arthropod brain. A dilemma. *In:* A.P. Gupta (ed.): *Arthropod Brain.* Wiley, New York, pp. 27–44.

Hanesch, U. Fischback, K.-F. and Heisenberg, M. (1989) Neuronal architecture of the central complex in *Drosophila melanogaster. Cell Tissue Res.* 257: 343–366.

Hanström, B. (1921) Über die Histologie und vergleichende Anatomie der Sehganglien und Globuli der Araneen. *Kungl. Svenska Vetensk. Acad. Handlingar.* 61: 1–39.

Hanström, B. (1923) Further notes on the central nervous system of arachnids, scorpions, phalangids and trap-door spiders. *J. Comp. Neurol.* 35: 249–272.

Hanström, B. (1924) Untersuchungen über das Gehirn, insbesondere die Sehganglien der Crustaceen. *Ark. Zool.* 16: 1–119.

Hanström, B. (1926) Untersuchungen über die relative Größe der Gehirnzentren verschiedener Arthropoden unter Berücksichtigung der Lebensweise. *Z. mikr.- anat. Forsch.* 7: 135–190.

Hanström, B. (1928a) *Vergleichende Anatomie des Nervensystems der wirbellosen Tiere.* Springer Verlag, Berlin, Heidelberg, New York.

Hanström, B. (1928b) Some points on the phylogeny of nerve cells and of the central nervous system of invertebrates. *J. Comp. Neurol.* 46: 475–492.

Hanström, B. (1928c) Die Beziehungen zwischen dem Gehirn der Polychäten und dem der Arthropoden. *Z. Morph. Ökol. Tiere* 11: 152–160.

Hanström, B. (1935) Fortgesetzte Untersuchungen über das Araneengehirn. *Zool. Jb. Anat.* 59: 455–478.

Harelson, A.L. and Goodman, C.S. (1988) Growth cone guidance in insects: fasciclin II is a member of the immunoglobulin superfamily. *Science* 242: 700–708.

Hennig, W. (1969) *Die Stammesgeschichte der Insekten.* W. Kramer, Frankfurt.

Holmgren, E. (1916) Zur vergleichenden Anatomie des Gehirns der Polychaeten, Onychophora, Xiphosuren, Arachniden, Crustaceen, Myriapoden und Insekten. *Kungl. Svenska. Vetensk. Handlingar.* 56: 1–303.

Homberg, U. (1985) Interneurones of the central complex in the bee brain (*Apis mellifera* L.). *J. Insect Physiol.* 31: 251–264.

Homberg, U. (1987) Structure and functions of the central complex in insects. *In:* A.P. Gupta, (ed.): *Arthropod Brain.* John Wiley & Sons, New York, pp. 347–368.

Jawlowski, H. (1936) Über den Gehirnbau der Käfer. *Z. Morph. Tiere* 32: 67–91.

Joly, R. and Deschamps, M. (1987) Histology and ultrastructure of the myriapod brain. *In:* A.P. Gupta (ed.): *Arthropod Brain.* John Wiley & Sons, New York, pp. 135–157.

Kent, K.S., Hoskins, S.G. and Hildebrand, J.G. (1987) A novel serotonin-immunoreactive neuron in the antennal lobe of the sphinx moth, *Manduca sexta* persists throughout postembryonic life. *J. Neurobiol.* 18: 451–456.

Kutsch, W. and Breidbach, O. (1994) Homologous structures in the nervous systems of arthropoda. *Adv. Insect Physiol.* 24: 1–113.

Kutsch, W., Urbach, R. and Breidbach, O. (1993) Comparison of motor patterns in larval and adult stage of a beetle, *Zophobas morio. J. Exp. Zoology* 267: 389–403.

Kuwada, J.W. (1984) Normal and abnormal development of an identified leech motor neuron. *J. Embryol. Exp. Morph.* 79: 125–137.

Kuwada, J.W. and Kramer, A.P. (1983) Embryonic development of the leech nervous system: Primary axon outgrowth of identified neurons. *J. Neurosci.* 10: 2098–2111.

Levine, R.B. (1986) Reorganization of the insect nervous system during metamorphosis. *Trends in Neurosci.* 6: 315–319.

Manton, S.M. (1977) *The Arthropoda.* Clarendon Press. Oxford.

Mobbs, P.G. (1985) Brain structure. *In:* G.A. Kerkut and L.I. Gilbert (eds). *Comprehensive Insect Physiology, Biochemistry, and Pharmacology. Vol. 5. Nervous System: Structure and Motor Function.* Pergamon Press, Oxford, pp. 288–370.

Muller, S.W. and Campbell, A. (1953) The relative number of living and fossil species of animals. *Systematic Zoology* 3: 168–170.

Nässel, D.R. and Elofsson, R. (1987) Comparative anatomy of the crustacean brain. *In:* A.P. Gupta (ed.): *Arthropod Brain.* John Wiley & Sons, New York, pp. 111–135.

Owen, R. (1848) *On the Archetype and Homologies of the Vertebrate Skeleton.* J. v. Voort, London.

Panov, A.A. (1959) Bau des Insektenhirnes während der postembryonalen Entwicklung II. Zentralkörper. *Entomol. Oboyr.* 38: 301–311.

Patel, N.H., Kornberg, T.B. and Goodman, C.S. (1989a) Expression of *engrailed* during segmentation in grasshopper and crayfish. *Development* 107: 201–212.

Patel, N.H., Martin-Blanco, E., Coleman, K.G., Poole, S.J., Ellis, M.C., Kornberg T.B. and Goodman C.S. (1989b) Expression of engrailed proteins in arthropods, annelids, and chordates. *Cell* 58: 955–968.

Pentreath, V.W. (1989) Invertebrate glial cells. *Comp. Biochem. Physiol.* 93A: 77–83.

Raper, J.A., Bastiani, M.J. and Goodman, C.S. (1984) Pathfinding by neuronal growth cones in grasshopper embryos. IV. The effects of ablating the A and P axons upon the behavior of the G growth cone. *J. Neurosci.* 4: 2329–2345.

Remane, A. (1956) *Die Grundlage des natürlichen Systems, der vergleichenden Anatomie und der Phylogenetik*. Akademische Verlagsgesellschaft Geest & Portig, Leipzig.

Retzius, G. (1892) Das Nervensystem der Lumbriciden. *Biol. Untersuch., N.F.* 3: 1–16.

Riehl, B. and Schule, W.R. (1989) Electronmicroscopical and histochemical studies on packet-, neuropile and connective glial cells of the leech *Hirudo medicinalis*. *Verh. Dtsch. Zool. Ges.* 82: 190.

Riehl, B., and Schlue, W.-R. (1990) Localization of carbonic anhydrase in identified glial cells of the leech central nervous system: application of histochemical and immunocytochemical methods. *J. Histochem. Cytochem.* 38: 1173–1178.

Salecker, I. and Distler, P. (1990) Serotonin-immunoreactive neurons in the antennal lobes of the American cockroach *Periplaneta americana*, light- and electron-microscopic observations. *Histochemistry* 94: 463–473.

Sandeman, D.C. (1982) Organization of the central ganglia. *In:* H.L. Atwood and D.C. Sandeman (eds): *The Biology of Crustacea, Vol. 3. Neurobiology: Structure and Function.* Academic Press, New York, pp. 1–61.

Sandeman, D.C., Sandeman, R.E. and Aitken, A.R. (1988) Atlas of serotonin-containing neurons in the optic lobes and brain of the crayfish, *Cherax destructor*. *J. Comp. Neurol.* 269: 465–478.

Sandeman, R.E., Sandeman, D.C. and Watson, A.H.D. (1990) Substance P antibody reveals homologous neurons with axon terminals among somata in the crayfish and crab brain. *J. Comp. Neurol.* 294: 569–582.

Schmidt-Ott, U. and Technau, G. (1992) Expression of *en* and *wg* in the embryonic head and brain of *Drosophila* indicates a refolded band of seven segment remnants. *Development* 116: 111–125.

Schmidt-Ott, U., González Gaitán, M., Jäckle, H. and Technau, G.M. (1993) Embryonic head phenotype of *Drosophila* "gap gene" mutants. *In:* N. Elsner and M. Heisenberg (eds): *Gene, Brain, Behaviour. Proceedings of the 21st Götingen Neurobiology Conference.* Thieme Verlag, Stuttgart, New York, p. 132.

Scholtz, G., Dohle, W., Sandeman, R.E. and Richter, S. (1993) Expression of *engrailed* can be lost and regained in cells of one clone in crustacean embryos. *Int. J. Dev. Biol.* 37: 299–304.

Schürmann, F.-W., Sandeman, R. and Sandeman, D. (1991) Dense-core vesicles and non-synaptic exocytosis in the central body of the crayfish brain. *Cell Tissue Res.* 265: 493–501.

Shaw, S.R. and Meinertzhagen, I.A. (1986) Evolutionary progression at synaptic connections made by identified homologous neurones. *Proc. Natl. Acad. Sci. USA* 83: 7961–7965.

Spörhase-Eichmann, U., Gras, H. and Schürmann, F.-W. (1987a) Patterns of serotonin-immunoreactive neurons in the central nervous system of the earthworm *Lumbricus terrestris* L. – I. Ganglia of the ventral nerve cord. *Cell Tissue Res.* 249: 601–614.

Spörhase-Eichmann, U., Gras, H. and Schürmann, F.-W. (1987b) Patterns of serotonin-immunoreactive neurons in the central nervous system of the earthworm *Lumbricus terrestris* L. – II. Rostral and caudal ganglia. *Cell Tissue Res.* 249: 625–632.

Snow, P.M., Zinn, K., Harrelson A.L., McAllister, L., Schilling, J., Bastiani, M.J., Makk, G. and Goodman, C.S. (1988) Characterization and cloning of fasciclin I and fasciclin II glycoproteins in the grasshopper. *Proc. Natl. Acad. Sci. USA* 85: 5291–5292.

Strauß, R. and Heisenberg, M. (1991) Genetic lesions in the central complex affect walking in *Drosophila melanogaster*. *In:* N. Elsner and H. Penzlin (eds): *Synapse-Transmission-Modulation. Proceedings of the 19th Göttingen Neurobiology Conference.* Thieme Verlag, Stuttgart, New York, p. 80.

Strauß, R. and Heisenberg, M. (1992) Function of the central complex as a control center for locomotor behaviour. *In*: N. Elsner and D.W. Richter (eds): *Rythmogenesis in Neurons and Networks. Proceedings of the 20th Göttingen Neurobiology Conference.* Thieme Verlag, Stuttgart, New York, p. 168.

Strauß, R. and Hofbauer, A. (1993) An immunocytochemical investigation on the structural effects of *Drosophila* central complex mutants. *In:* N. Elsner and M. Heisenberg (eds): *Gene-Brain-Behaviour. Proceedings of the 21st Göttingen Neurobiology Conference.* Thieme Verlag, Stuttgart, New York, p. 80.

Strausfeld, N.J. (1976) *Atlas of an Insect Brain.* Springer Verlag, Berlin, Heidelberg, New York.

Strausfeld, N.J. and Barth F.G. (1993) Two visual systems in one brain: neuropils serving the secondary eyes of the spider *Cupiennius salei. J. Comp. Neurol.* 328: 43–62.

Strausfeld N.J., Weltzien, P. and Barth F.G. (1993) Two visual systems in one brain: neuropils serving the principal eyes of the spider *Cupiennius salei. J. Comp. Neurol.* 328: 63–75.

Thomas, J.B., Bastiani, M.J., Bate, M. and Goodman, C.S. (1984) From grasshopper to *Drosophila*: A common plan for neuronal development. *Nature* 310: 203–207.

Urbach, R., Breidbach, O. and Kutsch, W. (1994). Comparative anatomy of muscle sets in larval and adult stages of *Zophobas morio* (Coleoptera, Tenebrionidae). *Zoomorphology.* 114: 47–57.

Wegerhoff, R. and Breidbach, O. (1992) Structure and development of the larval central complex in a holometabolous insect, the beetle *Tenebrio molitor. Cell Tissue Res.* 268: 341–358.

Weiß, N. (1994) *Die Entwicklung des kommissuralen Grundplanes im Käferhirn.* Thesis, Bonn.

Weltzien, P. (1988) *Vergleichende Neuroanatomie des Spinnenhirnes unter besonderer Berücksichtigung des "Zentralkörpers".* Ph.D. Thesis, Frankfurt.

Williams, J.L.D. (1975) Anatomical studies of the insect central nervous system. A groundplan of the midbrain and an introduction to the central complex in the locust *Schistocerca gregaria* (Orthoptera). *J. Zool.* 1976: 67–86.

Wilson, J.A. and Hoyle, G. (1978) Serially homologous neurons as concomitants of functional specialisation. *Nature* 274: 377–378.

Whitington, P.M., Leach, D. and Sandeman, R. (1993) Evolutionary change in neural development within the arthropods: axogenesis in the embryos of two crustaceans. *Development* 118: 449–461.

Whitington, P.M., Meier, T. and King, P. (1991) Segmentation, neurogenesis and formation of early axonal pathways in the centipede, *Ethmostigmus rubripes* (Brandt). *Roux's Archiv Develop. Biol.* 199: 349–363.

Zacharias, D., Williams, J.L.D., Meier, T. and Reichert, H. (1993) Neurogenesis in the insect brain: cellular identification and molecular characterization of brain neuroblasts in the grasshopper embryo. *Development* 118: 941–955.

The Nervous Systems of Invertebrates: An Evolutionary and Comparative Approach
ed. by O. Breidbach & W. Kutsch
© 1995 Birkhäuser Verlag Basel/Switzerland

The nervous systems of Echinodermata: Recent results and new approaches

J.L.S. Cobb

Gatty Marine Laboratory, The University, St. Andrews, Fife KY 16863B, Scotland, UK

Summary. The echinoderm nervous system is the least studied of all the major phyla. It does however have many unusual or unique features. There are two distinct and separate nervous systems, one of which may be mesodermal in origin. The nervous system is non-centralised and whole body co-ordination in complex behaviours appears to be able to be controlled by quite different parts of the nervous system depending upon circumstances. The action potential appears to be based on Ca^{++} ions and not Na^{+}, and connective tissue is mutable and innervated. Recently there has been an increase in activity looking at neuropeptides and other modulators. This review looks at what has been achieved and offers suggestions for future research.

Introduction

There have been a number of recent reviews of echinoderm nervous systems (Cobb, 1987–89, 90). Recently, study of the nervous system of this phylum has increased especially with the availability of immunoprobes against neuromodulators and transmitters. The main thrust of this review will be to look at problematical areas and assess what might be done to resolve them.

Background

Echinoderms are large and very obviously fascinating animals and it was to be expected that they attracted early morphologists. Their works are most excellently summarised in Hyman's volume on the Echinodermata (1955). This reference is absolutely invaluable to the neurobiologist for understanding the general and histological morphology of these animals. Since that time, the general echinoderm literature has never been adequately reviewed. Authors who rely on general texts on the invertebrates to understand the progress on function, in particular over the last 40 years, will be sadly misled and a really comprehensive review of general echinoderm biology is sorely needed.

There are six classes of extant echinoderms; one, the Crinoidea, is deemed the most primitive and has the opposite orientation to the rest and the nervous system reflects this. All the different classes of echinoderms have a nerve ring around the mouth called – appropriately – the circumoral nerve ring. Each arm then has a radial nerve cord that runs from the circumoral ring to the periphery. In most cases there are five radial nerve cords, with only a few asteroids having more than five arms. There are two main parts of the nervous system. The ectoneural nervous system occurs throught the animal and is concentrated into the ring and nerve cords. It is sensory, inter-neuronal and indirectly motor. The hyponeural is largely associated with substantial muscle blocks and is purely motor. In the echinoids, for instance, it is associated with the complex muscles of the lantern. It is highly developed in the nerve cord of the brittlestars associated with rapid locomotorary movements by the skeletomuscular system.

The odd group out is the Crinoidea – the feather stars. This class is considered ancestral to the rest. In fact, remarkably little is known about the function of crinoid nervous systems. The nerves are small and totally enclosed in a calcite skeleton. They have attracted some recent attention. Grimmer and Holland (1990) have described the nervous system of the stalkless crinoid *Holopus* and Welsch et al. (1990) have described parts of the nervous system of *Antedon*.

Smith adopted the nervous system of echinoderms as a major study and published a series of detailed and significant papers; see Smith (1966) for review. At that time, he largely relied on the use of modifications of the vital dye methylene blue. This dye does stain echinoderm nerves very effectively. Unknown to Smith (and only shown with the advent of electron microscopy) many muscles send long narrow processes – called muscle tails – to make synaptic contact with motor neurones at long distances from the main body of the muscles, and these also stain with vital dyes and thus misled Smith (see Cobb, 1987 for details). Because of this, much of Smith's work is flawed when re-analysed at the level of electron microscopy. Another aspect of Smith's work was to look at behaviour. Based on fairly simple experimentation he described a series of centres near or within the circumoral nerve ring in starfish responsible for co-ordinating whole-animal locomotion (and, by implication, other functions). Regrettably this idea too has not stood the test of more modern techniques. Nevertheless, Smith's work was pioneering at a time when the phylum was beginning to be ignored by neuroscientists, something which has sadly continued.

Another major controversy still not resolved is that introduced by Bullock (1965). Von Uexkull (1897) had described the nervous system of sea-urchins as a republic of reflexes. Bullock (1965) and Kinosita (1941) both worked on the behaviour of the peripheral organs (spines, tubefeet etc.) of these sea-urchins. Kinosita described these organelles as being connected by a nerve net of the classical coelenterate sort and thus rather primitive. Bullock was less sure and proposed a massive set of direct connections between the points of stimulus and the various effectors. There is no compelling evidence from morphology to date.

This author assumes direct connections to peripheral ganglia in the manner of Bullock's original suggestion. Recent studies by Cobb and Begbie (in press) show that each spine is run by a complex ganglion of ectoneural neurones which makes the idea of a nerve net even less likely.

A unique feature of echinoderms is a phenomenon that is being shown to be increasingly widespread. The connective tissue is mutable and able to change its constituancy from stiff to pliable very rapidly (see Wilkie 1988; Motokawa 1988). This appears to be under the control of the nervous system. There are special cells associated in ganglion-like aggregations which have been called juxtaligamental cells and in a number of sites these have been shown to be innervated by hyponeural motor nerves. It is likely that this phenomenon will be found in a very wide range of tissues associated with the skeleton. A largely accepted idea is that mutable connective tissue can be softened to allow muscles to move heavy appendages and parts of the main body to new positions where the connective tissue then hardens to fix the position. This energy-saving system may well underpin the entire evolutionary strategy of the echinoderms. The phenomenon has been demonstrated in many situations but most notably in sea-urchin spines, holothurian body wall and asteroid and particularly brittlestar arm movements. It is involved in autotomy of body parts. There have been a number of attempts at clarifying the mechanisms of the system. It has been known for some time that the system is calcium sensitive. Most recently, Matsuno and Motokawa (1992) have described storage cells for calcium among the catch connective tissue in sea-cucumbers. They envisage calcium release as changing binding properties of the glyco-protein matrix allowing the collagen bundles to slide past each other or remain fixed according to need. This implies that calcium is acting as a primary messenger. Whether calcium can be sufficiently sequestered in the extracellular space in this way in a calcium-rich environment remains to be seen. It is however worth bearing in mind that there have previously been suggestions of an excitatory factor as well as an inhibitory factor. Recently, and not before time, this unique phenomenon of a "smart" material has begun to attract some attention and it may be that a much clearer understanding will soon be achieved. At present, it seems most likely that the uniqueness lies in the properties of the connective tissue and not in the chemicals released from the juxtaligamental tissues. Notwithstanding this and Matsuno and Motokawa's description of calcium storage cells, there is an urgent need to study the neurochemistry of the juxtaligamental cells since it is by no means certain that there are not unusual release factors involved. There is now available a wide range of test preparations, some of which should be suitable for the sort of molecular biological analysis that is needed, in particular see Wilkie (1992).

Electron microscope morphology

Introduction

The advent of electron-microscopy radically changed the ideas on the layout of the nervous system since many echinoderm nerves are extremely small and not resolvable with the light microscope. Electron microscopy of a good quality is not easily achieved. As a general rule, marine tissue is more difficult to fix than freshwater, echinoderms are the most difficult marine tissue, nerves are harder to fix than other tissues and the developing nervous tissue is a nightmare. Very often a simple buffered osmium tetroxide produces less osmotic damage than a carefully prepared isotonic glutaraldehyde. A consequence of the difficulty in producing consistent results means that even identifying single nerves is difficult, let alone making pronouncements about cellular detail such as synaptic structure.

Sensory

Echinoderms have few specialised receptor organs. The best known is the eyespot in asteroids although even here its removal makes little difference to their photosensitivity and most echinoderms are probably very sensitive to light all over their integument. The same is true of the other modalities of reception. There is a complex chemosensitive and touch-sensitive hillock on the inside of the globiferous pedicellaria of sea urchins and there are tracts of individual modified ciliated receptors on other places; see Cobb (1987) for review. There is undoubted proprioception but no description of the morphology of the receptors (Moore and Cobb 1987). Thus there appears to be widespread receptor distribution in a range of modalities but no specialized concentration. This fits in with the idea that is emerging that the nervous system is not centralised. Since so far it has proved impossible to record unitary potentials from any sensory neurones, the description of various cellular structures as sensory is based on their morphology alone.

Nerve cord and rings

It is extraordinary that although much of the literature has assumed that the circumoral ring is a co-ordinating centre, there have been almost no morphological studies on it. The radial nerve cords of all the classes are segmental and clearly, on the evidence from the brittlestars, this segmentation is a series of similar ganglia linked together by interganglionic regions. Only in the brittlestars is there a detailed description involving individual neurones or tracts.

A valuable potential feature of the nerve cords is the fact that the neurone cell bodies are arranged in a largely two-dimensional array on the oral surface. In other words, there is a nice two-dimensional map which, at least in the brittlestar (and likely so in the other classes), is consistent in detailed cell positioning from one ganglion to the next. Maerkel and Roeser (1991) have recently described the morphology of the nerve cord in an echinoid.

What studies there have been on the circumoral nerve ring provide no evidence at all that it is any way co-ordinating. There are cell bodies and tracts running off to skeletomuscular systems around the mouth and jaws and also to the gut. For the main part it simply is a set of connections between the radial nerve cords. The idea that it acts as any sort of central control can be totally ruled out since it simply does not have the structure for this.

A significant feature of acculumations of neurones into segmental ganglia in the radial nerve cords is that there are no glial cells or glial-like cells. There are relatively few support cells. This lack of glial cells is extraordinary given the increasing importance that is being assigned to their co-operative function with neurones in other phyla. It should be borne in mind as well that echinoderms do not have a circulatory system of blood vessels that in any way resembles the blood capillaries or blood spaces of other organisms. The neurones then really do have to fend entirely for themselves as far as maintaining extracellular ion balance and taking up and removing nutrient and waste substances, as well as the manufacture and distribution of complex proteins and the like. It is worth commenting that another specific feature of the nerve cells is that a very large proportion (50% or more of total volume at times) is taken up with glycogen stores. This has the knock on effect that there is much less in the way of rough endoplasmic reticulum and so again echinoderms must maintain their axons and synapses apparently with proportionately less than half the protein translating machinery that most other phyla have in their neurones.

Chemical synapses

At the cytoplasmic level echinoderm neurones are characterised by typical microtubules, small mitochondria and various vesicles. Vesicles, as in other animal groups, show a wide range of size and content. To date there is no evidence linking a particular type of transmitter with a particular type of vesicle. At best small clear vesicles are associated with endings that have long been assumed to be cholinergic. This type of vesicle is mainly present at such neuromuscular junctions that have been described by nerves of hyponeural system (See Cobb, 1987 for details). There have been no reports of the small eliptical clear vesicles that are associated with the inhibitory transmitters. Many nerve endings are filled with vesicles of different types and appear to be varicose with no synaptic specialisations. They thus resemble endings in the vertebrate autonomic nervous system. Cobb (1985) reported a synapse onto juxtaligamental cells by hyponeural nerves with the typical pre- and post-synaptic specialisations of vertebrate central nervous tissue but these types of synapse are few and far between. Studying the detailed structure of synapses is seriously handicapped by the poor quality of fixation. There is a class of nerve-like processes in between the muscles of the ampullae and tubefeet that contain large granular vesicles. There is not "meant" to be a nerve plexus within these muscles and their function is obscure. There is a similar plexus of processes with large, dense, cored vesicles (as well as another class with typical small agranular ones) found in between the muscle of the gut and the detailed structure. The function of gut innervation, however, is obscure. Careful use of suitable immunoprobes may be the best way to start finding out about the layout of these putative nerve plexuses, which may then allow testable hypotheses to be made about their function.

Electrical synapses

To date there has been no description in any adult echinoderm tissue of structures resembling gap junctions. They have been diligently searched for in muscle tissue but not found. They might be presumed to be present between the giant fibres that make up the rapid through-conducting system but there is no morphological evidence for this, and intracellular dyefills show no evidence. It is of course possible that echinoderms have dispersed ion channels between the adjacent cells and have not formed them into the complexes normally found in gap junctions. The eggs and early larvae do show electrical connections between cells, so it remains unclear whether they have been lost in adult echinoderms or whether we still have not found them.

Motor innervation

There is either a completely separate hyponeural nervous system which directly innervates the muscles, or there is indirect innervation by ectoneural nerves with transmitter released – sometimes long distances – across the basement membrane (basal lamina) which underlies the ectodermal nervous tissue. The hyponeural nervous tissue is found associated with Aristotle's lantern in echinoids, the radial nerves and rings of asteroids and ophiuroids and the radial nerve cords of holothuria. It is itself innervated by transmitter released across a basement membrane. This strange inability of the ectoneural nervous system to make direct motor connections led this author to claim that the hyponeural nervous system must be mesodermal in original.

In the case of ectoneural nerves releasing transmitter across connective tissue basement membranes, the best descriptions are of the tubefeet and ampullae innervation and the innervation of the hyponeural tissue across the basement membrane in th brittlestar radial cord. Florey and Cahill (1980) showed in the case of the tubefeet that transmitter might be released very many microns from the target cells. The innervation of the hyponeural tissue by pre-synaptic ectoneural neurones in the brittlestar is well described (see Cobb, 1987 for detailed references) and there are physiological studies of this synapse (see below).

A strange exception to the rule of either non-direct innervation or a separate hyponeural innervation was the reported direct innervation of the muscles of the spines of sea-urchins described by Peters (1985). A re-investigation of this shows that there is the usual indirect connection onto a previously undescribed set of hyponeural neurones which then pass to directly innervate the muscles and juxtaligamental cells (Cobb and Begbie, in press).

Physiology

Extracellular

These studies started a long time ago and originally used wick electrodes to record summated potentials. These early studies showed very little (see Cobb, 1987 for references). Things changed when Brehm (1975) showed that it was possible to use plastic suction electrodes to record individual action potentials from the nerve cords of brittlestars. It has since been shown that the brittlestar *Ophiura ophiura* is a valuable preparation. All these records have however been obtained from large interneurones and direct recording from sensory nerves has not proved possible. The nearest anyone has come to this is in the summed electrical records made from sea-urchin spines by Berrios et al. (1985); though even here it may be on-going interneuronal information that is recorded rather than the sensory neurones.

We have little knowledge of how many different species of brittlestar can be recorded from usefully. There is no doubt that large specimens of *Ophiura ophiura* are especially fortuitous. It is, however, possible to record from *Ophiocomina nigra* and *Ophiothrix fragilis*. A survey of about 24 species of Caribbean brittlestars from a range of different families tried to relate speed of certain behaviours to size of radial cord interneurone (Cobb, unpublished studies). This project failed because the fixation of the material for electron microscopy on a remote field site was not good enough to make accurate measurements of the size of the nerve cords in all species. This study did show that none of these species had nerve fibres the size of those found in similar sized *Ophiura*, even though a few approached *Ophiura* for speed of righting behaviour or arm movement. It may be that the largest giant fibres will be found associated with cold water species where conduction velocity may be compromised by low temperatures. Extra-cellular recording is a good guide to giant fibre size and some simple recording experiments in marine laboratories in various parts of the world would indicate whether there are species other than *Ophiura ophiura* that are as good or better for recording single action potentials and thus useful for carrying out intracellular studies.

A series of papers by Moore and Cobb (1985a,b, 1987) showed that extracellular recording from brittlestar radial nerve cords – and therefore from classes of interneurones – was invaluable for monitoring modalities and levels of sensory reception. These studies unequivocally show that brittlestars (and by extrapolation other echinoderms) are exquisitely sensitive to changes in the environment even though they have no specialised sensory receptors. This simple technique still has considerable potential for understanding more about the way sensory information is integrated and passed on through the nervous systems of these animals and is easy to carry out.

Intracellular

Intracellular studies on excitable cells in adult echinoderms were begun using floating micro-electrodes on the muscles of the echinoid lantern (Cobb, 1968). There have been no intracellular recordings made since on muscle. Intracellular recordings have however been made on the giant fibres of the brittlestar *Ophiura*. It has proved possible to record from large ectoneural neurones as well as the hyponeural motor neurones. Recording from the ectoneural neurones and, under some circumstances, the hyponeural neurones, is done with a very minimum amount of dissection. The oral plate is removed and the sheaths of the epineural sinus dissected away leaving the cell bodies of the ectoneural interneurones laid out in a two-dimensional map. Dye fill experiments show that this map is consistent from segment to segment and the positions of a few identified cells are now known. The idea that each ganglion consists of a repeated, consistently positioned set of neurones is further confirmed by immunological probes to various neurochemicals.

Ectoneural studies

In favourable preparations it may be possible to see some cells with a dissecting microscope and these can then be impaled. Other impalements are done blind but, as an understanding has developed of the detail of the map, it has become possible to aim at particular regions with an increasing probability of finding a particular cell. One cell in particular can often be visually identified. This cell is the giant through-conducting neurone (GTC). The morphology of neurone which occurs in each segment – usually on alternate sides – has been described in a preliminary paper (Cobb and Ghyoot, 1993). The axon passes for at least six full segments of the nerve cord and always centrally. This neurone conducts in both directions. It is the only through-conducting neurone, so that once one fires the message is rapidly propagated throughout the animal by a chain of GTC neurones at a faster velocity than any other type of nerve fibre. There is a range of other interneurones identifiied so that their position and shape is documented but little is known of their function. There is for example an H-shaped neurone with its cell body on one side and four axon processes that run up to four segments (thus, one neurone has a field of at least eight segments). There are some neurones that run transversely within a single segment. Thin sections stained with toluidine blue confirm a very highly organised set of axon pathways in different orientations at different levels in the ganglion.

Hyponeural

A pool of muscle cells occurs on either side of the midline in each segmental ganglion (see Cobb, 1987 for references). In a large animal there are about 50 large cells and probably about 100 smaller cells in each ganglion on either side of the midline (see Cobb and Begbie, in press). They occur rather like a small bunch of grapes and are easy to impale with micro-electrodes. There are three main motor nerve branches which arise and a number of minor nerves, and two of the larger nerves can be dissected out from the ossicle in which they run and the axons are large enough to impale. It is necessary to desheath the neurones, and there is some evidence that this, and severing the motor nerves distally, causes the initiation of action potentials to fail even though synaptic activity is unimpaired. These motor cells normally show some spontaneous synaptic activity and occasionally substantial activity. Stimulation of the tip of the arm of the animal or the ectoneural radial nerve cord itself produces substantial, repeatable, and predictable synaptic activity. There are inhibitory and excitatory synaptic potentials which summate and during arm movements of certain types the inhibitory potentials on one side of the arm mirror image the excitatory potentials of the other side (Cobb, 1990). Many of these neurones are bipolar and the area of dendrites which make synaptic contact with the overlying ectoneural tissue are very extensive for each neurone. The dendrites from a single neurone thus contact the ectoneural tissue at the basement membrane for virtually half of each ganglion but very few if any pass from one ganglion to another or across the ganglion. The motor nerves pass into the ossicles and connective tissue and divide to innervate the inter-vertebral muscle as well as the juxtaligamental tissue associated with the innervation of the connective tissue (see below).

Until very recently the evidence suggested that the hyponeural nerves did not themselves inter-communicate and integrate motor responses further. However there are synapse-like structures between them (Cobb and Begbie, in press) and preliminary results suggest that there may be some neurochemical diversity (Ghyoot and Cobb, in press). Intracellular studies with two electrodes should make this phenomenon relatively easy to investigate.

Ionic basis

Brehm (1977) and Tuft and Gilly (1984) implied that some of the nervous system involved a calcium spike. These studies were carried out using ion substitution and drugs with recording using extracellular electrodes. Cobb and Moore (1988) used the intracellular capability in *Ophiura* to study the ionic basis of the action potential in that species. The results of Tuft and Gilly, where they found suggestions of both sodium and calcium ion involvement, were shown to be misleading due to the stimulatory effect of the choline chloride used as an ion substitute for sodium. Cobb and Moore used sucrose instead of choline chloride. These last studies presented evidence that the action potentials in all recordable neurones (and probably by extrapolation all neurones) in both the ectoneural and the hyponeural nervous system are based on a calcium spike without the apparent need for the involvement of sodium ions.

It must be admitted that the double electrode voltage clamp experiments needed to look at the actual currents have not been done. Double electrode studies in a single cell would be straightforward in hyponeural cells with high quality stable micro-electrode placement systems and certainly feasible in at least the GTC of the ectoneural nervous system. Even without this data the ion substitution experiments present very repeatable evidence with no obvious alternative explanation but a purely calcium based rising phase to the action potential.

Pharmacology

There is quite a substantial literature on the effect of a wide range of pharmacological agents on echinoderm preparations (see Cobb, 1988 for review). The problem with a great many of these studies is that it is never clear where the agents are acting. In the same way, histochemical techniques have shown widespread distribution of substances such as cholinesterase and catecholamines. Recently, attempts have been made to quantify these and map the distribution more precisely. Sloley and Juorio (1990) examined three species of starfish and showed high levels of dopamine and noradrenaline in all three. The tubefeet were rich in dopamine but not noradrenaline while the gut had much of both. 5 HT was noteworthy as being absent.

Immunoprobes

More recently there has been the application of immunoprobes, particularly against neuropeptides and related compounds. The most important has been the description of a new family of peptides. Elphick et al. (1991) used a radioimmunoassay based on a molluscan peptide and subsequently sequenced a new peptide from a starfish. This has been named SALMFamide (known as S1 with a related peptide being known as S2). Subsequently, Diaz-Miranda et al. (1992) used the same technique to isolate related novel peptides in a holothurian. Probes to some of these have been used to locate and map the distribution of the putative neuromodulators in various echinoderms. Thorndyke et al. (1992) used it on the larvae of the sand dollar *Dendroaster*, Moore and Thorndyke (1993) on the starfish, Heinzeller et al. (1993) on the crinoid *Antedon* and Ghyoot and Cobb (in press) on the brittlestar *Ophiura*. In the radial nerve cord of *Ophiura* a single giant cell has been identified in each segmental ganglion. Ghyoot and Cobb (in press) have also looked at the distribution of GABA, Glycine, 5 HT and the anthozoan peptide RF amide. The latter peptide shows interesting results in *Ophiura* since although it largely co-labels (but by no means entirely) with S1 (a SALMFamide) controls suggest this may represent co-localization of the two peptides and not cross-labelling. Martinez et al. (1993), however, used S1 and FRMFamide on the starfish gut but showed cross-reactivity. These authors also looked at other potentially regulating gastric neuropeptides and their results with S1 and FMRFamide show labelling not only in the basiplexus derived from the ectoneural tissue but also from the subcoelomic plexus, which is little understood. Garcia-Arraras et al. (1991) used probes to examine cholecystokinin-like activity in sea cucumbers. They showed that CCK and related peptides released basal muscle tension induced by ACH application. Welsch et al. (1989) used a range of probes against the crinoid *Antedon*. Ghyoot and Cobb (in press) have used anti Gaba, Glycine, 5 HT and RFamide as well as the S1 and S2 probes. This does show that the echinoderm nervous system has the same, precisely repeated neurochemical pattern as has been demonstrated using other morphological techniques with the layout of the neurones in the brittlestar nervous system.

Organ bath

There is quite a number of experiments of this sort which have been reviewed previously by Cobb (1987). The best that one can glean from them is that acetylcholine is excitatory as far as many muscles are concerned and that noradrenaline can be inhibitory. The evidence for the latter is rather poor. Cobb (1987) has previously argued that it is a direct effect on the muscle. As long as bath applications are made to fairly complicated neuro-effector preparations there is little chance of finding out very much about where these substances work and what their *in vivo* function is. (See also larval studies below).

Other morphological techniques

When Falck and Hillarp introduced the fluorescence technique to show the distribution of amines, echinoderms were subjected to this technique like every other group of animal! The most interesting detail which came out of these studies was that biogenic amines were only present in the ectoneural tissue and never in the hyponeural. There is a very substantial plexus in the gut directly joined to the circumoral nerve ring. This gut plexus lies among the secretory cells of the gut lining and does not extend into the outer circular and longitudinal muscles. The apple green colour of the fluorescence has suggested that there may be extensive dopamine. Organ bath experiments with the gut tentatively suggested that dopamine may be involved with mucus secretion (Cobb, unpublished). There is a differential distribution of noradrenaline and dopamine reported between the tubefeet and elsewhere in the nerve cord (Sloley and Juorio 1990).

The control of gamete release has progressed rapidly since the intitial description of a neurosecretory substance by Kanatani (for review see Cobb, 1988 and Shirai, 1991).

Behavior

Behavior is the most difficult field to review, since there are many studies but few say much original about the function of the nervous system.

Learning

Some limited evidence for associative learning was reviewed in Cobb (1987). The problem with this sort of research is that we still do not understand the organisation of the nervous system – how and where the stimuli are perceived and what significance the motivational state of the animal has on the outcome of the learning trials. It does appear from these studies that a learning phenomenon may be present but what we understood about it is at a very preliminary stage.

Central control

Some of the most important work on behavior was that carried out by Smith (1966) on starfish locomotion. Smith presented evidence which he interpreted as indicating there were central control centres. The evidence that has accumulated since that time has not supported this interpretation.

It is difficult to see how analysis of behavior can tell us much about the layout of the nervous system until we have a much better idea of what that layout actually is. Behavioral experiments need thinking out very carefully to test hypotheses about the nature of the layout. This has never been very effectively carried out.

An example where this approach has been valuable is in the response of the spines to stimulation of the test (skin) in a sea-urchin. This has been a controversial field for some time but recently Smith and Bullock (1992) made a mathematical analysis as to what the basic layout might be to produce the observed results. Although it does not answer the problem, it has at least opened this phenomena to a testable hypothesis.

Moore (1988) described hierarchical behavior in the brittlestar *Ophiura* and, as this is the species that has had far and away the most detailed study carried out on its nervous system, this might warrant a more detailed study. What is required is the sort of approach used by Smith and Bullock (1992) where alternatives in layout could be hypothesised and then ways to test the ideas developed.

Larval studies

A few echinoderms have a direct development but the vast majority has various forms of planktonic larvae. These larvae show a bilateral symmetry and the adult arises from them almost as a new organism. The work of Burke (see Burke et al., 1986) in a series of papers has described the morphology of the larval nervous system (see Cobb, 1989 for detailed references). There have more recently been a number of studies on various aspects of larval nervous systems but we still know remarkably little about the relationship of the larval to the adult nervous system. Indeed, there is some evidence that there is no relationship, with the adult nervous system developing *de novo*. Burke et al. stated in 1986 that the nervous system of the adult may not be related to that of the larvae. Nezlin et al. (1991) used fluorescence techniques to show distribution of monoamines and showed that the larval nervous system of *Asterias* totally degenerated as the adult metamorphosed.

Thorndyke et al. (1992) used immunological techniques to study the distribution of the newly described family of echinoderm peptides, the SALMFamides. These studies showed it was widespread and they suggest it may have diverse functions but there is no evidence as to how this relates to its adult distribution. They also looked at distribution of 5 HT and showed co-localisation with SALMFamides in the apical ganglion.

There have been a number of recent studies showing the effect of bath application of various neurochemicals on larval behaviour and in particular ciliary beating. Lacalli and Gilmour (1990) in sea-urchin pluteus suggested that catecholamines initially trigger the cholinergic response which follows. Mogami et al. (1992), again using a sea-urchin pluteus, showed bath applications of dopamine reduced swimming and that 5 HT increased it but suppressed backward swimming. However the effect of neurochemicals when applied to whole organisms at best only point vaguely at what may be going on in the nervous system.

Conclusions and the future

A basic understanding of how this non-centralised nervous system generates co-ordinated whole-animal behaviour is still not in place. The nervous system is so different in so many respects to that in other phyla that without this basic understanding interpretation of many aspects of their biology is fraught with risk. Anyone looking at rapid brittlestar behaviour will be impressed by the quality of the control and co-ordination. It is clear however that this behaviour is generated by a neural net work that can shift from different physical sets of neurones in different parts of the body. A tentative hypothesis at present is that the main function of GTC is to set up a neural net work in a particular anatomical region and then alert the rest of the animal that it has done so. Just how much of the nervous system is required to form a particular control network for a rapid co-ordinated whole-body response is not clear. Dyefill studies show that although each segmental ganglion is laid out in the same pattern, the fields of action are substantially overlapping. It certainly seems likely therefore that more than one ganglion is involved.

The intracellular work that has been achieved has been carried out with very basic student equipment now more than 25 years old. State-of-the-art manipulators and cell penetration equipment, used with persistence, would make serious studies of ectoneural giant interneurones possible. Recording from hyponerual cells is easy, even with multiple electrodes and low-grade equipment. Whether there are larger and hence even easier species of brittlestar elsewhere in the world is not clear. Invertebrate neuroscience has tended to move away from comparative studies to in depth analysis of some of the best preparations. It would be sad however if the extraordinary phenomena of the echinoderms remain unstudied.

Appendix

Potential of various preparations

Crinoidea
These are always going to be difficult to work with physiologically. The main importance will be comparative since it is from ancestors of this class that the others evolved, with a major re-arrangement of the nervous system. Comparisons at the molecular and genetic level with other phyla would need to include material from this class.

Holothuria
The most neglected class of the Eleuthereozoa (those that live mouth down, unlike the Crinoidea). There is no comprehensive EM study of the nervous system and, in particular, the structure of the radial nerve cord and the innervation of the substantial segmental muscles. Some of these animals are capable of very rapid body contractions and it may be possible that there are larger than average neurones present. It would seem at least possible in some species that extracellular electrodes might be used to record unitary potentials and possibly there are classes of giant neurones large enough to impale intracellularly.

Asteroidea
The nerve cord is external and directly accessible to recording electrodes. There are however no descriptions of really rapid moving species where there might be large neurones that could be recorded from. They are more robust and less inclined to arm autotomy than brittlestars and thus may be more valuable in behaviour experiments. The generally very small nerve fibres will always make them very difficult to study neurophysiologically.

Ophiuroidea
The giant fibres with cell bodies over 20 μm in the ectoneural and up to 30 μm in the hyponeural makes this the only echinoderm preparation

where detailed intracellular studies are possible. The great value of the nerve cord is that it can be exposed with a very minimum of dissection in a whole animal and that the neurones of each ganglion form a predictable pattern in two dimensions so that it is possible to impale the same set of neurones from one animal to another. At present this is only possible in large specimens of *Ophiura*. A disadvantage is that the nerve cord is oral and the animals live oral side down. The only easy way to expose the nerve cord for intracellular work is to have the animal upside down. Inverted extracellular electrodes can be used successfully on animals the right way up with no further dissection. It is possible to dissect through from the aboral surface and record intra-cellularly with the animal the right way up, but this requires very intrusive dissection of the segments adjacent to those recorded from. The ectoneural cells are then impaled through the hyponeural tissue – it is however perfectly possible and routinely used in this lab. Echinoderms have substantial powers of regeneration. This was reviewed by Cobb (1987) and involves the nervous system. The precise segmentation of the arm in brittlestars and the wealth of knowledge of the nervous system make arm regeneration in brittlestars a very valuable preparation. To date, remarkably little is known about even the most basic aspects of regeneration.

Echinoidea

The nerve cord being inside requires very damaging dissection to expose it. The nerves are very small and tightly packed and it is very unlikely that unitary potentials can be recorded. It is difficult to stimulate the nerve cord since the tight packing of nervous tissue in the cord requires damagingly high levels of current. The test however is an ideal preparation to settle the question of the type of peripheral nervous system present. Use of modern staining procedures should settle this controversial aspect of the layout of the nervous system. Aristotle's lantern is a remarkable structure and a much undervalued potential neuromuscular preparation. Wilkie et al. (1992) have recently drawn attention to its value in studying mutable connective tissues. The gut of the regular echinoids would almost certainly be the best to look at neuromuscular function (perhaps holothurians would be equally good). Even the most primitive level of study remains to be carried out on visceral function in echinoderms and is long overdue. Aristotle's lantern is thus a much undervalued organ.

References

Berrios, A., Brink, D., del Castillo, J. and Smith, D.S. (1985) Some properties of the action potential conducted in the spines of the sea-urchin *Diadema*. *Comp. Biochem. Physiol.* 81A: 15–23.

Brehm, P. (1977) Electrophysiology and luminescence of an ophiuroid radial nerve. *J. Exp. Biol.* 71: 213–227.

Bullock, T.H. (1965) Comparative aspects of conduction in echinoids and asteroids. *Amer. Zool.* 5: 545–562.

Burke, R.D., Brand, D.G. and Bisgrove, B.W. (1986) Structure of the nervous system of the auricularia larva of *Parastichopus californicus*. *Biol. Bull.* 170: 450–460.

Candia Carnevali, M.D., Wilkie, I.C., Lucca, E, Andrietti, F. and Melone, G. (1993) The Aristotle's lantern of the sea urchin *Stylocidaris affinis*: functional morphology of the musculo-skeletal system. *Zoomorph.* 113: 173–189.

Cobb, J.L.S. (1968) Observations on the electrical activity within the retractor muscle of the lantern of *Echinus esculentus*. *Comp. Biochem. Physiol.* 24: 311–324.

Cobb, J.L.S. (1985a) The motor innervation of the oral plate ligament in the brittlestar *Ophiura*. *Cell Tissue Res.* 242: 685–688.

Cobb, J.L.S. (1985b) The neurobiology of the ectoneural/hyponeural synapse. *Biol. Bull.* 168: 432–446.

Cobb, J.L.S. (1987) Neurobiology of the Echinodermata. *In:* M. Ali (ed.): *Invertebrate Nervous Systems*. Plenum Press, New York, pp. 483–526.

Cobb, J.L.S. (1988) Neurohumors and neurosecretion in echinoderms: A review. *Comp. Biochem. Physiol.* 91C: 151–158.

Cobb, J.L.S. (1989) Enigmas of echinoderm nervous systems. *In:* P.A.V. Anderson (ed.): *Evolution of the First Nervous System*. Plenum Press, New York, pp. 329–337.

Cobb, J.L.S. (1990) A physiological study of the correlation between motor output in the hyponeural nerves and arm movements in the brittlestar *Ophiura ophiura*. *Mar. Behav. Physiol.* 17: 147–157.

Cobb, J.L.S. and Begbie, K. (1994) Aspects of the hyponeural nervous system. *In:* B. David (ed.): *Proceedings of the 8th International Echinoderms Conference*. Balkema, Rotterdam; *in press*.

Cobb, J.L.S. and Moore, A. (1988) Studies on the ionic basis of the action potential of the brittlestar *Ophiura ophiura*. *Comp. Biochem. Physiol.* 91A: 821–827.

Diaz-Miranda, L. Price, D.A., Greenberg, M.D., Lee, T.D., Doble, K.E. and Garcia-Arraras, J.E. (1992) Characterization of the two novel peptides from the sea-cucumber *Holothuria glaberrima*. *Biol. Bull.* 182: 241–247.

Elphick, M.R., Price, D.A., Lee, T.D. and Thorndyke MC (1991) The SALMFamides: A new family of neuropeptides isolated from an echinoderm. *Proc. Roy. Soc Acad. B* 243: 121–127.

Florey, E. and Cahill, M.A. (1980) Cholinergic motor control of sea urchin tubefeet: Evidence for chemical transmission without synapses. *J. Exp. Biol.* 88: 281–292.

Garcia-Arraras, J.E., Enamorado-Ayala, T., Torres-Avillan, T. and Rivera, V. (1991) FMR-Famide-like immunoreactivity in cells and fibres of the holothurian nervous system. *Neuroscience Lett.* 132: 199–202.

Garcia-Arraras, J.E., Torres-Avillan, I. and Ortiz-Miranda, S. (1991) Cells in the intestinal system of holothurians express cholecystokinin-like immunoreactivity. *Gen. Comp. Endocr.* 83: 233–242.

Ghyoot, M. and Cobb, J.L.S. (1994) Immunocytochemical investigations of the radial nerve cord of *Ophiura ophiura*. *In:* B. David (ed.): *Proceedings of the 8th International Echinoderms Conference*. Balkema, Rotterdam; *in press*.

Grimmer, J.C. and Holland, N.D. (1990) The structure of a sessile stalkless crinoid *Holopus rangii*. *Acta Zool.* 71: 61–68.

Gustafson, T. (1989a) Pharmacological control of muscular activity in the sea urchin larva. I Effects of nicotinic and muscarinic agents. *Comp. Biochem. Physiol.* 94C: 1–14.

Gustafson, T. (1989b) Pharmacological control of muscular activity in the sea urchin larva. II Role of calcium in nicotinic stimulation and paralysis, and the modulatory role of muscarinic agents. *Comp. Biochem. Physiol.* 94C: 15–21.

Gustafson, T. (1991) Pharmacological control of muscular activity in the sea-urchin larva – IV Effects of monoamines and adenosine. *Comp. Biochem. Physiol.* 98C: 307–315.

Hernandez, Z.M., Morales, M., Smith, D.S. and del Castillo, J. (1987) Barium spikes are generated in the spines of the sea-urchin *Diadema*. *Comp. Biochem. Physiol.* 86A: 355–359.

Hyman, L.H. (1955) *The Invertebrates*. Vol. IV *Echinodermata*. McGraw-Hill, New York.

Kinosita, H. (1941) Conduction of impulses in the superficial nervous system of a sea urchin. Jap. *J. Zool.* 9: 221–245.

Lacalli, T.C. and Gilmour, T.H.J. (1990) Ciliary reversal and locomotory control in the pluteus larvae of *Lytechinus pictus*. *Phil. Trans. Roy Soc. B.* 330: 391–396.

Maerkel, K. and Roeser, U. (1991) Ultrastructure and organisation of the epineural canal and the nerve cord in the sea urchins. *Zoomorph.* 110: 267–280.
Martinez, A., Lopez, J., Montuenga, L. and Sesma, P. (1993) Regulatory peptides in gut endocrine cells and nerves in the starfish *Marthasterias glacialis. Cell Tissue Res.* 271: 375–380.
Matsuno, A. and Motokawa, T. (1992) Evidence for calcium translocation in catch connective tissue of the sea cucumber *Stichopus chloronotus. Cell Tissue Res.* 267: 307–312.
Mladvenov, P.V., Bisgrove, B., Asotra, S. and Burke, R.D. (1988) Mechanisms of an arm tip regeneration in the starfish *Leptasterias hexactii. Roux's Arch. Dev. Biol.* 198: 19–28.
Mogami, Y., Watanabe, K., Ooshima, C., Kawano, A. and Baba, S.A. (1992) Regulation of ciliary movement in sea urchin embryo: dopamine and 5 HT change the swimming behaviour. *Comp. Biochem. Physiol.* 101C: 251–254.
Moore, A. (1988) Neurophysiological evidence for hierarchical behaviour in an ophiuroid. *In:* R.D. Burke, P.V. Mladenow, P. Lambert and R.L. Parsley (eds): *Echinoderm Biology.* Balkema, Rotterdam, pp. 575–580.
Moore, A. and Cobb, J.L.S. (1985a) Neurophysiological studies on photic responses in *Ophiura ophiura. Comp. Biochem. Physiol.* 80A: 11–16.
Moore, A. and Cobb, J.L.S. (1985b) Neurophysiological studies on the detection of amino acids in *Ophiura ophiura. Comp. Biochem. Physiol.* 82A: 395–399.
Moore, A. and Cobb, J.L.S. (1987) Studies on tactile and vibratory stimuli in the brittlestar, *Ophiura. J. Mar. Biol. Ecol.* 104: 125–141.
Moore, S.J. and Thorndyke, M.C. (1993) Immunocytochemical mapping of the novel echinoderm neuropetide SALMFamide (S1) in the starfish *Asterias rubens. Cell Tissue Res.* 274: 609–618.
Morales, M, Del Castillo, J. and Smith, D.S. (1990) Acetylcholine sensitivity of the spine test articular capsule of the sea urchin *Eucidaris tribuloides. Comp. Biochem. Physiol.* 94C: 547–554.
Motokawa, T. (1988) Catch connective tissue – a key character for echinoderm success. *In:* R.D. Burke, P.V. Mladenow, P. Lambert and R.L. Parsley (eds): *Echinoderm Biology.* Balkema, Rotterdam, pp. 39–55.
Nezlin, L.P., Prodnikov, I.M. and Davydov, P.V. (1991) Catecholaminergic neurones in the larval nervous system of starfishes during metamorphosis. *Zhu. Evol. Biokh. Fiziol.* 27: 19–27.
Peters, B.H. (1985) The innervation of spines in the sea-urchin *Echinus esculentus. Cell Tissue Res.* 39: 219–238.
Sloley, B.D. and Juorio, A.V. (1990) Biogenic amines in the nervous system and other tissues and low levels of 5 hydroxytryptamine. *Biogenic Amines* 7: 341–350.
Shirai, H., Kondo, N, Bulet, P., Iosbe, M., Imai, K., Goto, T. and Kubota, I. (1986) Endocrine control of oocyte maturation and spawning in starfish. *In:* M. Porchet, J.-C. Andies and A. Dhainaut (eds): *Advances in Invertebrate Reproduction 4,* Elsevier, Amsterdam, pp. 342–383.
Smith, D.P.B. and Bullock, T.H. (1990) Model nerve net can produce rectilinear non diffuse propagation as seen in the skin plexus of sea-urchins. *J. Theor. Biol.* 143: 15–40.
Smith, J.E. (1966) The form and function of the nervous system. *In:* R.A. Boolootian (ed.): *Echinoderm Physiology.* Wiley, New York.
Thorndyke, M.C., Crawford, B.D. and Burke, R.D. (1992) Localisation of a SALMFamide neuropeptide in the larval nervous system of the sand dollar *Dendraster. Acta Zool.* 73: 207–212.
Tuft, P.J. and Gilly, W.F. (1984) Ionic basis of the action potential propagation along two classes of "giant axons" in the ophiuroid *Ophiopteris papillosa. J. Exp. Biol.* 113: 337–350.
Uexkull, J. von (1897) Über Reflexe bei den Seeigeln. *Z. Biol.* 37: 334–403.
Valentincic, T. (1991) Behavioural responses of the brittlestar *Ophiura ophiura* to amino acids, acetylcholine and related low-molecular weight compounds. *Chem. Senses* 16: 251–266.
Welsch, U., Heinzeller, T. and Cobb, J.L.S. (1989) Histochemistry and fine structural observations on the nervous tissue of *Antedon. Biomed. Res.* 10: 145–154.
Welsch, U., Heinzeller, T. and Cobb, J.L.S. (1990) Ultrastructure and innervation of ligament connective tissue in *Antedon bifida. In:* C. De Ridder et al. (eds): *Echinoderm Research.* Balkema, Rotterdam, pp. 255–261.
Wilkie, I.C. (1988) Design for disaster: the ophiuroid intervertebral ligament as a typical mutable connective tissue. *In:* R.D. Burke, P.V. Mladenow, P. Lambert and R.L. Parsley (eds): Proceedings 6th Echinoderm Conference, pp. 25–38.
Wilkie, I.C. (1992) Variable tensility of the oral plate ligaments of the brittlestar *Ophiura ophiura. J. Zool.* 228: 5–26.

The Nervous Systems of Invertebrates: An Evolutionary and Comparative Approach
ed. by O. Breidbach & W. Kutsch
© 1995 Birkhäuser Verlag Basel/Switzerland

The homology concept – still alive

M. Schmitt

Zoologisches Forschungsinstitut und Museum Alexander Koenig, Adenaueralle 160, D-53113 Bonn, Germany

Summary. A typical definition of homology which allows phylogenetic inferences is "homologous features in two or more organisms are those that stem phylogenetically from the same feature in the immediate common ancestor of those organisms".

Similarity of all kind is a proper (and possibly the only) valid empirical test of homology. "Similarity" should, therefore, not be part of the definition. Shared embryological fate is just one kind of similarity which is a strong, but by no means the only, indication of homology.

"Homology" and "synapomorhy" are separate concepts. They are empirically tested by different methods (similarity *versus* outgroup comparison). "Synapomorphy" does refer to monophyly of a certain taxon, while "homology" does not.

Within a phylogenetic approach, homologizing parts of the same individual or of individuals of the same species makes no sense. Thus, "serial homology" (homonomy) is of no help in phylogenetic systematics.

"Homology" can and may be defined in different ways. However, it is crucial that (1) the possible conclusions to be drawn from a statement on homology are limited by the definition, and (2) the diagnosis of homology – the empirical test – is kept strictly independent from the definition of homology.

Introduction

In biology, no comparison is possible without an understanding of "homology". Even when looking for non-homologous similarities a tool is needed in order to sort out homologous ones. However, from time to time it is stated that the homology concept is outdated or not a scientific one (cf., e.g., Sokal and Sneath, 1968; Peters and Gutmann, 1973).

Here, the attempt is made to show that the concept of homology and the method of recognizing homologies are legitimate elements of a scientific procedure. The coexistence of several, partly competing, definitions of homology reflects a variety of approaches and different aims in different domains of science rather than a profound confusion.

The aim of the present work is to present a logically consistent methodological framework for the practice of homologizing as well as a short historical review of the terms involved.

The definition of homology adopted here is that of Bock (1989): "homologous features (or condition of the features) in two or more organisms are those that stem phylogenetically from the same feature (or condition) in the immediate common ancestor of these organisms".

Definition of homology

The matter of the "correct" definition of homology has been intensively discussed (reviews in Holmes, 1980; Rieppel, 1988; De Pinna, 1991; Panchen, 1992). Owen defined (in 1943) homology as "the same organ in different animals under every variety of form and function". "Same" meant "according to the same plan"; of course, no idea of evolutionary transformation was involved. In 1847, Owen differentiated between "special homology," "general homology," and "serial homology". "Special homology" is defined as above. It is "determined by its relative position and connections, with a part or organ in a different animal, the determination of which homology indicates that such animals are constructed on a common type". In contrast, Owen defined "general homology" as the relation "in which a part or series of parts stands to the fundamental or general type ... Thus when the basilar process of the human occipital bone is determined to be the 'centrum' or 'body of the last cranial vertebra', its general homology is inunciated". By "serial homology" he meant a "repetition or representative relation in the segments of the same skeleton".

There were several attempts to introduce a general concept of homology which included all the types of homology listed by Owen (and possibly some more). In the German-speaking scientific community the best known definition is that of Osche (1973): "homolog sind... Strukturen, deren nicht zufällige Übereinstimmung auf gemeinsamer Information beruht" (homologous are structures whose non-incidental resemblances are based on shared information). Similarly comprehensive is the definition by Van Valen (1982) as "correspondence caused by a continuity of information". More recently, Roth, Wagner, and Haszprunar published general concepts of homology (Roth, 1984: basis of homology is the sharing of pathways in development, 1988: no neat congruence between genetics, development, gross morphology and evolution, 1991: homology and polymorphic populations; Wagner 1989a,b: hierarchical and cyclical inductive networks explain homology; there are three concepts of homology; Haszprunar, 1992: homology on different levels of organization). The special problem of homonomy (serial homology) is treated extensively by Minelli and Peruffo (1991).

While Bock (e.g., 1969, 1974, 1977, 1989) claimed that the homology concept had to be strictly phylogenetic, other authors explicity insisted on non-phylogenetic definitions (e.g., Zangerl, 1948; Patterson, 1982; Aboitiz, 1988). Advocating these latter positions, it is argued that the homology concept is older than the theory of evolution and, consequently, there was somewhat of a rule of priority speaking in favor of a definition without any reference to evolution. However, Holms (1980), Bock (e.g., 1989), and others have stated that since Darwin the generally accepted (and only reasonable) explanation of homology is the theory of descent with modification. Therefore, it would make no sense to apply non-phylogenetic definitions of homology.

As long as a definition meets the general requirements of a proper scientific definition there is no intrinsic argument speaking for a certain definition of homology. Which one to prefer must be decided on the basis of the underlying theory, which must be evaluated by means of its heuristic and explanatory power (Schmitt, 1989a).

A definition of homology must (1) allow valid conclusions, and (2) be independent of the empirical test of these hypotheses. In order to make a phylogenetic conclusion valid, the definition of homology has to be a phylogenetic one, since the definition acts as a premise in the syllogism. If we accept Bock's (1989) definition, we can draw the following conclusion: The paranota of Zygentoma (i.e., the primarily wingless dicondylian insects) and the wings of Pterygota are homologous. This means that their immediate common ancestor possessed a feature from which the paranota as well as the wings were derived. If conclusions other than phylogenetic ones are intended homology must be defined in the appropriate way. A "purely morphological" definition, e.g., "homologous are those structures which are composed according to a common plan," might of course allow only "purely morphological" conclusions, e.g., "Bat wings and horse forelimbs are homologous, so they are composed according to the same plan".

Consequently, a general definition of homology only allows general inferences, e.g., "characters a and b in the taxa A and B are based on shared information, respectively". Special phylogenetic inferences require specification(s) of the general definition (see Fig. 1). It does not matter if these specifications are made *a priori* and explicitly, or as a subsequent "explanation".

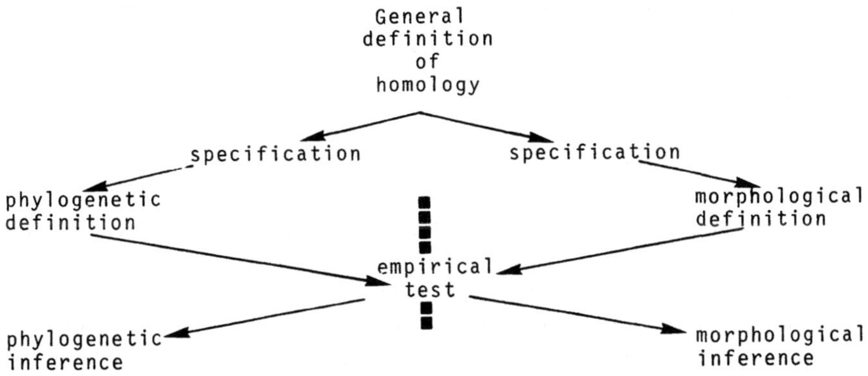

Figure 1. Pathways of drawings conclusions in homology research. A general (comprehensive) definition of homology might be possible, but specific inferences need specifications of this general definition. In order to draw phylogenetic inferences, a phylogenetic definition of homology is required and so on. As long as statements on homology are based on the same empirical test hardly any practical problems will arise.

Conditional phrase

From Bock's definition, it follows, if understood in a strict sense, that any two characters in any two animals or plants are homologous as long as we accept the monophyly of the Eukaryota. In order to avoid such trivial statements, Bock (e.g., 1977, 1989) introduced a "conditional phrase". This should provide "the required information about the exact relative nature" of a hypothesized homology, thus stating "as what" any two features are supposed to be homologous. Bird wings and bat wings are homologous "as tetrapod forelimbs," but they are not homologous "as tetrapod wings". Consequently, each conditional phrase is, at the same time, the denotation of a certain level of taxonomic inclusiveness: Tetrapods are more inclusive than are birds or bats. Basing on this "address to the level of the taxic hierarchy," Patterson (1982) as well as De Pinna (1991) and Rieppel (1992) treat "homology" as a synonym of "synapomorphy," which is a fault, as discussed on p. 432f.

What we homologize is, in fact, not a certain structure, but a specific pattern, generated by genetic and epigenetic processes responsible for the characters under consideration, with a single origin in evolution and passed from its origin through the generations. The crucial step in homology research is to identify this specific pattern. It is obvious that this pattern can decrease until complete disappearance. Thus a "homology" in two or more species can get lost in the course of their evolution. This means that there are still derivatives of the original character from which the homologous characters were derived, but the specific pattern which caused these characters to be distinct identifiable features got lost. Snakes may still possess four spots of tissue along their body sides which stem from the limbs of their lizard-like ancestors, but nothing remains that could be specifically identified as a limb-homologue.

Empirical test

Homology of characters is determined by (or "empirically tested against observations of") similarity of the characters in question.

"Use of shared similarities as the empirical test of postulated homologues is the only one available for testing independent homologous features" (Bock, 1989). Remane (1958) listed several "criteria" of homology. Since "criterion" means a necessary and sufficient condition, it is not the proper term to denote these operational sentences. They are rather "indications"of homology. Remane's three "main critera" are

1. corresponding position
2. correspondence of substructures
3. connecting intermediate steps.

His three "supporting criteria" are

4. occurence in many most similar species
5. congruence with other characters with equal distribution among taxa
6. probability of homology decreases with the frequency of occurrence of the characters under consideration in definitely not related species.

"Criterion " (6) is just the complement of (5). Moreover, it refers to an non-observational prerequisite, i.e., phylogenetic relationship. (3) through (5) are not independent empirical tests since they depend on other statements on homology. Moreover, they easily lead to circular statements when these other statements are based on the same "criteria". (1) and (2) remain as they are simply denotations of different kinds of similarity.

Especially "pattern cladists" (Patterson, 1982; De Pinna, 1991; Rieppel, 1992) emphasize "congruence" as a valid test of hypotheses on homology. To this "test" the same applies as is said about Remane's "criteria" (3)–(5): it is not a valid test of independent homologies (*vide* Bock, 1989) since it is based on a circular argument: the acceptance of the homology hypothesis pertaining to a certain character is dependent on the homology of all other (congruent) characters.

The frequently stressed definition of "structural homology" (e.g., Boyden, 1943), "topological homology" (e.g., Rieppel, 1992) or "transformational homology" (Patterson, 1982; De Pinna, 1991; Panchen 1992, p. 69: "two structures that differ in some significant way but are thought to be built on a common plan") is, in my opinion, not a proper definition of homology (i.e., not one which would allow reasonable and intended inferences), but rather the description of the empirical test of hypotheses on homology (which is called "taxic" homology by the above authors, but excludes serial homology).

Shared ontogenetic development is another type of similarity which can be empirically observed. Neither "similarity" nor "development" should be part of the definition of homology. Of course, these aspects can be used to define homology, but then the definition does not allow phylogenetic conclusions, and it is difficult to find an independent empirical test.

Shared developmental transformation of characters in two or more species is a strong indication of homology (see Dohle, 1976), but it is by no means a necessary argument in favor of homology as, for example, Kutsch and Breidbach (1994) point out in regard to homologous neurons. There are numerous examples of undoubtedley homologous features which develop through extremely different pathways (*vide* Baltzer, 1950; Dohle, 1988; Fioroni, 1987, p. 392). A striking example was provided by Spemann (1915; see Sander, 1989; Fig. 2): The formation of a lens in, for example, *Rana fusca* is induced by an eye-Anlage from the neural tube (a), so that the epidermis folds inwards (b),

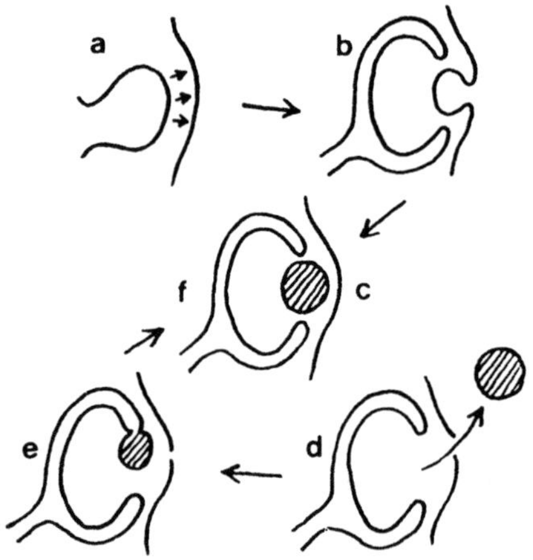

Figure 2. The problem of the homology of an anuran lens developed through induction by epidermal tissue (a → b → c) and a lens formed by a dorsal rim of the iris, i.e., neural tissue, after artifically removing the original lens (d → e → f). (From Sander (1989), after Spemann (1915).)

resulting in a lens (c). When this lens is removed artificially (d), a new lens is formed by the dorsal rim of the iris (e). Then the question arises whether the two lenses (f) and (c) are homologous (assuming that a → b → c took place in one species, e.g., *Rana fusca*, while d → e → f was carried out in another, e.g., *Rana lessonae*, just to avoid the problems of "intraspecific homologies"). Here, we see clearly that we do not homologize the physical structures, but rather the underlying (genetic and epigentic) pattern. By this it is demonstrated that different developmental origin does not disprove a hypothesis on homology. Some authors (e.g., Remane, 1952; Osche, 1973) made "complexity" a constitutive element of homology. In my opinion, this is not a necessary prerequisite for the validity of similarity as the empirical test of homology. Also, very simple characters can be homologous. But the more complex a character is the more reliable is our decision.

Basically, all empirical tests of homology are attempts to judge the probability of multiple independent evolutionary origin (or chance) *versus* single evolutionary origin (or conservatism, cf., Riedl, 1975). It is possible to do so by estimating the number of determination steps leading to the character under study. The estimation is identical to the examination of the number of independently variable elements constituting this character (cf., Dohle, 1989). The more evolutionary steps necessary to form the observed similarity, the more it is improbable that the similarity of the studied character was achieved two or more times in evolution. "Evolutionary steps" do not simply equal "mutations", but rather "phenotypic changes". A similar possibility to estimate the reliability of a hypothesis on homology is to weight the number of possible evolutionary answers to a certain requirement.

An example for the importance of thorough estimation of the kind and the degree of similarity is the new view of the arthropod brain illustrated by the chapters of Breidbach, Sandeman and Scholz, and Strausfeld et al. in the present volume.

Riedl (1975) provides an attempt to quantify, or at least semiquantify, these estimations of the probability of independent *versus* single origin.

The question of whether two neurons are homologous (as for example, interneurons of a certain type) can be approximately answered by determining all kinds of similarity between them: certain characteristics of their shape, their relative position, their ontogenetic development, their biochemical equipment, or their functional role. Campbell and Hodos (1970) list nine types of data which are "useful to establish homologies in the central nervous system: (1) experimentally determined fiber connections, (2) topology, (3) topography, (4) the position of reliably occurring sulci, (5) embryology, (6) morphology of individual neurons, (7) histochemistry, (8) electrophysiology, (9) behavioral changes resulting from stimulations, lesions, etc. The greater the degree of concordance among the characters (i.e., aspects of similarity, M.S.), the stronger becomes the justification for drawing the inference that structures in two different species may have been derived from corresponding structures in a common ancestor" (see Kutsch and Breidbach 1994, for further aspects).

Neither of these different aspects of similarity is an actual proof of their homology, nor does lack of any of them stand against the hypothesis of homology (see Meier and Reichert, this volume,).

The crucial question is not "how different may characters be unless we reject the hypothesis of homology," but "how similar must they be before we accept it". There is no satisfying answer to this crucial question (cf., Dohle, 1989, p. 371). As long as there is no conflicting evidence, we may accept the hypothesis of homology of two characters even if there are few distinctive correspondences, if not just one, i.e. even if the characters are simple. If conflicting evidence appears we have to weight characters, i.e., to balance the arguments speaking in favor and against homology, and then assume non-homology of the less complex character, i.e., complexity expressed in terms of "determination steps". In cladistic computer algorithms weighting of characters leads to an increase of tree length, i.e., to a higher number of necessary transformation steps (cf., Siebert, 1992).

An enhancement of the reliability of hypotheses on homology is to check all possible causes of independently evolved similarity. Rieger and Tyler (1979) gave several indications of independently acquired similarity (termed "analogy" by them): (1) influence of common selective pressures, (2) composition of similar material under the influence of similar environmental conditions, (3) if "they are likely to be the only possible means by which a particular function can be accomplished (the probability of analogy is inversely proportional to the number of structural solution for a given function)," (4) ontogenetic development from dissimilar origin, (5) influence of "similarity-dependent selective pressures" (i.e., "the influence of pressure to evolve mimicry").

Homology and phylogenetic systematics

In his basic works (1950, 1953, 1966), Hennig did not specifically clarify the role of the homology concept for phylogenetic systematics. In short, phylogenetic systematics is the method of detecting sistergroup-relations between monophyletic taxa or species, respectively. A taxon is monophyletic if it contains an ancestral species and all of its descendants.

Monophyly of a taxon can be determined by showing that all of its members possess at least one shared character in a uniquely derived state, which is called synapomorph (or -ic or -ous). The unaltered (primitive) alternative of this character state is called plesiomorph (or -ic or -ous), its occurence in several taxa is then termed "symplesiomorphy". The term "apomorh" means "derived", without saying whether the underlying transformation had taken place once or several times, i.e., convergencies are included. The meaning of syn- or autapomorph is a relative one: a certain character state is synapomorph on one taxic level, autapomorph on another one, and symplesiomorph on a third one. For example the possession of wings is autopomorph in the Pterygota (the primarily winged insects) and synapomorph in all of their subtaxa as compared to the Apterygota (the primarily wingless insects), but it is symplesiomorph in the flies as compared to the fleas (which are secondarily wingless). Several indications can be employed in order to determine character polarity (apomorh-plesiomorph), e.g., ontogeny, paleontology, functional analysis, etc. (cf., Stevens, 1980; Schmitt, 1985). The most widely applied method is the outgroup comparison (see Watrous and Wheeler, 1981): If of two states of a character one occurs only within a taxon under consideration, and the other one within and outside, than the latter is most probably the plesiomorph and the first is the apomorph state. The concept of "character" and "character state" is also a relative one: a feature is treated as a character state on a certain taxic level and, at the same time, as a character on another level (cf., Platnick, 1979).

As long as there is no conflicting evidence, any character whose state is determined as apomorph by one of the above methods can be regarded as a synapomorphy of all the taxa where it occurs, thus indicating monophyly of the taxon comprising all these taxa and their last common ancestor. When there is conflicting evidence, the principle of parsimony is applied and all excluded characters are interpreted as independently acquired apomorph character states (convergences, homoplasies or however they may be termed). "Sister-groups" (or "adelphotaxa," Ax, 1984, p. 15) are those monophyletic taxa (or "monophyla," Ax, 1984, p. 15) which stem from or are the immediate descendants of an ancestral species. "Outgroup" is an operational term which must not be confused with "sister-group". Any taxon possessing the character to be analyzed in one of two (or more) states may be used as an outgroup in an outgroup comparison (cf., Schmitt, 1989b).

Ax (1984, 1988, 1989) emphasized that it makes no sense to homologize "negative" characters, although these characters may be well identified as synapomorphies that substantiate a hypothesis on monophyly. This is certainly true; however, Panchen (1992, p. 74) and Patterson (1982) deem secondary lack of a character as apomorph *and* homologous. This discussion becomes meaningless if statements on homology include a conditional phrase (*sensu* Bock).

There is another, more important, difference between syn- (or aut-) apomporphy and homology: Any statement on synapomorphy does necessarily refer to monophyly of a concrete taxon. It is incomplete and useless to state that the wings are considered to be synapomorph without saying which monophyly hypothesis is based on that statement. So, "synapomorphy" (and related terms) is an exclusively phylogenetic concept which is applied in order to detect sistergroup relations. Of course, any statement on homology does necessarily mean that there is a taxon which could be shown to be monophyletic by means of this character. However, it is possible and legitimate to determine homology without detecting this taxon, even without intending to do so. It is a legitimate aim of morphology (or embryology or any other branch of biology) to ask for the homologous alternative of a given character without looking for the cladogenetic relations of the taxa where they occur (Schmitt, 1989a). Moreover, the methods to determine homology and synapomorphy are different: a search for similarities on the one hand, and outgroup comparison on the other. Therefore, these two concepts should be kept separate. McKitrick (1994) also suggests that synapomorphy is not equivalent to homology – mainly in order to keep phylogenetic analysis non-tautological.

Analogy, convergence and related terms

Several terms are often used when discussing homology. Since these, like all scientific terms, are subject to historical change as well as to different usage among different scientific schools, the following is a personal selection in order to present a consistent idea of homology.

(a) Analogy: was originally defined by Owen (1843, Glossary) as "a part or organ in one animal which has the same function as another part or organ in a different animal". This definition does not refer to homology or non-homology, thus leaving the possibility that homologous characters are analogous at the same time. Cracraft (1967) and other authors restricted "analogy" to "non-homologous features that have become similar in structure due to a similar interaction with the environment". I find no advantage in excluding homology from the definition of analogy and prefer Owen's original version.

(b) Convergence: is "the development of similar characters separately in two or more lineages without a common ancestry pertinent to the similarity but involving adaptation to similar ecological status" (Simpson, 1978). In most cases where characters are treated as being convergent the corresponding adaptation is not mentioned at all, thus making "convergent" synonymous with "non-homologous".

(c) Homoiology: independent parallel perfection of homologous characters (Plate, 1922, 1928), means basically the same as Simpson's "parallelism".

(d) Homoplasy: resemblance not traceable to common ancestry (Lankester, 1870), "it includes parallelism, convergence, analogy, mimicry, and chance similarity" (Simpson, 1961 "analogy" is defined here as in Cracraft, 1967).

(e) Parallelism: "the development of similar characters separately in two or more lineages of common ancestry" (Simpson, 1961). Normally used in the sense of 'homoiology'.

None of these terms is synonymous to "homology," some of them – at least "convergence" and "homoplasy" – explicitly exclude homology. Therefore, they are not "types of homology" as claimed by Campbell and Hodos (1970).

Iterative homology

Probably most puzzling is the term "serial homology". Its introduction into the homology concept caused much discussion and confusion and contributed a major part to the search for a "general" definition of homology. Owen (1847) wrote "The recognition of the same elements throughout the series of segments of the same skeleton I call 'the determination of the serial homology' ". Most remarkably, he explained on p. 333: *It will, of course, be obvious that the humerus is not 'the same bone' as the femur of the same individual in the same sense in which the humerus of one individual or species is said to be 'the same bone' as the humerus of another individual or species. In the instance of the serial homology above-cited, the femur, though repeating in its segment the humerus in the more advanced segment, it is not its namesake, not properly, therefore, its 'homologue'. I propose, therefore, to call the bones so related serially in the same skeleton 'homotypes', and to restrict the term 'homologue' to the corresponding bones in different species....*

The relation between "serially homologous" (or "homotypic") characters has been termed "homonomy" or "homonomous" by Bronn (1858; there, amazingly, termed "homonym" in its first use). In order to avoid confusion, I suggest to use these terms for "serial homology".

Scientists investigating protein sequences in different taxa are facing the analogous problem of finding similarities between proteins caused by splitting of genealogical lineages with subsequent retention of corresponding features or by iteration of a sequence with subsequent differentiation of the resulting units within one lineage. Fitch (1970) coined the terms "orthologous" and "paralogous" as "two subclasses of homology". 'Paralogous' refers to "the result of gene duplication so that both copies have descended side by side during the history of an organism".

"Orthologous" means "the result of speciation so that the history of the gene reflects the history of the species". This distinction could virtually contribute to the clarity of interpretations of relations between molecules. However, there seems to have been considerable misuse of the term "homology" in molecular studies, so that in 1987, Reeck et al. pointed out a strict separation of the homology concept and certain current usages in their field of science. Basically, this separation pertains to the distinction of homology as a phylogenetic concept and similarity as a tool to determine homology. In fact, "degrees of homology" expressed by a certain percentage of corresponding amino acids or base pairs are quantitative measurements of "similarity" and should be termed such (Gould, 1988). Using the term "homology" instead of "sequence similarity" would not only lead to profound confusion, but also to the need of a new term for the old homology.

As stated on p. 427, the preference for a certain definition of homology has to be based on its heuristic and its explanatory power within a certain theory. So, when we want to include "homonomy" in a comprehensive definition of homology, we should clearly state the aims and reasons thereof. Obviously, homonomy is meaningless in a phylogenetic framework since it does not help to uncover phylogenetic relations between taxa. It may, however, be useful to delimit the range of validity of certain statements, thus giving a frame for generalizations.

This may work according to the following model: "we found that neuron A was homonomous (or "homotypic") to neuron B (the "same" neurons in different segments of one individual). Since homonomous neurons are coded by the same part of the genome (equally transcribed in all the segments where homonomous neurons occur), we may expect that neuron B synthesizes the same neuropeptides as we found that neuron A does" (which would be a hypothesis to be tested, not a logical inference; cf., Agricola and Bräunig, this volume).

In any case, there must be accordance of definition and conclusion(s), and independence of definition and empirical test.

Conclusion

Regardless of the sometimes rather confusing and personal discussions about the "correct" definition or concept of homology, it is still the basic concept for all comparisons made in biology. Different definitions might coexist and they reflect different goals and approaches. As long as all statements on homology are based on the same empirical test (similarity), there will be little practical confusion.

There is no need for a new definition or methodology of homology (see also De Pinna, 1991). What we rather need are new methods to reveal new types of similarity, and we need more logical and methodological scrutiny of those who homologize. For example, the application of a strict method of homologizing would lead to both clear and discountable hypotheses on *the brain* or *the mushroom body* of the arthropods by indicating the kind and the degree of similarity on which each argument is based.

Acknowledgements
I am grateful to Dr. S. Naumann (Bonn) for her help with preparing the manuscript, to Ms. G. Uhl (Freiburg), Dr. Dr. O. Breidbach (Bonn), and an anonymous reviewer for critically reading and commenting on an earlier draft of this work.

References

Aboitiz, F. (1988) Homology: A comparative or a historical concept? *Acta Biotheoret.* 37: 27–29.

Ax, P. (1984) *Das phylogenetische System.* Gustav Fischer Verlag, Stuttgart and New York (English edition: *The Phylogenetic System.* John Wiley & Sons, Chichester and New York, 1987).

Ax, P. (1988) *Systematik in der Biologie.* Gustav Fischer Verlag, Stuttgart.

Ax, P. (1989) Homologie in der Biologie – ein Relationsbegriff im Vergleich von Arten. *Zool. Beitr. N.F.* 32: 487–496.

Baltzer, F. (1950) Entwicklungsphysiologische Betrachtungen über Probleme der Homologie und Evolution. *Rev. Suisse Zool.* 57: 451–477.

Bock, W.J. (1969) The concept of homology. *Annals New York Acad. Sci.* 167: 71–73.

Bock, W.J. (1974) Philosophical foundations of classical evolutionary classification. *Syst. Zool.* 22: 375–392.

Bock, W.J. (1977) Foundation and methods of evolutionary classification. *In:* M.K. Hecht, P.C. Goody and B.M. Hecht (eds): *Major Patterns in Vertebrate Evolution.* NATO Advanced Study Institute, Ser. A., Vol, 14. Plenum Press, New York, pp. 851–895.

Bock, W.J. (1989) The homology concept: Its philosophical foundation and practical methodology. *Zool. Beitr. N.F.* 32: 327–352.

Boyden, A.A. (1943) Homology and analogy: A century after the definition of "homologue" and "analogue" of Richard Owen. *Quart, Rev. Biol.* 18: 228–241.

Bronn, H.G. (1858) *Morphologische Studien über die Gestaltungsgesetze.* C.F. Winter, Leipzig, Heidelberg.

Campbell, C.B.G. and Hodos, W. (1970) The concept of homology and the evolution of the nervous system. *Brain Behav. Evol.* 3: 353–367.
Cracraft, J. (1967) Comments on homology and analogy. *Syst. Zool* 16: 355–359.
De Pinna, M.C.C. (1991) Concepts and tests of homology in the cladistic paradigm. *Cladistics* 7: 367–394.
Dohle, W. (1976) Zur Frage des Nachweises von Homologien durch die komplexen Zell- und Teilungsmuster in der embryonalen Entwicklung höherer Krebse (Crustacea, Malocastraca, Peracarida). *Sitzungsber. Ges. Naturf. Fr. Berlin N.F.* 16: 125–144.
Dohle, W. (1988) Book review: Søren Løvtrup, 1987, Darwinism: The Refutation of a Myth. Croom Helm Ltd., London. *J. Evol. Biol.* 1: 283–285.
Dohle, W. (1989) Zur Frage der Homologie ontogenetischer Muster. *Zool. Beitr. N.F.* 32: 355–389.
Fioroni, P. (1987): *Allgemeine und vergleichende Embryologie der Tiere*. Springer Verlag, Berlin, Heidlberg, New York.
Fitch, W.M. (1970) Distinguishing homologous from analogous proteins. *Syst. Zool.* 19: 99–113.
Gould, S.J. (1988) The heart of terminology. *Natural History* 2/88: 24–31.
Haszprunar, G. (1992) The types of homology and their significance for evolutionary biology and phylogenetics. *J. Evol. Biol.* 5: 13–24.
Hennig, W. (1950) *Grundzüge einer Theorie der phylogenetischen Systematik*. Deutscher Zentralverlag, Berlin.
Hennig, W. (1953) Kritische Bemerkungen zum phylogenetischen System der Insekten. *Beitr. Entomol.* 3 (Sonderh.), 1–85.
Hennig, W. (1966) *Phylogenetic Systematics*. Univ. Illinois Press, Urbana.
Holmes, E.B. (1980) Reconsiderations of some systematic concepts and terms. *Evol. Theor.* 5: 35–87.
Kutsch, W. and Breidbach, O. (1994) Homologous structures in the nervous systems of Arthropoda. *Advances in Insect Physiol.* 24: 1–113.
Lankester, E.R. (1870) On the use of the term homology in modern zoology, and the distinction between homogenetic and homoplastic agreements. *Annals and Magazine of Natural History Ser.* 4, 6: 35–43.
McKitrick, M.C. (1994) On homology and ontological relationship of parts. *Syst. Biol.* 43: 1–10.
Minelli, A. and Peruffo, B. (1991) Developmental pathways, homology and homonomy in metameric animals *J. Evol. Biol.* 4: 429–445.
Osche, G. (1973) Das Homologisieren als eine grundlegende Methode der Phylogenetik. *Aufs. Red. Senckenb. Naturf. Ges.* 24: 155–165.
Owen, R. (1843) *Lectures on the Comparative Anatomy and Physiology of the Invertebrate Animals*. Longman, Brown, Green and Longmans, London.
Owen, R. (1847) On the archetype and homologies of the vertebrate skeleton. *Brit. Assoc. Rep.* 1846: 169–340.
Panchen, A.L. (1992) *Classification, Evolution, and the Nature of Biology*. Cambridge Univ. Press, Cambridge, New York, Oakleigh.
Patterson, C. (1982) Morphological characters and homology. *In:* K.A. Joysey and A.E. Friday (eds): *Problems of Phylogenetic Reconstruction*. Academic Press, London, New York, pp. 21–74.
Peters, D.S. and Gutmann, W.F. (1973) Die Stichhaltigkeit des Homologiebegriffs. *Mathem. Naturwiss. Unterr.* 26: 274–279.
Plate, L. (1922) *Allgemeine Zoologie und Abstammungslehre*, Vol. 1. Gustav Fischer Verlag, Jena.
Plate, L. (1928) Über Vervollkommnung, Anpassung und die Unterscheidung von niederen und höheren Tieren. *Zool. Jb. Allg. Zool.* 45: 745–798.
Platnick, N.I. (1979) Philosophy and the transformation of cladistics. *Syst. Zool.* 28: 537–546.
Reeck, G.R., De Haën, C., Teller, D.C., Doolittle, R.F., Fitch, W.M., Dickerson, R.E., Chambon, P., McLachlan, A., Margoliash, E., Jukes, T.H. and Zuckerkandl, E. (1987) "Homology" in proteins and nucleic acids: A terminology muddle and a way out of it. *Cell* 50: 667.
Remane, A. (1958) *Die Grundlagen des natürlichen Systems, der verleichenden Anatomie und der Phylogenetik. Theoretische Morphologie und Systematik I*, Akademische Verlagsgesellschaft Geest & Portig, Leipzig.

Riedl, R. (1975) *Die Ordnung des Lebendigen, Systembedingungen der Evolution.* Paul Parey, Hamburg and Berlin (English edition: *Order in Living Organisms: A Systems Analysis of Evolution.* John Wiley & Sons, Chichester, New York, 1978).

Rieger, R.M. and Tyler, S. (1979) The homology theorem in ultrastructural research. *Amer. Zool.* 19: 655–664.

Rieppel, O.C. (1988) *Fundamentals of Comparative Biology.* Birkhäuser Verlag, Basel, Boston, Berlin.

Rieppel, O.C. (1992) Homology and logical fallacy. *J. Evol. Biol.* 5: 701–715.

Roth, V.L. (1984) On homology. *Biol. J. Linn. Soc.* 22: 13–29.

Roth, V.L. (1988) The biological basis of homology. *In:* C.J. Humphries (ed.): *Ontogeny and Systematics.* Columbia Univ. Press, New York, pp. 1–26.

Roth, V.L. (1991) Homology and hierarchies: Problems solved and unsolved. *J. Evol. Biol.* 4: 167–194.

Sander, K. (1989) Zum Geleit: Homologie und Ontogenese. *Zool. Beitr. N.F.* 32: 323–326.

Schmitt, M. (1985) Thesen zur Theorie und Methodologie der Phylogenetischen Systematik. *Zool. Beitr. N.F.* 29: 19–33.

Schmitt, M. (1989a) Das Homologie-Konzept in Morphologie und Phylogenetik. *Zool. Beitr. N.F.* 32: 505–512.

Schmitt, M. (1989b) Claims and limits of phylogenetic systematics. *Z. Zool. Syst. Evolutionsforsch* 27: 181–190.

Siebert, D.J. (1992) Tree statistics; trees and 'confidence'; consensus trees; alternatives to parsimony; character weighting; character conflict and its recolution. *In:* P.L. Forey, C.J. Humphries, I.J. Kitching, R.W. Scotland, D. Siebert and D.M. Williams (eds): *Cladistics – A Practical Course in Systematics.* Clarendon Press, Oxford, pp. 73–88.

Simpson, C.G. (1978) *Principles of Animal Taxonomy.* Columbia Univ. Press, New York.

Sokal, R.R. and Sneath, P.H.A. (1968) *Principles of Numerical Taxonomy.* Freeman, San Francisco.

Spemann, H. (1915) Zur Geschichte und Kritik des Begriffs der Homologie. *In:* P. Hinneberg (ed.): *Die Kultur der Gegenwart*, Teil 3, Abt. 4. G.B. Teubner, Leipzig, pp. 63–86.

Stevens, P.F. (1980) Evolutionary polarity of character states. *Annu. Rev. Ecol. Syst.* 11: 333–358.

Van Valen, L.M. (1982) Homology and causes. *J. Morphol.* 173: 305–312.

Wagner, G.P. (1989a) The origin of morphological characters and the biological basis of homology. *Evolution* 43: 1157–1171.

Wagner, G.P. (1989b) The biological homology concept. *Annu. Rev. Ecol. Syst.* 20: 51–69.

Watrous, L.E. and Wheeler, Q.D. (1981) The out-group comparison method of character analysis. *Syst. Zool.* 30: 1–11.

Zangerl, R. (1948) The methods of comparative anatomy and its contribution to the study of evolution. *Evolution* 2: 351–374.

The Nervous Systems of Invertebrates: An Evolutionary and Comparative Approach
ed. by O. Breidbach & W. Kutsch
© 1995 Birkhäuser Verlag Basel/Switzerland

Are the main grades of brains different principally in numbers of connections or also in quality?

T.H. Bullock

Department of Neurosciences, University of California, San Diego-La Jolla, CA 92093, USA

Summary. From left to right, no and yes. The premise is underlined that grades of complexity of brain and consequent behavior have evolved over an extremely wide span, if we compare major taxa from jellyfish to cuttlefish – not linearly, steadily or invariably and not necessarily correlated with survival value or improved adaptation. This relatively rare form of evolution, unequivocal only between some phyla, some classes and some orders, can be called vertical gradation and should be distinguished from lateral radiation that leads to species, genera and many families, dominated by adaptive specialization to a particular habit of life. More complex brains may have quantitatively more connections but are principally recognized by qualitatively different components. They can logically be called more advanced or "better" in this dimension, without implication of moral value or improved adaptation. Complexity is usefully measured by the number of kinds of components of structure, function and behavior, a usage quite distinct from a common application to highly specific or amazing achievements.

The inverebrate groups are reviewed to underline the saltations in grade of complexity of brains and behavior between many phyla and classes and some orders. The proposition is asserted that neuroscience has relatively neglected the study of such differences, to discern them, quantitate and evaluate them, using modern methods in each relevant domain – anatomy, physiology, chemistry, and behavior. It will be important to complete the detailed characterization of the major grades of complexity found among animals. This agenda is likely to uncover new principles in brain operation and in the biology of diversity.

For more than a century it has been clear that a wide disparity exists between the principal nervous concentrations of different phyla of invertebrates. Brains are extremely diverse. Between the marginal bodies of scyphomedusae and the quite respectable brain of polyclad flatworms there is a large difference in size and appearance; still greater is the difference between the flatworm brain and the vastly larger and more differentiated one of the garden snail and even more the great circumesophageal mass of cephalopods. Not only between phyla but sometimes within phyla the disparity can be great, as it is between the modest thickening of a subepithelial plexus that represents the brain of some polychaete annelids, like the sessile terebellids, to the impressive, insect-like brains of others, like the amphinomids and the scale worms. The same can be said comparing the meager circumesophageal ganglion of mussels, like *Mytilus*, to the substantial and relatively intricate one in pulmonate slugs, like *Ariolimax*. Within the extraordinarily diverse gastropods, and within the crustaceans, and among the families and orders of free-living flatworms there are striking differences between the brains, some being exceedingly modest and others well developed cerebral ganglia.

These differences among brains fascinate me. I believe they are greater than the differences among digestive systems, reproductive systems, muscles, skin or any other system in the animal body – and that the differences are highly consequential for behavior and all its sequelae. Note that nothing in these observations is said or assumed about the phylogenetic relations and no simple evolutionary ladder is inferred. Even before or without agreed-upon phylogenetic trees and sequences of sister groups, it must signify something that such drastic discrepancies in nervous system complexity turn up over and over again, between and within phyla.

To some it may seem esoteric to worry about differences between sandworm and earthworm brains, oyster and slug brains, or even between starfish and cuttlefish nervous systems. My assertion, however, is that these disparities are of the essence in several spheres. One is in testing theories on how brains work, where we can be led to quite implausible ideas if we know a good deal about the brain in only one or a few orders of animals. A second sphere is in recognizing what evolution has wrought and what principles of biology may still hide in the diversity of life. In this essay, rather than develop these two, or other spheres of significance, I will confine myself to a heuristic comment on the implications for neuroscience, in terms of research agenda and opportunities.

Given the great disparity between brains, the broad domain of inquiry that asks what dimensions, traits and features differ so much has been relatively neglected, compared to the domain that seeks basic common mechanisms (Bullock 1993). A first question is whether the disparities consist mainly of differences in brain size, cell number and number of connections or whether qualitative differences among the cells and connections play a substantial role. The main claim I make is that the latter is true and that diversity of brains between major groups of animals is principally in the degree of complexity.

It is important to distinguish this kind of evolution which is reflected in the overall grade of complexity of the brain and behavior, and can be called **vertical gradation**, from another which is reflected in adaptive specialization, often in one or a few domains of adaptation and can be called **lateral radiation** (Bullock 1992). These two distinct products of evolution overlap and are difficult to separate among lower taxa – species, genera and many families, where adaptive radiation is believed to be the commonest result of evolution and complexity cannot be said unequivocally to differ. Comparing higher categories – some phyla, some classes and some orders, it is easier to recognize distinct grades of complexity in the brain if we compare taxa that are not close together. In such cases the degree of adaptation to their lifestyle cannot be said unequivocally to differ. I am concerned in this essay with the grades of complexity of higher taxonomic categories.

Complexity can be usefully defined and measured by the number of different components at all levels. This translates, in our domain, into the number of kinds of anatomical parts, physiological processes and interdependencies, alternate behaviors, including sensory and social discriminations, and the response repertoire, whether motor, vocal, chemical, pigmentary, electrical or luminescent.

The great biological phenomenon I am speaking of is the fact that brains and, I believe, behaviors have more kinds of components in squid than in snails, in insects than in leeches, in mammals than in amphibians. The greater complexity is not necessarily or obviously more adapted, specialized or wonderful. The numbers of such major taxa obviously different in grade of brain complexity are far less than the numbers of taxa clearly different in adaptive specialization. The latter is the common form of evolution, producing millions of species, whereas the former is much less common, producing higher categories, only some of which are clearly different in brain complexity. In other words, I am not speaking about the accomplishments of bat echo location or moth olfaction or other amazing adaptations, but about comparisons of phyla and classes with respect to overall grades of complexity of the brain and behavior, as defined. Since in some taxa there is a wide range of complexity within the phylum or class, the contrast I am highlighting is best seen by choosing the most complex species of each such taxon to compare.

I can now rephrase the opening statement: it has long been clear that brains differ widely, both in relative numbers of cells and in qualitative features such as the microscopic anatomy. These combine to give us brains with many distinguishable parts. I shall sometimes use shorthand and call these higher, better or more advanced brains, meaning only that they have more different parts and might be capable of more different discriminations and behaviors, not that they are any more perfectly adapted to the lifestyle of the species than simpler brains are for their species. The latter, indeed, quite often have a longer record of survival and in some cases adaptability.

Size of the brain no doubt contributes something, in spite of the many counter examples where small animals have more advanced brains and presumably behavior than simpler but larger forms. The evidence on brain size, cell number and presumed number of connections, and the qualifications and alternative measures have been dealt with by other authors, mainly based on data from vertebrates (Jerison 1991; Ringo 1991; Deacon 1990; Ridet and Bauchot 1990; Pirlot 1987; Riska and Atchley 1985). I will only underline that the relevant parameters are presumably not simply brain mass in proportion to soft body mass but number of neurons and, even more importantly, connections among neurons. Many of the taxa I have just alluded to differ greatly in bulk of brain and number of cells and probably also in number of connections. This may have only a small effect, if any, for example between small and large species of octopus or species of insect (Rensch 1954). But we cannot exclude that it may in some cases have a larger significance for performance. If some crucial feature of connectivity is present, sheer numbers could, in principle, determine emergent features of organization and of assembly activity and cooperativity. One wishes it would be possible to evaluate how far this goes. Even if modelling and artificial "neural" nets could convincingly show, as McCulloch and Pitts (1943) claimed for a much simpler model, that assemblies of elements with a minimum of differentiation can imitate the performance of living brains, we would still want to know how nature has done it. Lacking the measures for evaluating the contribution of sheer numbers of connections, we can prospect for veins of ore among the qualitative differences between brains.

Many of the chapters in this volume demonstrate examples from histology, with and without special stains. These include varieties of cell types, concentrations of small cells, segregation of cells from fiber tracts, subdivisions of neuropiles according to texture, laminae, glomeruli, cartridges and other traits, each of which is characteristic of, or more developed in, certain taxa than in others. These low power features are suggestive of further differentiation in fine structure.

For the present purpose, we need not work out or agree on the phylogenetic history. Establishment of homology, interesting and important as it is – witness the attention it receives in this volume – is not the be all and end all of the biological importance of diversity. Indeed we could make the case that before coming to decisions about homology, we should first learn more about the actual differences between sister groups. For the goal of this essay it is important, even before that, to recognize the reality that some phyla, classes and orders exhibit a span of neural and behavioral complexity distinct from lateral radiation by adaptive specialization.

I must appeal to your mental imagery, to avoid burdening this volume with a gallery of pictures. Think of jellyfish and cuttlefish, flatworms and bees, ascarids and polynoids. Within the phylum, think of the most advanced pelecypods and gastropods, oligochaetes and polychaetes, branchipods and brachyurans. You may not know that marine polyclad brains are much more differentiated than triclad planarian brains; that polyplacophoran molluscs (chitons) have simpler brains than monoplacophorans; tetrabranchiates (*Nautilus*) than dibranchiates (octopods and decapods). Nemertean brains likewise come in simple and complex; the same can be said for prosobranch gastropods, for opisthobranchs, for bivalve (lamellibranch) molluscs, and for arachnids.

Of course, plausible "reasons" for the differences are sometimes a sessile or burrowing or parasitic life, or retrograde evolution or paedomorphosis. For those who would dismiss these differences, it matters not that plausibility is no assurance that an explanation is correct, or that such an explanation does not answer the question here posed, namely, in what anatomical, biochemical or physiological details do the differences consist? I am here concerned with facing the fact of a vast span of brain complexity without more than a superficial and preliminary list of the qualitative differences involved.

The opportunity, the challenge and the message of this piece is to work on that list as a major perspective on how brains operate at the diverse grades of advancement. It is already a significant list anatomically, and this is the frontier most likely to add new items, whereas chemical differences between orders, classes and phyla appear to be fewer, and physiological differences, although certainly important, are difficult to discern. Behavioral differences between major taxa, in my amateur opinion, are sure to be substantial and major but laborious to define. These assertions deserve some comment.

Anatomically more advanced brains are, to the eye, more differentiated, and here is our principal present knowledge of qualitative differences between major taxa. It is based on identifiable parts, subdivisions, boundaries, tissue textures, diversity of cell types, packed clusters of small cells with dense nuclei and little cytoplasm, and tracts of fibers, each representing an increase in complexity by our definition. This list must be far from complete and calls for quantitation of each feature. One suspects, for example, that the most advanced gastropod brain has a relatively smaller number of amacrine cells than dibranchiate cephalopods and the same may be true for other local neurons, including the "globuli" cells. Are the processes of similar or of average cells in gastropods less branched or less thorny or less studded with synapses, or are the synapses less segregated according to the presynaptic source? Are there fewer recurrent or reverberating circuits? Do they have fewer transmitters and modulators, or peptide-specific or multiple-peptide cells, or slower axon transport? Is the plan of having a large proportion of unique, identifiable cells relatively primitive and the plan of having larger numbers of each kind of cell, as in higher centers of vertebrates, cephalopods and arthropods, a more advanced one? Where do prosobranch, oligochaete and polychaete brains belong in this respect?

Golgi-defined cell types have been much less studied in molluscs, including cephalopods, than in insects; histology with axon transport tracers in cephalopods has hardly begun. It seems likely that these classical approaches will reveal new cell types and possibly novel relations among them. This volume testifies to a healthy interest in comparing species of gastropods but most of the prosobranchs have hardly been sampled. The same must be said for other major taxa such as pelecypods, chelicerates and polychaetes. The more advanced platyhelminths and nemerteans have been little studied with modern methods. Less advanced taxa tend to be neglected even more – amphineurans, oligochaetes, triclads, diplopods, lower crustaceans and insects, hemichordates, urochordates and cephalochordates. To end the litany of complaints, I must point out the sad tradition and continued tendency for even the best anatomical monographs to overlook the importance of comparing across major taxa.

I am not an anatomist but hear enough, as a disciple, to realize the range of questions and measurements now within reach which could expand our understanding of the grades of complexity. We can expect quantification of the textures of punctate substance, of raked-pile, coarse and fine thicket-like and hay-stack-like plexuses, neuropiles and fiber tracts and of transitions that may define boundaries of subdivisions. Recognition and tracing of perimeters and areas of profiles in electron micrographs can be applied to fibers as well as cells, giving counts and densities and histograms. Angularity, twigs or spines, branching patterns, local densifications, blebs, beads and varicosities, terminal arbors, even histochemically stained synapses can be enumerated and classified. Many features may have to remain for the present up to the trained observer to discern and characterize, for example the types and numbers of amacrine cells. the richness of recurrent collaterals or the relative development of reciprocal connections, the degree of profusion of dendrite-like ramifications and the degree of differentiation into miniatures, midgets and giants, Ts, Ys, and cones, climbing, linking and serpentining fibers, bushes, brushes, fans, nests and baskets, tangential, quasi-tangential and radial, small-field, wide-field, palmate, calyciform, wedge-shaped, club-shaped, anchor-like, chandelier-like or parasol-like, bi-stratified and tri-stratified arbors. Glomeruli are described by electron microscopy, but we can hope to learn whether they develop a characteristic 3-dimensional form, with varieties or gradations, for example, having more a intricate composition and distribution in bees than in annelids or in primitive insects with simpler brains.

The reason for this long list of traits is to point to the scarcity of comparisons across class and phylum lines in respect to specific, mensurable or estimable traits. We have hardly begun the job, as biologists, of really characterizing what evolution has created in the major categories of taxa, in the principal branches of biodiversity, as distinct from the twigs, in the most differentiated organ and its output – brain and behavior.

Special mention is due the nearly unique contributions of Johnson et al. (1994 and earlier) who have compared brain anatomy across many orders of mammals and defined 24 quantifiable characters, with their distribution among 152 species. This is what we need in other classes and phyla. The studies of Johnson et al. are limited, so far, to mammals and primarily to multicellular characters. Many more characters are doubtless available, including subcellular ones and cell types.

Classifying cell types, for a long time to come. will require human judgement and we can expect later workers to subdivide the categories of earlier workers, since good anatomists are splitters, not lumpers. Even so, they are unlikely to do full justice to the specifications inherent in nerve cells since members of the same subtype, even after categorizing by transmitters, modulators, immunological and cytochemical characters into still smaller subsets, will in general have receptive fields and projection fields that only partially overlap and hence are partially unique. Which part of the visual field or body surface or muscle a cell is connected to is essential for its normal function so that, unlike the liver and most other tissues, many neurons are not equivalent because their connections are partially nonoverlapping.

Besides the exact connectivity, another large domain of specifications characterizes cells, namely the physiological "personality". Anatomy and physiology together established the reality of identifiable neurons in many groups of invertebrates, and some workers have given informed guesses about how many such cells there are in a few species. Some years ago I undertook to extend the query, asking how many classes of cells have two or three or more fully equivalent members and how many distinct equivalence classes of whatever size may occur in several species (Bullock 1993, Chapter 7). This is the same as asking for the number of kinds of neurons there are, taking into consideration all criteria, chemical, physiological and anatomical – including the receptive and projective fields, the tendencies to adapt or rebound, to burst or to fire irregularly, to facilitate or potentiate and dozens of other alternatives. I cannot reveiw the procedure here for arriving at the estimates; they are large extrapolations from a few knowns, assumed to be representative. The procedure provides no obvious grounds for expecting a systematic over-estimate or under-estimate. Only two invertebrates were included. The estimate for a 400 g *Aplysia* was ca. 2×10^3 equivalence classes, including the identifiable neurons; for a similar sized lobster ca. 10 times more (for a carp of the same size ca. 4×10^4, only twice as many as the lobster; for a rat of similar size nearly 2×10^6; for the human $> 10^9$). The method is crude but even if the estimates are incorrect by a factor of 5 or 10 or more, they suggest that the span among taxa of qualitatively different neural elements is wide. One must realize, too, that these estimates compare only the cell types. A proper appreciation of the kinds of neural components would have to add the numbers of connected circuits, interactive assemblies or constellations of diverse cells.

The foregoing sample questions arise from the present, rather meager level of information. My claim is that explicit comparison of major taxa of quite distinct grades of complexity, using modern methods, will uncover new traits, each opening the door to new analytical study. Quantifiable variables should be sought across all disciplines. including the molecular, chemical, immunological, biophysical, pharmacological, and developmental.

Comparative neurophysiology is in particular need of imaginative new research. You may be surprised to hear that it lacks a recognized list of significant, measurable variables concerning operations and interactions that can be compared across major taxa. The following is only a tentative stage in formulating such a list. Are non-spiking neurons for example, distributed among taxa at random? Plateau potentials, diverse degrees of accommodation, homo- and hetero-facilitation and defacilitation, regular, irregular and patterned firing, cells requiring complex stimuli such as faces or species-specific songs, "recruiting" responses, epileptiform responses? Do some classes of animals have smaller or simpler long-latency event-related potentials to novel stimuli, or shorter or simpler post-tetantic potentiation, or simpler forms of delayed response learning, or fewer varieties of arousal and attention? Comparing measures of cooperativity among neurons, do we find synchrony less developed in some groups or coherence relatively low for each frequency band in the Fourier spectrum of the ongoing compound field potentials of a population of cells, computed between two loci a short distance apart? Is phase coupling between frequencies in the ongoing or evoked electrical activity of assemblies, such as that measured by the bicoherence correlated with other signs of advancement? Are ongoing, spontaneous brain rhythms or stimulus-induced rhythms marks of more advanced processing?

The assumption that we will basically understand the brain as soon as we know its circuitry and have a molecular explanation of the known properties of synapses and their plasticity is very far from justified. We still do not know how much signal processing and intercellular communication occurs extrasynaptically. We know almost nothing about when, where and how synchronization in various degrees appears in populations of cells, let alone what it does or means.

Particularly inadequate is our grasp of the significant variables at higher integrative levels, which are perhaps more likely to reward investigation with behaviorally correlated findings. A special field that has explosively expanded but only scratched the surface as yet is in new forms of analysis and display of neural population activity, particularly the correlational and cooperative properties of organized assemblies of cells. The usual static mental picture of the brain as a wired collection of cells should give way to a dynamic mental image of constant ongoing and stimulated activity that has a fine structure in space and time, quite possibly as characteristic of the species as is behavior. Learning how to record, display and describe the structure of activity in three dimensions, at appropriately fine resolution in space and time to reveal assembly-encoding schemes is a worthy challenge. Is there a range of variation among invertebrates in respect to measures of cooperative activity?

Behavioral comparative biology, including ethology, comparative learning, and the broad, overlapping area called comparative cognition, is also poorly developed in terms that recognize the evolution of complexity. Quantitative evaluations of comparable estimators of behavioral complexity across major taxa are scarce. You may feel – as I do – that appropriate measurements will surely confirm more complex behavior in an insect than a flatworm and perhaps still higher cognitive capacities in an octopus. Adequate data from a sufficient battery of quantitative measures to test this expectation are still, for the most part, to be gathered. Certainly the task is difficult and fraught with traps but I believe it is doable, with sufficient labor and ingenuity – and worth the effort. The first step is to recognize the importance of doing so, as a part of general biology.

I have elsewhere put forward an agenda of more than 25 practical measures of cognitive capacities – not easy to make, but feasible (Bullock, 1986). These should be added to species-specific ethograms to make at least preliminary comparisons possible, preferably between taxa sufficiently far apart so that the results are more different than the uncertainties of the methods. A third category of variables, besides the measures of cognitive capacities and the ethogram, is even more difficult; that is the category of knowledge or stored information. It is relevant here because it may be one of the principal variables distinguishing the more complex major taxa, and potentially capable of spreading species on a significant spectrum, once we find a way to estimate it.

Many of the higher level functional novelties in evolution, whether due to sheer number of connections or to qualitative features like spines or lamination can be called **emergents**. Saltatory conduction, face-selective neurons, the courtship songs of fruit flies are emergents in the sense that they could not have been predicted even from an extensive knowledge of their substrates. One way to express the message of this diatribe is to say that we have neglected the search for emergent features that distinguish the major taxa where our preliminary, short list of traits indicate different grades of complexity. Many such features are still out there, waiting to be discovered; some of them, we can be sure, are broad principles. Turned up, usually, by physiological natural history or "mere phenomenology," as with cognitive event-related potentials, kindling, computed maps, plateau potentials, posttetanic responses, and most of the other unpredicted properties, each emergent will open new windows for reductionist analysis. Brain and behavior must be the principal reservoir of biological emergents, and these must have increased greatly, in number and complexity, from the pre-Cambrian ancestors of corals and sea urchins, and at each subsequent advance in grade of complexity.

After a long period in the shadow of the biology of universal or common principles and basic mechanisms of life, biodiversity now has its supporters, even if the spokespersons tend to concentrate on the practical value due to molecular diversity. For biologists aiming to understand living nature, it is not enough to shout biodiversity, to protect it, name and catalog it. We need to study it, discover what is diverse and assess the differences with every disciplinary approach and at each level of organization up to the highest, that of brain and behavior.

References

Bullock, T.H. (1986) Suggestions for research on ethological and comparative cognition. *In*: R.J. Schusterman, J.A. Thomas and F.G. Wood (eds): *Dolphin Cognition and Behavior: A Comparative Approach*, Lawrence Erlbaum Assoc, Hillsdale NJ, pp. 207–219.

Bullock, T.H. (1992) Comparisons of major and minor taxa reveal two kinds of differences: "lateral" adaptions and "vertical" changes in grade. *In*: D.B. Webster, R.R. Fay and A.N. Popper (eds): *The Evolutionary Biology of Hearing*, Springer-Verlag, New York, pp. 15–19.

Bullock, T.H. (1993) How are more complex brains different? One view and an agenda for comparative neurobiology. *Brain Behav. Evol.* 41: 88–96.

Deacon, T.W. (1990) Problems of ontogeny and phylogeny in brain-size evolution. *Int. J. Primatol.* 11: 237–282.

Jerison, H.J. (1991) *Brain Size and the Evolution of Mind.* American Museum of Natural History, New York.

Johnson, J.I. Kirsch, J.A.W., Reep, R.L. and Switzer R.C.III (1994) Phylogeny through brain traits: more characters for the analysis of mammalian evolution. *Brain Behav. Evol.* 43: 319–347.

McCulloch, W.S. Pitts, W. (1943) A logical calculus for ideas immanent in nervous activity. *Bull. Math. Biophys.* 5: 115–133.

Pirlot, P. (1987) Contemporary brain morphology in ecological and ethological perspectives. *J. Hirnforsch.* 282: 145–211.

Rensch, B. (1954) The relations between the evolution of central nervous functions and the body size of animals. *In*: J. Huxley, A.C. Hardy and E.B. Ford (eds): *Evolution as a Process*, Allen & Unwin, London, pp. 181–200.

Ridet, J.M. and Bauchot, R. (1990) Analyse quantitative de l'encéphale des téléostéens: caractères evolutifs et adaptatifs de l'encéphalisation. I. Généralités et analyse globale. *J. Hirnforsch.* 31: 51–63.

Ringo, J.L. (1991) Neuronal interconnection as a function of brain size. *Brain Behav. Evol.* 38: 1–6.

Riska, B. and Atchley, W.R. (1985) Genetics of growth predict patterns of brain-size evolution. *Science* 229: 668–671.

Subject index

(The page number refers to the exact page within a chapter in which the keyword first occurs)

BioMethods

NEUROPHYSIOLOGY • NEUROPSYCHOLOGY

Methods in Comparative Psychoacoustics

Edited by

G.M. Klump, *Institue for Zoology, Techn. University of Munich, Germany*
R.J. Dooling, *Psychology Dept., University of Maryland, College Park, MD, USA*
R.R. Fay, *Parmly Hearing Institute, Loyola University, Chicago, IL, USA*
W.C. Stebbins, *Psychology Dept., University of Michigan, Ann Arbor, MI, USA*

1995. Approx. 350 pages. Hardcover • ISBN 3-7643-5079-2
BioMethods

Comparative psychoacoustics is an area of animal psychophysics which has seen remarkable advances in methodology over the past 25 years. The development of small laboratory computers has had a crucial impact on the field. Acoustic stimuli are now routinely generated using digital methods providing the researcher with unprecedented possibilities. The strategies and paradigms for data collection are becoming more refined as well, again due in large part to the use of computer technology.

This volume presents an array of methods designed to measure detection and discrimination of both simple and complex acoustic stimuli along with strategies designed to assess how acoustic stimuli are perceived, categorized or classified. Refinements in modern methodologies now make it possible to compare the hearing capabilities of diverse species tested under similar, if not identical, experimental conditions. Tests on nonverbal human infants fit well in this category and have much in common with comparative methodologies used in testing animals.

This book will serve as a reference source for graduate students and scientists interested in the comparative aspects of hearing.

Birkhäuser Verlag • Basel • Boston • Berlin

EXPERIENTIA SUPPLEMENTUM

Molecular Ecology and Evolution: Approaches and Applications

Edited by

B. Schierwater, B. Streit, *Zool. Institut d. J.W. Goethe-Univ., Frankfurt, Germany*
G.P. Wagner, *Dept. of Biology, Yale Univ., New Haven, USA*
R. DeSalle, *American Museum of Natural History, New York, NY, USA*

1994. 622 pages. Hardcover • ISBN 3-7643-2942-4 (EXS 69)

The past 25 years have witnessed a revolution in the way ecologists and evolutionary biologists approach their disciplines. Modern molecular techniques are now reshaping the spectrum of questions that can be addressed while studying the mechanisms and consequences of the ecology and evolution of living organisms.

Molecular Ecology and Evolution: Approaches and Applications describes, from a molecular perspective, several methodological and technical approaches used in the fields of ecology, evolution, population biology, molecular systematics, conservation, and development. Modern techniques are introduced, and older, more classic ones refined. The advantages, limitations, and potentials of each are discussed in detail, and thereby illustrate the widening range of cross-field research and applications which this modern technology is stimulating.

The book will serve as an important textbook for graduate and advanced undergraduate students, and as a key reference work for researchers.

Birkhäuser Verlag • Basel • Boston • Berlin

GPSR Compliance

The European Union's (EU) General Product Safety Regulation (GPSR) is a set of rules that requires consumer products to be safe and our obligations to ensure this.

If you have any concerns about our products, you can contact us on ProductSafety@springernature.com

In case Publisher is established outside the EU, the EU authorized representative is:

Springer Nature Customer Service Center GmbH
Europaplatz 3
69115 Heidelberg, Germany

Batch number: 10732324

Printed by Printforce, the Netherlands